Advances in Process Systems Engineering – Vol. 2

STOCHASTIC GLOBAL OPTIMIZATION

Techniques and Applications in Chemical Engineering

Advances in Process Systems Engineering

Series Editor: Gade Pandu Rangaiah
(National University of Singapore)

Vol. 1: Multi-Objective Optimization:
Techniques and Applications in Chemical Engineering
ed: Gade Pandu Rangaiah

Vol. 2: Stochastic Global Optimization:
Techniques and Applications in Chemical Engineering
ed: Gade Pandu Rangaiah

Advances in Process Systems Engineering – Vol. 2

STOCHASTIC GLOBAL OPTIMIZATION

Techniques and Applications in Chemical Engineering

editor

Gade Pandu Rangaiah

National University of Singapore, Singapore

World Scientific

NEW JERSEY • LONDON • SINGAPORE • BEIJING • SHANGHAI • HONG KONG • TAIPEI • CHENNAI

Published by

World Scientific Publishing Co. Pte. Ltd.

5 Toh Tuck Link, Singapore 596224

USA office: 27 Warren Street, Suite 401-402, Hackensack, NJ 07601

UK office: 57 Shelton Street, Covent Garden, London WC2H 9HE

British Library Cataloguing-in-Publication Data
A catalogue record for this book is available from the British Library.

Advances in Process Systems Engineering — Vol. 2
STOCHASTIC GLOBAL OPTIMIZATION
Techniques and Applications in Chemical Engineering
(With CD-ROM)

Desk Editor: Tjan Kwang Wei

ISBN-13 978-981-4299-20-6
ISBN-10 981-4299-20-0

Typeset by Stallion Press
Email: enquiries@stallionpress.com

Printed in Singapore by B & Jo Enterprise Pte Ltd

PREFACE

In Chemical Engineering, optimization plays a key role in the design, scheduling and operation of industrial reactors, separation processes, heat exchangers and complete plants. It is also being used on a larger scale in managing supply chains and production plants across the world. Furthermore, optimization is useful for understanding and modeling physical phenomena and processes. Without the use of optimization techniques, chemical processes would not be as efficient as they are now. Optimization has, in short, proven to be essential for achieving sustainable processes and manufacturing.

In many applications, the key is to find the global optimum and not just a local optimum. This is desirable as the former is obviously better than the latter in terms of the desired objective function. In some applications such as phase equilibrium, only the global optimum is the correct solution. Finding the global optimum is more challenging than finding a local optimum. Methods for finding the global optimum can be divided into two main groups: deterministic and stochastic (or probabilistic) techniques. Stochastic global optimization (SGO) techniques involve probabilistic elements and consequently use random numbers in the search for the global optimum. They include simulated annealing, genetic algorithms, taboo/tabu search and differential evolution. SGO techniques have a number of attractive features including being simple to understand and program, requiring no assumptions on the optimization problem, the wide range of problems they can solve, their ability to provide robust results for highly nonlinear problems even with many decision variables, and faster convergence towards global optimal solution.

Significant progress has been made in SGO techniques and their applications in the last two decades. However, there is no book devoted to SGO techniques and their applications in Chemical Engineering, which motivated the preparation of this book. The broad objective of this book is to provide an overview of a number of SGO techniques and their applications to Chemical Engineering. Accordingly, there are two parts in the book. The first part, Chapters 2 to 11, includes description of the SGO techniques and reviews of their recent modifications and Chemical Engineering applications. The second part, Chapters 12 to 19, focuses on Chemical Engineering applications of SGO techniques.

Each chapter in the book is contributed by well-known and active researcher(s) in the area. A brief resume and photo of each of the contributors to the book, are given on the enclosed CD-ROM. Each chapter in the book was reviewed anonymously by at least two experts and/or other contributors. Of the submissions received, only those considered to be useful for education and/or research were revised by the respective contributor(s), and the revised submission was finally reviewed for presentation style by the editor or one of the other contributors. I am grateful to my long-time mentor, Dr. R. Luus, who coordinated the anonymous review of chapters co-authored by me.

The book will be useful to researchers in academia and research institutions, to engineers and managers in process industries, and to graduates and senior-level undergraduates. Researchers and engineers can use it for applying SGO techniques to their processes whereas students can utilize it as a supplementary text in optimization courses. Each of the chapters in the book can be read and understood with little reference to other chapters. However, readers are encouraged to go through the Introduction chapter first. Many chapters contain several exercises at the end, which can be used for assignments and projects. Some of these and the applications discussed within the chapters can be used as projects in optimization courses at both undergraduate and postgraduate levels. The book comes with a CD-ROM containing many programs and files, which will be helpful to readers in solving the exercises and/or doing the projects.

I am thankful to all the contributors and anonymous reviewers for their collaboration and cooperation in producing this book. Thanks are also due to Mr. K.W. Tjan and Ms. H.L. Gow from the World Scientific, for their suggestions and cooperation in preparing this book. It is my pleasure to acknowledge the contributions of my postgraduate students (Shivom Sharma, Zhang Haibo, Mekapati Srinivas, Teh Yong Sing, Lee Yeow Peng, Toh Wei Khiang and Pradeep Kumar Viswanathan) to our studies on SGO techniques and to this book in some way or other. I thank the Department of Chemical & Biomolecular Engineering and the National University of Singapore for encouraging and supporting my research over the years by providing ample resources including research scholarships.

Finally, and very importantly, I am grateful to my wife (Krishna Kumari) and family members (Santosh, Jyotsna and Madhavi) for their loving support, encouragement and understanding not only in preparing this book but in everything I pursue.

Gade Pandu Rangaiah

CONTENTS

Chapter 1

INTRODUCTION

Gade Pandu Rangaiah

Department of Chemical & Biomolecular Engineering
National University of Singapore, Singapore 117576
chegpr@nus.edu.sg

1. Optimization in Chemical Engineering

Optimization is very important and relevant to practically all disciplines. It is being used both qualitatively and quantitatively to improve and enhance processes, products, materials, healthcare, and return on investments to name a few. In Chemical Engineering, optimization has been playing a key role in the design and operation of industrial reactors, separation processes, heat exchangers and complete plants, as well as in scheduling batch plants and managing supply chains of products across the world. In addition, optimization is useful in understanding and modeling physical phenomena and processes. Without the use of sophisticated optimization techniques, chemical and other manufacturing processes would not be as efficient as they are now. Even then, it is imperative to continually optimize the plant design and operation due to the ever changing technology, economics, energy availability and concerns on environmental impact. In short, optimization is essential for achieving sustainable processes and manufacturing.

In view of its importance and usefulness, optimization has attracted the interest of chemical engineers and researchers in both industry and academia, and these engineers and researchers have made significant contributions to optimization and its applications in Chemical Engineering. This can be seen from the many optimization books written by Chemical Engineering academicians (e.g. Lapidus and Luus, 1967; Beveridge and Schechter, 1970; Himmelblau, 1972; Ray and Szekely, 1973; Floudas, 1995 and 1999; Luus, 2000; Edgar *et al.*, 2001; Tawarmalani and Sahinidis, 2002; Diwekar, 2003; Ravindran *et al.*, 2006).

Optimization can be for minimization or maximization of the desired objective function with respect to (decision) variables subject to (process) constraints and bounds on the variables. An optimization problem can have a single optimum (i.e. minimum in the case of minimizing the objective function or maximum in the case of maximizing the objective function) or multiple optima (Fig. 1), one of which is the global optimum and the others are local optima. A global minimum has the lowest value of the objective function throughout the region of interest; that is, it is the best solution to the optimization problem. On the other hand, a local minimum has an objective function value lower than those of the points in its neighborhood but it is inferior to the global minimum. In some problems, there may be more than one global optimum with the same objective function value.

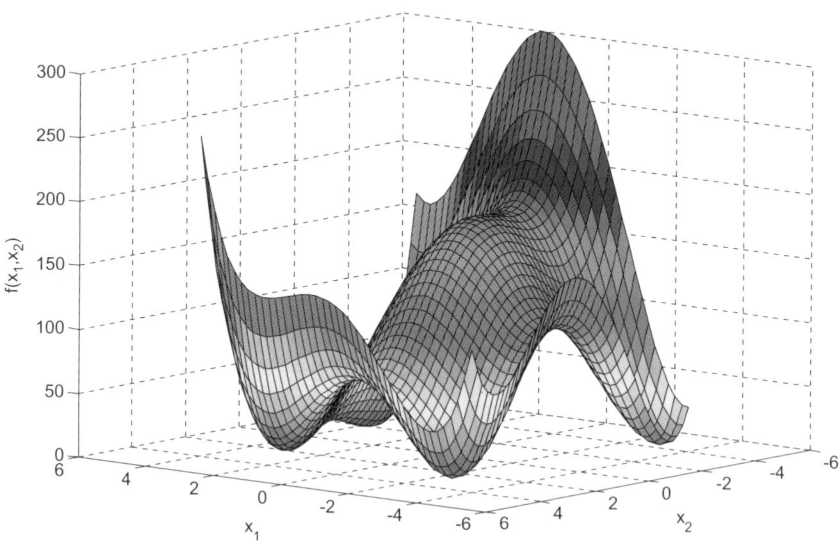

Figure 1. Three-dimensional plot of the modified Himmelblau function (Eq. (2)) showing four minima of which one is the global minimum.

In most applications, it is desirable to find the global optimum and not just the local optimum. Obviously, the global optimum is better than a local optimum in terms of the specified objective function. In some applications such as phase equilibrium, only the global optimum is the correct solution. Global optimization refers to finding the global optimum, and it encompasses the theory and techniques for finding the global optimum. As can be expected, finding the global optimum is more difficult than finding a local optimum. However, with the availability of cheap computational power, interest in global optimization has increased in the last two decades. Besides the need for global optimization, the application can involve two or more conflicting objectives, which will require multi-objective optimization (MOO). There has been increasing interest in MOO in the last two decades. This led to the first book on MOO techniques and its applications in Chemical Engineering (Rangaiah, 2009).

Many of the optimization books by chemical engineers cited above focus on optimization in general. Only two books: Floudas (1999) and Tawarmalani and Sahinidis (2002), are dedicated to global optimization, and they focus on deterministic methods. Besides these methods, however, many stochastic methods are available and attractive for finding the global optimum of application problems. Lack of a book on stochastic global optimization (SGO) techniques and applications in Chemical Engineering is the motivation for the book you are reading.

The rest of this chapter is organized as follows. The next section presents several examples having multiple minima, thus requiring global optimization. An overview of the global optimization methods is provided in Sec. 3. Scope and organization of the book are covered in the last section of this chapter.

2. Examples Requiring Global Optimization

2.1. Modified Himmelblau function

Consider the Himmelblau function (Ravindran *et al.*, 2006):

$$\text{Minimize} \quad f(x_1, x_2) = (x_1^2 + x_2 - 11)^2 + (x_1 + x_2^2 - 7)^2, \tag{1a}$$

with respect to x_1 and x_2,

$$\text{Subject to} \quad -6 \leq x_1, x_2 \leq 6. \tag{1b}$$

Here, $f(x_1, x_2)$ is the objective (or performance) function to be minimized; it is a function of two (decision) variables: x_1 and x_2. The feasible region is defined by the bounds on the variables (Eq. (1b)), and there are no other constraints in this problem. The above optimization problem has four minima with objective function value of 0 at $(x_1, x_2) = (3, 2)$, (3.584, −1.848), (−2.805, 3.131) and (−3.779, −3.283).

Himmelblau function has been modified by adding a quadratic term, in order to make one of these a global minimum and the rest local minima (Deb, 2002). The modified Himmelblau function is

$$\text{Minimize} \quad (x_1^2 + x_2 - 11)^2 + (x_1 + x_2^2 - 7)^2 + 0.1[(x_1 - 3)^2 + (x_2 - 2)^2], \tag{2a}$$

with respect to x_1 and x_2,

$$\text{Subject to} \quad -6 \leq x_1, x_2 \leq 6. \tag{2b}$$

With the addition of the quadratic term, the minimum at $(x_1, x_2) = (3, 2)$ becomes the global minimum with objective value of 0 whereas the other minima have positive objective values (Table 1 and Fig. 1). Note that the locations of the local minima of the modified Himmelblau function have changed compared to the minima of the Himmelblau function in Eq. (1).

2.2. Ellipsoid and hyperboloid intersection

Consider an ellipsoid and a hyperboloid in three dimensions. There can be four intersection points between these surfaces, one of which will be the farthest from the origin. Luus (1974) formulated a global optimization problem for finding this particular intersection, which is also considered in

Table 1. Multiple minima of the modified Himmelblau function (Eq. (2)).

No.	Objective function	Decision variables: x_1 and x_2
1	0.0	3.0, 2.0
2	1.5044	3.5815, −1.8208
3	3.4871	−2.7871, 3.1282
4	7.3673	−3.7634, −3.2661

Chapter 2 of this book. The global optimization problem can be expressed mathematically as:

$$\text{Maximize} \quad x_1^2 + x_2^2 + x_3^2, \tag{3a}$$

with respect to x_1, x_2 and x_3,

$$\text{Subject to:} \quad 4(x_1 - 0.5)^2 + 2(x_2 - 0.2)^2 + x_3^2 + 0.1x_1x_2$$

$$+0.2x_2x_3 = 16, \tag{3b}$$

$$2x_1^2 + x_2^2 - 2x_3^2 = 2. \tag{3c}$$

Here, the objective function (Eq. (3a)) is the square of the distance of a point (x_1, x_2 and x_3 in the three-dimensional space) from the origin, and Eqs. (3b) and (3c) are the equality constraints. Since Eqs. (3b) and (3c) represent respectively an ellipsoid and a hyperboloid in the three-dimensional space, any point satisfying these constraints corresponds to an intersection of the two surfaces. The global optimization problem (Eq. (3)) for finding the farthest intersection between the two surfaces has four maxima as shown in Table 2, of which only one is the global maximum and also the correct solution.

2.3. Reactor design example

We now consider a reactor design example that has two minima. In this problem, it is desired to find the optimal design of three continuous stirred tank reactors (CSTRs) wherein the following series-parallel reactions take place.

$$A + B \xrightarrow{k_1} X; \qquad X \xrightarrow{k_3} Y, \tag{4a}$$

$$A + B \xrightarrow{k_2} P; \qquad X \xrightarrow{k_4} Q. \tag{4b}$$

Table 2. Multiple optima for the optimization problem in Eq. (3).

No.	Objective function	Decision variables: x_1, x_2, x_3
1	8.7823	1.5682, −1.8007, −1.7551
2	9.5691	1.3763, −2.1262, 1.7761
3	10.473	1.0396, 2.4915, 1.7846
4	11.677	0.9884, 2.6737, −1.8845

Here, reactant A is expensive whereas reactant B is available in excess amount for reaction. The desired product is Y via the intermediate X, whereas P and Q are the products of side reactions. The above reactions are taken to be first order with respect to concentration of A (for the first two reactions) and X (for the last two reactions). The specific reaction rates are given by (Denbigh, 1958):

$$k_2/k_1 = 10{,}000\, e^{-3000/T}, \tag{5a}$$

$$k_3/k_1 = 0.01, \tag{5b}$$

$$k_4/k_1 = 0.0001\, e^{3000/T}, \tag{5c}$$

where T is the reaction temperature.

Component mass balances for A, X and Y around the nth reactor are:

$$C_A^{n-1} = C_A^n + (k_1^n + k_2^n)C_A^n\theta^n, \tag{6a}$$

$$C_X^{n-1} = -k_1^n C_A^n\theta^n + C_X^n + (k_3^n + k_4^n)C_X^n\theta^n, \tag{6b}$$

$$C_Y^{n-1} = -k_3^n C_X^n\theta^n + C_Y^n, \tag{6c}$$

where C is the concentration of a component (A, X and Y as indicated by the subscript), θ is the residence time in the CSTR and superscript n refers to the reactor number. Assume that concentrations in the feed to the first reactor are $C_A^0 = 1$, $C_X^0 = 0$ and $C_Y^0 = 0$. Optimal design of the three CSTRs is to find the values of T (involved in the rate coefficients) and θ for each reactor in order to maximize the concentration of the desired product Y from the last CSTR. In effect, the problem involves 6 design variables. For simplicity, the optimization problem is formulated in the dimensionless variables:

$$\alpha = \theta k_1 \quad \text{and} \quad \tau = k_2/k_1 = 10{,}000e^{-3000/T} \tag{7}$$

The mathematical problem for the optimal design of the three CSTRs is:

$$\text{Minimize} \quad -0.01x_1x_4 - 0.01x_5x_8 - 0.01x_9 \times \frac{[x_8 + x_7x_9/(1 + x_9 + x_9x_{10})]}{(1 + 0.01x_9 + 0.01x_9/x_{10})}, \tag{8a}$$

with respect to $x_1, x_2, \ldots, x_{10}$,

$$\text{Subject to} \quad x_3 = 1/(1 + x_1 + x_1x_2), \tag{8b}$$

$$x_4 = x_1x_3/(1 + 0.01x_1 + 0.01x_1/x_2), \tag{8c}$$

$$x_7 = x_3/(1 + x_5 + x_5 x_6), \tag{8d}$$

$$x_8 = (x_4 + x_5 x_7)/(1 + 0.01x_5 + 0.01x_5/x_6), \tag{8e}$$

$$0 \leq x_i \leq 2100 \quad \text{for } i = 1, 5, 9, \tag{8f}$$

$$0 \leq x_i \leq 4.934 \quad \text{for } i = 2, 6, 10, \tag{8g}$$

$$0 \leq x_i \leq 1.0 \quad \text{for } i = 3, 4, 7, 8, \tag{8h}$$

Here, the objective function is $[-C_Y^3]$, whose minimization is equivalent to maximizing C_Y^3 (i.e. concentration of the desired product Y in the last CSTR). Variables: x_1, x_5 and x_9 correspond to α^1, α^2 and α^3 (i.e. residence time multiplied by the rate coefficient k_1 in each of the three reactors); x_2, x_6 and x_{10} are the temperature as given by τ in each CSTR; x_3 and x_7 are the concentration of A in reactor 1 and 2 respectively; and x_4 and x_8 are the concentration of X in reactor 1 and 2 respectively.

The above problem for the CSTRs design is a constrained problem with 10 decision variables and 4 equality constraints besides bounds on variables. Alternatively, the equality constraints can be used to eliminate 4 decision variables (x_3, x_4, x_7 and x_8) and then treat the problem as having only 6 decision variables with bounds and inequality constraints. One solution to the design optimization problem is −0.50852 at (3.7944, 0.2087, 0.1790, 0.5569, 2100, 4.934, 0.00001436, 0.02236, 2100, 4.934), and another solution is −0.54897 at (1.3800, 0.1233, 0.3921, 0.4807, 2.3793, 0.3343, 0.09393, 0.6431, 2100, 4.934) (Rangaiah, 1985). The latter solution is the global solution and also better with higher concentration of the desired product in the stream leaving the third CSTR.

2.4. Stepped paraboloid function

Consider the two-variable, stepped paraboloid function synthesized by Ingram and Zhang (2009):

$$\text{Minimize} \quad 0.2(\lfloor x_1 \rfloor + \lfloor x_2 \rfloor) + [\text{mod}(x_1, 1) - 0.5]^2 + [\text{mod}(x_2, 1) - 0.5]^2, \tag{9a}$$

with respect to x_1 and x_2,

$$\text{Subject to} \quad -5 \leq x_1, x_2 \leq 5. \tag{9b}$$

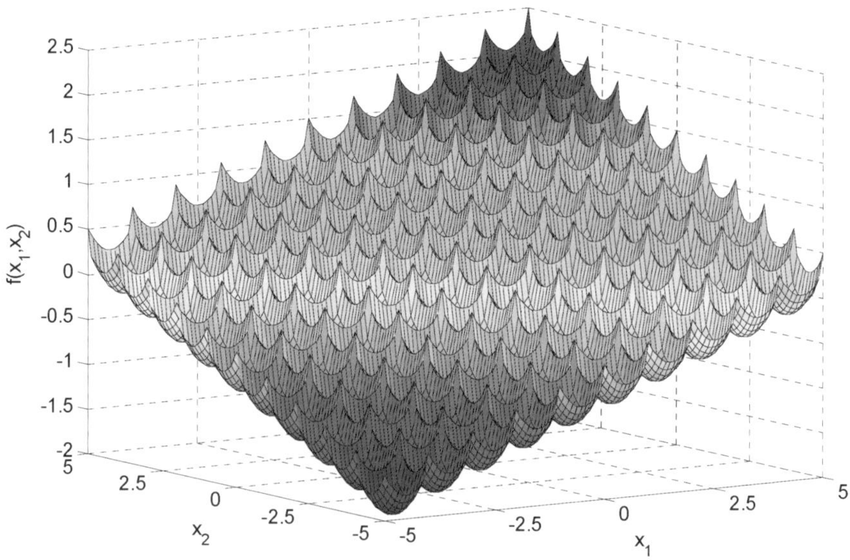

Figure 2. Three-dimensional plot of the discontinuous stepped paraboloid function (Eq. (9)) showing 100 minima of which one is the global minimum.

The notation $\lfloor x \rfloor$ denotes the floor function, which returns the largest integer less than or equal to x, and $\mathrm{mod}(x, y)$ is the remainder resulting from the division of x by y. Equation (9) contains 100 minima within the search domain, which are located at $(x_1, x_2) = (-4.5 + i, -4.5 + j)$ for i and $j = 0, 1, \ldots, 9$ (Fig. 2). In contrast to the examples considered thus far, there are discontinuities in both the function value and the function's derivative within the solution domain, specifically at $x_1 = -5, -4, \ldots, 4, 5$ and at $x_2 = -5, -4, \ldots, 4, 5$. The global minimum is located at $(x_1, x_2) = (-4.5, -4.5)$ and has an objective function value of -2. The problem can be extended to any number of variables and also can be made more challenging by decreasing the coefficient (0.2) in the first term of Eq. (9a), thus making the global minimum comparable to a local minimum.

The examples considered above have relatively simple functions, a few variables and constraints but still finding their global optimum is not easy. In general, optimization problems for many Chemical Engineering applications involve complex algebraic and/or differential equations in the constraints and/or for computing the objective function as well as numerous decision variables. Objective function and/or constraints in the application problems may not be continuous. Chemical Engineering problems generally involve continuous variables with or without integer variables. All these characteristics make finding the global optimum challenging. SGO techniques are well-suited for such problems. Hence, this book focuses on SGO techniques and their applications in Chemical Engineering.

3. Global Optimization Techniques

The goal of global optimization techniques is to find reliably and accurately the global minimum of the given problem. Many methods have been proposed and investigated for global optimization, and they can be divided into two main groups: deterministic and stochastic (or probabilistic) techniques. Deterministic methods utilize analytical properties (e.g. convexity) of the optimization problem to generate a deterministic sequence of points (i.e. trial solutions) in the search space that converge to a global optimum. However, they require some assumption (e.g. continuity of functions in the problem) for their success and provide convergence guarantee for problems satisfying the underlying assumptions. Deterministic methods include branch and bound methods, outer approximation methods, Lipschitz optimization and interval methods (e.g. see Floudas, 1999; Horst *et al.*, 2000; Edgar *et al.*, 2001; Biegler and Grossman, 2004; Hansen and Walster, 2004).

Stochastic global optimization (SGO) techniques, the subject of this book, involve probabilistic elements and consequently use random numbers in the search for the global optimum. Thus, the sequence of points depends on the seed used for random number generation. In theory, SGO techniques need infinite iterations to guarantee convergence to the global optimum. However, in practice, they often converge quickly to an acceptable global optimal solution. SGO techniques can be divided into four groups: (1) random search techniques, (2) evolutionary methods, (3) swarm intelligence methods and (4) other methods (Fig. 3).

Random search methods include pure random search, adaptive random search (ARS), two-phase methods, simulated annealing (SA) and tabu search (TS). ARS methods incorporate some form of adaptation including region reduction into random search for computational efficiency. Two-phase methods, as the name indicates, have a global and a local phase.

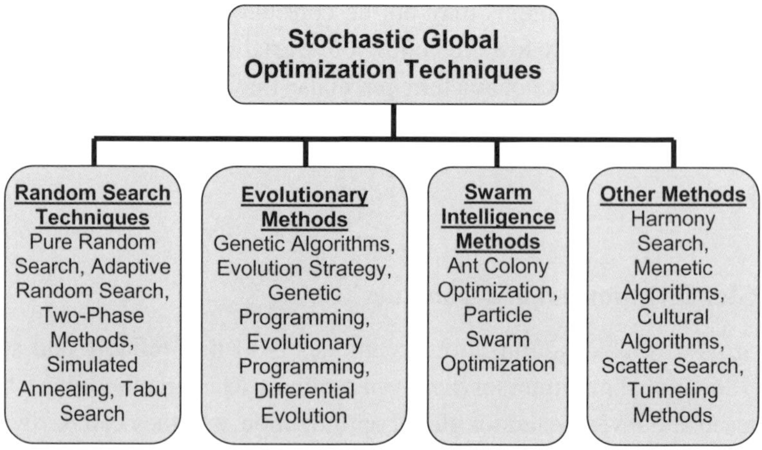

Figure 3. Classification of stochastic global optimization techniques.

Multi-start algorithms and their variants such as multi-level single-linkage algorithm belong to two-phase methods. SA is motivated by the physical process of annealing (i.e. very slow cooling) of molten metals in order to achieve the desired crystalline structure with the lowest energy. TS is derived from principles of intelligent problem solving such as tabu (i.e. prohibited) steps and memory. In this book, ARS methods are covered in Chapters 2, 3, 12 and 15, SA is presented in Chapter 3, and TS is described in Chapters 5 and 19.

Evolutionary methods/algorithms are population-based search methods inspired by features and processes of biological evolution. They have found many applications in Chemical Engineering. Genetic algorithms (GA), evolution strategy (ES), genetic programming, evolutionary programming and differential evolution (DE) belong to evolutionary methods. GA and ES are now quite similar although the former was originally based on binary coding compared to real coding used in ES. GA and its applications are discussed in Chapters 4, 16 and 18, and DE and its variants are the subject of Chapters 6, 13 and 14.

Ant colony optimization (ACO) covered in Chapter 7 and particle swarm optimization (PSO) presented in Chapter 8 are motivated by the *swarm intelligence* or social behavior. An application of ACO is described in Chapter 17. *Other SGO methods* include harmony search (HS, introduced in Chapter 9), memetic algorithms, cultural algorithms, scatter search and random tunneling methods. This book covers many SGO methods which have found applications in Chemical Engineering.

Many SGO techniques (such as SA, TS, GA, DE, PSO, ACO, HS, memetic algorithms, cultural algorithms and scatter search) are also known as meta-heuristic methods. A meta-heuristic guides a heuristic-based search in order to find the global optimum. On the other hand, a heuristic-based search such as a descent method is likely to converge to a local optimum.

SGO techniques have a number of attractive features. First, they are simple to understand and program. Second, they require no assumption on the optimization problem (e.g. continuity of the objective function and constraints), and hence can be used for any type of problem. Third, SGO methods are robust for highly nonlinear problems even with large number of variables. Fourth, they often converge to (near) global optimal solution quickly. Finally, they can be adapted for non-conventional optimization problems. For example, several SGO techniques have been modified for multi-objective optimization (Rangaiah, 2009).

Significant progress has been made in SGO techniques and their applications in the last two decades. However, further research is needed to improve their computational efficiency, to establish their relative performance, on handling constraints and for solving large application problems. More theoretical analysis of SGO techniques is also required for better understanding and for improving them.

4. Scope and Organization of the Book

The broad objective of this book is to provide an overview of a number of SGO techniques and their applications to Chemical Engineering. Accordingly, there are two parts in the book. The first part, Chapters 2 to 11, includes description of the SGO techniques and review of their recent modifications and Chemical Engineering applications. The second part, Chapters 12 to 19, focuses on Chemical Engineering applications of SGO techniques in detail. Each of these chapters is on one or more applications of Chemical Engineering using the SGO techniques described earlier. Each chapter in the book is contributed by well-known and active researcher(s) in the area.

Luus presents a simple and effective random search with systematic region reduction, known as Luus-Jaakola (LJ) optimization procedure in Chapter 2. He illustrates its application to several Chemical Engineering problems and mathematical functions, and discusses the effect of two parameters in the algorithm on a design problem. He also describes a way for handling difficult equality constraints, with examples.

In Chapter 3, Jeżowski *et al.*, describe in detail two SGO techniques, namely, a modified version of LJ algorithm and simulated annealing combined with simplex method of Nelder and Mead. They investigate the performance of these techniques on many benchmark and application problems as well as the effect of parameters in the techniques.

Chapter 4 deals with genetic algorithms (GAs) and their applications in Chemical Engineering. After reviewing the Chemical Engineering applications of GAs, Younes *et al.*, explain the main components of GAs and discuss implementation issues. Finally, they outline some modifications to improve the performance of GAs.

Tabu (or taboo) search (TS) for global optimization of problems having continuous variables is presented in Chapter 5 by Sim *et al.* After describing the algorithm with an illustrative example, they review TS methods for continuous problems, Chemical Engineering applications of TS and available software for TS. They also briefly describe TS features that can be exploited for global optimization of continuous problems.

In Chapter 6, Chen *et al.*, describe differential evolution (DE) including its parameter values. They summarize the proposed modifications to various components of DE and provide an overview of Chemical Engineering applications of DE reported in the literature. In particular, DE has found many applications for parameter estimation and modeling in addition to process design and operation.

Ant colony optimization (ACO) for continuous optimization problems is illustrated with an example, by Shelokar *et al.* in Chapter 7. They also review ACO for combinatorial optimization, multi-objective optimization and data clustering. Performance of ACO for test and application problems is presented and discussed in the later sections of the chapter.

Particle swarm optimization (PSO) motivated by the social behavior of birds and fishes, is the subject of Chapter 8. Jarboui *et al.* describe a basic PSO algorithm and its modifications that include global best and local best algorithms. They evaluate the performance of six PSO algorithms for solving nonlinear and mixed-integer nonlinear programming problems.

In Chapter 9, Ingram and Zhang introduce harmony search (HS), which is motivated by the improvisation process of musicians, describe its basic algorithm for global optimization and summarize many modifications to the basic algorithm. They also review HS applications, mention the available software and provide an illustrative example and programs for the HS algorithm.

Younes *et al.*, discuss many issues in the evaluation and reporting of SGO techniques, in Chapter 10. These include performance measures (of solution quality, efficiency and robustness), test problems, experiment design, parameter tuning, presentation and discussion of performance results.

Constraints are common in Chemical Engineering applications, and need to be tackled in solving the optimization problems by SGO techniques. In Chapter 11, Durand *et al.* present an overview of five approaches for handling constraints. Then, they describe a hybrid strategy involving Karush-Kuhn-Tucker conditions for optimality, for handling constraints in SGO techniques, and evaluate its performance on selected nonlinear and mixed-integer nonlinear programming problems.

In Chapter 12, Luus illustrates the use of LJ procedure to model reduction, parameter estimation and optimal control applications, and also investigates the potential of line search in the LJ procedure.

Bonilla-Petriciolet *et al.*, in Chapter 13, apply DE and TS, each in conjunction with a local optimizer, to phase stability and equilibrium calculations in reactive systems, which are formulated using transformed composition variables.

Srinivas and Rangaiah describe two versions of DE with tabu list in Chapter 14. They demonstrate their performance and compare them with DE and TS on benchmark and phase stability problems.

In Chapter 15, Poplewski and Jeżowski describe industrial water (usage) networks and the formulation of optimization problems for them. They then solve three water network problems with equality constraints and numerous binaries, by the modified LJ algorithm described in Chapter 3.

Bochenek and Jeżowski consider the difficult and yet important problem of heat exchanger network retrofitting in Chapter 16. They employ a two-level optimization with GA in both outer level (for structural optimization) and inner level (for parameter optimization) for solving two retrofitting problems.

Finding classification rules in measured data by ACO is described in Chapter 17 by Shelokar *et al.* One ACO algorithm for classification and another for feature selection are presented. The former was tested on a number of data sets, and the two algorithms together were used on two data sets for simultaneous classification and feature selection.

In the pen-ultimate Chapter 18, Kotecha *et al.*, apply GA and Constraint Programming (CP) for a job scheduling problem and a sensor network design problem, and compare the performance of the two techniques. Prior to the application, they describe CP that reduces the search domain for optimization, mainly based on constraint propagation.

Lin and Miller, in the last Chapter 19, describe schemes for developing parallel SGO algorithms and illustrate them for solving heat exchanger network synthesis and computer aided molecular design problems using parallel TS. The fully documented code for the first example is provided on the CD accompanying the book.

Each chapter in this book is comprehensive and can be read by itself with little reference to other chapters. Introduction to, description, algorithm, illustrative examples and programs of the SGO techniques given in the first part of this book are useful to senior undergraduates and postgraduates doing courses and projects related to optimization. Reviews of modifications and Chemical Engineering applications of the techniques in a number of chapters are of particular interest to researchers and engineers. Applications covered in the second part of this book and programs/files on the accompanying CD are valuable to many readers of this book.

References

Beveridge, G.S.G. and Schechter, R.S. (1970). *Optimization: Theory and Practice*, McGraw Hill, New York.

Biegler, L.T. and Grossman, I.E. (2004). Part II. Future perspective on optimization. *Computers and Chemical Engineering*, **28**, p. 1193.

Deb, K. (2002). *Optimization for Engineering Design*, Prentice Hall of India, New Delhi.

Denbigh, K.C. (1958). Optimal temperature sequences in reactors. *Chemical Engineering Science*, **8**, p. 125.

Diwekar, U.M. (2003). *Introduction to Applied Optimization*, Kluwer Academic, Norwell, Mass.

Edgar, T.F., Himmelblau, D.M. and Lasdon, L.S. (2001). *Optimization of Chemical Processes*, Second Edition, McGraw-Hill, New York.

Floudas, C.A. (1995). *Nonlinear Mixed-integer Optimization: Fundamentals and Applications*, Oxford University Press, New York.

Floudas, C.A. (1999). *Deterministic Global Optimization: Theory, Methods and Applications*, Kluwer Academic, Boston.

Hansen, G. and Walster, W. (2004). *Global Optimization Using Interval Analysis*, Marcel Dekker, New York.

Himmelblau, D.M. (1972). *Applied Nonlinear Programming*, McGraw-Hill, New York.

Horst, R., Pardalos, P.M. and Thoai, N.V. (2000). *Introduction to Global Optimization*, Kluwer Academic, Boston.

Ingram, G. and Zhang, T. (2009). Personal Communication.

Lapidus, L. and Luus, R. (1967). *Optimal Control in Engineering Processes*, Blaisdell, Waltham, Mass.

Luus, R. (1974). Two-pass method for handling difficult equality constraints in optimization. *AIChE Journal*, **20**, p. 608.

Luus, R. (2000). *Iterative Dynamic Programming*, Chapman & Hall, Boca Raton.

Rangaiah, G.P. (1985). Studies in constrained optimization of chemical process problems. *Computers and Chemical Engineering*, **4**, p. 395.

Rangaiah, G.P. (Ed.) (2009). *Multi-Objective Optimization: Techniques and Applications in Chemical Engineering*, World Scientific, Singapore.

Ravindran, A., Ragsdell, K.M. and Reklaitis, G.V. (2006). *Engineering Optimization: Methods and Applications*, Second Edition, John Wiley, New Jersey.

Ray, W.H. and Szekely, J. (1973). *Process Optimization with Applications in Metallurgy and Chemical Engineering*, Wiley, New York.

Tawarmalani, M. and Sahinidis, N.V. (2002). *Convexification and Global Optimization in Continuous and Mixed-integer Nonlinear Programming: Theory, Algorithms, Software and Applications*, Kluwer Academic, Dordrecht.

Exercises

(1) Find the global optimum of the modified Himmelblau function (Eq. (2)) and the geometric problem (Eq. (3)) using a local optimizer and/or programs provided on the attached CD. Try different initial estimates for

the decision variables and/or parameters in the optimization program. Is it easy to find the global optimum of these two problems?

(2) Develop the optimization problem (Eq. (8)) for the design of CSTRs based on the description and equations provided in this chapter. Note that this requires Chemical Engineering background.

(3) Solve the optimization problem (Eq. (8)) for the design of CSTRs using a local optimizer and/or programs provided on the attached CD. Try different initial estimates for the decision variables and/or parameters in the optimization program. Is it easy to find the global optimum of this problem?

(4) Find the global optimum of the stepped, paraboloid function (Eq. (9)) using a local optimizer and/or programs provided on the attached CD. Ensure that the floor and mod functions in Eq. (9) are correctly implemented in your program. Try different initial estimates for the decision variables and/or parameters in the optimization program. Present and discuss the results obtained.

Chapter 2

FORMULATION AND ILLUSTRATION OF LUUS-JAAKOLA OPTIMIZATION PROCEDURE

Rein Luus

Department of Chemical Engineering
University of Toronto, 200 College Street
Toronto, ON M5S 3E5, Canada
rein.luus@utoronto.ca

1. Introduction

We consider the general problem of maximizing (or minimizing) a real-valued scalar function of n variables, written as the performance index

$$I = f(x_1, x_2, \ldots, x_n), \tag{1}$$

subject to the set of inequality constraints

$$g_j(x_1, x_2, \ldots, x_n) \leq 0, \quad j = 1, 2, \ldots, p, \tag{2}$$

and the equality constraints

$$h_i(x_1, x_2, \ldots, x_n) = 0, \quad i = 1, 2, \ldots, m, \tag{3}$$

where the number of equality constraints m is less than the number of variables n. We assume that f, g_j, and h_i are continuous functions of the continuous variables $x_1, x_2, \ldots, x_n$. The p inequalities specify the feasible region over which $x_1, x_2, \ldots, x_n$ may be chosen.

2. LJ Optimization Procedure

To introduce the Luus-Jaakola optimization procedure (LJ optimization procedure) let us take the case where there are no equality constraints. We consider handling the equality constraints later in this chapter. The direct search optimization procedure suggested by Luus and Jaakola (1973) involves taking randomly chosen points over a reasonable region and then making the search more intensive around the best point by reducing the region size in a systematic fashion. There are basically only three steps:

(1) Choose some reasonable initial point (this point does not have to be a feasible point) $\mathbf{x}^*$ and a reasonable region size vector $\mathbf{r}$. Then choose a number of random points R in the n-dimensional space around this point through the equation:

$$\mathbf{x} = \mathbf{x}^* + \mathbf{D}\mathbf{r}^j, \tag{4}$$

where $\mathbf{D}$ is a diagonal matrix, where randomly chosen diagonal elements lie in the interval $[-1, +1]$, and $\mathbf{r}^j$ is the region size vector for the jth iteration.

(2) Check the feasibility of each such randomly chosen point with respect to the inequality constraint (2). For each feasible point evaluate the performance index I in Eq. (1), and keep the best $\mathbf{x}$-value.

(3) An iteration is defined by Steps 1 and 2. At the end of each iteration, $\mathbf{x}^*$ is replaced by the best feasible $\mathbf{x}$-value obtained in step 2, and the region size vector $\mathbf{r}^j$ is reduced by γ through

$$\mathbf{r}^{j+1} = \gamma\,\mathbf{r}^j, \tag{5}$$

where γ is a region contraction factor such as 0.95, and j is the iteration index. This procedure is continued for a number of iterations and the results are examined. The procedure is straightforward and Fortran programs using LJ optimization procedure are given by Luus (1993, 2000a).

The procedure is straightforward, but the user must specify the initial center of the region $\mathbf{x}^*$, the initial region size vector $\mathbf{r}^1$, the region reduction factor γ and decide on the number of random points R to be used in each iteration. The importance of the reduction factor was illustrated by Spaans and Luus (1992). Michinev *et al.* (2000), showed that for a large number of optimization problems the efficiency, as measured in terms of the number of function evaluations, can be increased quite substantially by reducing the number of random points R whenever there is an improvement in the performance index. They found that the reduction factor 0.6 worked very well for a number of problems. The reliability of obtaining the global optimum for nonunimodal systems can be improved by incorporating simple tunnelling into the LJ optimization procedure as was shown by Wang and Luus (1987, 1988). Bojkov *et al.* (1993) showed that the LJ optimization procedure was not restricted to low dimensional problems, but can be used in optimal control problems, where after parametrization, one can be faced with a very high dimensional optimization problem. In fact, they solved successfully a 960 variable optimization problem. The LJ optimization was found to be a good procedure by Luus and Hennessy (1999) for checking the results obtained for fed-batch reactor optimization by other methods.

An effective way of choosing the region size over which the random points are chosen to improve the convergence rate was presented by Luus (1998). We consider such improvements of the basic LJ procedure later. Now, however, let us illustrate the application of the basic LJ optimization procedure by taking a simple 7-food diet problem.

2.1. Example of an optimization problem — diet problem with 7 foods

Let us consider the problem of selecting a snack which gives the maximum amount of satisfaction, and which also provides some nutritional value, as considered by Luus (2000a). Let us suppose that while attempting to gain the maximum satisfaction, we are concerned about cost, calories, protein, and iron, and there are 7 foods from which to choose the snack. The nutritional values of these 7 foods, levels of satisfaction, and constraints are given in Table 1.

Each food is rated on a satisfaction scale from 1 to 100, called utility, where, for example beer is given a utility value of 95 and hot dog 19, and it is assumed that satisfaction is proportional to the amount of food consumed.

Table 1. Diet problem with 7 foods.

	Food	Utility	Cost dollars	Calories kcal	Protein g	Iron mg	Maximum allowed
1	Apple	35	0.70	70	0.0	0.4	1
2	Beer, 12-oz bottle	95	1.25	150	1.0	0.0	2
3	Hamburger	25	0.99	245	21.0	2.7	2
4	Hot dog	19	1.10	270	8.6	1.3	1
5	Milk, 250 mL	40	0.75	157	8.0	0.1	1
6	Muffin, bran	75	1.00	86	3.0	1.3	2
7	Pizza, sector	50	0.90	185	7.0	0.7	3
	Constraints		5.00	800	25	3.50	

These utility values will change from one individual to another, and from day to day. By choosing these values, we are saying that we prefer a bottle of beer 5 times more than a hot dog, and we prefer a hamburger, having given a utility of 25, slightly more than a hot dog on that particular day. The nutritional values for the foods have been taken from the report of the Health and Welfare Canada (1979) and the costs are estimated.

The goal is to choose the amount of each food to maximize the total utility, obtained by summing the amount of utility obtained from each food, and to meet the constraints given as the last row and the last column in Table 1. The total cost should not exceed \$5, and we do not want to consume more than 800 calories (food calories which are really kcal). However, we would like to obtain at least 25 g protein and 3.5 mg of iron. As indicated in the last column of Table 1, there is an upper limit placed upon the amount of each food to be eaten.

Let us denote by x_i the amount of food i to be consumed. Although some of these variables should have integer values, we consider the case where these are taken as continuous variables, and fractional amount of each food is allowed. By using the utility values in Table 1, the total amount of satisfaction is then given by the performance index:

$$I = 35x_1 + 95x_2 + 25x_3 + 19x_4 + 40x_5 + 75x_6 + 50x_7. \tag{6}$$

The goal here is to choose the amount of each food x_i to maximize this performance index, subject to the four inequality constraints related to the nutritional requirements:

$$0.7x_1 + 1.25x_2 + 0.99x_3 + 1.1x_4 + 0.75x_5 + x_6 + 0.9x_7 \leq 5 \quad \text{(cost)}, \tag{7}$$

$$70x_1 + 150x_2 + 245x_3 + 270x_4 + 157x_5 + 86x_6 + 185x_7 \leq 800 \quad \text{(calories)}, \tag{8}$$

$$x_2 + 21x_3 + 8.6x_4 + 8x_5 + 3x_6 + 7x_7 \geq 25 \quad \text{(protein)}, \tag{9}$$

$$0.4x_1 + 2.7x_3 + 1.3x_4 + 0.1x_5 + 1.3x_6 + 0.7x_7 \geq 3.5 \quad \text{(iron)}, \tag{10}$$

It is noted that the inequalities for protein and iron are reversed, since it is desired to have the snack contribute toward the minimum nutritional requirements for these two minerals. The x_i cannot be negative, and we also have upper bounds on the amount of each food as given in the last column of Table 1. Therefore, there are the additional inequality constraints:

$$0 \leq x_1 \leq 1, \tag{11}$$

$$0 \leq x_2 \leq 2, \tag{12}$$

$$0 \leq x_3 \leq 2, \tag{13}$$

$$0 \leq x_4 \leq 1, \tag{14}$$

$$0 \leq x_5 \leq 1, \tag{15}$$

$$0 \leq x_6 \leq 2, \tag{16}$$

$$0 \leq x_7 \leq 3. \tag{17}$$

Here all the mathematical expressions are linear and the variables are positive semi-definite, so linear programming is best suited for maximizing the performance index in Eq. (4). To solve this problem by linear programming, slack variables, surplus variables and artificial variables have to be introduced to put the problem into a suitable form. However, in spite of the additional variables, the optimization is accomplished very fast with linear programming (Luus, 2000a, pp. 38–42). In fact, to solve this problem with the linear programming computer program listed in the appendix of reference (Luus, 2000a, pp. 287–290), we obtain $I = 338.12747$ with $x_1 = 0$,

$x_2 = 1.439$, $x_3 = 0.455$, $x_4 = 0$, $x_5 = 1$, $x_6 = 2$, and $x_7 = 0$ in negligible computation time (less than 0.06 s). In many problems, instead of linear relations, we are faced with nonlinear expressions, and linear programming would be more difficult to use, since linearization would be required and the global optimum would not be assured.

Let us now consider this optimization problem as described by Eqs. (6)–(17) by LJ optimization procedure. As a starting point let us choose $\mathbf{x}^* = [0.5\ \ 0.5\ \ 0.5\ \ 0.5\ \ 0.5\ \ 0.5\ \ 0.5]^T$, the region size vector for the first iteration $\mathbf{r}^1 = [2.0\ \ 2.0\ \ 2.0\ \ 2.0\ \ 2.0\ \ 2.0\ \ 2.0]^T$, the region reduction factor $\gamma = 0.95$, and let us specify the total number of iterations to be performed as 301. We consider this as a mathematical problem to illustrate the LJ optimization procedure and attempt to get 8-figure accuracy for the optimum value of the performance index. The above algorithm yielded the results in Table 2, where the problem was run for different number of

Table 2. Diet problem with 7 foods solved by LJ optimization procedure, showing the performance index as a function of iteration number and number of randomly chosen points per iteration R.

	Number of random points R					
Itn no.	100	1,000	10,000	20,000	50,000	100,000
1	278.74555	308.20355	330.47691	330.47691	330.47691	334.55093
21	316.20312	333.80802	334.47839	336.59327	337.03895	337.37480
41	331.54830	336.63188	336.69458	337.86878	337.79660	338.04237
61	334.71816	336.97549	337.56456	338.04817	337.95275	338.04715
81	335.65701	337.52988	337.58666	338.04817	338.09620	338.12285
101	336.35799	337.97786	337.65449	338.09751	338.11635	338.12285
121	336.43562	337.97786	337.67202	338.11966	338.12477	338.12455
141	336.52514	337.98307	337.68489	338.12369	338.12706	338.12669
161	336.52514	337.98318	337.69031	338.12603	338.12706	338.12693
181	336.52770	337.98568	337.69048	338.12699	338.12728	338.12733
201	336.52932	337.98568	337.69111	338.12734	338.12738	338.12743
221	336.53040	337.98599	337.69140	338.12744	338.12746	338.12746
241	336.53043	337.98599	337.69154	338.12744	338.12746	338.12747
261	336.53059	337.98605	337.69154	338.12746	338.12747	338.12747
281	336.53064	337.98608	337.69161	338.12746	338.12747	338.12747
301	336.53064	337.98609	337.69161	338.12746	338.12747	338.12747
CPU (s)	0.11	0.88	7.91	15.87	39.60	79.20

random points R on a PentiumIII/600 digital computer. The PentiumIII/600 computer was found to be about 3 times slower than Pentium4/2.4 GHz.

The procedure is easy to program and, with the availability of very fast personal computers, a reasonable amount of computational inefficiency can be tolerated. One of the great advantages of the method is that no auxiliary variables are required, so that the user is closer to the problem at hand. As can be seen in Table 2, the optimum $I = 338.12747$, with $x_1 = 0$, $x_2 = 1.43943$, $x_3 = 0.45527$, $x_4 = 0$, $x_5 = 1$, $x_6 = 2$, $x_7 = 0$, is obtained with the use of $R = 50{,}000$ randomly chosen points at each iteration after 261 iterations, and after 241 iterations with $R = 100{,}000$. This solution is the same as obtained earlier with linear programming. Despite the large number of points required, the computation time on a PentiumIII/600 was only 1 min. For higher dimensional problems, however, it is desirable to improve the efficiency of the basic algorithm for the LJ optimization procedure. Thus, recently some research effort has been directed to improving the efficiency.

To increase the efficiency of this optimization procedure and to make the LJ optimization procedure viable for high-dimensional optimization problems, it was found by Luus *et al.* (1995) in solving an 84-dimensional cancer chemotherapy scheduling optimization problem, that the use of a multi-pass procedure, in which a relatively small number of randomly chosen points is used in each iteration, improved the computational efficiency quite substantially. In the multi-pass method the three-step procedure is repeated after a given number of iterations, usually with a smaller initial region size than the one used at the beginning of the previous pass. Recently, Luus (1998) showed that the strategy of choosing the region size at the beginning of the next pass to be the range over which the variables have changed during the current pass is an effective way of increasing the efficiency. Let us illustrate the use of such a multi-pass procedure for this same 7-food diet problem.

For the multi-pass procedure, as a starting point let us choose again $\mathbf{x}^* = [0.5\ 0.5\ 0.5\ 0.5\ 0.5\ 0.5\ 0.5]^T$, the region size vector $\mathbf{r} = [2.0\ 2.0\ 2.0\ 2.0\ 2.0\ 2.0\ 2.0]^T$, the region reduction factor $\gamma = 0.95$, and let us specify the total number of iterations to be performed at each pass as 21. Then after each pass the region size is put equal to the change of each variable during the pass. In order to avoid the collapse of the region size, if the change is less than 10^{-6} then the region size is put equal to 10^{-6}.

Table 3. Diet problem with 7 foods solved by LJ optimization procedure using a multi-pass procedure, showing the performance index as a function of pass number and number of randomly chosen points per iteration R.

	Number of random points R				
Pass no.	100	500	1,000	2,000	5,000
1	316.20312	335.06610	333.80802	336.49133	334.81180
5	333.97033	337.18234	336.38848	338.12742	338.12730
10	335.11970	337.19169	336.44332	338.12747	338.12747
15	335.16068	337.25475	336.44362	338.12747	338.12747
20	335.16083	337.72014	336.46736	338.12747	338.12747
25	335.16083	338.03720	338.10305	338.12747	338.12747
30	335.16084	338.03720	338.12746	338.12747	338.12747
CPU (s)	0.77	1.43	3.13	4.17	9.34

As can be seen in Table 3, where the computation time on a PentiumIII/600 for the 30 passes is given in the last row, the efficiency in getting the optimum to 8 figures is increased quite substantially by using a smaller number of points, but numerous passes. A substantial improvement in computation time can be obtained by noting that after a small number of iterations some of the inequality constraints become active and then use these as equalities, as was suggested by Luus and Harapyn (2003). If a feasible solution is not readily available at the beginning of the iterations, then continuation can be used effectively (Luus *et al.*, 2006). Thus the LJ optimization procedure keeps the user close to the problem and enables complex optimization problems to be solved with reasonable computational effort.

One may wonder if the efficiency of the optimization procedure can be increased if the center of the region is taken as the newly calculated point immediately, rather than waiting for the end of the iteration to make the change to the best point. For this problem, as is seen in Table 4, it appears that the original formulation is better than this "greedy" approach. When the number of random points is chosen as 2,000 or 5,000, there is not much difference, and the global optimum is established in 8 passes (versus 7 passes with the original version). However, if 1,000 or fewer points are taken, then the original version is better.

Table 4. Diet problem with 7 foods solved by LJ optimization procedure using a multi-pass procedure, showing the performance index as a function of pass number and number of randomly chosen points per iteration R, where the center of the region is immediately taken as the best value once obtained.

	Number of random points R				
Pass no.	100	500	1,000	2,000	5,000
1	323.83416	330.00494	331.07071	331.39871	333.15635
5	331.46914	332.01188	336.46777	338.11506	338.12536
10	331.47498	334.04033	336.85513	338.11506	338.12747
15	331.47498	334.04453	336.90429	338.12747	338.12747
20	331.47498	334.04461	336.90430	338.12747	338.12747
25	331.47499	334.04462	336.90431	338.12747	338.12747
30	331.47499	334.04462	336.90431	338.12747	338.12747
CPU (s)	0.77	1.48	2.37	4.12	9.33

2.2. Example 2 — Alkylation process optimization

We consider the alkylation process described by Payne (1958) and used for optimization with sequential linear programming by Sauer *et al.* (1964), and also by Bracken and McCormick (1968). The problem involves determining 10 variables, subject to 20 inequality constraints and 7 equality constraints. Luus and Jaakola (1973) arranged the equality constraints, so that 7 variables could be eliminated from the optimization by solving for the values in a seriatim fashion. We therefore have the following optimization problem: Maximize

$$P = 0.063x_4x_7 - 5.04x_1 - 0.35x_2 - 10x_3 - 3.36x_5, \tag{18}$$

subject to the inequality constraints

$$\alpha_i \le x_i \le \beta_i, \quad i = 1, 2, \ldots, 10, \tag{19}$$

where $\alpha = [0.01\ 0.01\ 0.01\ 0.01\ 0.01\ 85\ 90\ 3\ 1.2\ 145\ ]^T$ and $\beta = [2{,}000\ 16{,}000\ 120\ 5{,}000\ 2{,}000\ 93\ 95\ 12\ 4\ 162\ ]^T$, and the equality constraints

$$x_4 = x_1[1.12 + x_8(0.13167 - 0.006667x_8)], \tag{20}$$

$$x_5 = 1.22x_4 - x_1, \tag{21}$$

$$x_2 = x_1 x_8 - x_5, \tag{22}$$

$$x_6 = 89 + [x_7 - (86.35 + x_8(1.098 - 0.038x_8))]/0.325, \tag{23}$$

$$x_{10} = -133 + 3x_7, \tag{24}$$

$$x_9 = 35.82 - 0.222x_{10}, \tag{25}$$

$$x_3 = \frac{0.001x_4x_6x_9}{98 - x_6}. \tag{26}$$

It is seen that the equality constraints are arranged in a special order, so that when values for the variables x_1, x_8, and x_7 are given, then the remaining seven variables are calculated directly through the seven equations. Thus, there are only 3 independent variables for optimization. The variables have the following meaning:

x_1 = olefin feed (barrels/day)
x_2 = isobutane recycle (barrels/day)
x_3 = acid addition rate (thousands of pounds/day)
x_4 = alkylate yield (barrels/day)
x_5 = isobutane makeup (barrels/day)
x_6 = acid strength (weight percent)
x_7 = motor octane number
x_8 = external isobutane-to-olefin ratio
x_9 = acid dilution factor
x_{10} = F-4 performance number

For optimization, we took as initial center of the region $x_1^{(0)} = 1{,}500$, $x_7^{(0)} = 93$, and $x_8^{(0)} = 10$, and initial region sizes for the first pass of 200, 1, and 1, respectively. After the first pass the initial region size was determined by the extent of the variation of the variables during the previous pass. To avoid the collapse of the region size, the region collapse parameter $\epsilon = 10^{-6}$ was used. As was shown by Luus (2003), the choice of the region collapse parameter affects the convergence rate and a good approach is to choose initially a value such as 10^{-4} and reduce its value in a systematic way and terminate the optimization when it reaches a low value such as 10^{-11} (Luus *et al.*, 2006). But for simplicity, here we used a constant value of 10^{-6}. For optimization we used $R = 7$ random points per iteration and

201 iterations per pass. Convergence to $P = 1{,}162.03$ was very rapid, being achieved after the fourth pass, with $x_1 = 1{,}728.37$, $x_2 = 16000.0$, $x_3 = 98.13$, $x_4 = 3{,}056.04$, $x_5 = 2{,}000.0$, $x_6 = 90.62$, $x_7 = 94.19$, $x_8 = 10.41$, $x_9 = 2.616$, and $x_{10} = 149.57$. The computation time for six passes on the PentiumII/350 (which is about 6 times slower than Pentium4/2.4 GHz) was 0.11 s, as read from the computer clock. It is noted that x_2 and x_5 are at the upper bounds of the inequality constraints.

This example provides good insight into plant expansion. Suppose we contemplate expanding the isobutane recycle capacity and of isobutane makeup. Taken individually, we see from Fig. 1 that approximately \$43 benefit is expected if we expand the isobutane recycle handling capacity by 25%. From Fig. 2 we see that approximately \$46 benefit is expected if we expand the isobutane makeup capacity by 25%. However, if we expand *both* of them by 25% simultaneously, then a benefit of about \$210 is obtained. This type of synergistic effect is difficult to predict from one variable at a time consideration.

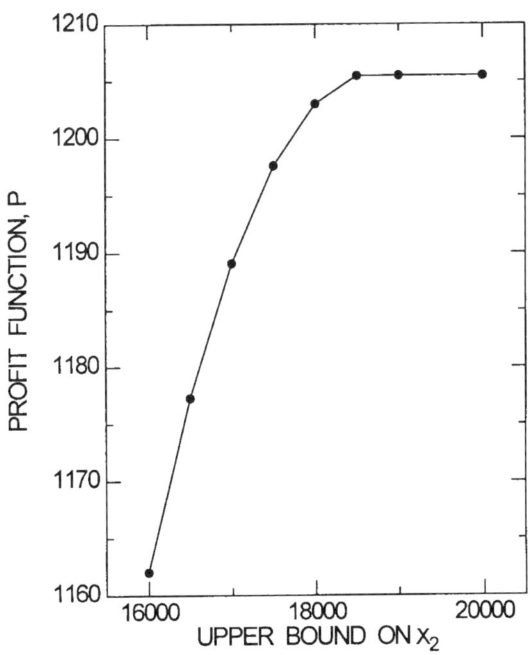

Figure 1. Effect of relaxing the upper bound on isobutane recycle.

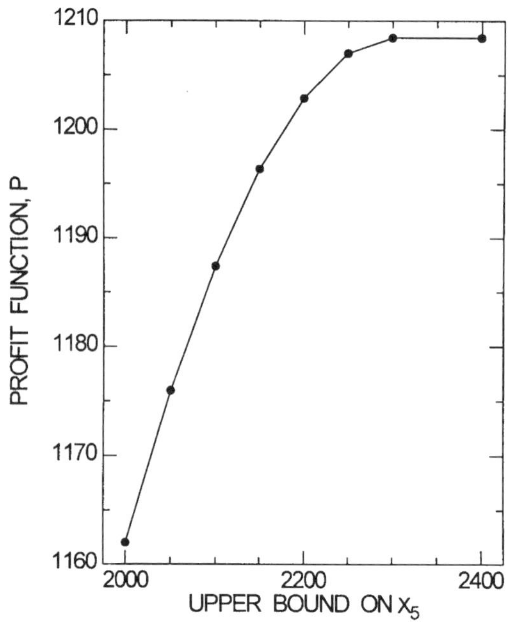

Figure 2. Effect of relaxing the upper bound on isobutane makeup.

2.3. Example 3 — Gibbs free energy minimization

At equilibrium the Gibbs free energy takes on the minimum value. Therefore, minimization of the Gibbs free energy should yield the composition of the phases at equilibrium. This approach was used by White *et al.* (1958) for determining the equilibrium composition of a mixture of chemical species in a single phase. The same problem was considered also by Luus and Jaakola (1973), Smith and Missen (1982), and by Luus (2000c).

To minimize $G/(RT)$ involves minimization of the performance index

$$I = \sum_{j=1}^{10} n_j \left[c_j + \ln P + \ln \left(\frac{n_j}{\sum_{i=1}^{10} n_i} \right) \right], \tag{27}$$

where n_j is the number of species j and c_j for each of the species is given by White *et al.* (1958) and by Smith and Missen (1982): $c_1 = -10.021$, $c_2 = -21.096$, $c_3 = -37.986$, $c_4 = -9.846$, $c_5 = -28.653$, $c_6 = -18.918$, $c_7 = -14.640$, $c_8 = -28.302$, $c_9 = -30.954$, and $c_{10} = -26.111$. The three mass balances are arranged into a form to eliminate three variables

$$n_{10} = 2 - n_1 - 2(n_2 + n_3) - n_6, \tag{28}$$

$$n_8 = 1 - n_4 - 2n_5 - n_6, \tag{29}$$

$$n_9 = 0.5(1 - n_{10} - n_3 - n_7 - n_8), \tag{30}$$

so the number of independent variables for the minimization of the performance index I is reduced from 10 to 7.

As starting conditions we chose $n_i = 0.2$, and $r_i = 0.2$ for $i = 1, 2, \ldots, 7$, and used region reduction factor $\gamma = 0.95$ and 101 iterations in each pass. After every pass the region size was put equal to the extent to which each variable had changed during the pass, using the region collapse parameter $\epsilon = 10^{-6}$. Convergence to the equilibrium composition was readily obtained in a small number of passes. The optimum compositions as obtained for two values of pressure that have been used in the literature are given in Table 5. It is noted that the optimum compositions and the performance index are quite sensitive to the value of pressure that has been used. For the pressure used by White *et al.* (1958), the optimum value of the performance index is −47.761090859 and for the pressure of 51.0 atm, the performance index is −47.761376550.

Table 5. Moles of chemical species at equilibrium for Example 3, as obtained with two values of pressure P with comparisons to literature values.

i	Species	Eqm. values obtained with $P = \exp(3.932)$	Eqm. values reported by White *et al.* (1958)	Eqm. values obtained with $P = 51.000$	Eqm. values reported by Smith and Missen (1982)
1	H	0.0406681	0.040668	0.0406727	0.0406728
2	H_2	0.1477303	0.147730	0.1477374	0.1477372
3	H_2O	0.7831534	0.783153	0.7831415	0.7831418
4	N	0.0014142	0.001414	0.0014143	0.0014143
5	N_2	0.4852466	0.485247	0.4852462	0.4852462
6	NH	0.0006932	0.000693	0.0006932	0.0006932
7	O	0.0179473	0.017947	0.0179494	0.0179494
8	NO	0.0273993	0.027399	0.0274000	0.0274000
9	O_2	0.0373144	0.037314	0.0373164	0.0373164
10	OH	0.0968713	0.096872	0.0968763	0.0968760

For convergence rate, we examine the sum of absolute deviations from the equilibrium composition

$$S = \sum_{i=1}^{10} \left| n_i - n_i^0 \right|, \tag{31}$$

where n_i^0 are the number of moles of species i at equilibrium, as given in Table 5. The use of S for examining convergence rate, rather than the performance index, is more practical due to the low sensitivity of the performance index to changes in composition: $I = -47.761090859$ corresponds to $S = 2.8330 \times 10^{-7}$, whereas $I = -47.761090853$ corresponds to $S = 5.0416 \times 10^{-4}$.

It is natural to expect faster convergence in terms of the number of passes if a larger number of random points is used in each iteration. However, the gain in terms of computational effort is no longer achieved when more than 100 random points are used. This is shown in Table 6 where a wide range of the number of random points is used for three different values for the region collapse parameter ϵ. Even with $R = 1{,}000$, the computation time is negligible. Pentium/120 was found to be about 15 times slower than pentium4/2.4 GHz.

As is shown in Table 7, the use of 201 iterations in each pass decreases the number of passes required if more than 50 random points are used in each

Table 6. Number of passes required to get $S < 5.0 \times 10^{-7}$ with 101 iterations per pass; computation time (in s on Pentium/120) is given in parentheses.

	Region collapse parameter		
Number of random points, R	$\epsilon = 10^{-7}$	$\epsilon = 10^{-6}$	$\epsilon = 10^{-5}$
11	18 (1.21)	12 (0.82)	12 (0.83)
15	14 (1.10)	11 (0.88)	9 (0.66)
25	9 (0.93)	10 (1.04)	22 (2.36)
50	12 (2.09)	9 (1.54)	17 (2.97)
100	8 (2.41)	8 (2.41)	11 (3.40)
200	21 (12.74)	13 (7.69)	8 (4.51)
500	7 (9.28)	6 (7.80)	5 (6.26)
1,000	15 (42.57)	6 (15.49)	6 (15.44)

Table 7. Number of passes required to get $S < 5.0 \times 10^{-7}$ with 201 iterations per pass; computation time (in s on Pentium/120) is given in parentheses.

	Region collapse parameter		
Number of random points, R	$\epsilon = 10^{-7}$	$\epsilon = 10^{-6}$	$\epsilon = 10^{-5}$
11	28 (2.80)	16 (1.64)	18 (1.86)
15	15 (1.87)	19 (2.42)	18 (2.20)
25	10 (1.75)	9 (1.59)	8 (1.49)
50	12 (3.90)	8 (2.53)	3 (0.88)
100	4 (2.31)	4 (2.31)	3 (1.59)
200	4 (4.34)	3 (3.13)	3 (3.18)
500	4 (10.60)	3 (7.58)	3 (7.69)
1,000	4 (21.09)	3 (15.16)	3 (15.21)

Table 8. Number of passes required to get $S < 5.0 \times 10^{-7}$ with 51 iterations per pass; computation time (in s on Pentium/120) is given in parentheses.

	Region collapse parameter		
Number of random points, R	$\epsilon = 10^{-7}$	$\epsilon = 10^{-6}$	$\epsilon = 10^{-5}$
11	37 (1.81)	33 (1.66)	NC
15	24 (1.37)	26 (1.49)	NC
25	NC	50 (3.75)	48 (3.41)
50	20 (2.15)	14 (1.48)	NC
100	42 (7.69)	25 (4.50)	NC
200	10 (3.02)	10 (3.02)	24 (7.75)
500	27 (20.43)	15 (10.93)	21 (15.65)
1,000	15 (21.42)	12 (16.86)	12 (16.76)

iteration, but there is no computational advantage. Similarly, as is shown in Table 8, there is no computational advantage in reducing the number of iterations per pass to 51. The rate of convergence with 101 iterations per pass, as is shown in Fig. 3, is systematic and rapid.

When there are more than one phase, the equations describing the system become more complex, but as has been shown by Lee *et al.* (1999) fairly complex situations can be handled by LJ optimization procedure.

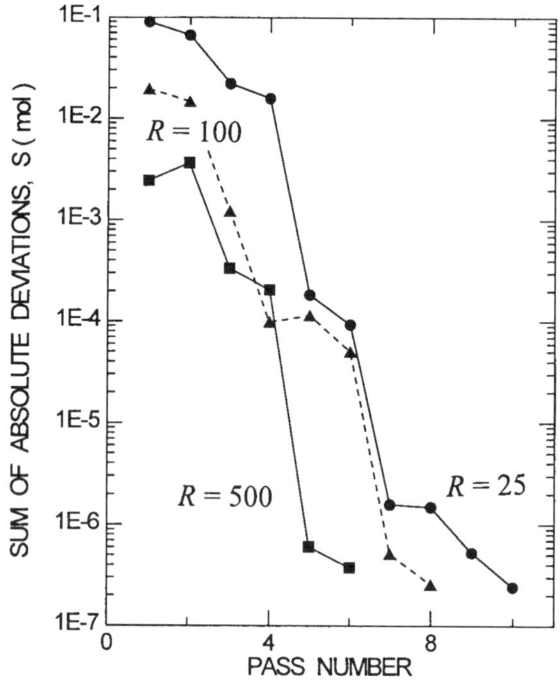

Figure 3. Convergence to the minimum value of the Gibbs free energy, showing the effect of the number of random points per iteration.

3. Handling Equality Constraints

In many optimization problems equality constraints are present. Frequently, the equalities can be arranged so that the optimization problem is actually simplified by reducing the number of free variables, as has been illustrated here with the optimization of the alkylation plant. The seven equality constraints were rearranged to reduce the dimensionality of the optimization problem from 10 to 3. Also, in the optimization problem involving minimization of the Gibbs free energy, the chemical equations were used to simplify the optimization problem. The question arises about what to do in cases where we have *difficult* equality constraints, i.e. equality constraints that cannot be arranged to enable elimination of variables.

Replacing each equality constraint by two inequality constraints has been used by some, for example, Himmelblau (1972), Miettinen *et al.* (2003), and Jeżowski *et al.* (2005). Although such "relaxation" of equality constraints by a small amount like 0.01 may not appear large, the resulting effect on the performance index can be very large. In the alkylation process optimization such a procedure yielded an increase of 50% in the profit expectation.

An effective way to handle difficult equality constraints, without introducing errors of approximation, is to use penalty functions. To outline this approach,we consider the situation where, in addition to the inequality constraints in Eq. (2), we also have m equality constraints

$$h_i(x_1, x_2, \ldots, x_n) = 0, \quad i = 1, 2, \ldots, m, \tag{32}$$

where these equality constraints are "difficult" in the sense that they cannot be used to solve for some particular variables in seriatim fashion.

Although a two-pass method to deal with equality constraints (Luus, 1974) was effective to solve optimization problems involving recycle streams (Luus, 1975), the general approach for handling equality constraints with LJ optimization procedure has not been solved satisfactorily, until quite recently. Hartig *et al.* (1995) showed that penalty functions can be used very effectively in direct search optimization. The work was extended by Luus and Wyrwicz (1996), and now it appears that the use of a quadratic penalty function incorporating a shifting term is a good way of dealing with difficult equality constraints (Luus, 1996). We consider the augmented performance index:

$$J = I + \theta \sum_{i=1}^{m} (h_i - s_i)^2, \tag{33}$$

where θ is a positive penalty function factor, and a shifting term s_i is introduced for each equality constraint so that θ can be reasonably small and the equality constraints would be accurately satisfied. Luus and Storey (1997) showed that there is usually a wide range for θ for which accurate results are obtained. To solve the optimization problem, LJ optimization procedure is used in a multi-pass fashion, where at the beginning of each pass consisting of a number of iterations, the region sizes are restored to a fraction of the sizes used at the beginning of the previous pass. The shifting terms s_i are updated after every pass simply by adjusting the values at the beginning of pass $(q+1)$ based on the deviation of the left hand side in Eq. (32) from zero; i.e.

$$s_i^{q+1} = s_i^q - h_i, \quad i = 1, 2, \ldots, m, \tag{34}$$

where q is the pass number. Upon optimization the product $-2\theta s_i$ gives the Lagrange multiplier associated with the ith equality constraint, yielding useful sensitivity information (Luus and Storey, 1997).

Another approach to dealing with equality constraints is to solve numerically the difficult algebraic equations at each iteration (Luus, 2000b, 2001). Even a large set of algebraic equations can be solved very efficiently (Luus, 1999, 2000a). This approach has the advantage of reducing the number of variables over which optimization is carried out, and we illustrate this by considering a simple example.

3.1. Example 4 — Geometric problem

Let us consider the geometric problem used by Luus (1974) to show how to handle equality constraints in LJ optimization procedure. The same example was used again (Luus, 1996), (Luus and Storey, 1997) to show the use of shifting terms in optimization. The problem is to find the maximum distance from the origin to the intersection of an ellipsoid with a hyperboloid. For mathematical convenience we consider the square of the distance. Therefore it is required to maximize

$$I = x_1^2 + x_2^2 + x_3^2, \tag{35}$$

subject to the two equality constraints

$$\begin{aligned} h_1 = {} & 4(x_1 - 0.5)^2 + 2(x_2 - 0.2)^2 + x_3^2 + 0.1x_1x_2 \\ & + 0.2x_2x_3 - 16 = 0, \end{aligned} \tag{36}$$

and

$$h_2 = 2x_1^2 + x_2^2 - 2x_3^2 - 2 = 0. \tag{37}$$

We construct the augmented performance index

$$J = I - \theta[(h_1 - s_1)^2 + (h_2 - s_2)^2], \tag{38}$$

where θ is a positive penalty function factor and the shifting terms s_1 and s_2 are initially put to zero and then, after ever pass q, are updated by

$$s_i^{q+1} = s_i^q - h_i, \quad i = 1, 2. \tag{39}$$

To observe the convergence of the iterative scheme, we look at the performance index I and the norm of the equality constraint violation defined by

$$S = |h_1| + |h_2|. \tag{40}$$

To solve this problem with LJ optimization procedure, we took the initial value for each x_i to be zero, and the region size for each variable to be 10. We used the region contraction factor of $\gamma = 0.95$, and allowed 5 iterations per pass, with $R = 25$ random points per iteration. For the first five passes the region size was restored to 0.80 of its value at the beginning of the previous pass ($\eta = 0.80$), and for the rest of the passes the region size was determined by the extent of the variation of the variables, with the region collapse parameter $\epsilon = 10^{-6}$, to prevent collapse of the region if the change in a variable is less than ϵ. The use of three different values for θ gives widely different convergence profiles, as is seen in Fig. 4, so the convergence profile is dependent on θ, but the end result after 100 passes was the same. The value of the performance index in each case was $I = 11.67664$ and the norm of the constraint violation was less than 10^{-6}. As is shown in Table 9, the converged values of the shifting terms times the penalty function factor yield the same result for each case. In fact, $2\theta s_i$ gives the Lagrange multiplier for the constraint i. Thus we conclude that the Lagrange multiplier associated with the first constraint is -0.67341 and with the second constraint, 0.21106. As was shown by Luus (1996), by choosing different starting points or different parameters in the optimization procedure, it is possible to get the other three local maxima with their associated Lagrange multipliers.

Luus (2000c) showed that the use of penalty functions to handle the equality constraints in the optimization of a realistic reactor-extractor system considered by Lee (1969), which was used for optimization studies by Luus (1975).

Instead of using penalty functions to handle equality constraints, a better way is to solve the algebraic equations numerically in each iteration by the procedure recommended by Luus (1999, 2000a). By using the approach

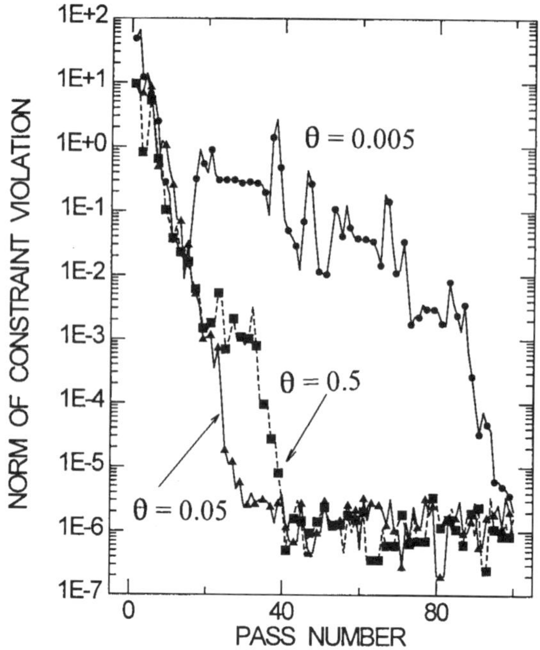

Figure 4. Convergence profile of the equality constraint violation norm with different values of the penalty function factor θ.

Table 9. Results after 100 passes with different values of θ for the geometric problem.

Penalty factor, θ	Performance index, I	Shifting terms		Variables		
		s_1	s_2	x_1	x_2	x_3
0.005	11.67664	−67.34144	21.10645	0.98842	2.67366	−1.88446
0.050	11.67664	−6.73414	2.110645	0.98842	2.67366	−1.88446
0.500	11.67664	−0.673415	0.211065	0.98842	2.67366	−1.88446

of solving the two difficult equality constraints numerically at each iteration Luus (2000b) reduced the problem to a one-dimensional search, where the global optimum was obtained in less than a second of CPU time on a PentiumIII/600. This same approach yielded the optimum operating conditions in a fluidized catalytic cracking unit (Iyer *et al.*, 2001; Luus *et al.*, 2002), where 12 difficult equality constraints were encountered.

To illustrate the computational advantages of this approach, we consider the case where Eq. (36) and Eq. (37) are solved numerically to obtain x_1 and x_2. Then optimization is simplified to finding the optimum value for x_3. Since there are four local maxima, it is important to solve the problem several times to ensure getting the global maximum. We therefore ran the optimization problem 50 times, starting with randomly chosen values for x_3 between -5 and 5. For optimization, we used $R = 10$ random points per iteration, 10 iterations per pass, and allowed 10 passes.

For the beginning of the first pass, the region size was chosen as 5. For subsequent passes, the region size at the beginning of the pass was put equal to the change in x_3 during the previous pass. If the region size so determined was less than the region collapse parameter $\epsilon = 10^{-6}$, then the region size was put equal to 10^{-6}. To solve the two algebraic equations, the method suggested by Luus (1999) was used with a tolerance of 10^{-12}. The starting values for x_1 and x_2 for the first iteration of the first pass were chosen at random between -5 and 5. For subsequent iterations, the best values for x_1 and x_2 obtained in the previous iteration were used as initial values for solving the algebraic equations. After every iteration the region size was reduced by $\gamma = 0.95$. If the performance index after 10 passes was not equal to the performance index after 9 passes, then convergence was not ensured, and the case was labelled as nonconvergence in Table 10. As can be seen, convergence to the global optimum $I = 11.67664$, with $x_1 = 0.98842$, $x_2 = 2.67366$, $x_3 = -1.88446$, was obtained 20 times, and nonconvergence resulted only once. The fourth local optimum $I = 8.78227$ was not obtained. The total computation time for the 50 cases was 0.44 s on a PentiumIII/600.

Table 10. Results from 50 different starting points for the geometric problem, by solving the two algebraic constraint equations numerically.

		Variables		
Performance index, I	Frequency	x_1	x_2	x_3
11.67664	20	0.98842	2.67366	−1.88446
10.47324	6	1.03949	2.49161	1.78454
9.56909	23	1.37627	−2.12617	1.77606

3.2. Example 5 — Design of columns

Let us consider a simply supported vertical solid column that is linearly tapered symmetrically from the center to its ends. Since the weight of the column is neglected, the largest cross-sectional area occurs at the middle of the column to provide the largest moment of inertia for resisting buckling. The problem is to determine the optimum linear tapering of the column so that the volume of material is minimized to withstand the applied axial load without buckling. This problem was solved with the LJ optimization procedure by Dinkoff *et al.* (1979). Here we wish to show the preliminary work that is necessary in getting the problem into the form so that optimization can be used.

Let $x = 0$ denote the bottom of the column, L be the length of the column, and $W(x)$ be the cross-sectional distance, or width, of the column. Then the volume of the column is

$$V = 2\omega \int_0^{\frac{L}{2}} (W(x))^2 dx, \tag{41}$$

where ω is the dimensionless area of the cross section chosen. For a square cross section $\omega = 1$, for a circular cross section $\omega = \pi/4$, and for an equilateral triangle, $\omega = \sqrt{3}/4$. For linear tapering

$$W(x) = W_0 + ax, \quad 0 \le x \le L/2. \tag{42}$$

where W_0 is the width of the column at the bottom and at the top. The maximum width occurs at the center of the column at $x = L/2$, where the width is $W_1 = W_0 + aL/2$.

Integration of Eq. (41) gives

$$V = \omega L \left(W_0^2 + \frac{LW_0 a}{2} + \frac{L^2 a^2}{12} \right), \tag{43}$$

which can be written in terms of the width at the center as

$$V = \omega L \left(W_1^2 - \frac{LW_1 a}{2} + \frac{L^2 a^2}{12} \right). \tag{44}$$

The axial deflection y of the slender column on the verge of buckling is given by

$$EI(x)\frac{d^2y}{dx^2} + Py = 0 \tag{45}$$

where E is the Young's modulus, P is the axial load and the moment of inertia is

$$I(x) = \xi(W(x))^4. \tag{46}$$

For square cross section $\xi = 1/12$, for circular cross section $\xi = \pi/64$ and for equilateral triangle $\xi = \sqrt{3}/96$.

The boundary conditions on Eq. (45) are

$$y = 0, \quad \text{at } x = 0, \tag{47}$$

$$\frac{dy}{dx} = 0, \quad \text{at } x = L/2. \tag{48}$$

Substituting Eq. (42) and Eq. (46) into Eq. (45) gives the differential equation

$$\left(\frac{W_0}{a} + x\right)^4 \frac{d^2y}{dx^2} + \alpha^2 y = 0, \tag{49}$$

where

$$\alpha^2 = \frac{P}{E\xi a^4}. \tag{50}$$

The substitution

$$x = 1/\theta \tag{51}$$

converts Eq. (49) into the Bessel equation

$$\frac{d^2y}{d\theta^2} + \frac{2}{\theta}\frac{dy}{d\theta} + \alpha^2 y = 0. \tag{52}$$

Further substitution

$$z = y\theta \tag{53}$$

converts the Bessel equation into a linear equation with constant coefficients

$$\frac{d^2z}{d\theta^2} + \alpha^2 z = 0. \tag{54}$$

Equation (54) is now solved with the given boundary conditions in Eqs. (47)–(48) to yield

$$\tan\left[\frac{k}{W_1 a} - \frac{k}{a(W_1 - aL/2)}\right] - \frac{k}{aW_1} = 0, \tag{55}$$

where

$$k = \sqrt{\frac{P}{\xi E}}. \tag{56}$$

We therefore have the equality constraint

$$h = \frac{k}{aW_1} - \frac{k}{a(W_1 - aL/2)} + \pi - \tan^{-1}\left(\frac{k}{aW_1}\right) = 0. \tag{57}$$

The optimization problem is to minimize V as given by Eq. (44) subject to the constraint in Eq. (57).

We therefore introduce the augmented performance index

$$J = V + \theta_1(h - s)^2 \tag{58}$$

where θ_1 is a positive penalty function factor and s is a shifting term which is updated after every pass q of the LJ optimization procedure by

$$s^{q+1} = s^q - h. \tag{59}$$

By taking $P = 1.0 \times 10^4$ N, $L = 2.50$ m, $E = 1.0 \times 10^{10}$ Pa, $\omega = 1.0$, $\xi = 1/12$, $\theta_1 = 10^2$, $\gamma = 0.95$, $R = 100$, 50 iterations per pass, in 50 passes, requiring 0.88 s of computation time on PentiumII/350 personal computer, we obtained $V = 6.178 \times 10^{-3}\,\text{m}^3$ with $|h| = 5.6 \times 10^{-6}$, $a = 0.0209234$, and $W_1 = 0.062213$ m. With these values, the ratio of the widths is $W_0/W_1 = 0.5796$, as reported by Dinkoff *et al.* (1979). For the initial values, we chose $a = 0.01$ and $W_1 = 0.05$. For the first three passes the initial region size was chosen to be 20% of the initial values of the corresponding variables, with a restoration factor $\eta = 0.70$. For the remainder of the passes, the adaptive region size was used.

Dinkoff *et al.* (1979) solved the optimization procedure slightly differently, by choosing the augmented performance index by using the Lagrange multiplier λ to append the equality constraint

$$J_2 = V + \lambda h, \tag{60}$$

and using the stationary conditions for the minimum of J_2:

$$\frac{\partial J_2}{\partial a} = 0, \tag{61}$$

and

$$\frac{\partial J_2}{\partial W_1} = 0. \tag{62}$$

This formulation thus requires the solution of three nonlinear algebraic equations Eq. (57), Eq. (61) and Eq. (62). The solution of these algebraic equations is readily accomplished by choosing a reasonable number of starting points at random and using the procedure suggested by Luus (1999).

For various combinations of P, L, E, and cross section, Dinkoff *et al.* (1979) found the following relationships for the optimum linear tapering:

$$\frac{W_0}{W_1} = 0.5796, \tag{63}$$

and

$$W_1 = 1.1849 \sqrt[4]{\frac{PL^2}{\xi \pi^2 E}}. \tag{64}$$

By using optimal linear tapering there is a material saving of 10.4%. It is interesting to note that the material saving is independent of P, L, E, and cross-section type. Further details are given in the paper by Dinkoff *et al.* (1979).

4. Effect of Parameters

To show the effect of the choice of two important parameters in the LJ optimization procedure, we consider the transformer design function minimization problem of Sarma (1990), that was used to illustrate the importance of the rate of search region reduction by Spaans and Luus (1992). The region reduction factor γ that is used to reduce the region size after every iteration is very important. Of equal importance is the region restoration factor η that is used in multi-pass procedure, where we wish to repeat the iterations but with a smaller initial region size. The best way to obtain a good feel for these parameters is to consider several examples. Here we consider a relatively simple, yet nontrivial, example. The problem is to minimize the transformer design function considered by Spaans and Luus (1992):

$$\begin{aligned} I = {} & x_1 x_4 \left(0.0204 + 0.0607 x_5^2\right)(x_1 + x_2 + x_3) \\ & + x_2 x_3 \left(0.0187 + 0.0437 x_6^2\right)(x_1 + 1.57 x_2 + x_4), \end{aligned} \tag{65}$$

subject to the constraints:

$$x_i > 0, \quad i = 1, \ldots, 6, \tag{66}$$

$$1 - (A + B) x_5^2 > 0, \tag{67}$$

where

$$A = 0.00062 x_1 x_4 (x_1 + x_2 + x_3), \tag{68}$$

$$B = 0.00058 (x_1 + 1.57 x_2 + x_4)(2.07)^2 \times 10^6 / x_1^2 x_2 x_3 x_4^2, \tag{69}$$

$$4AB < 1, \tag{70}$$

$$1/(2A) + C > x_5^2 > 1/(2A) - C, \tag{71}$$

where

$$C = \frac{[1 - 4AB]^{0.5}}{2A}, \tag{72}$$

$$0.16225 x_1 x_2 > 1, \tag{73}$$

$$x_3 > \frac{x_1 + x_2}{0.16225 x_1 x_2 - 1}, \tag{74}$$

$$x_4 > \frac{2x_1 + 2.57 x_2 + x_3}{0.16225 x_1 x_2 x_3 - (x_1 + x_2 + x_3)}, \tag{75}$$

and

$$x_6 = \frac{2.07 \times 10^3}{x_1 x_2 x_3 x_4 x_5}. \tag{76}$$

Although there are six variables, the presence of the simple equality constraint in Eq. (76) reduces the problem to a five-dimensional search. We chose the initial region sizes for these five variables to be the ranges

$$1 \leq x_1 \leq 10, \tag{77}$$

$$1 \leq x_2 \leq 10, \tag{78}$$

$$5 \leq x_3 \leq 15, \tag{79}$$

$$7 \leq x_4 \leq 17, \tag{80}$$

$$0.73 \leq x_5 \leq 1.27, \tag{81}$$

and the initial point at the center of the region, namely (5.5, 5.5, 10, 12, 1). By using a single pass consisting of 241 iterations, the minimum values for the performance index shown in Table 11 were obtained for different values of γ and R.

Although $R = 10{,}000$ appears to be a very large number of random numbers to be used at each iteration, the computation time on a

Table 11. Minimum performance index as a function of region reduction factor γ and the number of random points used per iteration R.

	Number of random points per iteration, R				
Reduction factor, γ	$R = 100$	$R = 250$	$R = 500$	$R = 1{,}000$	$R = 10{,}000$
1.00	138.75618	140.16045	137.09263	136.50329	135.91620
0.99	135.17358	135.20815	135.21686	135.15955	135.11182
0.98	135.08493	135.07967	135.07874	135.07767	135.07642
0.97	135.07719	135.07636	135.07611	135.07600	135.07599
0.96	135.07972	135.07600	135.07609	135.07597	135.07596
0.95	135.08123	135.07627	135.07603	135.07598	135.07596
0.94	135.07750	135.08088	135.07609	135.07597	135.07596
0.93	135.07864	135.07616	135.07638	135.07609	135.07596
0.92	135.08822	135.07645	135.07756	135.07603	135.07596
0.91	135.09007	135.07805	135.07633	135.07624	135.07596
0.90	135.12332	135.07629	135.07621	135.07622	135.07596
0.85	135.07830	135.08094	135.07844	135.07631	135.07596
0.80	135.31193	135.07730	135.08009	135.07666	135.07596
0.75	135.17052	135.13162	135.08872	135.08090	135.07597
0.70	135.08493	138.97426	135.08589	135.07651	135.07603

PentiumII/400 was only 8 s for 241 iterations, yielding the minimum value of $I = 135.07596$ for $0.80 \le \gamma \le 0.96$. So there is a wide range for γ to obtain the minimum.

The optimum values for the variables are $x_1 = 5.33257$, $x_2 = 4.65672$, $x_3 = 10.43293$, $x_4 = 12.08268$, $x_5 = 0.75260$, and $x_6 = 0.87865$.

To test whether convergence could be obtained faster by using a multi-pass method where the region size was set equal to the extent of the variation of a variable after each pass, we performed a series of runs consisting of 30 passes, each consisting of 41 iterations. The results in Table 12 show that there is no noticeable improvement in reducing computational effort by using this approach for this problem.

Another scheme of obtaining the initial region sizes in multi-pass method is to restore the region size to a factor η times the region size used in the previous pass. Here we illustrate the application of this idea by using 5 passes, each consisting of 241 iterations, and using a reduction factor $\gamma = 0.95$. Table 13 shows that this approach works very well with this example.

Table 12. Minimum performance index as a function of region reduction factor γ and the number of random points used per iteration R using 30 passes, each consisting of 41 iterations; the region size at the beginning of all passes after the first pass was put equal to the change in the corresponding variable in the previous pass.

	Number of random points per iteration, R				
Reduction factor, γ	$R = 20$	$R = 50$	$R = 100$	$R = 200$	$R = 2,000$
1.00	135.18905	135.07817	135.07598	135.09384	135.07596
0.99	135.12555	135.10038	135.08251	135.07663	135.07598
0.98	135.11597	135.07619	135.07656	135.07866	135.07599
0.97	135.07758	135.07651	135.07671	135.07752	135.07597
0.96	135.08249	135.07922	135.07636	135.07598	135.07597
0.95	135.12966	135.10139	135.08304	135.07664	135.07598
0.94	135.13423	135.08344	135.08468	135.07646	135.07597
0.93	135.19553	135.08063	135.10500	135.07604	135.07649
0.92	135.08789	135.07787	135.07851	135.07691	135.07597
0.91	135.34644	135.08955	135.08654	135.07695	135.07597
0.90	135.08379	135.08686	135.08502	135.07683	135.07602

Table 13. Minimum performance index as a function of region restoration factor η and the number of random points used per iteration R using 5 passes, each consisting of 241 iterations; the region size at the beginning of all passes after the first pass was put equal to η times the value used in the previous pass.

	Number of random points per iteration, R				
Restoration factor, η	$R = 20$	$R = 50$	$R = 100$	$R = 200$	$R = 2{,}000$
1.00	135.07817	135.07626	135.07731	135.07642	135.07596
0.95	135.08228	135.07623	135.07656	135.07619	135.07597
0.90	135.07864	135.07622	135.07633	135.07604	135.07597
0.85	135.07697	135.07629	135.07757	135.07613	135.07596
0.80	135.07696	135.07597	135.07688	135.07624	135.07597
0.75	135.07684	135.07622	135.07868	135.07617	135.07596
0.70	135.07787	135.07622	135.07686	135.07641	135.07596
0.65	135.07656	135.07606	135.07655	135.07600	135.07596
0.60	135.07990	135.07608	135.07653	135.07605	135.07597
0.55	135.07676	135.07627	135.07634	135.07637	135.07597
0.50	135.08006	135.07610	135.07696	135.07605	135.07597

These six examples have illustrated the application of the LJ optimization procedure in an introductory way. This optimization procedure has been found very useful in parameter estimation, model reduction and optimal control. These applications are illustrated in Chapter 12.

There are numerous optimization procedures which have arisen out of the basic idea of starting with a large search region and then making the search for the optimum more intensive by reducing the size of the region. An unbiased comparison amongst the methods is difficult. An attempt was made by Liao and Luus (2005). They found that there was no advantage in using the genetic algorithm rather than the LJ optimization procedure to optimize typical Chemical Engineering problems. Recently, Luus (2007) showed that by including a simple line search in the LJ optimization procedure, convergence rate is improved significantly and the optimum can be obtained very accurately. It is easier to apply than the gradient method suggested by Luus and Brenek (1989). This is especially important in model reduction, parameter estimation, and optimal control. This is presented in detail in Chapter 12 of this book, but here we wish to show the use of line search after every pass in minimization of the Rosenbrock function.

4.1. Example 7 — Minimization of Rosenbrock function

The Rosenbrock function has been widely used as a benchmark problem. The problem is to find $x_i, i = 1, 2, \ldots, n$ to minimize the performance index

$$I = \sum_{i=1}^{n-1} [(x_i - 1)^2 + 100(x_{i+1} - x_i^2)^2]. \tag{82}$$

In using the LJ optimization procedure the initial values for x_i were chosen to be random numbers between 0 and 5, and initial region size for each x_i to be 2.5. We specified 20 iterations per pass, region reduction factor of 0.95 after every iteration, and chose the region restoration factor of 0.85 for the first 10 passes. After that, we chose the region sizes from the amount that each variable has changed during the pass, with the region collapse parameter $\epsilon = 10^{-3}$ chosen initially. If the performance index did not change by more than ϵ in 3 passes, then ϵ was halved. The use of line search in the direction in which the variables have changed during the pass, using 10 function evaluations as suggested by Luus (2007), after every pass was tested. When ϵ became less than 10^{-11}, then the iterations were stopped.

We first considered the case $n = 100$ to show the effect of the line search. From 5 different starting points chosen at random with different seeds for the random number generator, convergence to I less than 10^{-14} was readily obtained with and without the line search as is shown in Table 14. However, the use of line search reduced the number of passes required for convergence quite substantially and reduced the average computation time per case from 8.8 s to 6.1 s with an AMD Athlon 3,800/2.4 GHz personal computer, which is about 1.5 times faster than Pentium4/2.4 GHz computer. Evaluation of line search was also carried out with $n = 3{,}000$, from two randomly chosen initial values as before. The results in Table 15 show that with and without line search convergence was obtained, and the use of line search speeds up the computations. The average computation time without line search was

Table 14. Effect of line search using $n = 100$ for the Rosenbrock function with the use of $R = 15$ random points per iteration and 20 iterations per pass.

	Without line search		With line search	
Case number	Performance index I	Passes	Performance index I	Passes
1	1.129×10^{-15}	3,073	1.376×10^{-15}	2,434
2	3.950×10^{-16}	2,610	1.935×10^{-15}	2,129
3	1.644×10^{-15}	2,602	5.134×10^{-15}	2,183
4	9.175×10^{-16}	3,466	3.292×10^{-15}	1,961
5	1.901×10^{-15}	3,247	1.057×10^{-15}	2,523

Table 15. Effect of line search using $n = 3{,}000$ for the Rosenbrock function with the use of $R = 15$ random points per iteration and 20 iterations per pass.

	Without line search		With line search	
Case number	Performance index I	Passes	Performance index I	Passes
1	4.479×10^{-13}	73,976	7.758×10^{-13}	44,735
2	4.678×10^{-13}	64,240	9.402×10^{-13}	50,155

5,726 s and with line search the time was reduced to 3,843 s. Therefore, line search was used for all the subsequent runs.

We next considered the Rosenbrock function with $n = 1{,}000$. From 5 different starting points, convergence to I less than 10^{-12} was readily obtained with both $R = 15$ and $R = 25$ randomly chosen points per iteration. The results in Table 16 show that, although the use of $R = 15$ requires a larger number of passes, computationally it is faster than the use of $R = 25$, requiring an average of 795 s per case as compared to 873 s per case with $R = 25$. For each case, the value 1.00000 was obtained for each x_i. The convergence profiles are shown in Figs. 5 and 6, where we see that the use of different seeds for the random number generator has no effect on the results of the optimization, since I less than 10^{-11} guarantees that each variable has attained the value 1.00000.

The use of $n = 10{,}000$ in the Rosenbrock function with $R = 15$ presented no difficulties as is shown in Fig. 7 where two different seeds

Table 16. Results with n = 1,000 with different values of R for the Rosenbrock function minimization.

	$R = 15$		$R = 25$	
Case number	Performance index I	Passes	Performance index I	Passes
1	6.234×10^{-14}	15,918	4.319×10^{-13}	10,469
2	5.312×10^{-13}	14,354	1.613×10^{-14}	11,451
3	7.012×10^{-14}	14,572	7.786×10^{-14}	10,761
4	2.034×10^{-13}	14,952	9.483×10^{-14}	10,140
5	1.338×10^{-13}	14,393	2.385×10^{-13}	10,152

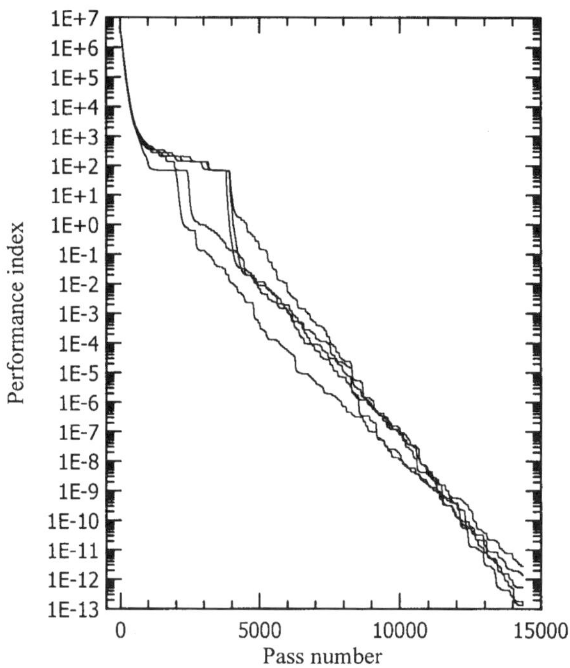

Figure 5. Convergence profile with n = 1,000 with the use of 15 random points per iteration.

were used in the random number generator. Convergence to $I = 5.451 \times 10^{-12}$ with 164,473 passes and $I = 6.554 \times 10^{-12}$ with 162,303 passes resulted with the average computation time of 44,177 s and all the variables were equal to 1.00000. As the dimensionality of the problem increases, the

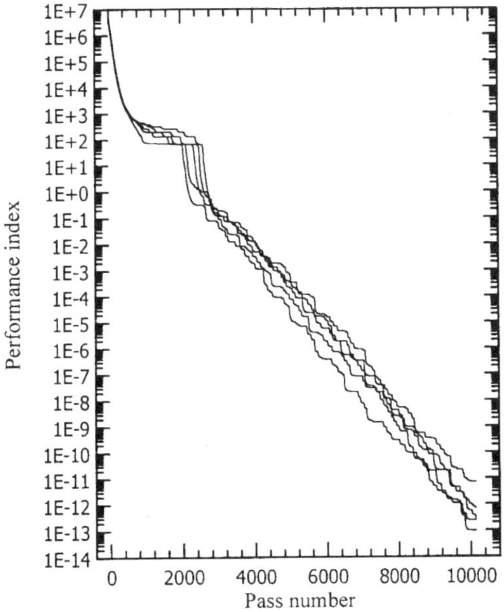

Figure 6. Convergence profile with $n = 1{,}000$ with the use of 25 random points per iteration.

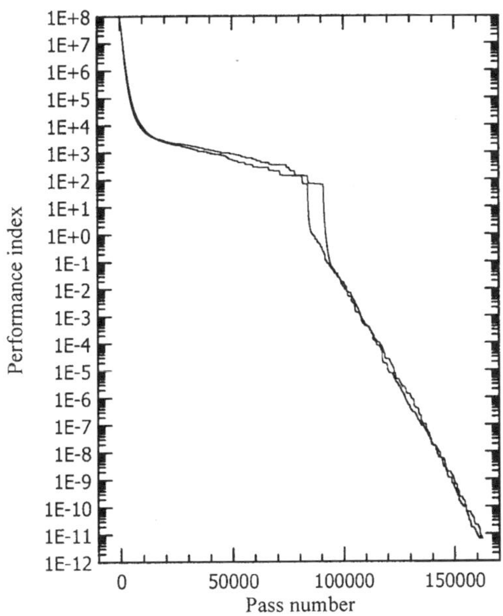

Figure 7. Convergence profile with $n = 10{,}000$ with the use of 15 random points per iteration.

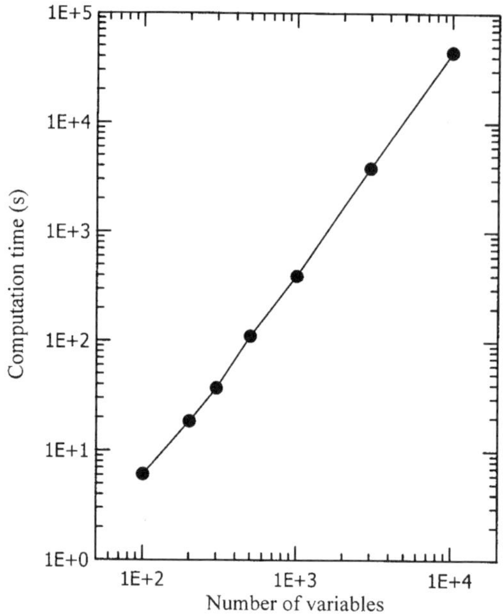

Figure 8. Effect of the number of variables n on the computation time on AMD Athlon/3,800 to obtain convergence.

computation time increases as is shown in Fig. 8. In each case $R = 15$ random points were used per iteration. It should be noted that the number of random points here does not have to be increased beyond 15 regardless of the dimensionality of the problem.

4.2. Example 8 — Maximization of the Shubert function

The Shubert function provides a good benchmark problem and has been used recently by Jeżowski *et al.* (2005) to test some optimization procedures. The problem is to find $x_i, i = 1, 2$ to maximize the performance index

$$I = \prod_{i=1}^{2} \sum_{j=1}^{5} j \cos[(j+1)x_i + j], \tag{83}$$

with

$$0 \leq x_i \leq 10, \quad i = 1, 2. \tag{84}$$

In using the LJ optimization procedure the initial values for x_i were chosen to be random numbers between 0 and 10, and initial region size for each x_i to be 5.0. As in the previous example, we specified 20 iterations per pass, region reduction factor of 0.95 after every iteration, and chose the region restoration factor of 0.85 for the first 10 passes. After that, we chose the region sizes from the amount that each variable has changed during the pass, with the region collapse parameter $\epsilon = 10^{-3}$ chosen initially. If the performance index did not change by more than ϵ in 3 passes, then ϵ was halved. The use of line search in the direction in which the variables have changed during the pass, using 10 function evaluations as suggested by Luus (2007), after every pass was also tested. When ϵ became less than 10^{-11}, then the iterations were stopped.

There was no difficulty in getting the global optimum $x_i = 5.48286$, $i = 1, 2$ with $I = 210.48229$. This is marginally better than 210.4820 reported by Jeżowski *et al.* (2005). In 100 runs, the success rate was over 50%, as is shown in Table 17. As is seen, the success rate was improved when R was increased from 15 to 100 random points per iteration. In each case convergence with $\epsilon < 10^{-11}$ was obtained in 30–42 passes with line search and 30–41 passes with the line search. There was no noticeable advantage in using the line search for this example. With the use of $R = 100$, besides the global optimum, only the local optimum with $I = 57.381399$ was obtained when one of the variables was zero and the other was 4.85806. However, with $R = 15$ we also obtained $I = 52.427358$ when one variable was 5.48286 and the other was 9.56321, and $I = 19.875836$ when both variables were 0. The computation time of 4.34 s for the 100 runs with $R = 100$ is negligible.

Table 17. Effect of the number of random points per iteration on the success rate in 100 cases in obtaining $I = 210.48229$ with $x_i = 5.48286$, $i = 1, 2$ for Shubert function maximization.

Number random points, R	Without line search		With line search	
	Number of successes	CPU time, s	Number of successes	CPU time, s
15	59	0.83	55	0.82
25	62	1.32	67	1.32
100	90	4.34	87	4.34

It was noticed that in most runs 8-figure accuracy was obtained for the performance index already in 15–18 passes, so to specify the limit $\epsilon < 10^{-11}$ may be too restrictive. For this problem we do not need 11-figure accuracy, and we could have reduced the computation time in half without sacrificing accuracy in the reported values. In the accompanying CD are Fortran programs for sample runs of optimization of the Rosenbrock function and Shubert function, and the output files.

5. Conclusions

The LJ optimization procedure has been found to be very effective in solving a wide range of optimization problems. When used in multi-pass fashion the efficiency is increased. Choosing the size of the region at the beginning of each pass from the amount by which each variable has changed during the previous pass, along with the collapse parameter ϵ if the change in some variable is less than ϵ, is an effective means of determining the region sizes. Then by reducing ϵ each time when the performance index changes by less than ϵ during 3 passes allows accurate convergence to be obtained. When ϵ becomes less than some value such as 10^{-11}, then iterations are stopped. This scheme provides a convenient termination condition. As was shown with the Rosenbrock function minimization, this works very well, and small number of random points per iteration can be used to get accurate convergence, even if the dimensionality of the problem is 10,000.

With the use of Rosenbrock function minimization, it was shown that the use of line search after every pass speeds up computations quite significantly, and is expected to be very useful for unconstrained optimization problems, or where inequality constraints are not active. When the inequality constraints are active, then these can be converted into equality constraints, and a simpler optimization problem can be solved with a reduced number of variables.

The LJ optimization procedure does not require a feasible point at the start, provided that within the specified number of iterations a feasible point is eventually obtained. If such is not the case, then continuation can be used, where the problem is changed slightly to enable a feasible solution to be obtained. Then the problem is changed gradually until the original problem is reached, and the optimal solution is then obtained for the original problem. This is illustrated by Luus *et al.* (2006). Whether the global optimum can always be obtained with continuation is not quite clear at present, and further research is required in this area.

References

Bojkov, B., Hansel, R. and Luus, R. (1993). Application of direct search optimization to optimal control problems. *Hungarian J. Ind. Chem.*, **21**, pp. 177–185.

Bracken, J. and McCormick, G.P. (1968). *Selected Applications of Nonlinear Programming*, Wiley, New York.

Dinkoff, B., Levine, M. and Luus, R. (1979). Optimum linear tapering in the design of columns. *Trans. ASME*, **46**, pp. 956–958.

Hartig, F., Keil, F.J. and Luus, R. (1995). Comparison of optimization methods for a fed-batch reactor. *Hungarian J. Ind. Chem.*, **23**, pp. 141–148.

Health and Welfare Canada (1979). *Nutrient Value of Some Common Foods*, Supply and Services Canada, Cat. No. H58-28/1979E.

Himmelblau, D.M. (1972). *Applied Nonlinear Programming*, McGraw-Hill, New York.

Iyer, R.S., Luus, R. and Woo, S.S. (2001). Optimization of a model IV fluidized catalytic cracking unit. *Canadian J. Chem. Eng.*, **79**, pp. 542–547.

Jeżowski, J., Bochenek, R. and Ziomek, G. (2005). Random search optimization approach for highly multi-modal nonlinear problems. *Advances in Engineering Software*, **36**, pp. 504–517.

Lee, E.S. (1969). Optimization of complex chemical plants by gradient technique. *AIChE Journal*, **15**, pp. 393–400.

Lee, Y.P., Rangaiah, G.P. and Luus, R. (1999). Phase and chemical equilibrium calculations by direct search optimization. *Comput. Chem. Eng.*, **23**, pp. 1183–1191.

Liao, B. and Luus, R. (2005). Comparison of the Luus-Jaakola optimization procedure and the genetic algorithm. *Engineering Optimization*, **37**, pp. 381–398.

Luus, R. (1974). Two-pass method for handling difficult equality constraints in optimization. *AIChE Journal*, **20**, pp. 608–610.

Luus, R. (1975). Optimization of multistage recycle systems by direct search. *Canadian J. Chem. Eng.*, **53**, pp. 217–220.

Luus, R. (1993). Optimization of heat exchanger networks. *Ind. Eng. Chem. Res.*, **32**, pp. 2633–2635.

Luus, R. (1996). Handling difficult equality constraints in direct search optimization. *Hungarian J. Industrial Chemistry*, **24**, pp. 285–290.

Luus, R. (1998). Determination of the region sizes for LJ optimization procedure. *Hungarian J. Industrial Chemistry*, **26**, pp. 281–286.

Luus, R. (1999). Effective solution procedure for systems of nonlinear algebraic equations. *Hungarian J. Industrial Chemistry*, **27**, pp. 307–310.

Luus, R. (2000a). *Iterative Dynamic Programming*, Chapman & Hall/CRC, London.

Luus, R. (2000b). Handling difficult equality constraints in direct search optimization. Part 2. *Hungarian J. Industrial Chemistry*, **28**, pp. 211–215.

Luus, R. (2000c). Luus-Jaakola optimization procedure. *Recent Res. Devel. Chemical Engng.*, Transworld Research Network, India, **4**, pp. 45–64.

Luus, R. (2001). Direct search Luus-Jaakola optimization procedure, LJ optimization procedure. In *Encyclopedia of Optimization*, Floudas, C.A. and Pardalos, P.M. (eds.), Kluwer Academic Publishers, Dortrecht, The Netherlands, **1**, pp. 440–444.

Luus, R. (2003). Effect of region collapse parameter on convergence rate of LJ optimization procedure. *Proceedings of IASTED International Conf. on Intelligent Systems and Control*, Salzburg, Austria, pp. 51–56.

Luus, R. (2007). Use of line search in the Luus-Jaakola optimization procedure. *Proceedings of the 3rd IASTED International Conference on Computational Intelligence CI2007*, July 2–4, 2007, Banff, Alberta, Canada, pp. 128–135.

Luus, R. (2008). Optimal control of oscillatory systems by iterative dynamic programming. *Journal of Industrial and Management Optimization*, **4**, pp. 1–15.

Luus, R. and Brenek, P. (1989). Incorporation of gradient into random search optimization. *Chem. Eng. Technol.*, **12**, pp. 309–318.

Luus, R. and Harapyn, I. (2003). Two-step method for active inequality constraints for direct search optimization. *Proceedings of IASTED International Conf. on Intelligent Systems and Control*, Salzburg, Austria, pp. 443–448.

Luus, R., Hartig, F. and Keil, F.J. (1995). Optimal drug scheduling of cancer chemotherapy by direct search optimization. *Hungarian J. Industrial Chemistry*, **23**, pp. 55–58.

Luus, R. and Hennessy, D. (1999). Optimization of fed-batch reactors by the Luus-Jaakola optimization procedure. *Ind. Eng. Chem. Res.*, **38**, pp. 1948–1955.

Luus, R., Iyer, R.S. and Woo, S.S. (2002). Handling equality constraints in direct search optimization. *Recent Developments in Optimization and Optimal Control in Chemical Engineering*, Luus, R. (ed.), Research Signpost, Kerala, India, pp. 129–153.

Luus, R. and Jaakola, T.H.I. (1973). Optimization by direct search and systematic reduction of the size of search region. *AIChE Journal*, **19**, pp. 760–766.

Luus, R., Sabaliauskas, K. and Harapyn, I. (2006). Handling inequality constraints in direct search optimization. *Engineering Optimization*, **38**, pp. 391–405.

Luus, R. and Storey, C. (1997). Optimal control of final state constrained systems. *Proc. IASTED International Conference on Modelling, Simulation and Control*, Singapore, August 11–14, pp. 245–249.

Luus, R. and Wyrwicz, R. (1996). Use of penalty functions in direct search optimization. *Hungarian J. Industrial Chemistry*, **24**, pp. 273–278.
Michinev, I., Stoyanov, S. and Keil, F.J. (2000). Investigation and modification of the Luus-Jaakola global optimization procedure, *Hungarian J. Industrial Chemistry*, **28**, pp. 231–239.
Miettinen, K., Makela, M.M. and Toivanen, J. (2003). Numerical comparison of some penalty-based constraint handling techniques in genetic algorithms. *J. Global Optimization*, **27**, pp. 427–446.
Payne, P.E. (1958). Alkylation — What you should know about this process. *Petroleum Refiner*, **37**, pp. 316–329.
Sarma, M.S. (1990). On the convergence of the Baba and Dorea random optimization methods. *JOTA*, **66**, pp. 337–343.
Sauer, R.N., Colville, A.R. and Burwick, C.W. (1964). Computer points the way to more profits. *Hydrocarbon Processing Petroleum Refiner*, **43**, pp. 84–90.
Smith, W.R. and Missen, R.W. (1982). *Chemical Reaction Equilibrium Analaysis: Theory and Algorithms*, Wiley, New York, pp. 137–139.
Spaans, R. and Luus, R. (1992). Importance of search-domain reduction in random optimization. *JOTA*, **75**, pp. 635–638.
Wang, B.C. and Luus, R. (1997). Optimization of non-unimodal systems. *Int. J. Num. Methods in Eng.*, **11**, pp. 1235–1250.
Wang, B.C. and Luus, R. (1998). Reliability of optimization procedures for obtaining global optimum. *AIChE Journal*, **24**, pp. 619–626.
White, W.B., Johnson, S.M. and Dantzig, G.B. (1958). Chemical equilibrium in complex mixtures. *J. Chemical Physics*, **28**, pp. 751–755.

Exercises

(1) Solve the 7 food diet problem by increasing the satisfaction (utility) of apple from 35 to 55. (*Ans.* $I = 338.53523$).

(2) (a) Solve the alkylation optimization problem by relaxing simultaneously the upper bound on isobutane recycle (x_2) from 16,000 to 17,600 barrels/day and the upper bound on isobutane makeup (x_5) from 2,000 to 2,200 barrels/day.

(b) Run this problem by increasing the upper bounds on x_2 and x_5 to 20,000 and 2,500 barrels/day, respectively.

(c) How do your results compare with what you would have expected from Figs. 1 and 2?

(3) Show that the given substitutions into Eq. (45) yield Eq. (49) and that the further substitution of Eq. (51) converts Eq. (49) into the Bessel Eq. (52)

which is easily converted to a second order differential equation with constant coefficient (Eq. (54)).

(4) Carry out the minimization of the Rosenbrock function with $n = 2{,}000$, making sure that adequate convergence is obtained so that I becomes less than 10^{-11} and all the variables become 1.00000.

Chapter 3

ADAPTIVE RANDOM SEARCH AND SIMULATED ANNEALING OPTIMIZERS: ALGORITHMS AND APPLICATION ISSUES

Jacek M. Jeżowski*, Grzegorz Poplewski and Roman Bochenek

Department of Chemical and Process Engineering
Rzeszów University of Technology, Rzeszów, Poland
**ichjj@prz.edu.pl*

1. Introduction and Motivation

Various stochastic optimization approaches (also called recently as "metaheuristic" ones) were applied and are expected to be widely used in the future in chemical and process engineering as well as in other engineering sciences. The methods that are applied in process engineering are as follows: random search (also called adaptive random search, ARS), simulated annealing (SA), genetic algorithms or more generally evolutionary algorithms (EA), tabu search, ant colony approach and particle swarm optimization. Stochastic methods usually provide only a general framework, and thus, they are often implemented in various versions of detailed programs. The key point in the successful application of a stochastic method is the "good" choice of control parameter setting. The choice of the right

*Corresponding author.

stochastic optimizer for the problem at hand is important too. This chapter addresses algorithms and parameter settings of two stochastic optimization techniques: ARS and SA with the simplex algorithm.

ARS is most likely the oldest stochastic technique having its peak of popularity in chemical and process engineering in the 1960s and 1970s. SA is also a mature optimization technique. Modern, often population-based, optimization approaches are more popular at present. However, the advantage of both ARS and SA for the potential user is that they are relatively well tested. The analysis of the literature shows that they proved their efficiency for an ample set of various tasks. The use of ARS for the challenging problem of water network optimization is described in Chapter 15 of this book. Also, some information on parameter settings, though still insufficient, can be found more easily for both SA and ARS techniques than for more recent optimization methods.

The rest of this chapter is organized in the following way. First, a version of ARS algorithm which is based on the concepts of a method of Luus and Jaakola (1973) is described in the next section. Then, we will proceed to explaining SA technique. We will focus on a method that embeds Simplex (Nelder–Mead) algorithm within SA general framework. The results of tests for both techniques will be analyzed next with special focus on parameter setting. Important application issues such as methods of dealing with constraints will be discussed, too. Summary section will conclude the chapter. The computer software for both optimization methods is available on the disc accompanying the book.

2. Adaptive Random Search Approach

2.1. Introduction

The ARS method is aimed at nonlinear programming (NLP) problems though there are some versions that tried to deal with mixed-integer nonlinear programming (MINLP) problems, for instance in Salcedo (1992) and Bochenek *et al.* (1999). However, it is possible to conclude from the examples solved in these works that the versions are able to solve cases with a small number of binary variables. At present, there is an opinion that ARS optimization is unable to cope with large problems. However, this opinion seems to be valid rather for MINLP formulation. Zabinsky (1998) stated that, for certain ARS algorithms, the computation load does not increase sharply with NLP problem size. Similar conclusions can be found in Hendrix *et al.* (2001). The works by Luus (1996), Lee *et al.* (1999) and Luus *et al.* (2002) to mention a few as well as the solution to water network optimization problem addressed in Chapter 15 support these conclusions. Liao and Luus (2005) reported recently superiority of ARS optimization over genetic algorithms for some benchmark global optimization tasks. Several authors, e.g. Edgar *et al.* (2001) in the textbook on optimization, pointed out the usefulness of ARS techniques for finding good initialization for NLP deterministic optimization approaches. ARS technique can also be applied in hybrid approaches to speed up computation — see e.g. Lima *et al.* (2006). Hence, literature information and the experience of the authors allow claiming that ARS optimization is efficient and easy to use for (at least) small and medium size NLP problems.

There exist several versions of ARS optimization method. Here, we limit ourselves to list those that have been presented in the literature on chemical and process engineering problems, usually together with applications, for instance the works of Luus (1973), Luus and Jaakola (1973), Jaakola and Luus (1974), Luus (1974), Luus (1975), Gaines and Gaddy (1976), Heuckroth *et al.* (1976), Campbell and Gaddy (1976), Kelahan and Gaddy (1977), Wang and Luus (1978), Martin and Gaddy (1982), Rangaiah (1985), Mihail and Maria (1986), Wang and Luus (1997), Luus and Brenek (1989), Salcedo *et al.* (1990), Salcedo (1992), Banga and Seider (1996), Banga *et al.* (1998), Bochenek *et al.* (1999), Michinev *et al.* (2000), Li and Rhinehart (2000), Bochenek and Jeżowski (2000), Jeżowski and Bochenek (2000; 2002), Luus *et al.* (2002), Jeżowski and Jeżowska (2003), Jeżowski *et al.* (2005), Ziomek *et al.* (2005).

The basic common concept of all these approaches is concentration of search in a small region around the current best solution. However, ARS algorithms proposed to date differ in regard to many other aspects, such as procedure of generating random trial points or pace of search region reduction. The method by Luus and Jaakola (1973), abbreviated to LJ algorithm in the following, is simple and its applications are numerous and well documented. We will present in the following two modified versions of the LJ method, starting with the original LJ method. The presentation in the following is for the minimization of unconstrained nonlinear problems but with bounds on variables, defined by (1) and (2), if not stated otherwise.

$$\min F(x), \tag{1}$$

subject to

$$x_i^l \le x_i \le x_i^{up}; \quad i = 1, \ldots, p. \tag{2}$$

Let x^0 denotes the starting point that should be provided by the user. The LJ algorithm generates certain number of random search points from the even distribution. The number of points will be called "the number of internal loops — *NIL*". These search points are generated in cycles. The number of cycles will be referred to as "the number of external loops — *NEL*". Symbol δ denotes size of search region. Initial sizes $(\delta_i^0, i = 1, \ldots, p)$ should be given in the input. They can be directly calculated from the lower and upper bounds on variables used in inequalities (2). The flow diagram of the LJ method is given in Fig. 1. Equations (3) and (4) in this flow sheet are given by:

$$x_i^k = x_i^* + r_i \delta_i^{k-1}; \quad i = 1, \ldots, p, \tag{3}$$

where r_i is random number of uniform distribution from the range (−0.5, 0.5) and

$$\delta_i^k = \beta \delta_i^{k-1}. \tag{4}$$

The LJ algorithm requires only three control parameters: β — contraction coefficient of search region size, *NEL* — number of external loops, *NIL* — number of internal loops. It is an important advantage over some other versions of ARS optimization. Note that the sizes of search regions in the LJ procedure are diminished at the same rate for each variable according to Eq. (4). The first modification to the LJ algorithm proposed in Jeżowski and Bochenek (2002) was to apply a pace of reduction dependent on the variable in order to reduce search region with different rate for each variable. This version was called M–LJ algorithm (for modified LJ algorithm). These authors proposed to use final size of search region for a variable given in the data instead of parameter β. For the given final sizes δ_i^f,

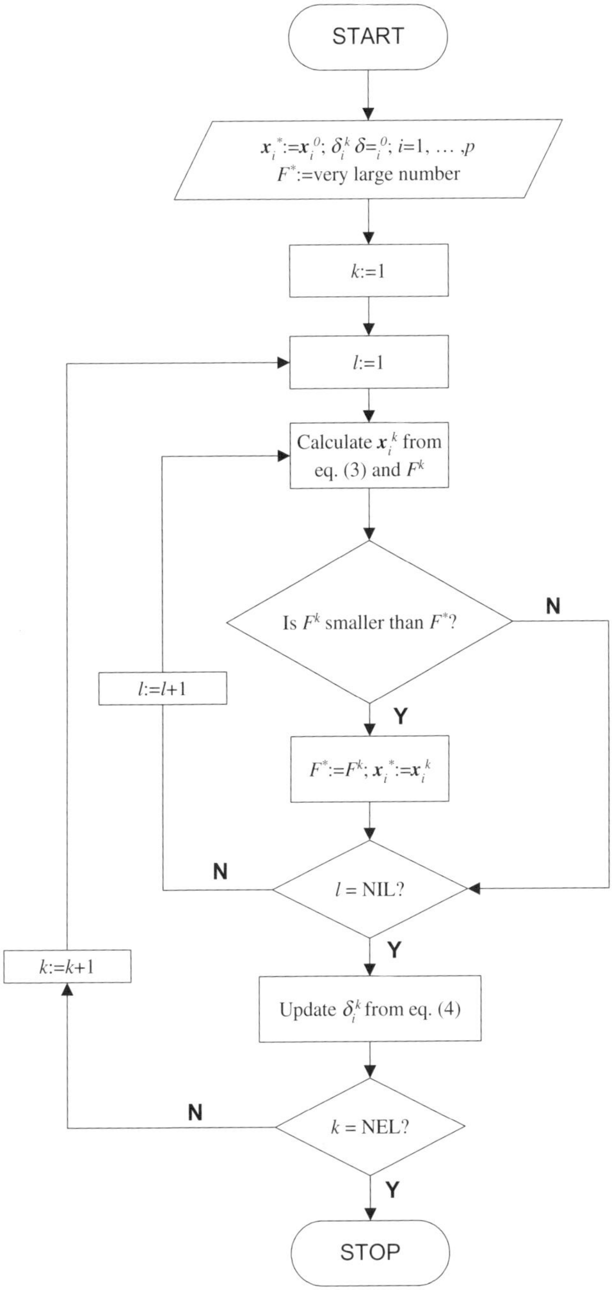

Figure 1. Flow chart of the LJ algorithm.

the values of contraction parameter β_i for variables x_i are calculated from:

$$\beta_i = \left(\frac{\delta_i^f}{\delta_i^0}\right)^{1/NEL}. \tag{5}$$

Alternatively, the search region size can be updated according to Eq. (7) instead of relation (4) and, thus, contraction parameter β_i^k for variable i after k external iterations is calculated from formula (6) instead of (5):

$$\beta_i^k = \left(\frac{\delta_i^f}{\delta_i^0}\right)^{k/NEL}, \tag{6}$$

$$\delta_i^k = \beta_i^k \delta_i^0. \tag{7}$$

One reason of applying the final region sizes in M–LJ method is that they are often easy to assess from "physical" interpretation of variables in an optimization problem and the knowledge of initial sizes of regions. Additionally, their application causes an effect of scaling the variables. Initial hyper-box search space is gradually diminished to a very small final hyper-box in such a manner that every edge (i.e. every variable) may change with different rate.

Jeżowski and Bochenek (2002) showed, via numerical experiments, that good results are also obtained when using one value (δ^f) for final sizes independent of variables as shown by relation (8), with recommended δ^f from the range $(10^{-1}, 10^{-4})$.

$$\delta_i^f = \delta^f; \quad i = 1, \ldots, p. \tag{8}$$

It is worth noting that in spite of the use of one final size δ^f, M–LJ method yields in majority of problems different values of contraction parameters β_i $(i = 1, \ldots, p)$ because of different initial sizes. Jeżowski and Bochenek (2002) found that the application of final sizes as control parameters in M–LJ version instead of one fixed parameter β in the LJ algorithm increases reliability of locating the global optimum by 10–20% in average for both unconstrained and constrained optimization problems with up to 20 variables.

The second change in the LJ algorithm is aimed mainly at increasing reliability of solving highly multi-modal NLP problems. This version is called LJ–MM method (LJ algorithm for multi-modal problems) and relies on change of reduction rate profile for search region sizes. In the LJ and M–LJ methods, the profile of reduction rate in terms of external loop number ($k = 1, \ldots, NEL$) is very steep in the initial phase of optimization and almost constant in the final stage in which many iterations are performed. This is illustrated in Fig. 2 for the LJ algorithm for parameters $\delta_i^0 = 1$ and $\delta_i^f = 10^{-3}$ for all variables. Due to a sharp reduction in search region at the beginning of calculations, the global optimum region can be cut off and the LJ and M–LJ algorithms can be trapped into a local optimum. This can be expected for highly multi-modal functions with local optima of similar values. Also, a bad starting point can cause a similar effect even for regular functions. To eliminate this potential pitfall, Jeżowski *et al.* (2005) applied a size reduction rate that features as close resemblance to Gaussian (normal) distribution as possible (see Fig. 3).

For normal distribution, size contraction coefficients β_i^k should conform:

$$\beta_i^k = \exp\left(-\left[\frac{z_i}{\sigma_i}\right]^2\right). \tag{9}$$

Additionally, these parameters must be kept in the range (0, 1) and the reduction scheme should follow the key idea of M–LJ version, i.e. size of the search regions of the last major iteration (*NEL*) has to be equal to the given final sizes. To meet these conditions, Jeżowski *et al.* (2005) derived

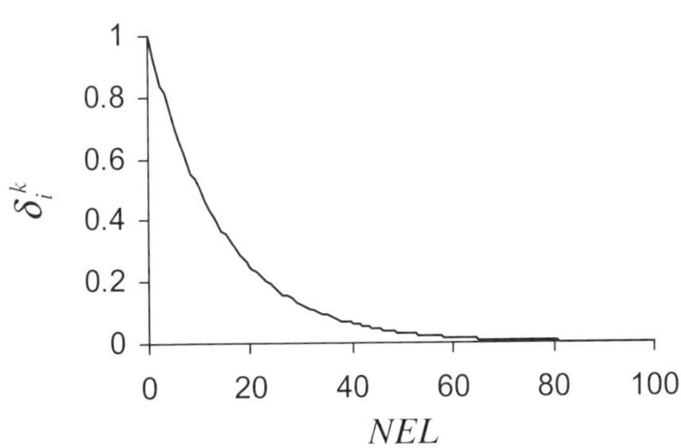

Figure 2. The profile of search region reduction rate in LJ or M–LJ method.

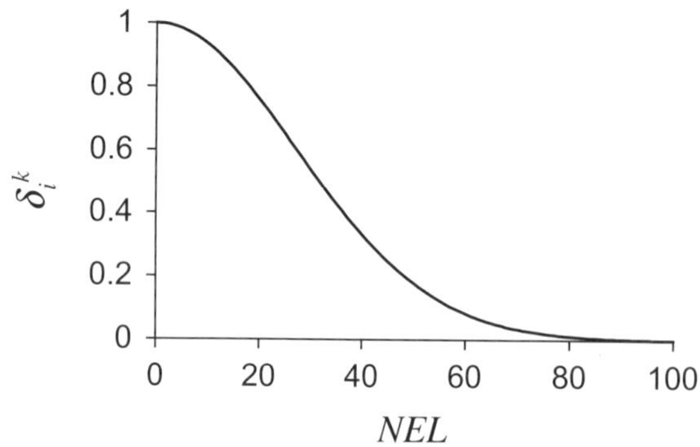

Figure 3. The profile of search region reduction rate in LJ–MM method.

the following formula for calculating parameters β_i^k:

$$\beta_i^k = \exp\left\{\left(\frac{k}{NEL}\right)^2 \cdot \ln\left(\frac{\delta_i^f}{\delta_i^0}\right)\right\}. \tag{10}$$

Note that these contraction coefficients require updating of region sizes using Eq. (7). As shown in Fig. 3, the rate of search region size reduction is slow in the initial phase of optimization by LJ–MM procedure, and hence this feature should provide a high reliability of locating the global optimum. This is, however, achieved by diminishing the number of iterations in final steps. Hence, a precise location of global optimum can be difficult, particularly for a flat goal function. Thus, LJ and LJ–MM algorithms can perform better in some cases if very precise location of the optimum is not required. The tests reported in Sec. 4 prove the validity of this analysis. This potential drawback does not seem serious in majority of industrial applications where very precise solution is not required.

Equation (10) can be rearranged into:

$$\beta_i^k = \left(\frac{\delta_i^f}{\delta_i^0}\right)^{\left[\frac{k}{NEL}\right]^2}. \tag{11}$$

This equation features a close resemblance to formula (6) applied in M–LJ version. Hence, semi-empirical Eq. (12) with heuristic parameter α is proposed for determination of region size reduction parameter:

$$\beta_i^k = \left(\frac{\delta_i^f}{\delta_i^0} \right)^{\left[\frac{k}{NEL}\right]^\alpha}. \tag{12}$$

Formula (12) provides the common framework for both M–LJ and LJ–MM algorithms since the former uses parameter $\alpha = 1.0$ while the latter sets α at 2.0. Generally, the use of Eq. (12) requires the good setting of α, and this issue will be addressed in Sec. 4.

3. Simulated Annealing with Simplex Method

3.1. Introduction

Simulated annealing (SA) is more recent optimization method than ARS approach, and its beginning can be set at 1983 year — the publication of seminal work by Kirkpatrick *et al.* (1983). The first monograph by Arst and Korst was published in the year 1989. SA mimics certain thermodynamic principles of producing an ideal crystal. It makes use of the algorithm of Metropolis *et al.* (1953) developed for simulating thermal moves of molecules at certain temperature T. This temperature is the crucial parameter of SA that influences both reliability and efficiency of optimization. To produce a crystal, the temperature has to be decreased. A rapid decrease of T gives rise to irregularities in crystal structure, which is similar to failure to locate the global minimum. Extremely slow decrease would result in prohibitive computation time. A choice of cooling scheme, i.e. the procedure of decreasing T will be addressed in the following. At certain temperature T, particularly at a high temperature, some molecules do not reach minimum energy level available at this temperature. To simulate this phenomenon, SA optimization mechanism allows accepting a worse solution but acceptance probability should drop with a decrease of T. Acceptance criterion is also important for SA optimization. Higher than zero probability of accepting currently worse solution has been the unique feature of SA. More recent tabu search optimization has also this feature which allows escaping of local minima traps.

Basic SA algorithm for unconstrained minimization problem defined by (1) and (2) is illustrated in Fig. 4. Equation (13) and conditions (14a)

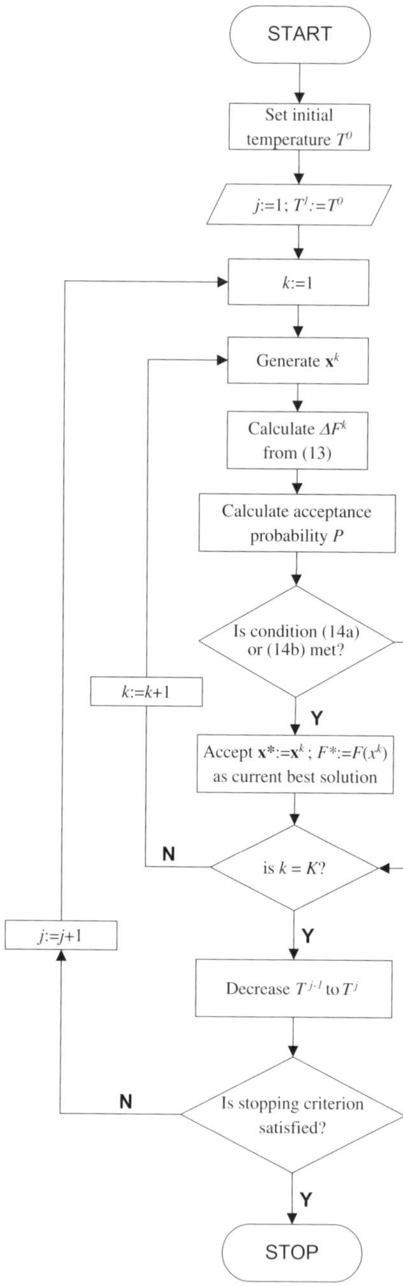

Figure 4. Flow chart of basic SA optimization method.

and (14b) in this figure are:

$$\Delta F^k = F(X^k) - F(X^{k-1}), \tag{13}$$

$$\Delta F^k \leq 0, \tag{14a}$$

$$\Delta F^k > 0 \quad \text{and} \quad P > RND, \tag{14b}$$

where P is the probability of accepting a feasible point and RND is a random number from uniform distribution between 0 and 1.

Various versions of the SA were proposed to date. Here, we limit ourselves to list some of them due to space limitations: Vanderbilt and Louie (1984), Corana *et al.* (1987), Press and Teukolsky (1991), Press *et al.* (1992), Cardoso *et al.* (1996; 1997), Maier and Whiting (1998), Sahin and Ciric (1998), Locatelli (2000), Chen and Su (2002), Chapter 8 in Spall (2003), and Őzcelik and Őzcelik (2004). For similar reason we mention only some applications of the SA in chemical and process engineering. The method is most often used for combinatorial problems or those that can be directly formulated as optimization problems with discrete-valued variables. Many papers address the application of the SA to scheduling and design of batch processes (e.g. Das *et al.* (1990), Ku and Karimi (1991), Patel *et al.* (1991), Murakami *et al.* (1997), Xi-Gang and Zhong-Zhou (1997), Hanke and Li (2000)), in combinatorial chemistry and molecular design (e.g. Ourique and Telles (1998), Marcoulaki and Kokossis (1998, 2000a,b)). Also, many papers deal with use of SA for design of processes, apparatus and process systems, for instance works by Dolan *et al.* (1989, 1990), Floquet *et al.* (1994), Chaudhuri and Diwekar (1997a,b), Maia *et al.* (1995), Cordero *et al.* (1997), Athier *et al.* (1997), Maia and Qassim (1997), Zhu *et al.* (2000).

Relatively few works deal with the application of the SA method for nonlinear problems (NLP) with continuous variables. As Press *et al.* (1992) pointed out, the possible reason is the deficiency of typical SA approaches where successive point $x^{i+1} \in R^p$ is randomly generated from the previous one x^i according to a scheme suggested in Vanderbilt and Louie (1984) or similar ones. This makes the optimization method inefficient for some characteristics of functions such as deep, narrow valleys. Hence, to eliminate this drawback, Press and Teukolsky (1991) proposed embedding the simplex algorithm (in particular, Nelder and Mead (NM) procedure) into the general

SA algorithm. Such an approach combines the robust, deterministic local optimizer for unconstrained problems with global optimization properties of the SA approach. Note that the NM method is used in every temperature level to find trial points for SA, and not just to refine the optimum in the final step. The method referred to as the SA–simplex (SA–S) in the following, was coded by Cardoso *et al.* (1996), with some extensions, as the subroutine called SIMPSA and was aimed at constrained NLP problems. Then, the application of SIMPSA was extended in Cardoso *et al.* (1997) for MINLP problems — M–SIMPSA algorithm. Both subroutines were successfully applied to various benchmark NLP problems and also small-scale MINLP problems. We adopted certain ideas from SIMPSA algorithm and coded the subroutine called SA–S/1 presented as follows.

3.2. SA–S/1 algorithm

To explain basic concepts of SA–S/1 method in detail we employ the description of the general SA from Sec. 3.1 as the background. The most important changes are as follows. Due to the application of NM procedure, the simplex with $p + 1$ vertices (for p variables) is used instead of a point x. Also, the block "Generate x^k" of the flow chart in Fig. 4 is modified to the following points:

(a) Generate a simplex using NM algorithm and perturb the goal functions of the simplex according to Eq. (15). Note that the perturbed vertices have higher values of the goal function (F') than those of the original simplex (F).

$$F_i' = F_i - T \cdot \log(RND_i) \tag{15}$$

where index i denotes vertex i of the simplex.

(b) select the vertex with maximum value of F_l' and reflect the simplex according to Nelder–Mead procedure.

(c) Perturb goal functions of reflected point "r" of the simplex according to Eq. (16). This results in the improvement of the goal function of vertex r of reflected simplex.

$$F_r' = F_r + T \cdot \log(RND) \tag{16}$$

In SA–S/1 algorithm, we applied an additional stochastic mechanism called "inverse movement". The simplex with perturbed values of simplex vertices according to Eqs. (15) and (16) is allowed to move uphill in case of minimization or downhill for maximization in step (4'b) of SA–S/1 method. Hence, the goal function may worsen. These inverse movements are controlled by parameter *INV*. If the value of *INV* is higher than random number *RND* of uniform distribution from the range (0,1) then the simplex is reflected to inverse direction. Condition (17) defines these reflections.

$$\text{if} \begin{cases} INV > RND & \text{then uphill reflection} \\ INV \leq RND & \text{then downhill reflection} \end{cases}. \tag{17}$$

This mechanism is expected to enable the search to escape from local optima traps at low values of the temperature, where SA–S/1 algorithm works as the original deterministic NM procedure.

Reliability and efficiency of the SA method depends largely on the cooling scheme, the choice of the initial simplex, the equilibrium criterion, the convergence criterion and on the initial temperature. In the following paragraphs, we describe mechanisms applied in SA–S/1 algorithm. We do not present an overview of other solution techniques that were proposed in the literature. The reader is referred to numerous papers for this, particularly papers with ample literature review such as Press *et al.* (1992), Cardoso *et al.* (1996, 1997), Chen and Su (2002), and Schmidt and Thierauf (2003).

3.3. Important mechanisms of SA–S/1 algorithm

3.3.1. Initial simplex generation

Coordinates ($x_i^s = i, \ldots, p$) of the starting vertex (denoted by superscript s) of initial simplex are generated from the uniform distribution according to:

$$x_i^s = x_i^l + RND_i(x_i^u - x_i^l); \quad i = 1, \ldots, p. \tag{18}$$

Coordinates of other vertices are calculated on the basis of the starting vertex from:

$$x_i = x_i^s + ((0.5 - RND_i) \cdot (x_i^u - x_i^l)); \quad i = 1, \ldots, p. \tag{19}$$

If bounds are not satisfied then the variable is set at the lower or upper bound.

3.3.2. Determination of the initial temperature

In SA–S/1 algorithm, we applied the iterative method from Cardoso *et al.* (1996, 1997), which is also similar to that of Chen and Su (2002) and Meier and Whiting (1997). It requires a number of simplex reflections at a high value of parameter T. We calculated this temperature as $T = 10^6 \cdot F^s$, where F^s is the goal function value of the initial vertex. The following parameters are calculated when performing the algorithm:

m_1 — the number of the reflections having improvement of the goal function
m_2 — the number of the reflections having no improvement of the goal function

The average increase of the goal function value is calculated from:

$$\Delta f^+ = \frac{\sum_{k=1}^{m_2}(F_k - F_{k-1})}{m_2}, \tag{20}$$

where F_k and F_{k-1} are values of the goal function of two successive feasible solutions for such simplex reflection which yielded an improvement of the goal function.

It is assumed that the acceptance probability (Px) of a solution at temperature T is defined by:

$$Px = \frac{m_1 + m_2 \cdot \exp\left(\frac{-\Delta f^+}{T}\right)}{m_1 + m_2}. \tag{21}$$

For fixed Px value we can determine the initial temperature T^0 from Eq. (22), such that the acceptance probability of the first point is not less than Px.

$$T^0 = \frac{-\Delta f^+}{\ln((Px \cdot (m_1 + m_2) - m_1)/m_2)}. \tag{22}$$

In the SA–S/1 algorithm we fixed acceptance probability Px at 0.95.

3.3.3. Acceptance criterion

From among various formulae suggested in the literature we have chosen the popular Metropolis criterion (Metropolis *et al.*, 1957). This criterion is given by:

$$P = \begin{cases} 1 & \text{if } \Delta F < 0 \\ \exp\left(-\dfrac{\Delta F}{T}\right) & \text{if } \Delta F \geq 0 \end{cases} \tag{23}$$

3.3.4. Cooling scheme — Temperature decrease

Often the simple exponential cooling scheme defined by Eq. (24) is used in SA:

$$T^{i+1} = T^i \cdot \Delta, \tag{24}$$

where $\Delta < 1$ and should be in the range (0.9, 0.99), as suggested in the literature. However, some researchers suggested that formula (25) called the adaptive cooling scheme or Aarts & van Laarhoven equation produces more reliable results than the exponential scheme for similar number of goal function evaluations (NFE).

$$T^{i+1} = \frac{T^i}{1 + \dfrac{T^i \cdot \ln(1+\gamma)}{3\sigma^i}}, \tag{25}$$

where σ^i is the standard deviation of the goal function at T^i defined by

$$\sigma^i = \sqrt{\frac{\sum_{k=1}^{ip}(F_k - F_a)^2}{ip}}. \tag{25a}$$

Here, *ip* is the number of feasible solutions found in temperature T^i, F_k is the value of the goal function of solution k at temperature T^i, F_a is the arithmetic mean value of the goal function for all feasible solutions calculated at temperature T^i.

Parameter γ in Eq. (25) is the control parameter of the cooling scheme. It takes values greater than zero. For small values, less than 1.0, the cooling is slow and, in consequence, optimization reliability increases but computational load too. Formula (25) has adaptive features, i.e. it accounts for "history" of the optimization by correlating a rate of cooling in terms of distance from the equilibrium state — the higher is the standard deviation (the system is far away from the equilibrium) the higher the rate of cooling. The exponential cooling scheme does not have such a feature. Because the reported literature does not give clear conclusion as to the choice of the cooling scheme, we decided to include both schemes in the SA–S/1 solver as the options. They have been tested and the results are given in Sec. 4.

3.3.5. Equilibrium criterion

The logical condition "is $k = K$?" in the flow chart of basic SA in Fig. 4 is usually called "equilibrium criterion". This criterion requires specification of parameter K. We applied K equal to the maximum number of generations given in input data. Parameter K is, thus, the control parameter of SA–S/1 method. Its influence will be discussed in Sec. 4.1.

3.3.6. Stopping (convergence) criterion

The simplest and commonly applied criterion is to stop the calculations if the temperature value is smaller than the given number $T^{\min}$, as shown by inequality (26). We also employed this condition.

$$T \leq T^{\min}. \tag{26}$$

Figure 5 shows the flow chart of the SA–S/1 optimization procedure for problems with inequality constraints (a more detailed description of the method of dealing with these constraints is addressed in Sec. 4.4).

4. Tests, Control Parameters Settings and Important Application Issues

4.1. Tests — Test problems and results

Both optimization approaches, i.e. algorithms LJ–MM and SA–S/1, were tested on many benchmark problems for (global) optimizers and also on several chemical and process engineering problems that are commonly used in the literature. The former are listed in Appendix A while the latter in Appendix B. The problems in Appendix A are both unconstrained (denoted by F) and constrained (with symbol E) formulations. They were considered

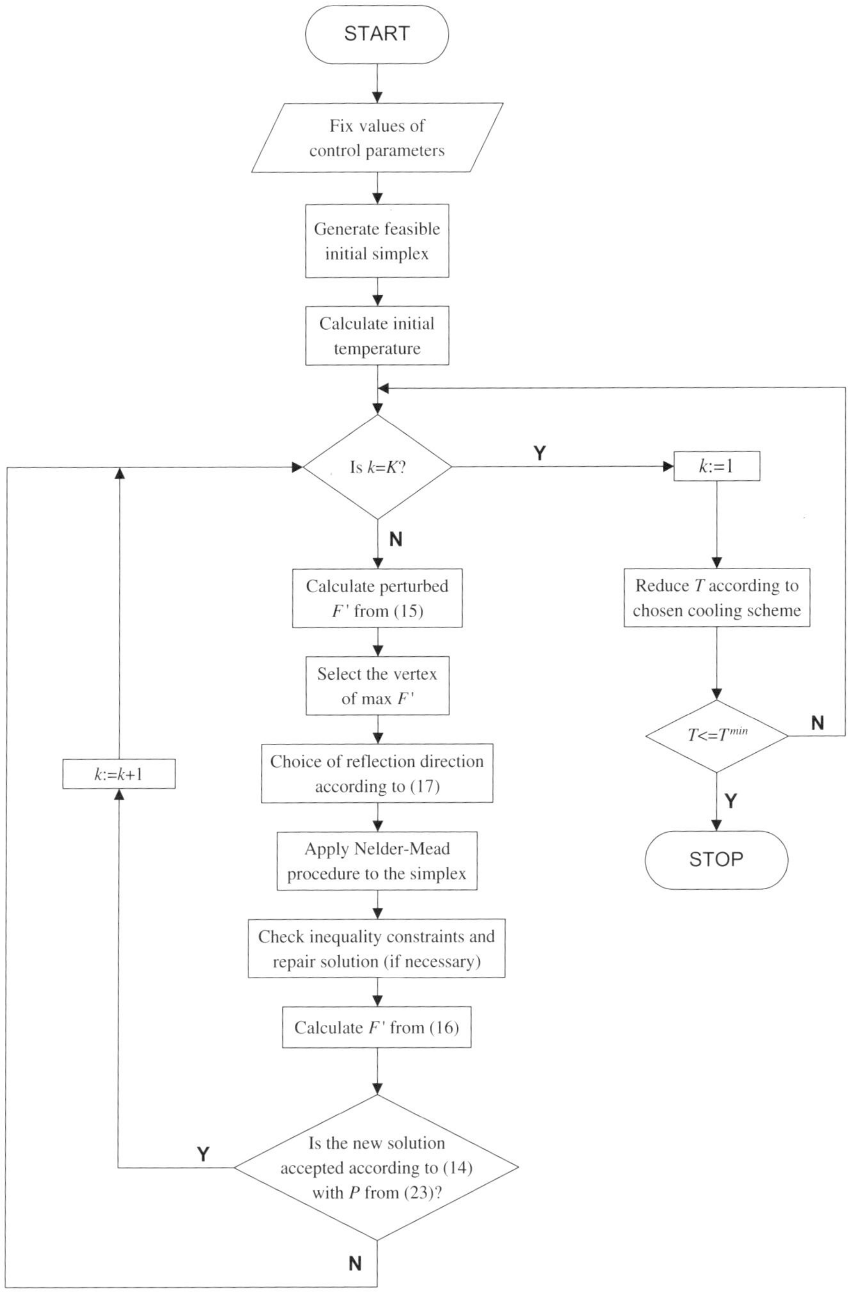

Figure 5. Flow chart of SA–S/1 algorithm.

in many works such as Floudas and Pardalos (1990), Michalewicz (1996), Ali *et al.* (1997), Michinev *et al.* (2000), Rajesh *et al.* (2000), Mathur *et al.* (2000), Siarry and Dognon (2001), Michalewicz and Fogel (2002), and Trafalis and Kasap (2002). Chemical and process engineering problems from Appendix B (with symbol B) can also be found in Salcedo *et al.* (1990), Visweswaran and Floudas (1990), Salcedo (1992), Vaidyanathan and El-Halwagi (1994), Ryoo and Sahinidis (1995), Floudas (1995), Biegler *et al.* (1997), Zamorra and Grossmann (1998) and de Gouvea and Odloak (1998). The main objectives of tests were to check efficacy and reliability of the algorithms and to estimate good parameter settings. The obvious conclusion from "no free lunch" theorem, e.g. Ho and Pepyne (2002), is that each optimizer does not perform equally well for all types of optimization problems. Additionally, one cannot expect that there exist universal and good parameter settings for all problems. Hence, we attempted to select an ample set of test problems though manageable in reasonable period of time. All test problems are of NLP type since both algorithms: LJ–MM and SA–S/1, particularly the latter, are for this class of optimization problems. However, LJ–MM subroutine can be used for MINLP formulation but with special treatment of binary variables as shown in Chapter 15.

We found that the performance of optimizers depends on whether a problem is multimodal with numerous sharp peaks of similar values, which are called "multi-modal" problems, or has relatively flat functions, particularly near the optimum — referred to as "flat" ones. Problems from F1 to F10, problems F14 and E8 belong to multi-modal ones. As an example, the plot of function F3 is shown in Fig. 6. The others belong to the class of "flat" problems (see Fig. 7).

Because both LJ–MM and SA–S/1 are stochastic optimizers, several runs of the subroutine are necessary with the same values of the parameters to ensure reliable results for the analysis. We run the optimizers from 100 up to 1,000 times depending on CPU time needed for the problem. To evaluate the performance of the optimization methods, two performance indices defined below are employed.

F^{mean} — mean value of the goal function for the problem found in all runs.
SR (Δe) — success rate calculated as the ratio of runs with the results that differ less than Δe from the best known result for the problem (treated as the global optimum) to the total number of runs.

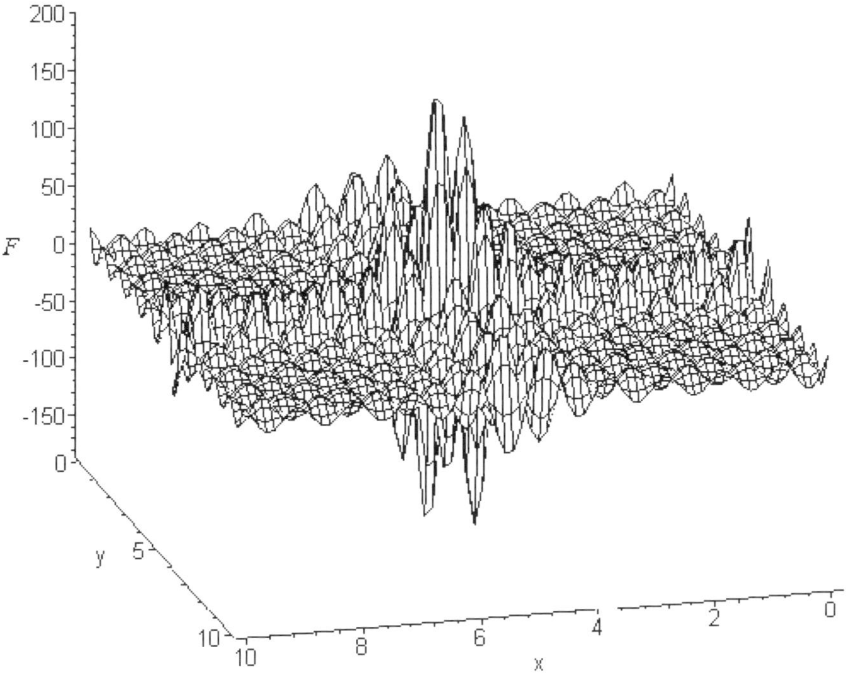

Figure 6. The profile of function F3.

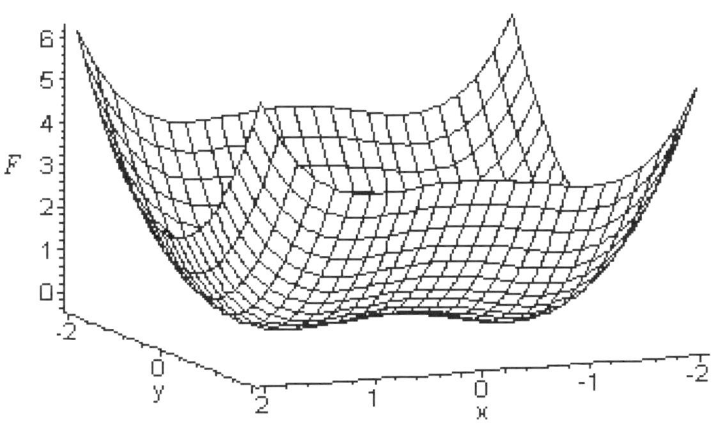

Figure 7. The profile of function F11 ($-2.0 \leq x_i \leq 2.0$).

The results were divided into classes according to the relative errors. The class Δe is defined as the range of results with the relative errors from 0.0 to Δe (in %). The relative error was calculated from:

$$\text{abs}\left[\frac{F^* - F}{F^*}\right] \cdot 100\% \ (F^* \text{ denotes the optimal value}).$$

For instance, class $\Delta\text{e} = 1.0$ contains the results with relative errors less than or equal to 1.0%. For each class Δe we calculated the success rate denoted by SR(Δe). This is the fraction of solutions from all runs that fall into the class Δe. For instance, "SR($\Delta\text{e} = 0.1$) $= 0.1$" means that the fraction of solutions that have the relative error less than or equal to 0.1% is 0.1 (i.e. 10%). Such representation of the results gives deeper insight into reliability of the optimization since one can observe the distribution of the results among classes of various relative errors. Additionally, the number of function evaluation (NFE) values and CPU times are reported for LJ–MM algorithm, particularly for example E8. For SA–S/1 algorithm we were not able to report NFE values since they can not be calculated directly from the values of parameters applied as it is the case of LJ–MM approach. However, CPU times are given to compare them with those for LJ–MM.

The rest of the material in the following is structured as follows. Results of tests for SA–q/1 and LJ–MM are analyzed in Secs. 4.2 and 4.3 respectively, with focus on parameters setting. The influence of the parameters on the optimization will be illustrated by figures or in tables for selected cases. Application issues are addressed in Sec. 4.4. The last Sec. 4.5 gives the results and the comparison of efficiency of both algorithms for the difficult problem E8, particularly for large number of variables. Due to the fact that the major aim of the tests was to find rules for parameter settings and, also, to estimate how the algorithms perform for large number of variables (Sec. 4.5), we did not try to locate the global optima for all test problems and both optimizers. Also, we did not use knowledge of the problem (e.g. active constraints) to make some modifications of solution procedures. The optimizers were generally initialized from mid-points calculated from the input data; the exceptions are very few (e.g. E8 with large number of variables). We believe that such a way of performing tests mimics the typical solution procedure of non-expert users.

4.2. Parameter settings for SA–S/1 algorithm

During the tests carried out for SA–S/1 algorithm, we have investigated the influence of the following.

- Cooling scheme: both cooling schemes defined by formulae (24) and (25, 25a), respectively, were investigated. For the exponential scheme, parameter Δ in Eq. (24) was varied in the range $(0.9, 0.99)$. For the adaptive cooling scheme parameter γ was varied from 0.02 to 1.5.
- Values of control parameter K in the equilibrium criterion.
- Value of parameter *INV* that controls the inverse movements according to formula (17).
- Value of final temperature $T^{\min}$ in the convergence criterion (26).

4.2.1. Cooling scheme

There is an opinion in the literature, based on experiences with typical SA algorithms for discrete-valued problems (and without simplex method), that the adaptive cooling scheme is better than the exponential one. The results of our tests for SA–S/1 showed that this is also valid for the SA algorithm with the simplex. For all the test problems the adaptive cooling scheme required smaller (or comparable at worse) CPU time than the exponential one to give solutions of similar F^{mean} and SR. Furthermore, a profile of dependency of current best solutions on temperature (T) shows a more or less steady improvement of the goal function value with temperature decrease for the adaptive cooling scheme. It means that this scheme provides consistent search such that solutions are improved steadily. In case of the exponential cooling scheme, often a sharp change was observed in a narrow region of high or low temperature range. The improvements beyond this region are small or even negligible. This is illustrated in Fig. 8 for problem E1. The character of plots does not depend on settings of other parameters such as final temperature or parameter *INV*. In the following, we present the results and the analysis only for the adaptive cooling scheme.

4.2.2. Influence of parameter *INV*

The higher the value of parameter *INV* (in the range (0, 1)) the higher is the chance of the inverse (uphill for minimization or downhill in maximization)

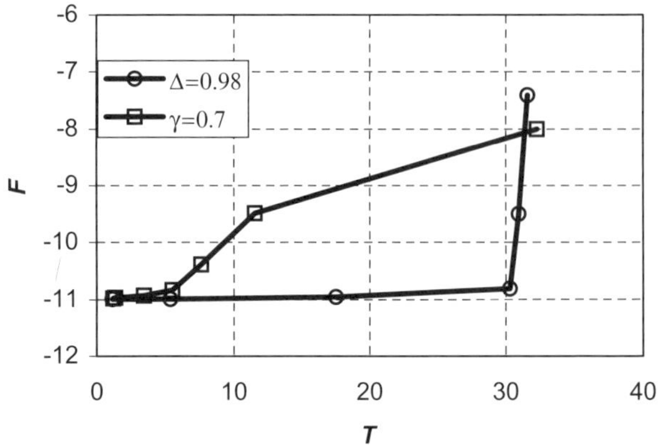

Figure 8. Profiles of the goal function changes in solving E1 by SA–S/1 algorithm: circles denote the solutions generated with the exponential cooling scheme for $\Delta = 0.98$ and squares the solutions for the adaptive cooling scheme for $\gamma = 0.7$.

movement of a simplex. The test results allowed concluding that the use of parameter *INV* higher than zero as proposed in SA–S/1 produces improvements though the scale of this effect is problem dependent. However, the general character does not depend significantly on the settings for other parameters. The influence of parameter *INV* depends on whether the problem is highly multi-modal or flat. For the flat functions, an increase of parameter *INV* from zero to a certain value denoted by INV^{lim} generates better (or at least not worse) SR for all meaningful classes of (Δe). The typical influence of parameter *INV* on indices $\text{SR}(\Delta\text{e} = 0.1)$ and $\text{SR}(\Delta\text{e} = 0.01)$ is illustrated in Fig. 9 for problem E8 with number of variables equal to 20. Fig. 10 shows the influence of *INV* on F^{mean} for example E1.

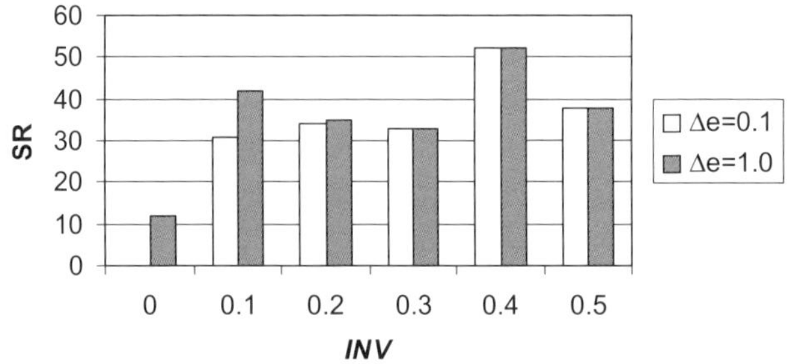

Figure 9. Influence of parameter *INV* on the performance index SR for problem E8 ($p = 20$).

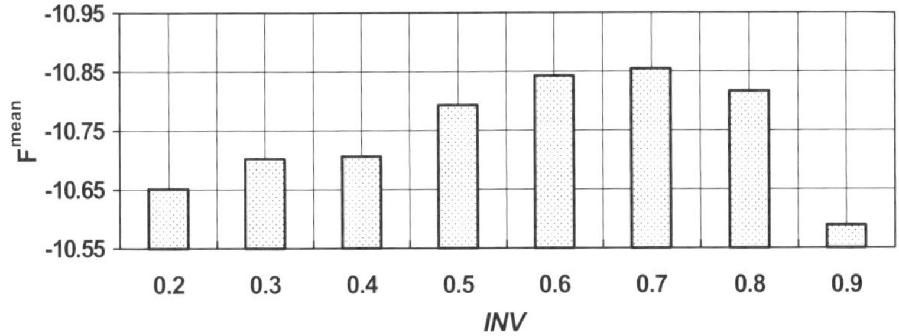

Figure 10. Influence of parameter *INV* on the performance index F^{mean} for problem E1.

However, for many highly multi-modal problems, particularly those with a large number of variables, the effect of *INV* is more complex. For some cases, *INV* equal to zero improved SR in comparison to that with *INV* greater than zero but only for very small Δe. However, for Δe of the order 0.5, values of SR are significantly higher for *INV* greater than zero. Results of such accuracy are sufficient in practice. This type of dependence is illustrated by the results for problem F9 in Table 1. High SR values were reached for $\Delta e = 0.5$ and $\Delta e = 1.0$ with parameter *INV* from the range (0.1, 0.3).

To conclude the analysis of the effect of parameter *INV*, the use of *INV* higher than zero increases the reliability of SA–S/1. The increase of *INV* has a positive effect up to some limiting value INV^{lim} beyond which the reliability decreases. This limiting value (INV^{lim}) is slightly problem dependent but setting INV^{lim} in the range 0.4–0.6 is safe.

4.2.3. Influence of parameter *K* in the equilibrium criterion

It seems obvious that SR and F^{mean} should improve with an increase of K but CPU time will soar, which necessitates some compromise. Cardoso *et al.* (1996, 1997) suggested the use of K of the order 100, independent of the problem. Our tests showed that for K higher than 60–70 both F^{mean} and SR($\Delta e \geq 0.1$) increase very slightly and, in some cases, no effect was noticed. Such type of influence was observed for flat problems with number of variables less than 10. Figure 11 illustrates this for test problem E1. Such influence of parameter K was observed for all tested settings of parameters.

Table 1. Influence of parameter *INV* on SR index for various classes Δe for problem F9.

	INV						
Δe [%]	0	0.1	0.2	0.3	0.4	0.5	0.6
0.01	16	0	0	0	0	0	0
0.05	16	0	0	0	0	0	0
0.10	16	4	1	0	0	0	0
0.50	16	24	23	22	11	11	5
1.00	16	34	42	59	43	29	18
2.00	16	36	54	69	67	68	46
5.00	17	36	57	71	72	75	71
higher than 5	83	64	43	29	28	25	29
F^*	−571.7583	−571.3261	−571.3217	−570.7633	−571.1881	−570.9129	−570.1998

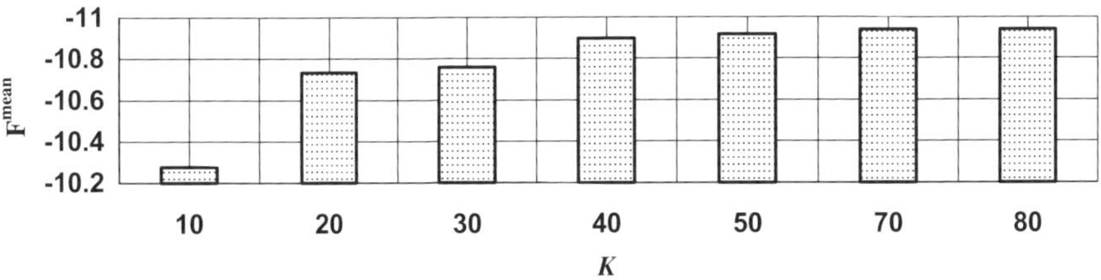

Figure 11. Influence of parameter K in SA–S/1 algorithm on index F^{mean} for problem E1.

However, for multi-modal problems with large number of variables, K of the order 70 may be insufficient. Deeper analysis showed that the value of K should be related to the number of variables (or the number of degrees of freedom for problems with easy equality constraints). Generally, we recommend using $K = \Phi p$ where p is the number of variables or the number of degrees of freedom and coefficient Φ should be from the range (5, 15). Larger values of K are necessary only for some difficult multi-modal problems particularly if reaching a high SR for $\Delta\text{e} \leq 0.1$ is of importance.

4.2.4. Influence of parameter γ in the adaptive cooling scheme

One can expect that small values of γ in Eq. (25), lower than 1.0 in particular, should improve reliability but at the cost of CPU time increase. We found that γ lower than 0.2 or so has practically no effect on SR (for Δe of order 0.1 or larger) and F^{mean}. This does not depend on values of other parameters. Generally, small γ values produce high SR values for small Δe. Hence, γ of order 0.2 can be treated as the lower bound in most cases particularly for flat functions and for reaching Δe of the order 0.1 or larger. Figure 12 illustrates the above statements for problem E2.

The general guideline is: start SA–S/1 subroutine with parameter γ set at 1.0, then decrease it gradually in steps of 0.1 and analyze the effect. If the results are not better or improvements are small stop the trials. For highly multi-modal problems, low γ values smaller than 0.2, are often necessary.

4.2.5. Influence of parameter T^{min}

The results of the tests proved that there is a lower limit on T^{min} below which its influence on reliability is not significant. Up to this limiting value,

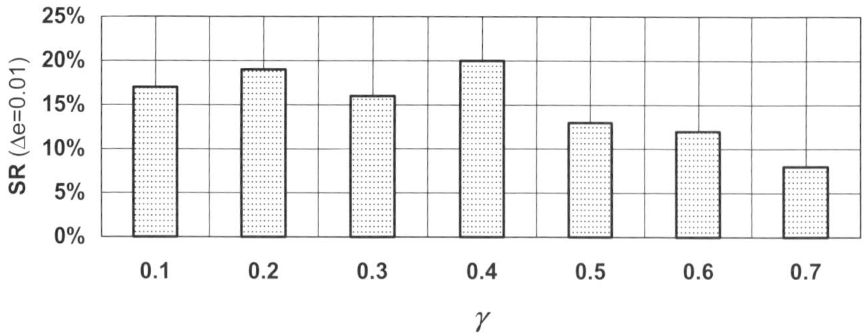

Figure 12. Influence of parameter γ on index SR for problem E2.

both performance indices increase with decreasing values of $T^{\min}$. The lower limit on $T^{\min}$ depends on the problem and also on values of control parameters: K, *INV* and γ. Furthermore, T^0 value calculated from the procedure described in Sec. 3 influences on the selection of good $T^{\min}$ value since the latter should be substantially smaller than T^0, usually by the order of 10^3. We did not find a simple correlation among lower bounds on all these parameters that ensure good performance. The useful observation from the tests is that $T^{\min}$ value does not influence CPU time largely. It was observed that the decrease of $T^{\min}$ by factor of 10 increases CPU time by factor of about 2. Hence, we conclude that the use of low values of $T^{\min}$ (below 0.1) is recommended if a high reliability is important or if parameter T^0 is low. In general, we recommend $T^{\min}$ below 1.0 but not much lower than 0.1 with values of the rest of control parameters set at the values recommended above. However, smaller $T^{\min}$ values are required if T^0 parameter is low. The experience gained shows that the use of another termination criterion may be useful. We plan to test convergence criteria similar to those used in deterministic approaches with some modifications accounting for stochastic nature of the SA method.

4.3. Results and analysis of tests for LJ–MM algorithm

It is obvious from the description in Sec. 2 that the key parameters of LJ–MM method are the numbers of external (*NEL*) and internal (*NIL*) iterations. The product of *NEL* and *NIL* can be considered to be the number of function evaluations. In fact it is equal to NFE if death penalty for inequality constraints is not used; else, NFE is smaller than this product since goal function is not calculated for cases when even one inequality is not met. Our tests revealed that the settings for *NEL* and *NIL* depend mainly on problem size. However, there are likely more complex dependencies such as for multi-modal versus flat problems, which are difficult to estimate. As a rule of thumb, we recommend that the product of *NEL* and *NIL* should be more than 10^5–10^6 if the number of degrees of freedom exceeds about 12–15, and *NEL* should not be higher than *NIL*. An *NEL* value more than *NIL* may be advantageous for large, highly multi-modal problems. The final choice of both parameters should be made after some trials.

To give some idea of NFE values, we show average values of *NEL* and *NIL*, in Table 2, for solving the problems in Appendix A. They were sufficient to reach reasonable SR values (of order 70%) even for small errors Δe (of order 0.1). Parameters and results for problem E8 will be analyzed in Sec. 4.5.

In regard to the setting for final sizes (δ^f), we applied the same value for each variable since the majority of the test problems has no "physical" interpretation. Generally, values of the order 10^{-2} to 10^{-4} produced the best results independent of problem type and settings of other parameters. The effect of the final size on SR for the multi-modal problem F9 is illustrated in Fig. 13.

Table 2. Values of parameters *NEL* and *NIL* used in the tests.

	Unconstrained problems															
	F1	F2	F3	F4	F5	F6	F7	F8	F9	F10	F11	F12	F13	F14	F15	F16
NEL	10	30	50	30	70	80	60	200	200	100	10	30	70	80	20	200
NIL	15	20	60	20	90	60	50	230	200	100	15	40	80	100	30	220

	Constrained problems (Except E8)						
	E1	E2	E3	E4	E5	E6	E7
NEL	80	300	140	140	200	55	100
NIL	80	500	100	100	250	60	140

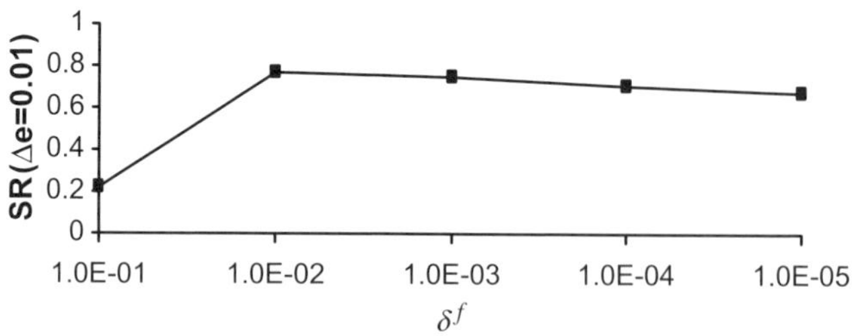

Figure 13. SR(Δe $= 0.01$) versus δ^f for problem F9: $x_i^0 = 11.0$; $x_i^* = 14.2076$.

Parameter α should be set in the range from 1.0 (equivalent to using M–LJ) to 2.0 (this is for LJ–MM). Also, slightly higher values, but not larger than 3.0, usually produce reasonable results. We found that α has more complicated influence than the values of final sizes, and the dependency is also problem specific. However, the character of this influence is similar for all settings of other parameters. Generally, higher α values are more advantageous for multi-modal functions than for flat ones. Parameter α of order 1.0 is sufficient for flat functions to give high SR, particularly for small Δe of order 0.01. However, also for flat functions, higher α values from the range (2.0, 2.5) allow reaching high SR for errors Δe from 0.1 to 1.0. For multi-modal problems, such high α values are advantageous to reach small errors Δe. Hence, as the rule of thumb, we advise to use α of order 2.0 for all kinds of optimization problems. The influence of α is illustrated in Fig. 14 for multi-modal function F9 and Fig. 15 for flat problem F11.

4.4. Selected application issues

The other important issues for the application of LJ–MM and SA–S/1 are the selection of starting point and dealing with constraints. The starting point does not influence the performance of SA–S/1 since the initial simplex is generated randomly using the lower and upper bounds on the variables. This is advantage in many cases but may be drawback if the user has some good understanding of the problem and is able to find a better initial point.

The ARS technique (LJ–MM) does not require a feasible initial solution, particularly for problems of moderate scale. However, the choice of

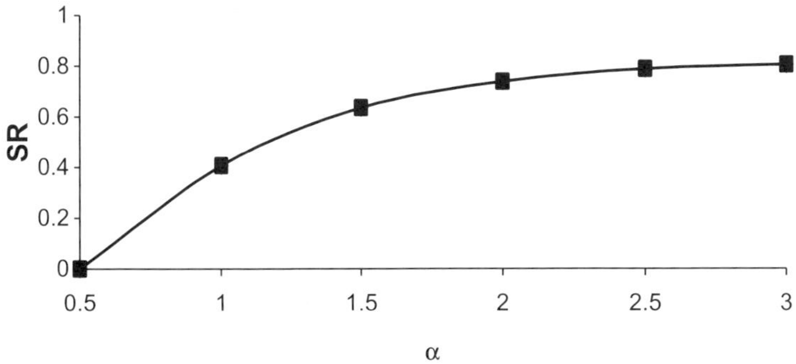

Figure 14. The dependence of SR ($\Delta e = 0.01$) on α parameter for problem F9.

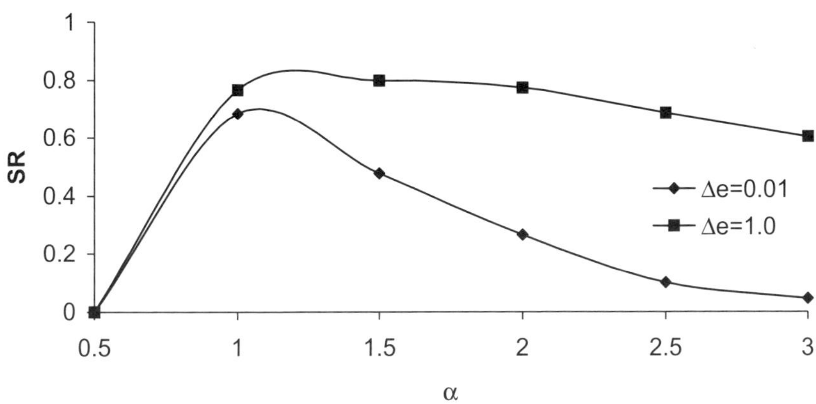

Figure 15. The dependence of SR ($\Delta e = 0.01$ and $\Delta e = 1.0$) on α parameter for problem F11.

an initial search space should be performed with great care to find possibly small one without cutting off the optimum. An example of determination of the initial search space is presented in Chapter 15 for water network design problem. Algorithm LJ–MM is relatively insensitive to the choice of initial point, particularly in comparison with classical deterministic methods and, also with the M–LJ version. Advantage of LJ–MM in this respect is particularly strong for multi-modal problems. This is illustrated in Fig. 16 for problem F11, which shows the influence of starting point on SR($\Delta e = 0.01$) for LJ–MM and LJ. Note that LJ algorithm performed better in regards to SR for small Δe, when it was initialized from a point around the optimum.

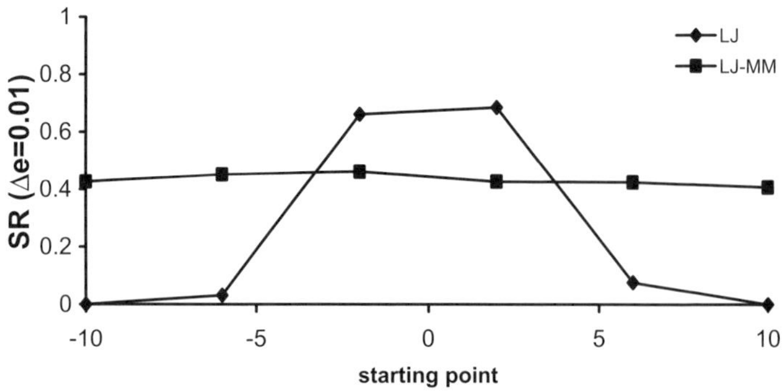

Figure 16. SR ($\Delta e = 0.01$) versus starting point for problem F11 ($x^* = -1.0497, 0.0008$).

For relatively high Δe values of 0.1 and larger, LJ–MM method is clearly more reliable than the LJ even if both are initialized near the optimum. Similar effect was noted for flat functions, LJ–MM yields better SR values for classes of results with Δe of 0.1 or larger. The improvements are, however, smaller than for multi-modal problems.

In regard to dealing with constraints, we can state, based on the tests, that different mechanisms should be applied to equality and inequality constraints. Also we implemented different techniques for dealing with inequality constraints in LJ–MM and SA–S/1. These are discussed below.

4.4.1. Dealing with inequality constraints

There are two types of inequality constraints: limits on variables, often called *explicit* constraints and typical inequalities: $g(x) \leq 0$, called *implicit* inequality constraints. Different methods were applied to deal with them in SA–S/1 subroutine. In a solution generated by simplex movement, if variable x_i is beyond the upper or lower bound then it is simply set at the violated limit. In case of *implicit* inequality constraints, several techniques are available including penalty terms in augmented goal function or death penalty commonly used in genetic algorithms. In SA–S/1, we applied the procedure taken from Cardoso *et al.* (1996, 1997), which is similar to repairing solutions in Genetic Algorithms. The reparation is performed by generating randomly points till all constraints are met. In case of LJ–MM method, we found that elimination of a point that does not meet inequality constraints is often the efficient and simple way to handle inequalities. Penalty terms in the augmented goal function perform well if penalty coefficients are properly tuned what is not easy task in general.

4.4.2. Dealing with equality constraints

Our experience is that equality constraints can be handled in both algorithms in the same manner. We suggest the use of direct solution as much as possible. For this, the set of all variables is divided into a subset of decision variables and second subset of dependent ones. The values of the former are generated randomly by ARS or SA–S/1 algorithm while the latter are calculated from equations that are linear with respect to them (for generated values of the decision variables). To show the method, consider the constraints:

$$\sum_{i=1}^{p} x_{i,m} y_{i,m} = 0; \quad m = 1, \ldots, M. \tag{27}$$

By choosing x_i or y_i to be the decision variables, constraints (27) can be transformed into linear equations with respect to y_i or x_i respectively. Note that constraints (27) are sums of bilinear terms $(x_i y_i)$. Such terms are common in process engineering problems since they occur in mass and energy balances as products of flow rate and concentration or flow rate and temperature. The examples of applications are in Chapters 15 and 16, respectively for water and heat exchanger network design.

Additional possibilities for applying the direct solution of equality constraints result from correctly organized sequential procedure of solving the constraints. Such sequential procedure can be coded in stochastic optimization approaches in contrast to deterministic optimization methods. Consider the following illustrative example.

$$\min\,(x_1 x_2 x_3), \tag{28}$$

s.t.

$$x_1 x_2 = 0, \tag{29}$$

$$x_3 \ln(x_2) = 0. \tag{30}$$

Let x_1 is chosen the decision variable and stochastic optimizer generates its value. Then, x_2 can be calculated from equality (29) and x_3 can be determined from constraint (30), which became linear with respect to x_3. The example contained three variables and two equations, and hence has one degree of freedom. It is thus possible to reduce the number of decision variables to the number of degrees of freedom. A substantial reduction of the number of decision variables allows solving even larger optimization problems with stochastic optimization methods in reasonable CPU time. The reader is referred to Luus *et al.* (2002) for additional examples. Also, these authors showed that another version of the LJ algorithm is able to handle difficult cases, which cannot be transferred into linear equations.

4.5. Problem size effect

Both ARS and SA–S/1 algorithms can be applied for solving large problems. However, one can expect that there is a limit in regard to problem scale, the number of variables in particular, where the ARS and/or SA–S strategy become inefficient. To test the influence of problem dimensionality on their performance, we have chosen problem E8 with number of variables increasing from 12 to 50. We also employed two genetic algorithm solvers: GENOCOP III by Michalewicz (1996) taken from http://www.cs.adelaide.edu.au/~zbyszek/evol-systems.html, and Gen-Com by Bochenek and Jeżowski.

As stated in Michalewicz and Fogel (2002), problem E8 is difficult for all types of optimization methods. The first constraint (see Appendix A) is active at the origin of the goal function where the global optimum is located. To exploit this knowledge, a specific initialization procedure and specific genetic operators were developed by Michalewicz to solve example E8 with genetic algorithm. The best results from those computations are reported, for various p values, in the Appendix A. The optima reported in the Appendix A were calculated by Luus (2008) using the LJ optimization procedure in multi-pass fashion, and using region size determination as suggested by Luus *et al.* (2006).

Additionally, large numbers of generations together with large population size were applied in the genetic algorithm. For instance, for $p = 50$, the number of generations was set at 30,000 and population size was 30.

Hence, to find "good" results with SA–S/1 and LJ–MM, large NFE should be employed if no enhancements from the knowledge of the problem are used. The basic aim of the present tests was to compare the performances of both methods. We decided to use the testing procedure that was used for the other examples. It means that the optimizers were run for a certain CPU time limit. We did not exploit the insights provided by Michalewicz and Fogel (2002). The midpoint calculated from the given ranges for variables was used as the starting point. However, other initialization was applied in very few cases, as stated in the text. Preliminary tests showed that SA–S/1 was able to locate a "good" solution in shorter CPU time than LJ–MM for practically all numbers of variables. The difference increased with the increase of p, and p of 20 was a threshold value. Hence, we report first the results for p from 12 to 20 and, then, for larger p. However, as shown in Table 3, LJ–MM gave better results than the other two methods, and the results are very close to the optima given in Appendix A. Note that for $p = 20$, the result is slightly better than 0.803553 reported in Michalewicz (1996) and in Michalewicz and Fogel (2002).

The tests for all values of p were organized in the following way. First, SA–S/1 method was run and the results of this method: the best goal function ($F^{*,\mathrm{SA}}$) and average CPU time — $\mathrm{CPU}^{\mathrm{SA}}$, were used as reference for other optimizers, namely, the control parameters influencing computation time for them were adjusted in such a way so as to reach similar goal function values. However, we set a limit on CPU time — it was not more than $\mathrm{CPU}^{\mathrm{SA}}$ multiplied by a factor of 15 for $p < 20$ and for greater p. The parameters of both algorithms applied to reach the results in Table 3 are presented in Table 4 together with CPU times.

The success rate for $p = 20$ with LJ–MM algorithm was lower than with SA–S/1. Better performance was reached with starting point of $x_i = 1.0$ for

Table 3. Comparison of F^* for example E8 with the number of variables p from 12 to 20.

p	LJ–MM	SA–S/1	GENECOP III/GenCom
12	0.76256	0.76243	0.75038
15	0.78244	0.78238	0.78205
20	0.80362	0.80361	0.79651/0.80362

Table 4. Parameters and CPU times for the optimizers to calculate results in Table 3.

p	LJ–MM	SA–S/1
12	average time = 62 secs/run $NEL = 500$ $NIL = 20{,}000$ $\delta^f = 1\text{E-}3$	average time = 7 secs/run $INV = 0.4$ $\gamma = 0.1$ $K = 120$ $T^0 = 4\text{E-}2$ $T^{\min} = 1\text{E-}4$
15	average time = 70 secs/run $NEL = 500$ $NIL = 20{,}000$ $\delta^f = 1\text{E-}3$	average time = 13 secs/run $INV = 0.4$ $\gamma = 0.1$ $K = 150$ $T^0 = 4\text{E-}2$ $T^{\min} = 1\text{E-}4$
20	average time = 2700 secs/run $NEL = 5{,}000$ $NIL = 40{,}000$ $\delta^f = 1\text{E-}3$	average time = 25 secs/run $INV = 0.4$ $\gamma = 0.1$ $K = 200$ $T^0 = 4\text{E-}2$ $T^{\min} = 1\text{E-}6$

$i = 1, \ldots, p$. To reduce computation time for 100 runs we have applied this starting point in tests for $p \geq 20$ with LJ–MM. The results from GENOCOP III can be most likely improved by better adjustment of parameters. This solver uses many control parameters and options which are difficult to tune. CPU times for GENOCOP III (not shown here) were regularly higher than those for LJ–MM. The difference was not substantial and decreased with increase in number of variables. CPU time needed by GenCom solver was similar to that used for GENOCOP III. Note that best result found by GenCom for $p = 20$ is the same as LJ–MM.

The results for problem E8 with larger values of p ($p = 30, 40, 50$) are shown in Table 5. The results obtained with SA–S/1 are quite close to the optimal solutions given in Appendix A. To get accurate results, 9-figure accuracy is required in F. With SA–S/1, we were able to get only 3-figure accuracy in x. Table 6 contains values of parameters of both algorithms applied to achieve results close to the optima. CPU times are also given there.

Table 5. Comparison of F^* for problem E8 with $p = 30, 40$ and 50.

n	LJ–MM	SA–S/1
30	0.81273	0.82188
40	0.80720	0.83101
50	0.80334	0.83525

Table 6. Parameters and CPU times for the optimizers to give results in Table 5.

p	LJ–MM	SA–S/1
30	average time = 2900 secs/run $NEL = 15{,}000$ $NIL = 15{,}000$ $\delta^f = 1E\text{-}3$	average time = 120 secs/run $INV = 0.5$ $\gamma = 0.1$ $K = 300$ $T^0 = 2E\text{-}2$ $T^{\min} = 1E\text{-}6$
40	average time = 6,000 secs/run $NEL = 20{,}000$ $NIL = 15{,}000$ $\delta^f = 1E\text{-}3$	average time = 170 secs/run $INV = 0.5$ $\gamma = 0.1$ $K = 300$ $T^0 = 2E\text{-}2$ $T^{\min} = 1E\text{-}6$
50	average time = 13,300 secs/run $NEL = 25{,}000$ $NIL = 25{,}000$ $\delta^f = 1E\text{-}3$	average time = 530 secs/run $INV = 0.5$ $\gamma = 0.05$ $K = 200$ $T^0 = 2E\text{-}2$ $T^{\min} = 1E\text{-}6$

The results from the tests for problem E8 showed that SA–S/1 performed better than LJ–MM for p larger than 20 (see Table 5). The superiority of the former increased with the increase of p. This effect can be expected because LJ–MM is more "random" in nature than SA–S/1. The latter makes use of efficient deterministic method to generate trial points while the former uses a kind of blind search. This is the evidence of efficiency of ARS technique for large scale problems. We plan to investigate why LJ–MM is not so efficient for problem E8 with high p even though it performs well for small and medium size cases.

5. Summary

The chapter presents two stochastic algorithms: LJ–MM that is of ARS type and SA–S/1. The latter embeds Nelder–Meade simplex procedure into SA framework. The methods were tested for many problems listed in Appendices A and B. In almost all cases, they showed good performance. Also, they are easy to use and simple, particularly LJ–MM algorithm requires small number of parameters that are easy to tune in most cases. The tests performed allow concluding that reliability of LJ–MM is slightly better than that of SA–S/1 for small problems. The difference is not important in view of the very small CPU time for both methods. However, LJ–MM appears to be less efficient than SA–S/1 method for problem E8 with 30 and more variables. Special enhancements such as good initialization, initial region size adjustment and so on are required to improve its reliability for large problems with active constraints.

Symbols

F — goal function in optimization
F^{mean} — optimization performance index, mean value of goal function in a series of runs
INV — parameter controlling inverse movement of simplex
K — parameter of equilibrium criterion
P — probability of accepting a feasible point
x — variable
$\boldsymbol{x}$ — vector of variables
n — number of variables in optimization problem E8
NEL — number of external loops
NIL — number of internal loops
p — number of variables
r, RND — random number from uniform distribution from the range (0, 1)
SR — optimization performance index, success rate in a series of runs
$T/T^{\min}/T^0$ — temperature, parameter of SA optimization /minimal value of T applied as termination criterion/initial temperature

α — parameter of LJ–MM algorithm
β — coefficient for reducing search region size in ARS
γ — parameter of the adaptive cooling scheme in SA — Eq. (25)
δ — search region size
Δ — parameter of the exponential cooling scheme in SA — Eq. (24a)
Δe — relative error of optimization algorithm

Superscripts

f — final value
l — lower bound
max/min — maximum/minimum
u — upper bound
$*$ — optimum or best result
0 — initial value or initial condition

Acronyms

ARS — adaptive random search
LJ — Luus-Jaakola (optimization method)
LJ–MM — Luus-Jaakola (algorithm) for multimodal problems
NLP — nonlinear programming
MINLP — mixed-integer nonlinear programming
M–LJ — modified Luus-Jaakola (algorithm)
SA — simulated annealing
SA–S — simulated annealing with simplex method
SA–S/1 — simulated annealing with simplex method — version of the authors

References

Aarst, E. and Korst, J. (1989). *Simulated Annealing and Boltzmann Machines — A Stochastic Approach to Combinatorial Optimization and Neural Computers*, John Wiley and Sons, New York.

Ali, M.M., Torn, A. and Vitanen, S.A. (1997). Numerical comparison of some modified controlled random search algorithms. *Journal of Global Optimization*, **11**, pp. 377–385.

Amarger, R.J., Biegler, L.T. and Grossmann, I.E. (1992). An automated modeling and reformulation system for design optimization. *Computers and Chemical Engineering*, **16**(7), pp. 623–636.

Andre, J., Siarry, P. and Dognon, T. (2001). An improvement of the standard genetic algorithm fighting premature convergence in continuous optimization. *Advances in Engineering Software*, **32**, pp. 49–60.

Athier, G., Floquet, P., Pibouleau, L. and Domenech, S. (1997). Process optimization by simulated annealing and NLP procedures. Application to heat exchanger network synthesis. *Computers and Chemical Engineering*, Suppl, 21, pp. S475–S480.

Banga, J.R. and Seider, W.D. (1996). Global optimization of chemical processes using stochastic algorithms. *State of Global Optimization*, Floudas, C.A. and Pardalos, P.M. (eds.), Kluwer Avademic Publishers, The Netherlands, pp. 563–583.

Banga, J.R., Irizarry-Rivera, R. and Seider, W.D. (1998). Stochastic optimization for optimal and model-predictive control. *Computers and Chemical Engineering*, **22**(4/5), pp. 603–612.

Biegler, L.T., Grossmann, I.E. and Westerberg, A.W. (1997). *Systematic Methods of Chemical Process Design*, Prentice Hall PTR, New Jersey.

Bochenek, R. and Jeżowski, J. (2000). Optymalizacja metodą adaptacyjnego przeszukiwania losowego w projektowaniu procesów. II. Przykłady zastosowań w inżynierii procesowej. *Inżynieria Chemiczna i Procesowa*, **21**, pp. 465–487.

Bochenek, R., Jeżowski, J., Poplewski, G. and Jeżowska, A. (1999). Studies in adaptive random search optimization for MINLP problems. *Computers and Chemical Engineering*, Suppl., pp. S483-S486.

Campbell, J.R. and Gaddy, J.L. (1976). Methodology for simultaneous optimization with reliability: Nuclear PWR example. *AlChE Journal*, **6**(22), pp. 1050–1055.

Cardoso, M.F., Salcedo, R.L. and de Azevedo, S.F. (1996). The simplex-simulated annealing approach to continuous non-linear optimization. *Computers and Chemical Engineering*, **20**(9), pp. 1065–1080.

Cardoso, M.F., Salcedo, R.L., de Azevedo, S.F. and Barbosa, D. (1997). A simulated annealing approach to the solution of MINLP problems. *Computers and Chemical Engineering*, **21**(12), pp. 1349–1364.

Chaudhuri, P.D. and Diwekar, U.M. (1997a). An automated approach for the optimal design of heat exchangers. *Industrial Engineering Chemistry and Research*, **36**, pp. 3685–3693.

Chaudhuri, P.D. and Diwekar, U.M. (1997b). Synthesis under uncertainty with simulators. *Computers and Chemical Engineering*, **7**(21), pp. 733–738.

Chen, T.-Y. and Su, J.-J. (2002). Efficiency improvement of simulated annealing in optimal structural design. *Advances in Engineering Software*, **33**, pp. 675–680.

Corana, A., Marchesi, M., Martini, C. and Ridella, S. (1987). Minimizing multimodal functions of continuous variables with the simulated annealing algorithm. *ACM Transactions on Mathematical Software*, **13**, pp. 262–280.

Cordero, J.C., Davin, A., Floquet, P., Pibouleau, L. and Domenech, S. (1997). Synthesis of optimal reactor networks using Mathematical programming and simulated annealing. *Computers and Chemical Engineering*, Suppl, **21**, pp. S47–S52.

Das, H., Cummings, P.T. and Le Van, M.D. (1990). Scheduling of serial multiproduct batch processes via simulated annealing. *Computers and Chemical Engineering*, **14**(12), pp. 1351–1362.

Dolan, W.B., Cummings, P.T. and Le Van, M.D. (1989). Process optimization via simulated annealing: application to network design. *AlChE Journal*, **35**, pp. 725–736.

Dolan, W.B., Cummings, P.T. and Le Van, M.D. (1990). Algorithmic efficiency of simulated annealing for heat exchanger network design. *Computers and Chemical Engineering*, **14**, pp. 1039–1050.

Edgar, T.E., Himmelblau, D.M. and Lasdon, L.S. (2001). *Optimization of Chemical Processes*. Boston, USA, McGraw Hill International Edition (2nd ed.)

Floquet, P., Pibouleau, L. and Domenech, S. (1994). Separation sequence synthesis: How to use simulated annealing procedure. *Computers and Chemical Engineering*, **18**, pp. 1141–1148.

Floudas, C.A. (1995). *Nonlinear and Mixed-Integer Optimization. Fundamentals and Applications*, Oxford University Press, New York.

Floudas, C.A. and Pardalos, P.M. (1990). A collection of test problems for constrained global optimization algorithms. *Lecture Notes in Computer Science*, Springer-Verlag, Berlin.

Gaines, L.D. and Gaddy, J.L. (1976). Process optimization by flow sheet simulation. *Industrial and Engineering Chemistry, Process Design and Development*, **1**(15), pp. 206–211.

Hanke, M. and Li, P. (2000). Simulated annealing for the optimisation of batch distillation processes. *Computers and Chemical Engineering*, **24**, pp. 1–8.

Hendrix, E.M.T., Ortigosa, P.M. and Garcia, I. (2001). On success rates for controlled random search. *Journal of Global Optimization*, **21**, pp. 239–263.

Heuckroth, M.W., Gaddy, J.L. and Gaines, L.D. (1976). An examination of the adaptive random search technique. *AlChE Journal*, **22**(4), pp. 744–750.

Ho, Y.C. and Pepyne, D.L. (2002). Simple explanation of the no free lunch theorem of optimization. *Cybernetics and Systems Analysis*, **38**(2), pp. 292–298.

Jaakola, T.H.I. and Luus, R. (1974). A note on the application of nonlinear programming to chemical-process optimization. *Operations Research*, **22**(2), pp. 415–417.

Jeżowski, J. and Bochenek, R. (2002). Experiences with the use of the Luus-Jaakola algorithm and its modifications in optimization of process engineering problems. *Recent Developments in Optimization and Optimal Control in Chemical Engineering*, Luus, R. (ed.), Research Signpost, Trivandrum, India, pp. 89–114.

Jeżowski, J. and Bochenek, R. (2000). Optymalizacja metodą adaptacyjnego przeszukiwania losowego w projektowaniu procesów. I. Opis metody optymalizacji. *Inżynieria Chemiczna i Procesowa*, **21**, pp. 443–463.

Jeżowski, J. and Jeżowska, A. (2003). Adaptive random search optimization procedures in solver Opti-Sto. *Inżynieria Chemiczna i Procesowa*, **24**(2), pp. 261–279.

Jeżowski, J., Bochenek, R. and Ziomek, G. (2005). Random search optimization approach for highly multi-modal nonlinear problems. *Advances in Engineering Software*, **36**(8), pp. 504–517.

Kelahan, R.C. and Gaddy, J.J. (1977). Synthesis of heat exchange networks by mixed integer optimization. *AlChE Journal*, **23**(6), pp. 423–435.

Kirkpatrick, S., Gellat, C.D. and Vecchi, M.P. (1983). Optimization by simulated annealing. *Science*, **200**, pp. 671–680.

Ku, H. and Karimi, I. (1991). An evaluation of simulated annealing for batch process scheduling. *Industrial and Engineering Chemistry Research*, **30**, pp. 163–169.

Lee, Y.P., Rangaiah, G.P. and Luus, R. (1999). Phase and chemical equilibrium calculations by direct search optimization. *Computers and Chemical Engineering*, **23**, pp. 1183–1191.

Li, J. and Rhinehart, R. (1998). Heuristic random optimization. *Computers and Chemical Engineering*, **3**(22), pp. 427–444.

Liao, B. and Luus, R. (2005). Comparison of the Luus-Jaakola optimization procedure and the genetic algorithm. *Engineering Optimization*, **37**(4), pp. 381–398.

Lima, R.M., Salcedo, R.L. and Barbosa, D. (2006). SIMOP: Efficient reactive distillation optimization using stochastic optimizers. *Chemical Engineering Science*, **61**, pp. 1718–1739.

Locatelli, M. (2000). Convergence of a simulated annealing algorithm for continuous global optimization. *Journal of Global Optimization*, **18**, pp. 219–234.

Luus, R. (1973). A direct approach to optimization of a complex system. *AlChE Journal*, **19**(3), pp. 645–646.

Luus, R. (1974). Two-pass method for handling difficult equality constraints in optimization. *AlChE Journal*, **20**(3), pp. 608–610.

Luus, R. (1975). Optimization of multistage recycle systems by direct search. *Canadian Journal of Chemical Engineering*, **53**, pp. 217–220.

Luus, R. (1996). Numerical convergence properties of iterative dynamic programming when applied to high dimensional systems. *Chemical Engineering Research and Design* (*Transactions of the Institution of Chemical Engineers*), **74**, (Part A), pp. 55–62.

Luus, R. (2008). Personal Communication.

Luus, R. and Brenek, P. (1989). Incorporation of gradient into random search optimization. *Chemical Engineering and Technology*, **12**, pp. 309–318.

Luus, R. and Jaakola, T.H.I. (1973). Optimization by direct search and systematic reduction of the size of search region. *AIChE Journal*, **19**(4), pp. 760–766.

Luus, R., Iyer, R.S. and Woo, S.S. (2002). Handling equality constraints in direct search. in *Recent Developments in Optimization and Optimal Control in Chemical Engineering*, Luus, R. (ed.), Research Signpost, Trivandrum, India, pp. 120–178.

Luus, R., Sabaliauskas, K. and Harapyn, I. (2006). Handling inequality constraints in direct search optimization. *Engineering Optimization*, **38**(4), pp. 391–405.

Maia, L.O.A. and Qassim, R.Y. (1997). Synthesis of utility systems with variable demands using simulated annealing. *Computers and Chemical Engineering*, **21**, pp. 947–950.

Maia, L.O.A., de Carvalho, L.A. and Qassim, R.Y. (1995). Synthesis of utility systems by simulated annealing. *Computers and Chemical Engineering*, **19**(4), pp. 481–488.

Maier, R.W. and Whiting, W.B. (1998). The variation of parameter settings and their effects on performance for the simulated annealing algorithm. *Computers and Chemical Engineering*, **23**(1), pp. 47–62.

Marcoulaki, E.C. and Kokossis, A.C. (1998). Molecular design synthesis using stochastic optimisation as a tool for scoping and screening. *Computers and Chemical Engineering*, Suppl, **22**, pp. S11–S18.

Marcoulaki, E.C. and Kokossis, A.C. (2000a). On the development of novel chemicals using a systematic synthesis approach. Part I. Optimisation framework. *Chemical Engineering Science*, **55**, pp. 2529–2546.

Marcoulaki, E.C. and Kokossis, A.C. (2000b). On the development of novel chemicals using a systematic synthesis approach. Part II. Solvent design. *Chemical Engineering Science*, **55**, pp. 2547–2561.

Martin, D.L. and Gaddy, J.L. (1982). Process optimization with the adaptive randomly directed search. *AIChE Symposium Series*, **78**(214), pp. 99–107.

Mathur, M., Karle, S.B., Priye, S., Jayaraman, V.K. and Kulkarni, B.D. (2000). Ant colony approach to continuous function optimization. *Industrial and Engineering Chemistry Research*, **39**, pp. 3814–3822.

Metropolis, N., Rosenbluth, A., Rosenbluth, M., Teller, A. and Teller, E. (1953). Equation of state calculations by fast computing machines. *Journal of Chemical Physics*, **21**, pp. 1087–1092.

Michalewicz, Z. (1996). *Genetic Algorithms + Data Structures = Evolution Programs*, Springer-Verlag, Berlin.

Michalewicz, Z. and Fogel, D.B. (2002). *How to Solve it: Modern Heuristics*, Springer-Verlag, Berlin.

Michinev, I., Stoyanov, S. and Keil, F.J. (2000). Investigation and modification of the Luus-Jaakola global optimisation algorithm. *Hungarian Journal of Industrial Chemistry*, **28**, pp. 231–239.

Mihail, R. and Maria, G. (1986). A modified Matyas algorithm (MMA) for random process optimization. *Computers and Chemical Engineering*, **10**(6), pp. 539–544.

Murakami, Y., Uchiyama, H., Hasebe, S. and Hashimoto I. (1997). Application of repetitive S.A. method to scheduling problems of chemical processes. *Computers and Chemical Engineering*, Suppl, **21**, pp. S1087–S1092.

Ourique, J.E. and Telles, A.S. (1998). Computer-aided molecular design with simulated annealing and molecular graphs. *Computers and Chemical Engineering*, Suppl, **22**, pp. S615–S618.

Őzcelik, Y. and Őzcelik, Z. (2004). Solving mixed integer chemical engineering problems via simulated annealing approach. *Chemical and Biochemical Engineering Quarterly*, **18**(4), pp. 329–335.

Patel, A.N., Mah, R.S.H. and Karimi, I.A. (1991). Preliminary design of multiproduct noncontinuous plants using simulated annealing. *Computers and Chemical Engineering*, **15**(7), pp. 451–469.

Press, W.H. and Teukolsky, S.A. (1991). Simulated annealing optimization over continuous spaces. *Computational Physics*, Jul/Aug, pp. 426–429.

Press, W.H., Teukolsky, S.A., Velterling, W.T. and Flannery, B.P. (1992). *Numerical Recipes in C. The Art of Scientific Computing*, 2nd edition, Cambridge University Press, Cambridge.

Rajesh, J., Jayaraman, V.K. and Kulkarni, B.D. (2000). Taboo search algorithm for continuous function optimization. *Chemical Engineering Research and Design (Transactions of the Institution of Chemical Engineers)*, **78**(Part A), pp. 845–848.

Rangaiah, G.P. (1985). Studies in constrained optimization of chemical process problems. *Computers and Chemical Engineering*, **9**(4), pp. 395–404.

Ryoo, H.S. and Sahinidis, N.V. (1995). Global optimization of nonconvex NLPs and MINLPs with applications in process design. *Computers and Chemical Engineering*, **19**(5), pp. 551–566.

Sahin, K.H. and Ciric, A.R. (1998). A dual temperature simulated annealing approach for solving bilevel programming problems. *Computers and Chemical Engineering*, **23**, pp. 11–25.

Salcedo, R.L. (1992). Solving nonconvex nonlinear programming and mixed-integer nonlinear programming problems with adaptive random search. *Industrial and Engineering Chemistry Research*, **31**, pp. 262–273.

Salcedo, R.L., Goncalves, M.J. and de Azevedo, S.F. (1990). An improved random search algorithm for non-linear optimization. *Computers and Chemical Engineering*, **14**(10), pp. 1111–1126.

Sauer, R.N., Colville, A.R. Jr and Burwick, C.W. (1964). Hydrocarbon process. *Petroleum Refinery*, **43**, pp. 84.

Schmidt, H. and Thierauf, G. (2005). A combined heuristic optimization technique. *Advances in Engineering Software*, **36**, pp. 11–19.

Spall, J.C. (2003). *Introduction to Stochastic Search and Optimization*, John Wiley & Sons, Hoboken, New Jersey.

Trafalis, Bt. and Kasap, S. (2002). A novel metaheuristics approach for continuous global optimization. *Journal of Global Optimization*, **23**, pp. 171–190.

Tvrzska de Gouvea, M. and Odloak, D. (1998). A new treatment of inconsistent quadratic programs in a SQP-based algorithm. *Computers and Chemical Engineering*, **22**(11), pp. 1623–1651.

Vaidyanathan, R. and El-Halwagi, M. (1994). Global optimization of nonconvex nonlinear programs via interval analysis. *Computers and Chemical Engineering*, **10**(18), pp. 889–897.

Vanderbilt, D. and Louie, S.G. (1984). A Monte Carlo simulated annealing approach to optimization over continuous variables. *Journal of Computational Physics*, **56**, pp. 259–271.

Visweswaran, V. and Floudas, C.A. (1990). A global optimization algorithm (GOP) for certain cases of nonconvex NLPs — II. Application of theory and test problems. *Computers and Chemical Engineering*, **14**(12), pp. 1419–1434.

Wang, B.C. and Luus, R. (1978). Reliability of optimization procedures for obtaining global optimum. *AIChE Journal*, **24**(4), pp. 619–626.

Wang, B.C. and Luus, R. (1997). Optimization of non-unimodal systems. *International Journal for Numerical Methods in Engineering*, **11**, pp. 1235–1250.

Xi-Gang, Y. and Zhong-Zhou, Ch. (1997). A hybrid optimization method for design of batch chemical processes. *Computers and Chemical Engineering*, Suppl. **21**, pp. S685–S689.

Zabinsky, Z.B. (1998). Stochastic methods for practical global optimization. *Journal of Global Optimization*, **13**, pp. 433–444.

Zamorra, J.M. and Grossmann, I.E. (1998). Continuous global optimization of structured process systems models. *Computers and Chemical Engineering*, **12**(22), pp. 1749–1770.

Zhu, Y., Wen, H. and Xu, Z. (2000). Global stability analysis and phase equilibrium calculations at high pressures using the enhanced simulated annealing algorithm. *Chemical Engineering Science*, **55**, pp. 3451–3459.

Ziomek, G., Kaspereit, M., Jeżowski, J., Seidel-Morgenstern, A. and Antos, D. (2005). Effect of mobile phase composition on the SMB processes efficiency. Stochastic optimization of isocratic and gradient operation. *Journal of Chromatography A*, **1070**, pp. 11–24.

Exercises

(1) Develop an algorithm for solving the constraints as a set of linear equations in problems B1, B2, B3. Follow the explanation in Sec. 4.4. Did you achieve the reduction of independent variables number to the number of degrees of freedom in all cases? Try to develop more than a single solution scheme for test problem B2. How many versions of linearization are possible for this task?

(2) Remove the first constraint in example E8 and try to solve it for $n = 12$. Did you notice any changes in comparison with keeping the constraint? Then, remove the second constraint. Is the solution to unconstrained problem better? If so, explain the reason.

(3) Consider the results of selecting decision variables for example B3 in exercise 1. You should achieve the problem with 2 independent variables. Therefore, it is possible do draw the plot the goal function and the constraint for two variables to visualize the optimization problem. Draw the profile of the goal function and constraint using x_5 and x_6 as two independent variables. Why is this problem difficult? Hint: find local optima and compare their values.

(4) Find a function that approximates investment cost in [€/m^2] of insulating a flat brick wall with foamed polystyrene as a function of its thickness. (Hint: use least squares optimization model). The cost data are provided in Table 7.

Table 7. Cost of foamed polystyrene versus thickness.

Insulation thickness [mm]	Insulation price [€/m^2]
50	1.91
80	3.06
100	3.82
120	4.59
150	5.74

Fixed cost of installation = 4.41 [€/m^2]. Try quadratic function. Is the quadratic approximation necessary? Is linear relation of sufficient accuracy?

(5) Develop and solve optimization problem of calculating optimal thickness of insulation for the case of flat wall in exercise 4. As the goal function, use the annual profit calculated as:

$$\begin{aligned}&(\text{profit caused by applying insulation})\\&\quad-(\text{annual investment cost of insulation})\end{aligned}$$

The first term is proportional to the difference between heat losses through no insulated wall (bricks) and insulated wall (bricks with foamed polystyrene). Investment cost should be calculated on the basis of the equation developed in exercise 4. Use the data given below:

- Thermal conductivity of polystyrene, $\lambda_{\text{insulation}} = 4.2 \times 10^{-5}$ [kW/(m·K)]); thermal conductivity of bricks, $\lambda_{\text{brick}} = 4 \times 10^{-4}$ [kW/(m·K)]; brick wall thickness = 0.15 [m]
- For heating energy cost, use the data for natural gas: Unit price = 0.412 [€/m^3]; calorific value = 37 [MJ/m^3]; thermal efficiency = 0.8
- convective heat-transfer coefficients: inside = 20 [kJ/(h·m^2·K)]; outside = 32.7 [kJ/(h·m^2·K)]
- average temperature difference = 20 [K]; average time of heating season 600 [h/year]; depreciation time = 15 [year]
  Hint: consult example 3.3 (pp. 89–91) *in the book by Edgar et al.* (2001).

(6) The problem called Branin function has the form:

$$\min F(x_1, x_2) = \left(x_2 - \frac{5.1}{4\pi^2}x_1^2 + \frac{5}{\pi}x_1 - 6\right)^2 + 10\left(1 - \frac{1}{8\pi}\right)\cos(x_1) + 10$$

The region of interest is: $-5 \le x_1 \le 10$; $0 \le x_2 \le 15$. Calculate all three global optima applying the software at the disc. Hint: the solutions are: $[-\pi, 12.275]$; $[\pi, 2.275]$; $[3\pi, 2.475]$ with $F^* = 5/(4\pi)$

Appendix A

Unconstrained test problems

F1 : max $F(\mathbf{x}) = -\sum_{i=1}^{2}(x_i^2 - \cos(18x_i^2))$ (Rastrigin function)

$-0.625 \le x_1 \le 0.125$; $-0.250 \le x_2 \le 0.500$; $F^* = 2.0000$; $x^* = [0.0, 0.0]$

F2 : max $F(\mathbf{x}) = \sum_{i=1}^{2} x_i^2 + 40\prod_{i=1}^{2}\sin(x_i)$

$0.000 \le x_1 \le 7.000$; $-3.000 \le x_2 \le 6.000$; $F^* = 86.8057$; $x^* = [4.9278, 4.9278]$

F3 : max $F(\mathbf{x}) = \prod_{i=1}^{2}\sum_{j=1}^{5} j\cos[(j+1)x_i + j]$ (Shubert 2-variable function)

$0.000 \le x_i \le 10.000$; $F^* = 210.4820$; $x^* = [5.4827, 5.4827]$

F4 : max $F(\mathbf{x}) = \sum_{i=1}^{3}(x_i^2 - \cos(18 \cdot x_i^2))$

$-0.625 \le x_1 \le 0.125$; $-0.250 \le x_2 \le 0.500$; $-0.300 \le x_3 \le 0.450$; $F^* = 3.0000$; $x^* = [0.0, 0.0, 0.0]$

F5 : max $F(\mathbf{x}) = \prod_{i=1}^{3} x_i \sin(x_i)$

$6.0 \le x_i \le 16.0$; $F^* = 2846.6093$; $x_i^* = 14.2076$ for $i = 1, \ldots, 3$.

F6 : max $F(\mathbf{x}) = \sum_{i=1}^{3} x_i^2 + 40\sin x_1 \sin x_2 + 40\sin x_1 \sin x_3 + 40\sin x_2 \sin x_3$

$0.000 \le x_1 \le 7.000$; $-3.000 \le x_2 \le 6.000$; $-4.000 \le x_3 \le 5.000$; $F^* = 188.3367$; $x_i^* = 4.8344$ for $i = 1, \ldots, 3$.

F7 : max $F(\mathbf{x}) = -\sum_{i=1}^{5}(x_i^2 - \cos(18 \cdot x_i^2))$

$-1.0000 \le x_i \le 1.0000$; $F^* = 5.0000$; $x_i^* = 0.0$ for $i = 1, \ldots, 5$.

F8 : $\max F(\mathbf{x}) = 0.001 \prod_{i=1}^{5} \sum_{j=1}^{5} j \cos[(j+1)x_i + j]$

$0.000 \le x_i \le 10.000$; $F^* = 642.7431$; $x_i^* = 5.4827$ for $i = 1, \ldots, 5$

F9 : $\max F(\mathbf{x}) = 0.001 \prod_{i=1}^{5} x_i \sin(x_i)$

$6.000 \le x_i \le 16.000$; $F^* = 571.7690$; $x_i^* = 14.2076$ for $i = 1, \ldots, 5$

F10 : $\max F(\mathbf{x}) = \prod_{i=1}^{3} \sum_{j=1}^{5} j \cos[(j+1)x_i + j]$

$0.000 \le x_i \le 10.000$; $F^* = 3053.6724$; $x_i^* = 5.4827$ for $i = 1, \ldots, 3$

F11 : $\min F(x) = \frac{x^4}{4} - \frac{x^2}{2} + \frac{x}{10} + \frac{y^2}{2}$; (Quartic function)

$-10.000 \le x, y \le 10.000$; $F^* = -0.35239$

F12 : $\min F(x) = -\sum_{i=1}^{4} c_i \exp\left[-\sum_{j=1}^{3} a_{ij}(x_j - p_{ij})^2\right]$ (Hartman 4-variables function)

$-1.000 \le x_j \le 1.000$; $F^* = -3.86278$;
$x^* = [0.114614, 0.555649, 0.852547]$

I	a_{ij}			c_i	p_{ij}		
	$j=1$	$j=2$	$j=3$		$j=1$	$j=2$	$j=3$
1	3.0	10.0	30.0	1.0	0.36890	0.1170	0.2673
2	0.1	10.0	35.0	1.2	0.46990	0.4387	0.7470
3	3.0	10.0	30.0	3.0	0.10910	0.8732	0.5547
4	0.1	10.0	35.0	3.2	0.03815	0.5743	0.8828

F13 : $\min F(x) = -\sum_{i=1}^{19} \left[(x_i^2)^{(x_{i+1}^2+1)} + (x_{i+1}^2)^{(x_i^2+1)}\right]$; (Brown 3 function)

$-1.00 \le x_i \le 4.00$; $F^* = 0.000$

F14 : $\min F(\mathbf{x}) = \dfrac{1}{0.1 + \sum_{i=1}^{10} \frac{x_i^2}{4000} - \prod_{i=1}^{10} \cos \frac{x_i}{\sqrt{i}} + 1}$

$-5.12 \le x_i \le 5.12$; $F^* = 5.0000$

F15 : $\min F(x) = \sum_{k=1}^{n-1}[(1-x_k)^2 + 100(x_{k+1} - (x_k)^2)^2]$; $n = 3$ (Rosenbrock 3-variables function)

$$-1.2 \le x_k \le 1.2; \quad F^* = 0.00; \quad x^* = [1, 1, 1]$$

F16 : $\min F(x) = \sum_{k=1}^{n-1}[(1-x_k)^2 + 100(x_{k+1} - (x_k)^2)^2]$; $n = 6$; (Rosenbrock 6-variables function)

$$-1.2 \le x_k \le 1.2; \quad F^* = 0.00; \quad x^* = [1, 1, 1, 1, 1, 1]$$

Constrained test problems

E1 : $\min F(x, y) = 6.5x - 0.5x^2 - y_1 - 2y_2 - 3y_3 - 2y_4 - y_5$

s.t.:

$x + 2y_1 + 8y_2 + y_3 + 3y_4 + 5y_5 \le 16$ $\quad 0.2x + 2y_1 + 0.1y_2 - 4y_3 + 2y_4 + 2y_5 \le 12$

$-8x - 4y_1 - 2y_2 + 2y_3 + 4y_4 - y_5 \le -1$ $\quad -0.1x - 0.5y_1 + 2y_2 + 5y_3 - 5y_4 + 3y_5 \le 3$

$2x + 5y_1 + 0.2y_2 - 3y_3 - y_4 - 4y_5 \le 24$

$y_3 \le 1 \quad y_4 \le 1$ and $y_5 \le 2$ $\quad x_i \ge 0 \quad y_i \ge 0$; for $1 \le i \le 5$

$F^* = -11$; $(x^*, y^*) = [0.0, 6.0, 0.0, 1.0, 1.0, 0.0]$

E2 : $\min F(x, y) = 5x_1 + 5x_2 + 5x_3 + 5x_4 - 5\sum_{i=1}^{4} x_i^2 - \sum_{i=1}^{9} y_i$

s.t.:

$2x_1 + 2x_2 + y_6 + y_7 \le 10$ $\quad 2x_1 + 2x_3 + y_6 + y_8 \le 10$

$2x_2 + 2x_3 + y_7 + y_8 \le 10$ $\quad -8x_1 + y_6 \le 0$

$-8x_2 + y_7 \le 0$ $\quad -8x_3 + y_8 \le 0$

$-2x_4 - y_1 + y_6 \le 0$ $\quad -2_2 - y_3 + y_7 \le 0$

$-2y_4 - y_5 + y_8 \le 0$ $\quad 0 \le x_i \le 1 \quad i = 1, 2, 3, 4$

$0 \le y_i \le 1 \quad i = 1, 2, 3, 4, 5, 9$ $\quad 0 \le y_i \quad i = 6, 7, 8$

$F^* = -15$; $(x^*, y^*) = [1, 1, 1, 1, 1, 1, 1, 1, 1, 3, 3, 3, 1]$

$$\textbf{E3} : \min F(x) = -(75.196 - 3.8112x_1 + 0.126948x_1^2 - 2.0567\ 10^{-3}x_1^3 + 1.0345\ 10^{-5}x_1^4 - 6.8306x_2 + 0.030234x_1x_2 - 0.00128134x_2x_1^2 + 3.5256\ 10^{-5}x_2x_1^3$$

$$
\begin{aligned}
&- 2.266\ 10^{-7} x_2 x_1^4 + 0.25645 x_2^2 - 0.0034604 x_2^3 \\
&+ 1.3514\ 10^{-5} x_2^4 - 28.106/(x_2+1) - 5.2375\ 10^{-7} x_1^2 x_2^2 \\
&- 6.3\ 10^{-8} x_1^3 x_2^2 + 7\ 10^{-10} x_1^3 x_2^3 + 0.00034054 x_1 x_2^2 \\
&- 1.6638\ 10^{-6} x_1 x_2^3 - 2.8673 \exp(0.0005 x_1 x_2))
\end{aligned}
$$

s.t.:

$x_1 x_2 - 700 \geq 0; \quad x_2 - 0.008 x_2^2 \geq 0; \quad (x_2 - 50)^2 - 5(x_1 - 55) \geq 0;$
$0 \leq x_1 \leq 75; \quad 0 \leq x_2 \leq 65$
$F^* = -7.80057; x^* = [13.554, 51.645]$

E4: $\min F(\mathbf{x}) = x_1^{0.6} + x_2^{0.6} - 6x_1 - 4x_3 + 3x_4$
s.t.:
$x_1 + 2x_3 \leq 4; \quad x_2 + 2x_4 \leq 1; \quad x_2 - 3x_1 - 3x_3 = 0$
$0 \leq x_1 \leq 3; \quad 0 \leq x_2 \leq 4; \quad 0 \leq x_3 \leq 4; \quad 0 \leq x_4 \leq 1$
$F^* = -4.5142; x^* = [4/3, 4, 0, 0]$

E5: $\min F(\mathbf{x}) = x_1^{0.6} + x_2^{0.6} + x_3^{0.4} - 4x_3 + 2x_4 + 5x_5 - x_6$
s.t.:
$x_4 - x_6 = 0; \quad -3x_1 + x_2 - 3x_4 = 0; \quad -2x_2 + x_3 - 2x_5 = 0;$
$x_1 + 2x_4 \leq 4; \quad -0.01 x_2 x_5 \leq 4; \quad x_3 + x_6 \leq 6;$
$0 \leq x_1 \leq 3; \quad 0 \leq x_2 \leq 3; \quad 0 \leq x_3 \leq 4;$
$0 \leq x_4 \leq 2; \quad 0 \leq x_5 \leq 2; \quad 0 \leq x_6 \leq 12$
$F^* = -13.4123$

E6: $\max F(\mathbf{x}) = x_1^2 + x_2^2 + x_3^2$
s.t.:
$4(x_1 - 0.5)^2 + 2(x_2 - 0.2)^2 + x_3^2 = 0.1 \quad x_1 x_2 + 0.2 x_2 x_3 \leq 16$
$-(2x_1^2 + x_2^2 - 2x_3^2) \leq -2 \quad -2.3 \leq x_i \leq 2.7 \quad i = 1, 2, 3$
$F^* = 11.67664 \quad x^* = [0.988, 2.674, -1.884]$

E7: $\min F(\mathbf{x}) = -9x_5 - 15x_9 + 6x_1 + 16x_2 + 10x_6$
s.t.:
$x_1 + x_2 - x_3 - x_4 = 0 \quad x_3 - x_7 - x_5 = 0$
$x_4 + x_8 - x_9 = 0 \quad x_7 + x_8 - x_6 = 0$

$$3x_1 + x_2 - x_{10}(x_3 + x_4) = 0 \quad x_3x_{10} + 2x_7 - 2.5x_5 \leq 4$$
$$x_{10}x_4 + 2x_6 - 1.5x_9 \leq 0$$
$$[0, 0, 0, 0, 0, 0, 0, 0, 0, 1] \leq x$$
$$\leq [300, 300, 100, 200, 100, 300, 100, 200, 200, 3];$$
$$F^* = -400.000;$$
$$x^* = [0, 100, 0, 100, 0, 100, 0, 100, 200, 100]$$

Remark: 5 variables were calculated from equality constraints.

E8:

$$\max F(\mathbf{x}) = \left| \frac{\sum_{i=1}^{p} \cos^4(x_i) - 2\prod_{i=1}^{p} \cos^2(x_i)}{\sqrt{\sum_{i=1}^{p} i x_i^2}} \right|$$

s.t.:

$$\prod_{i=1}^{p} x_i \geq 0.75; \quad \sum_{i=1}^{p} x_i \leq 7.5p$$
$$0 < x_i \leq 10 \quad \text{for} \quad i = 1, 2, \ldots, p$$

Luus (2008), using LJ optimization procedure as suggested by Luus *et al.* (2006), obtained the following for global minimum of E8.

$$p = 12\text{: } F^* = 0.762564133$$
$$x^* = [3.14637, 3.09385, 3.04190, 2.98916, 1.512005, 0.50086, 0.49172, 0.48333, 0.47559, 0.46840, 0.46167, 0.45535]$$

$$p = 15\text{: } F^* = 0.782444960$$
$$x^* = [3.14786, 3.10156, 3.05570, 3.00921, 2.96081, 2.90869, 0.49537, 0.48539, 0.47648, 0.46841, 0.46101, 0.45418, 0.44782, 0.44188, 0.43628]$$

$$p = 20\text{: } F^* = 0.803619104$$
$$x^* = [3.16246, 3.12833, 3.09479, 3.06145, 3.02793, 2.99383, 2.95867, 2.92184, 0.49482, 0.48836, 0.48232, 0.47665,$$

0.47130, 0.46623, 0.46142, 0.45684, 0.45246, 0.44827,
0.44425, 0.44038]

$p = 30$: $F^* = 0.821884397$

$x^* = $ [6.26484, 3.15035, 3.12809, 3.10609, 3.08424, 3.06244,
3.04057, 3.01856, 2.99628, 2.97361, 2.95042, 2.92652,
0.47667, 0.47309, 0.46964, 0.46631, 0.46310, 0.45998,
0.45695, 0.45402, 0.45117, 0.44840, 0.44571, 0.44308,
0.44053, 0.43803, 0.43560, 0.43322, 0.43090, 0.42862]

$p = 40$: $F^* = 0.831015627$

$x^* = $ [6.27655, 3.16258, 3.14555, 3.12870, 3.11201, 3.09542,
3.07888, 3.06236, 3.04581, 3.02920, 3.01247, 2.99558,
2.97848, 2.96109, 2.94337, 2.92524, 0.48364, 0.48087,
0.47816, 0.47552, 0.47296, 0.47046, 0.46803, 0.46565,
0.46332, 0.46105, 0.45884, 0.45666, 0.45453, 0.45245,
0.45040, 0.44839, 0.44644, 0.44451, 0.44262, 0.44076,
0.43893, 0.43714, 0.43538, 0.43365]

$p = 50$: $F^* = 0.835262348$

$x^* = $ [6.28358, 3.16994, 3.15607, 3.14236, 3.12887, 3.11527,
3.10185, 3.08848, 3.07515, 3.06179, 3.04844, 3.03504,
3.02159, 3.00803, 2.99440, 2.98055, 2.96659, 2.95240,
2.93800, 2.92328, 0.48825, 0.48593, 0.48366, 0.48152,
0.47935, 0.47723, 0.47515, 0.47315, 0.47117, 0.46921,
0.46728, 0.46542, 0.46360, 0.46179, 0.45998, 0.45829,
0.45656, 0.45488, 0.45322, 0.45157, 0.44998, 0.44839,
0.44684, 0.44531, 0.44379, 0.44231, 0.44082, 0.43939,
0.43795, 0.43656]

$p = 75$: $F^* = 0.841752435$

$p = 100$: $F^* = 0.845044789$

$p = 200$: $F^* = 0.8499890369$

Appendix B

B1: Optimization of alkylation process

$$\max F = 0.063x_4x_7 - 5.04x_1 - 0.035x_2 - 10x_3 - 3.36x_5$$

s.t.

$x_4 = x_1(1.12 + 0.13167x_8 - (x_8)^2)$; $\quad x_7 = 86.35 + 1.098x_8 - 0.038(x_8)^2 + 0.325(x_6 - 89)$

$x_9 = 35.82 - 0.222x_{10}$; $\quad x_{10} = 3x_7 - 133$

$x_8 = (x_2 + x_5)/x_1$; $\quad x_4 = x_1 + x_5 - 0.22x_4$

$1000 \cdot x_3 = x_4x_6x_9/(98 - x_6)$;

$$\begin{aligned} F^* &= 1161.336694; \\ x^* &= [1728.310416, 16000.00, 98.133457, 3055.992144, \\ &\quad 2000.000, 90.618812, 94.189777, 10.414796, \\ &\quad 2.615609, 149.569330] \end{aligned}$$

B2: Chemical reaction equilibrium

$$\min F(\mathbf{x}) = \sum_{i=1}^{10} x_i \left(c_i + \ln \frac{x_i}{x_1 + \cdots + x_{10}} \right)$$

s.t.:

$x_1 + 2x_2 + 2x_3 + x_6 + x_{10} = 2$, $\quad x_4 + 2x_5 + x_6 + x_7 = 1$

$x_3 + x_7 + x_8 + 2x_9 + x_{10} = 1$, $\quad x_i \geq 0.000001$; $(i = 1, \ldots, 10)$,

$c_1 = -6.089$ $\quad c_2 = -17.164$

$c_3 = -34.054$ $\quad c_4 = -5.914$

$c_5 = -24.721$ $\quad c_6 = -14.986$

$c_7 = -24.100$ $\quad c_8 = -10.708$

$c_9 = -26.662$ $\quad c_{10} = -22.179$

$$\begin{aligned} F^* &= -47.760765; \\ x^* &= [0.04035, 0.15387, 0.77497, 0.00167, 0.48469, 0.00069, \\ &\quad 0.02826, 0.01849, 0.03850, 0.10128] \end{aligned}$$

Local optimum −47.761106 at

$$x^* = [0.040668; 0.147730, 0.783153; 0.001414; 0.000693, 0.485247, 0.027399, 0.017947, 0.037314, 0.096872]$$

B3: Reactor train optimization

$$\max[x_4]$$

s.t.:

$$x_1 - c_{A0} + k_1 x_1 x_5 = 0; \quad x_3 + x_1 - c_{A0} + k_3 x_3 x_5 = 0$$

$$x_2 - x_1 + k_2 x_2 x_6 = 0; \quad x_4 - x_3 + x_2 - x_1 + k_4 x_4 x_6 = 0$$

$$\sqrt{x_5} + \sqrt{x_6} \leq 4$$

$$x_4^* = 0.3888143$$

$$x^* = [0.771462, 0.51699247, 0.204234, 0.3888143, 3.0355687, 5.09672619]$$

Chapter 4

GENETIC ALGORITHMS IN PROCESS ENGINEERING: DEVELOPMENTS AND IMPLEMENTATION ISSUES

Abdunnaser Younes and Ali Elkamel

Department of Chemical Engineering, University of Waterloo
Waterloo, Ontario, Canada

Shawki Areibi

School of Engineering, University of Guelph
Guelph, Ontario, Canada

1. Introduction

Over the last two decades, genetic algorithms (GAs) have been successfully applied in many fields, such as engineering, robotics, economics, chemistry, genetics, operations research, art, and social sciences.

GAs are stochastic search techniques that seek improved performance by sampling promising regions of the solution space, i.e. regions with a high probability for leading to good solutions. These algorithms manipulate candidate solutions in a manner similar to the mechanisms of natural selection. The basic idea is to start from an initial, usually randomly created, set of solutions. In every iteration, new solutions are generated from the existing set using mechanisms modeled after those currently believed to apply in natural evolution of organisms. Some solutions are retained for the subsequent iteration, in which a new cycle of genetic operations are performed. The algorithm, thereby, proceeds in an evolutionary manner where the fittest individuals survive.

The idea of applying principles of natural evolution to design search techniques dates back to several papers in the 1950s (Spall, 2003). Later on, three general approaches were developed independently: *evolutionary programming* by Fogel *et al.* (1966), *evolution strategies* by Rechenberg (1973), and *genetic algorithms* by Holland (1975).

In evolutionary programming, the individuals or "organisms" are finite state machines. Organisms that best solve some target function get the opportunity to reproduce. In reproduction, parent organisms are mutated to create offspring.

Evolution strategies (ES) are a sub-class of nature-inspired direct search methods which use mutation and selection applied to a population of individuals containing candidate solutions in order to evolve iteratively better and better solutions. Two examples of ES are the $(\mu + \lambda)$-ES and (μ, λ)-ES. These two models differ the creation of the new population with either totally new offsprings or a combination of both the best parents and offspring. Recombination operators for evolutionary strategies also tend to differ from Holland-style crossover, allowing operations such as averaging parameters, for example, to create an offspring.

The terms *evolutionary computing* and *evolutionary algorithms* (EAs) are used interchangeably to include the previously mentioned approaches and many other search methods that employ population-based techniques mimicking some principles of natural evolution. GAs represent one of the most (if not the most) widely used classes of EAs; however, they are not necessarily the best alternative all the time. In fact, many EA-based techniques are specifically designed for applications that GAs cannot handle easily. Moreover, current state-of-the-art implementations are in fact hybrids of several classes of EAs, which make it difficult to pinpoint the implementation to a certain class. In many publications, the authors even use the terms GA and EA interchangeably.

There are some characteristics of GAs to which their success as search procedures is attributed. First, GAs manipulate a population (set) of solutions simultaneously; thus, the probability of getting trapped in local optima is reduced compared with methods that proceed from point to point in the solution space. This characteristic also makes them suitable for parallelization. Second, GAs use operators that can act on a coding of the parameter space rather than the parameters themselves, which simplifies implementation. Third, they can work with almost any kind of objective function as they do not require continuity or differentiability of the function to be optimized, which makes them suitable for problems with complex objectives and for those evaluated by simulation. However, GAs are not without limitations. For example, they can suffer from premature convergence and could be less effective for fine tuning the search. Consequently, solutions could deteriorate if not handled properly. Thus, a successful GA implementation should aim to alleviate these difficulties.

The main objective of this introductory chapter is to help engineering students and practitioners understand the principles, the merits, the limitations, and the implementation issues associated with GAs. More insight into evolutionary algorithms can be gained by consulting other references, such as Beasley *et al.* (1993a,b) for an introduction to GAs; Michalewicz (1992) and Mitchell (1996) for more details on genetic operators; Gen and Cheng (1999), Haupt and Haupt (1998), and Michalewicz and Fogel (2004) for implementation and practical issues; and Reeves and Rowe (2002) and Whitley (2001) for theoretical aspects.

A review of Chemical Engineering applications in the past few years will be introduced in Sec. 2. This will be followed by an explanation of the main components of a traditional GA in Sec. 3. In Sec. 4 we focus on implementation issues such as encoding mechanisms, effective evaluation mechanisms and parameter tuning of GAs. Finally in Sec. 5 we describe a few advanced techniques that are used to improve the performance of GAs.

2. Review of Chemical Engineering Applications

In researching GA-based publications in the field of chemical engineering, one notices their large number and their wide range of applications. In this section, we present a quick review of the use of GAs in this field.

Recent publications cover different types of optimization problems in a variety of disciplines of Chemical Engineering: They range from parameter estimation to plant design, from single objective optimization to multiple-objective optimization, from pure integer programming to mixed-integer-nonlinear-programming (MINLP), from optimization under well-defined conditions to optimization under uncertainty, from problems

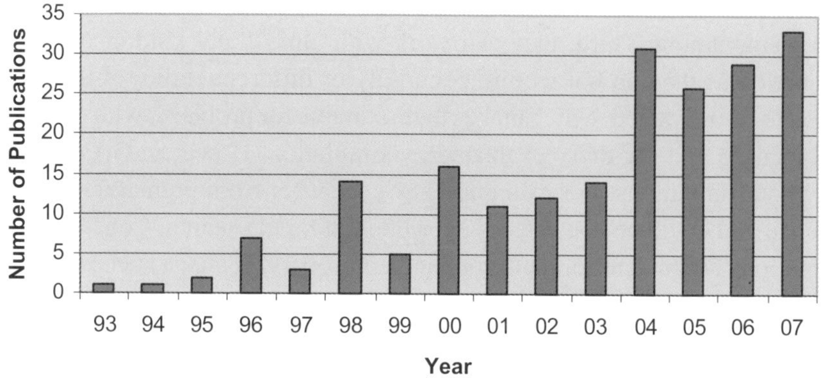

Figure 1. GAs in Chemical Engineering. Publications in international journals. (Based on engineering village databases, University of Waterloo, Canada, 15/03/2008)

with explicit objective functions to problems with simulation-determined objective functions. Figure 1 clearly shows a trend of increasing interest in the use of GAs.

Restricting our attention to journal articles that have been published during the past two years, we present some of these articles in Table 1. Here, it should be noted that this table is by no means comprehensive; It is worthwhile to conduct a survey on the use of GAs and possibly with other EAs in Chemical Engineering. The examples given in the table are merely mentioned to give the reader an idea of the popularity of GAs and the scope of their applications.

3. The Basic Genetic Algorithm

As a search technique, GAs examine and manipulate a set of possible solutions simultaneously at each *generation* (or iteration of the algorithm). The set of solutions pop_j is referred to as the *population* of the jth generation.

Each candidate solution is represented by a string of symbols called a chromosome (in some applications, a solution is represented by more than one chromosome) as seen in Fig. 2. Thus, at the outset, there must be a code or scheme that allows for a representation of possible solutions to the problem, and a suitable function to assess the fitness of any solution. The population can then evolve for a predetermined total number of generations under the application of evolutionary rules called *genetic operators*. These

Table 1. Examples of GA applications in chemical engineering.

Ref.	Application	Objective	Method
Wu *et al.* 2007	Distillation of acetone-methanol	Molecular design for extractive distillation	GA with improved encoding and operators
Yang *et al.* 2007/03/	FCC gasoline secondary reaction	Optimization of fuzzy neural network	Real-coded GA
Masoori *et al.* 2007	Process design and reaction kinetics	Rates of reaction in reacting systems	Binary-coded GA vs. real-coded GA
Tao and Wang 2007	Chemical engineering processes	Parameter estimation	RNA based GA
Cao *et al.* 2007	Water-using networks	Minimum freshwater consumption	Multi-agent GA
Jeżowski *et al.* 2007	Heat exchanger networks	Superstructure optimization of retrofit problems	Dual alternating GAs: structure and parameter optimization
Young *et al.* 2007	Chemical process	MINLP in the design of chemical systems	Entropy-guided GA
Elliott *et al.* 2006	Combustion of aviation fuel	A reduced mechanism to simulate fuel combustion	Multi-objective real GA
Agrawal *et al.* 2006	Chemical reactors	Two-objective low-density polyethylene reactor	Non-dominated sorting binary GA with diversity-maintaining jumping genes

(*Continued*)

Table 1. (*Continued*)

Ref.	Application	Objective	Method
Istadi and Amin 2007	Chemical reactors	ANN-based model for plasma reactor	Weighted-sum multi-objective real GA
He and Hui 2007	Chemical plants	Large multi-stage multi-product scheduling	Permutation-coded GA with heuristic rules penalty methods
Jung *et al.* 2007	Chemistry of materials	Search for blue phosphors in seven cation systems	GA-assisted combinatorial chemistry
Till *et al.* 2007	Chemical batch scheduling	Two-stage stochastic integer programs	Hybrid EA and mixed-integer programming
Cubillos *et al.* 2007	Real-time process optimization	Process optimization of grey-box neural models	Hybrid GA and nonlinear programming

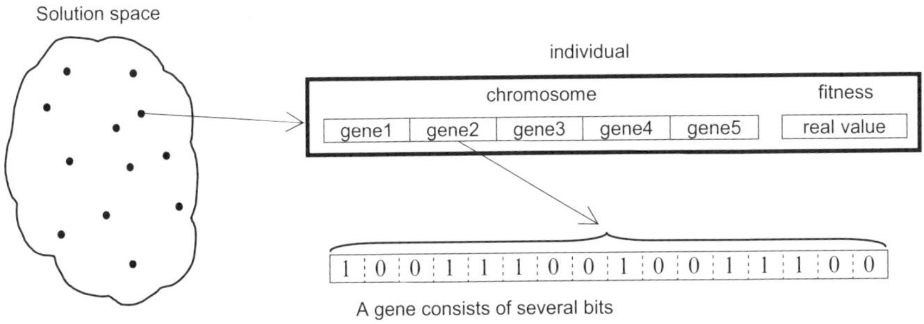

Figure 2. Solution representation. A candidate solution is mapped into an individual that consists of one or more chromosomes.

```
Procedure GeneticAlgorithm
    // initiate generations counter
  g = 0;
    // generate initial population;
  Generate(pop);
    // assign fitness values to individuals in the population
  Evaluate( pop );
  repeat
      g = g+1;
        // select individuals for reproduction
      pop(1) = Select( pop );
        // recombine individuals if crossover condition is satisfied
      pop(2) = Cross( pop(1) );
        // mutate individuals if mutation condition is satisfied
      pop = Mutate( pop(2) );
      Evaluate( pop );
  until terminating condition is satisfied
  return best individual;
```

Figure 3. A simple generic genetic algorithm.

rules include techniques for selecting solutions for recombination, and transformation functions that create new individuals from selected parent solutions. Figure 3 shows the main modules of a traditional GA. In the next few sections, we will cover each GA module in more detail.

3.1. Encoding

Traditional GAs do not work directly on solutions but on *chromosomes*. Each candidate solution is represented by a string of symbols constituting one or more chromosomes. The encoded version of each parameter of a candidate solution is termed a *gene*, and the encoding of the entire solution is termed *genotype*. Thus, a gene is the basic unit of the genotype. In natural selection, a genotype consists of several chromosomes, each consisting of a number of genes. GAs, however, often use a single chromosome to represent a candidate solution. Hence, the terms *individual*, chromosome, and genotype are often used interchangeably in GAs. Strictly speaking, we should distinguish between the terms *phenotype* and genotype. Whereas genotype refers to the representation of the solution in the chromosome(s), phenotype refers to the problem version of the genotype, i.e. the actual solution.

To evaluate individuals, the user needs to define a decoding function that maps chromosomes (consisting of genes) back to solutions (consisting of decision variables). A decision variable x_i is represented by a gene g_i that consists of l_i binary bits. Using direct conversion from binary bits to decimal value, the decimal value of the gene is

$$\upsilon_i = \sum_{m=1}^{l_i} b_{i,m}\, 2^{m-1}, \tag{1}$$

where $b_{i,m}$ is the mth bit in gene g_i. The decimal value of the gene g_i is linearly mapped to the value of the decision variable x_i by

$$x_i = X_i^{lo} + \frac{X_i^{hi} - X_i^{lo}}{2^{l_i} - 1}\,\upsilon_i, \tag{2}$$

where x_i^{lo} and x_i^{hi} are the lower and the upper limits of x_i respectively.

The precision of the approximation inflicted by this transformation is determined from $\frac{X_i^{hi} - X_i^{lo}}{2^{l_i}-1}$. Thus, the number of bits l_i is chosen so that the required precision is obtained.

Example 1

Consider an example (Michalewicz, 1992) of an optimization problem in which we wish to maximize

$$f_1(x_1, x_2) = 21.5 + x_1 . \sin(4\pi x_1) + x_2 . \sin(20\pi x_2), \quad (3)$$

where $-3.0 \leq x_1 \leq 12.1$ and $4.1 \leq x_2 \leq 5.8$.

The number of bits to be used for each variable is first decided according to the required precision. Suppose that the user wishes to have a precision of four decimal places for each variable. In this case, l_i must satisfy the inequality $(X_i^{hi} - X_i^{lo})/(2^{l_i} - 1) \leq 10^{-4}$. Accordingly, the computed values of l_1 and l_2 are 18 and 15, respectively. That is, 18 bits are required to represent x_1, and 15 bits are required to represent x_2, and the total length of the chromosome is $18 + 15 = 33$.

The string $<$ **100001011001110101, 101001010100110** $>$ would be a chromosome representing one possible solution $< x_1, x_2 >$. The value of x_1 is determined by decoding the first gene (first 18 bits): Eq. (1) is used to compute the integer $v_1 = dec((100001011001110101)_2) = 136821$. Then, the mapping Eq. (2) is used to find $x_1 = -3.0 + \frac{12.1-(-3.0)}{2^{18}-1} * 136821 = 4.8811$. Similarly, the last 15 bits are decoded to $x_2 = 5.1977$.

3.2. Fitness evaluation

In applying a genetic algorithm to any optimization problem one has to devise a particular type of objective function called a *fitness function* or an *evaluation function*. This function evaluates and ranks individuals in the population according to their *fitness* (i.e. how good their solutions are) so that individuals producing the best solutions in the population are retained.

It is important to distinguish between the term fitness and the term *fitness landscape*. The latter is a metaphor commonly used to visualize the relation between fitness values of all candidate solutions and distances between them as a structure of fitness hills and valleys.

In simple optimization problems the fitness function is more or less a modification of the objective function. The fitness function in Example 1 is the objective function itself. Hence, the fitness associated with the string $<$ **100001011001110101, 101001010100110** $>$ can be directly obtained by substituting the decoded variables into the objective function (3), i.e.

$f_1(4.8811, 5.1977) = 21.5 + 4.8811 \sin(4\pi * 4.8811) + 5.1977 \sin(20\pi * 5.1977) = 15.8847$.

Most introductory books give simple examples, similar to this one, on fitness evaluation. The simplicity serves to keep the reader focused on the main topic; however, it should not be taken as an indication that fitness evaluation is a trivial task. Indeed, in real world problems the task of evaluating a candidate solution is often challenging and time consuming. Different techniques to estimate the evaluation or reduce its cost are discussed in more detail in Sec. 4.

3.3. Initial population

The algorithm starts by generating individuals that constitute the initial population, which should eventually lead to the best final results in terms of solution quality and running time. Several approaches can be used to generate the initial population. In this section, we discuss three general approaches that are commonly used: random, uniform, and biased.

In random initialization, we merely assign a random value to each variable in each chromosome such that this value falls within any known limits of the variable. This random initialization aims to sample the search space and to avoid biasing the population around particular regions in the search space.

One drawback in random sampling is that it does not ensure a fair sampling of the search space, especially when the population size is small. For this reason, some implementations use specific procedures to ensure that the population is distributed uniformly over the search space, e.g. by dividing the search space into subregions and ensuring that each subregion is represented by one individual in the initial population.

If the user has some prior knowledge about some good solutions or about regions that are likely to contain high quality solutions, then the user should bias the initial population towards promising regions by injecting the initial population either with known good solutions or with solutions picked from promising regions.

The three approaches described earlier are illustrated in Fig. 4. Note, however, that these are not the only approaches that can be used. In fact,

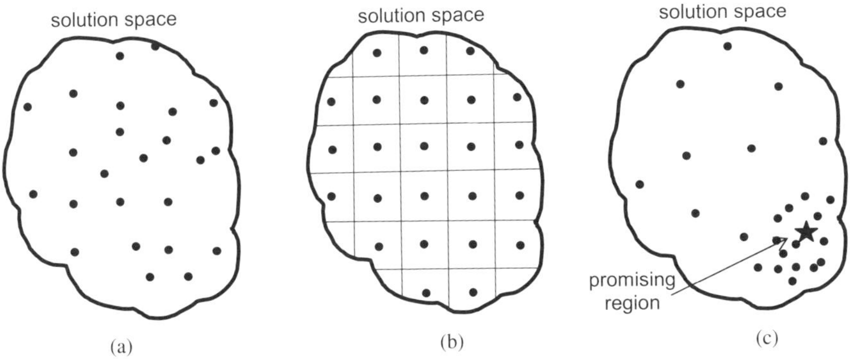

Figure 4. Population initialization. (a) random (b) uniform (c) biased.

adopting hybrid approaches can generally be more effective than using primitive approaches.

3.4. Selection

In GAs, selection operators perform a role equivalent to Charles Darwin's *natural selection*; that is, they determine which individuals in the current population survive and reproduce offspring. Individuals are selected for mating according to their fitness: those with greater fitness are awarded more offspring than those with lesser fitness. There exist many selection schemes in the literature, such as roulette wheel selection (RWS), fitness scaling, ranking selection, and tournament selection. This section will be devoted to RWS, the earliest of these schemes, and briefly discuss the other ones. The interested reader can find more details on these schemes and several others in Michalewicz (1992) and Mitchell (1996).

3.4.1. Fitness proportionate selection

As the name implies, RWS simulates a roulette wheel having a spoke that spins and lands on a sector (an individual) in the wheel, as illustrated in Fig. 5. In this method, the probability of selecting an individual is equal to its fitness relative to the average fitness in the population. Although RWS is still one of the most used methods, it is not the ideal scheme. RWS is

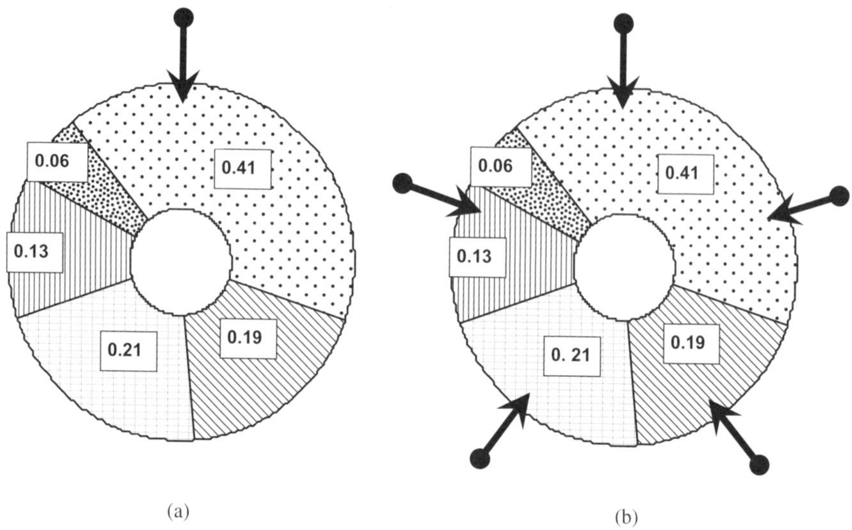

Figure 5. Fitness proportionate selection. (a) Roulette wheel sampling. (b) Stochastic universal sampling.

essentially a sampling with replacement scheme; thus, the actual number of selected copies of an individual can drastically vary from the expected value. Stochastic universal sampling (SUS) (Baker, 1987) solves this problem by employing a sampling without replacement scheme. SUS can be viewed as a RWS with multiple spokes (Fig. 5b).

Example 2

Assume we seek to optimize the objective function given in Example 1, and that we have created an initial population consisting of five individuals $\{S_1, \ldots, S_5\}$. Suppose that their fitness values are found to be $\{10.92, 6.73, 3.31, 21.03, 9.68\}$ using Eq. 3. To apply RWS or SUS to this population, we first compute the relative fitness of each individual from the relation $\bar{f}(S_i) = \frac{f(S_i)}{\sum_{i=1}^{5} f(S_i)}$, e.g. $\bar{f}(S_1) = (10.92)/(10.92 + 6.73 + 3.31 + 21.03 + 9.68) = 0.21$. The resulting five relative fitness values $\{0.21, 0.13, 0.06, 0.41, 0.19\}$ are then used to compute accumulative fitness values $\phi(S_i) = \sum_{k=1}^{i} \bar{f}(S_k)$, e.g. $\phi(S_1) = 0.21$ $\phi(S_2) = 0.21 + 0.13 = 0.34$. The resulting set of accumulated fitness values $\{0.21, 0.34, 0.40, 0.81, 1.00\}$ is used to simulate a roulette wheel with five unequal segments. A spin of the wheel is simulated by generating a random number r between 0 and 1, and selecting the individual that has the smallest accumulated fitness larger or equal to r. Repeating the process of generating random numbers and selecting the corresponding individuals five times, we end up with five individuals. Suppose that the generated random numbers are 0.515, 0.534, 0.015, 0.853, and 0.163. The corresponding individuals are then S_4, S_4, S_1, S_5, and S_1 respectively.

Since RWS and SUS increase the probability of selecting an individual in proportion to its fitness, they cannot work directly on minimization problems, where the probability of selection should decrease when the objective function decreases. Therefore, these problems are transformed into maximization ones, usually by taking the reciprocal of the objective function. That is,

$$f = \frac{1}{a + f_{\text{minimization}}}, \tag{4}$$

where a is a positive constant (usually 1) added to avoid division by zero.

Regardless of the type of the objective function (minimization or maximization), the main disadvantage of RWS and SUS is inherent to their reliance on fitness proportionates. In early generations, fitness values are typically small and can be quickly dominated if a relatively good solution is discovered, and thus the population can converge prematurely towards this solution. On the other hand, fitness values are typically large toward the end of the run, and hence their differences relative to their average are small, allowing for little exploitation of good individuals.

3.4.2. Other selection schemes

Fitness scaling (Goldberg, 1989) is proposed to alleviate the problems of proportionate selection, by converting the fitness values to new values that receive the sample emphasis throughout the run (the emphasis is not magnified or diminished by the absolute values of fitness). A commonly used fitness scaling function uses linear transformation of the original function. That is,

$$f = af_{\text{original}} + b, \tag{5}$$

where a and b are the parameters that determine the extent of scaling. The main problem with this method is that it requires continuous re-scaling of the fitness.

Ranking methods avoid scaling the population by first sorting the individuals according to their fitness values. The probability of selecting an individual is then determined by its rank in the ordered population. However, ranking methods have the disadvantage of the extra cost required for sorting the population, which can be very expensive in large populations.

Tournament selection is yet another popular approach used by many researchers. It works in a manner similar to ranking selection (which avoids the pitfalls of proportionate selection methods), but is easier to implement and less time consuming. The idea is simple: The population is divided into subgroups (pairs in the case of the most common binary tournament). The best individual (the one with highest fitness or lowest cost) in each group is selected. The process is repeated until the desired number of individuals to be injected in the new population is reached. With small populations, it is recommended to implement selection without replacement to avoid errors due to stochastic biases.

3.5. Crossover

Crossover, also called *recombination* operator, is used to recombine genetic material in parent chromosomes (usually two) to produce one or more offsprings that share characteristics of parents. However, in so doing crossover operators disrupt the original chromosomes. Thus, crossover has a dual effect on the search process: First, it has a recombination effect as it passes traits from parents to offspring. Second, it has an explorative effect, as it directs the search into new regions. These effects can be better illustrated by studying three types of crossover that are used extensively in binary encoded GAs.

A *single-point crossover* is a type of crossover in which the parent strings are bisected at a randomly chosen point simultaneously. A child string can be produced by joining the leftmost portion of one parent and the rightmost portion of the other parent. If a second child is to be produced, it is constructed from the unused portion of both parents. Figure 6 illustrates the operation of a single-point crossover. The second type is the more

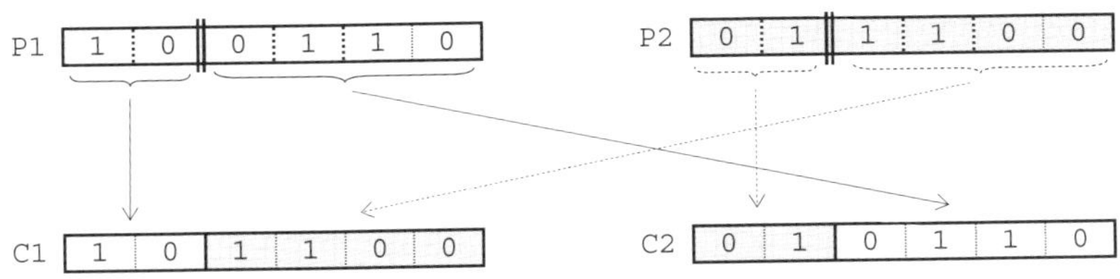

Figure 6. Single-point crossover.

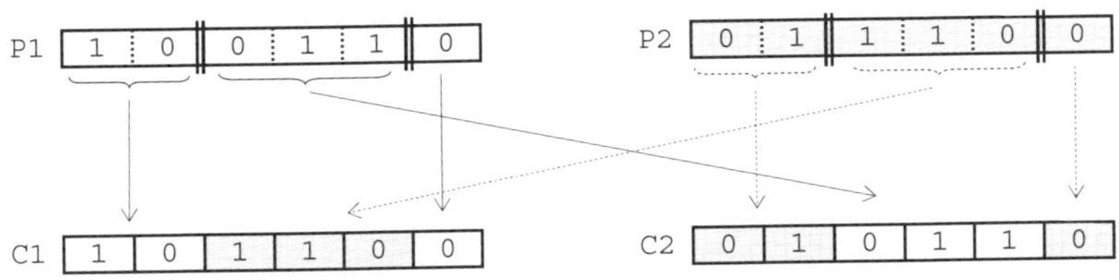

Figure 7. Two-point crossover.

general *n-point crossover*. For example in 2-point crossover each parent chromosome is bisected at two cut points, as shown in Fig. 7. The outer portions of one chromosome are then recombined with the inner portion of the other chromosome to produce one child chromosome. A second child chromosome can be constructed by joining the unused portions of the parents. The third type of crossover is called *uniform crossover*, see Fig. 8. When uniform crossover is applied to a bit string, a random binary mask is created first. A child chromosome is then constructed bit by bit: A bit is copied from the first parent if the corresponding bit in the mask is 0 or from the second parent if the corresponding mask bit is 1. If a second child is to be created, then it is generated by simply copying the unused bits of both parents. Comparing the three crossover types, single-point crossover

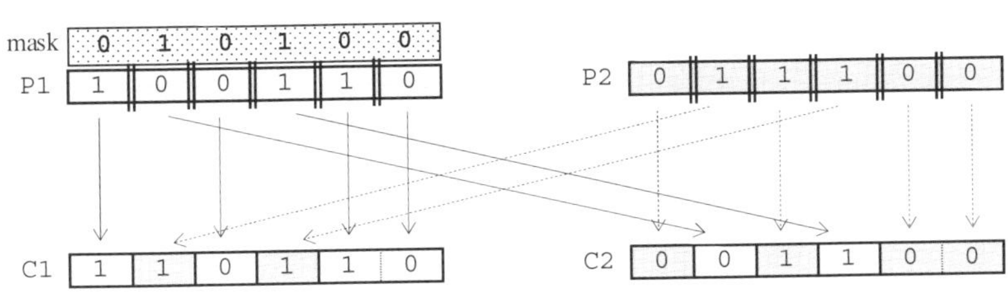

Figure 8. Uniform crossover.

tends to produce the minimum explorative effect whereas uniform crossover produces maximum exploration. In fact, uniform crossover can be viewed as a multiple-point crossover. However, these three crossover techniques are unsuitable for non-binary encodings since they produce infeasible solutions and exchange variables between parents without changing their values in real-coded GAs.

Real-coded GAs use several forms of crossover that basically produce linear combinations of the parental variables. The most used form is the blending crossover of Eshelman and Schaffer (1993). In this operator, a child chromosome is produced gene by gene from two parent chromosomes. Each gene y_i is chosen randomly in the range between $x_i^1 - \alpha(x_i^2 - x_i^1)$ and $x_i^2 + \alpha(x_i^2 - x_i^1)$ where x_i^1 and x_i^2 are the corresponding genes in the parent chromosomes, and α is a parameter that controls the limits of this range. Thus, a formula for computing a possible value for a child gene can be given as

$$y_i = \gamma_i(x_i^2) + (1 - \gamma_i)(x_i^1), \tag{6}$$

where $\gamma_i = (1 + 2\alpha)r_i - \alpha$ and r_i is a random variable between 0 and 1.

For permutation based representation several crossover techniques can be used. *Order crossover* (Davis, 1985), *partially mapped crossover* (Goldberg and Lingle, Jr., 1985), and *cycle crossover* (Oliver *et al.*, 1987) are examples of operators that are specifically developed to recombine solutions to create feasible children. A detailed description of these operators with many more can be found in Michalewicz (1992).

3.6. Mutation

The main objective of applying the *mutation* operators is to ensure that the population is supplied with new genetic material throughout the search process. Mutation enables the GA to maintain diversity in the population and introduces some random search behavior necessary to explore new (unvisited) regions of the search space that may not be reached by applying crossover only. Thus, mutation plays a complementary role to that of crossover, which works on material already present in the population and thus cannot introduce new genetic material. In other words, without a

Figure 9. Simple random mutation.

mutation operator, genetic material nonexistent in the initial population or lost with discarded individuals can never be developed.

Mutation is applied as a unary operator that transforms one parent chromosome into a different child chromosome. In binary representations, the mutation operator replaces each bit in the string by a randomly selected bit if a probability test is passed (see Fig. 9). In real-coded GAs, an individual can be mutated by slightly increasing or decreasing the value of some of its genes. Gaussian Mutation is a common method that adds a normally (Gaussian) distributed error to the gene.

$$y_i = x_i + e(i). \tag{7}$$

In permutation representations, mutation can be accomplished by swapping two or more genes in the chromosome to produce a different solution.

3.7. Theoretical aspects

Researchers have proposed several theories to explain the behavior of GAs and their theoretical convergence. Three theorems in particular have greatly contributed to the understanding of GAs, despite the debate that surrounds the extent of their applicability: the schema theorem (Holland, 1975), the *building block hypothesis* (Goldberg, 1989), and the no free-lunch (NFL) theorem (Wolpert and Macready, 1997). In this section, we discuss these theorems briefly. For a more refreshing analysis, we refer the interested reader to Reeves and Rowe (2002).

Holland's schema theorem is the earliest attempt to explain how GAs work; it is also the most cited GA-related theorem. Holland introduced the term *schema* to explain his theorem in binary representation. A schema is a template made from the alphabet {0 1 #}, with the "#" symbol matching 0 or 1. For example, the schema {0 1 1 # 0 0} matches the two strings {0 1 1 0 0 0} and {0 1 1 1 0 0}, and the schema {# 1 1 0 # 0} matches the strings {0 1 1 0 0 0}, {0 1 1 0 1 0}, {1 1 1 0 0 0}, and {1 1 1 0 1 0}. Holland defined the *order* of the schema as the number of 0's and 1's it contains, and the *defining length* of the schema as the distance between the two outer most non-# symbols in the schema. With these terms, the schema theorem can be stated as

> *At a certain generation of the genetic algorithm, existing schemata that are of short defining length, low order, and above the average fitness receive exponentially increasing trials in subsequent generations.*

Goldberg gives a similar explanation as to how GAs work in his *Building Block Hypothesis* (Goldberg, 1989). He compares the performance of a genetic algorithm in its search of an optimal solution, to the arrangements made by a child to construct a big castle from small building blocks.

Both explanations assume that high quality solutions consist of good "building blocks" where a good building block is a few bits in the chromosome that contribute to the high fitness value of an individual. Thus, the success of a GA in finding good solutions is primarily attributed to its ability to discover, emphasize, and recombine good building blocks from different chromosomes to construct final solutions. At one time, these explanations were seen to be satisfactory, and many algorithms were designed to exploit the building blocks and increase the number of good schemata. However, both explanations are criticized by many researchers today. A contemporary view, that is shared by many researchers, is that building blocks should be viewed as a dynamic entity that changes over the search process. Under this view, the quest to detect and combine good building blocks makes little sense, since what can be a good block at one generation might become less effective in the next generation.

The NFL theorem is not exclusively concerned with GAs, but rather with search algorithms in general. This theorem states that the performance of all search algorithms is the same when averaged over all possible functions. That is, no search method can be the best technique for all problems.

The details of NFL nor its counter criticism will be discussed any further. The main interest from a practical point of view is in its implications. One implication of this theorem is that claims of effectiveness or superiority of an algorithm should be supported by specifying a set of problems that define the scope of these claims. Another major implication of this theorem is that we need to integrate as much problem specific information as possible in the GA. This can be included in the selected representation, the generation of the initial population, the design of the genetic operators and the handling of the constraints, and externally in the form of hybridizing the GA with other problem specific heuristics.

3.8. General characteristics

Before implementing a GA, we must ensure that GAs are the right choice for the problem at hand. This is not a trivial task, but a good approach starts by understanding the advantages and disadvantages of GAs.

3.8.1. Advantages

GAs have some characteristics to which their success as search procedures is attributed. First, GAs use several transformation operators to manipulate a population of candidate solutions at each generation, making the associated search process global and multiple directional. Thus, they sample the search space more effectively and are less apt to getting trapped in local optima, compared with methods that proceed from point to point in the solution space, as illustrated in Fig. 10. Hence, they are effective in solving constrained and multi-modal objective functions.

Second, genetic operators make use of a coding of the parameter space rather than the parameters themselves (only objective function information is used), which simplifies implementation, and makes them applicable to a wide range of applications and able to handle a variety of decision variables involving binary, integer, or continuous representations.

Third, GAs are not limited by characteristics of objective functions and can work even with a partial or approximate evaluation. With the appropriate

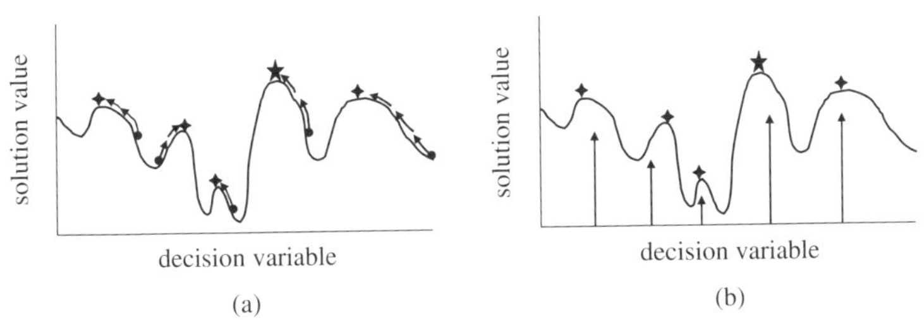

Figure 10. A point-based search versus a population heuristic. (a) A local search method that inspects a single point at a time ends at a local optimum that depends on the starting point. (b) Population-based heuristics sample several regions of the search space simultaneously.

choice of selection operators, GAs need only a method to rank solutions rather than completely evaluate them. Thus they are suitable for problems with complex objective functions, such as those evaluated by simulation.

Finally, parallel implementations of GAs are relatively easy allowing for great savings in run time.

3.8.2. Disadvantages

GAs can be less effective in some applications, they can suffer from premature convergence, and they are generally poor at fine tuning the search process. These disadvantages can be alleviated by adding specific procedures to maintain population diversity and by hybridizing with other algorithms. These procedures are discussed in Secs. 5.1 and 5.2.

However, their main disadvantage, which they share with other meta-heuristic approaches, is the inability to guarantee optimality. Actually, in optimization problems these algorithms do not recognize the optimum even after they find it. For this reason, if there is an exact method or a simple heuristic that can solve the problem at hand in reasonable and acceptable time, then we should not use GAs or any other meta-heuristic approach for this problem. Nevertheless, most real-world problems are too complex to be solved exactly and in short time, and hence GAs remain good candidates for many of such problems.

3.9. When should we use GAs?

The real power of the GA comes from the combined effect of exploration and exploitation: exploration of the search space by crossover and mutation and exploitation of good solutions by selection.

GAs are particularly well-suited to problems where the space of all potential solutions is truly huge — too vast to search exhaustively in any reasonable amount of time. Most problems that fall into this category are known as *nonlinear*. In a linear problem, the fitness of each component is independent, thus an improvement to one component will result in an improvement of the system as a whole. Needless to say, few real-world problems are like this. Nonlinearity is the norm, where changing one component may have ripple effects on the entire system, and where multiple changes that individually are detrimental may lead to much greater improvements in fitness when combined.

One important factor to consider is epistasis, the interaction between different genes in a chromosome. On one extreme, little or no epistasis means that each gene contributes independently to the fitness of the chromosome, making the task of the GA simple. Problems that can be represented in this way can be easily solved by simple hill climbers. On the other extreme, excessive epistasis means that the contribution of a gene to the fitness of the chromosome depends greatly on the values of the other genes. Problems with such representations are better solved by random search. Between both extremes lie many engineering problems that can be represented with chromosomes having moderate epistasis. GAs are known to be effective for such problems.

GAs are also proven to be efficient in the field of multi-objective optimization and many researchers have investigated their use for such applications. The majority of evolutionary algorithms targeting multi-objective problems make use of the pareto-dominance concept to create a set of non-dominant solutions called the pareto front. The pareto-front of any multi-objective optimization problem is the set of solutions in the objective space that represents the tradeoffs for the different objectives, as shown in Fig. 11. Thus another motivation of the use of GAs is their ability to handle multi-objective problems, which constitute most practical problems.

In summary, the ease of parallelization of GAs make them particularly good candidates for problems with complex, time-consuming, objective

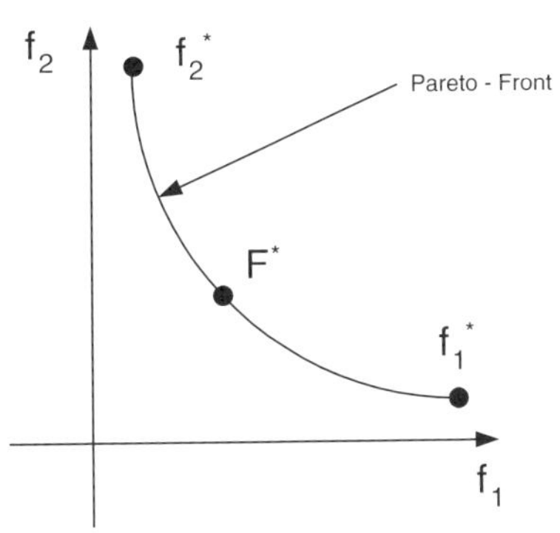

Figure 11. The concept of the pareto-front.

functions. As well, the population-to-population approach is useful for finding several Pareto solutions for multiple objective optimization problems and for finding several alternative solutions for dynamic optimization problems and optimization under uncertainty.

4. Implementation Issues

The simple generic GA introduced in the previous section is usually not effective for solving complex problems. For example, many optimization problems in Chemical Engineering are modelled as MINLP. They are characterized by two features that make them potentially hard to solve: First, they deal with integer and continuous variables simultaneously, which make the problem both combinatorial and open ended. Second, the presence of non-linearity in objective function and the constraints give rise to non-convexity issues. In addition, the evaluation of candidate solutions is often complex and time-consuming.

Therefore, a successful GA implementation involves several decisions, including the selection of encoding schemes and genetic operators, and many other decisions that influence almost all components of the GA. Moreover, for any given problem, GAs can be implemented in many ways, giving users a margin for integrating problem specific domain information, their own experience, and intuitions in the implementation.

4.1. Primary decisions

4.1.1. Encoding

Traditional GAs use binary encoding; however, nowadays GAs use real values, integers, and permutations, in addition to binary, for solution representation. Deciding on a scheme for representing solutions is one of the earliest decisions taken in the course of a GA implementation. It also greatly determines the success of the implementation because it influences the topology and the size of the search space, and the choice and the effectiveness of the GA operators.

In applying the traditional binary encoding, for example, the user has to decide upon the number of bits, as described in Sec. 3.1. High precision of variable representation can be obtained by increasing the number of bits.

However, an increase in the number of bits increases the size of the search space and possibly the runtime. Hence, precision should be limited to the value appropriate to the problem at hand.

A question that is often raised is whether to use fixed or floating point representation for real valued decision variables. There seems to be no clear-cut answer to this question even though many researchers advocate floating point representation. On the one hand, fixed point (binary) representation of a particular problem requires a smaller search space than that used by floating point representation for the same problem. On the other hand, binary representation of real variables results in loss of precision because the infinite number of values that the original continuous variable can take are mapped into a finite number of integer values.

In many situations, a specific encoding arises naturally. For example, in problems involving yes/no decisions, such as knapsack problems, binary encoding is the natural choice, and for problems involving permutation, such as scheduling problems, permutation is the natural choice. However, such choices are not necessarily the best alternatives. The user has to review relevant publications and investigate commonly used representations to decide on the best representation for the problem under consideration.

4.2. Complex evaluations

Real-world problems often involve complex, time consuming, ambiguous, or conflicting objectives. In particular, applications of Chemical Engineering are dominated by computationally expensive evaluation functions that rely on lengthy, complex simulations. These characteristics can make a GA run prohibitively expensive, since GAs need a large number of evaluations to produce an optimal or an acceptable solution. Consequently, special procedures are necessary to reduce the time spent in function evaluation, by reducing the total number of evaluations or the cost of individual evaluation. Some of the commonly used procedures are listed below.

4.2.1. Reducing the total number of evaluations

- Ensure that the initial population contains distinct solutions only. An inexpensive way to achieve this is by using unique patterns that deterministically set the value of some bits according to a specific pattern and set the remaining bits randomly. The pattern is designed so that no two sets of values of the selected bits are identical in any two chromosomes.

- Avoid reevaluating the same individual. The population is checked prior to evaluation for the existence of identical copies of the same chromosome. Extra copies are then either discarded or assigned the objective values of the originals. This procedure is computationally expensive as it might involve comparing several chromosomes bit by bit.
- Avoid recombining identical solutions. Identical parents if recombined, do not produce any new solutions. Their child solutions are also duplicates of the parent solutions. Indeed, recombination of parent solutions should be avoided all the time — even when the cost of evaluation is not large.
- Store visited solutions in a list that is checked whenever a new candidate solution is considered. This naive approach is very expensive, since the list of visited solutions becomes extremely long towards the end of the run. Hence, we do not recommend this approach in practical large problems.

4.2.2. Reducing the cost of individual evaluation

- Perform incremental evaluation. That is, the difference in fitness between a new individual and its parent is computed, rather assessing the new individual completely. The fitness of the new individual can then be determined by simply adding the computed difference to the fitness of the parent. This procedure can be very effective; however, it is not always applicable.
- Estimate or partially evaluate the fitness of a candidate solution, rather than completely evaluate it. Look for solution traits that allow quick ranking of solutions, without complete evaluation. This technique can be very useful; however, it is problem dependent and requires a deep understanding of the problem at hand. To reduce errors from estimations or incomplete evaluations, these techniques are applied in the early generations only whereas more accurate evaluations are performed in the subsequent generations.

It is important to note that some of the above procedures are expensive and hence should be performed only when their benefits surpass their costs.

4.3. Constraint handling

Most real world problems are subject to some kind of constraints. Hence, GAs used to solve such problems have to handle infeasible solutions during the search process. A particular difficulty here is that solutions (feasible or infeasible) are not guaranteed to evolve into feasible solutions. That is, the mutation or the recombination of perfectly feasible solutions might produce infeasible offspring. The literature contains several methods of constraint handling: Some discard infeasible solutions, some try to prevent them, some try to repair them, and others just penalize them.

Discard infeasible solutions The simplest method to handle infeasible solutions is discarding them. In this way, the need to evaluate infeasible solutions is eliminated and the search space is effectively reduced. However, discarding infeasible solutions also means discarding the connectors between feasible regions of the search space, and hence this method frequently fails when the feasible region is not convex and when the optimal solution is located close to the problem constraints.

Prevent infeasible solutions In many COPs, problem specific operators (i.e. crossover and mutation) are designed to produce feasible solutions only. For example, order based crossover and edge recombination crossover are used in TSP to produce feasible offspring.

Repair infeasible solutions Many popular techniques work by repairing infeasible solutions. That is, infeasible solutions are transformed into feasible ones by replacing genes that cause infeasibility. These techniques use two general approaches: In the first approach, the fitness of the repaired solution is assigned to the infeasible one, and in the other approach, the infeasible chromosome is completely replaced with the repaired chromosome. These techniques are easy to apply in many problems. They are well suited and often used with COPs such as TSP and knapsack problems. However, in heavily constrained problems, repair can be both challenging and time consuming — it can be as hard as solving the original problem.

Penalize infeasible solutions Penalty methods constitute another popular approach in which an infeasible solution is made less attractive to the selection operators by adding a penalty term to its cost function. A penalty term is made sufficiently large so that any infeasible solution is not better than the worst feasible solution in the population. The term increases with the number of violated constraints and with the depth of constraint violation. Often, the term is reduced with the number of generations, as the degree of constraint violation decreases.

Note that as in most aspects of GAs, the best approach to handle infeasibility is problem dependent, and that a hybrid approach that uses more than one method is likely to produce the best results.

4.4. Genetic operators and parameters

Once a decision has been made on a specific representation and an evaluation criteria has been set, there remains other practical considerations that influence the implementation of GAs. The user has to decide on genetic operators and their parameters, and answer questions like: Which operator to apply and how often? What is the 'optimal' population size? when to terminate the search?

4.4.1. Genetic operators

The selected encoding scheme helps in selecting some operators. Actually, it helps us to decide what not to use, leaving us several operators that can be used with the current encoding. For example, binary encoding permits the three well-known types of crossover, but the user has to decide upon the best technique that would minimize the disruption of the genetic material of good solutions. The general view now is that two-point crossover is better than single-point crossover. Some researchers advocate uniform crossover; however, we need to be more careful with this type as it causes larger disruptions in the chromosome structure. Mutation, also, can be implemented in many ways. For example, instead of using a completely random mutation, one can use random increments that pushes the individuals away from the population center to increase population diversity. For selection, we strongly recommend tournament selection, as it avoids the pitfalls of fitness proportionate selection and those of ranking selection. As well, tournament selection can be modified so that the fitter individual is selected at a user-specified probability that permits the selection of the less fit individual from time-to-time. Thus, the user can control selection pressure by adjusting the probability parameter. Irrespective of the selection operator, the user must ensure that an elitism strategy is followed. That is, the best found solution at any time during the run is saved to ensure that the best solution does not deteriorate if crossover or mutation produce inferior solutions.

4.4.2. Parameter tuning

Running a GA requires setting a number of parameters. However, finding settings that work well on a specific problem is not a trivial task. Poor settings lead to inferior results; good settings require time-consuming trials to find. Some of the methods used to tune parameters include (1) using values recommended in the literature by other researchers working on similar problems; (2) using two different sets of test problems, one for parameter tuning and one for reporting the final results; (3) adaptively changing the parameters in response to some statistical measures that represent the search status (Adaptive settings will be discussed in Sec. 5.1.1).

4.4.3. Termination criteria

GAs can in principle evolve without stopping. Hence, some strategy for terminating the run has to be set. The most commonly used criterion is to stop the algorithm once a pre-specified number of generations is reached. Other criteria terminate the GA run when a best-so-far solution does not improve for a specific number of generations, a pre-determined solution quality is obtained, the population reaches a lower limit of diversity that indicate convergence, or simply a maximum length of CPU time is reached.

5. Advanced Topics

The literature of GAs is vast, containing numerous advances over the basic GA of Sec. 3. A few techniques that have been used by the authors will be briefly discussed. These techniques are sufficiently general to be of interest to engineering students.

5.1. Maintaining population diversity

The degree of success of GAs on a given problem depends largely on their ability to strike a balance between *exploration* and *exploitation* during the search process. Population diversity plays an important role in achieving this balance, since high diversity directs the search toward the exploration of unvisited regions of the search space whereas low diversity focuses the search on specific regions to exploit possibly good solutions. Hence, many GA implementations employ some measure of population diversity to assess the search status (between exploration vs exploitation), and then use this measure to manipulate genetic parameters to drive the search toward either more exploration or more exploitation.

One way to measure diversity is to sum the distances between each individual and the rest of the individuals in the population. High computational costs associated with this measure can be reduced by limiting the population size to a few individuals; however, small populations might hinder the search process. Another way of measuring diversity is to use a single aggregation point as a representative of the whole population. Since GAs are designed to converge around the population-best, it is appropriate to measure population diversity in terms of distances from the population-best solution rather than distances from an average point. As well, this measure has the advantage of greatly reducing costs of computing diversity without imposing unnecessary limitations on the population size. By reserving individual v_n for the population-best, the aggregated genotypic measure (d) of a population of size n can be given as

$$d = \sum_{i=1}^{n-1} dist(v_i, v_n). \tag{8}$$

Diversity measures are normalized so that a 'zero' diversity means all individuals in the population are similar (ideally identical) to each other whereas a diversity value in the vicinity of 'one' indicates that they are very different from each other.

Several techniques for controlling diversity are used in the literature. Here we discuss two widely used techniques: adaptive parameter tuning and island based approaches.

5.1.1. Adaptive parameter setting

The goal of using adaptive parameters is twofold. First, it controls population diversity and hence balances exploration and exploitation, as mentioned earlier. Second, it reduces the efforts spent on parameter tuning and makes it more systematic and meaningful.

Mutation and crossover have great influence on population diversity. Generally, a GA population gradually converges with the number of generations, and the rate of convergence decreases as the rate of mutation increases. At extensively high mutation, diversity does not decrease at all. Figure 12(a) illustrates typical effects of mutation rate on population diversity.

The effect of crossover on population diversity is similar to that of mutation rate. It is important to note that crossover is less diversifier than mutation, as can be seen in Fig. 12(b). Therefore, by changing the rates of crossover and mutation, population diversity — and consequently the search status between exploration and exploitation — can be manipulated.

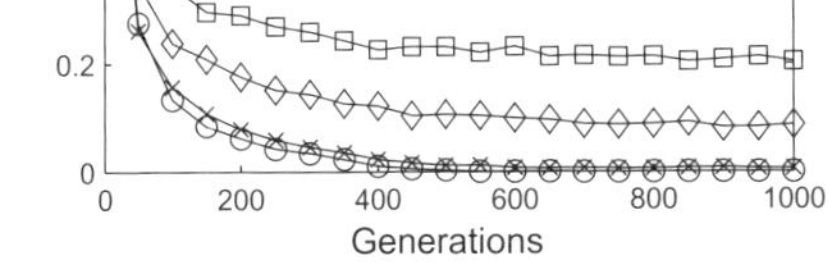

(a) Mutation (per thousand)

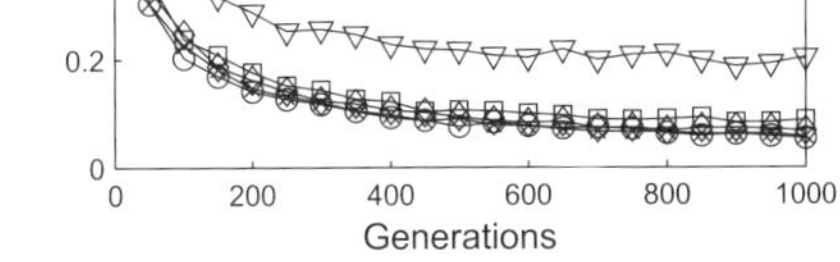

(b) Crossover (per ten)

Figure 12. Effect of genetic operators on population diversity on a typical problem. (a) Mutation rate is varied while crossover rate is fixed. Rates are per thousands; e.g. M005 means mutation rate of 0.005. (b) Crossover rate is varied while mutation rate is fixed. Rates are per tens; e.g. C06 means crossover rate of 0.6.

In the diversity control method adopted by Younes *et al.* (2007), population diversity $d(g)$ at generation g is measured and compared to two reference values, an upper limit d_h and a lower limit d_l. Mutation rate $\mu(g)$ for the current generation can then be computed according to this comparison, as follows:

$$\mu(g) = \begin{cases} Z_l \cdot \left(\overline{\mu} - \mu(g-1)\right) + \mu(g-1), & d(g) < d_l \\ \mu(g-1) - Z_h \cdot \left(\mu(g-1) - \overline{\mu}\right), & d(g) > d_h \\ \mu(g-1), & d_l \leq d(g) \leq d_h \end{cases}, \quad (9)$$

where $Z_l = \min\left\{\frac{d_l - d(g)}{d_h - d_l}, 1\right\}$, $Z_h = \min\left\{\frac{d(g) - d_h}{d_h - d_l}, 1\right\}$. Similar formulas can be used for adaptive crossover rate (and for adaptive selection probability if modified tournament is used).

To implement this method, one needs to specify two values for population diversity (d_l and d_h), and two values (a minimum limit and maximum limit) for each of the adaptive genetic parameters. Diversity is measured at the end of each generation and then used to update genetic parameters of the subsequent generation.

5.1.2. Island-based genetic algorithms

Island genetic algorithm (IGA) is a popular method that divides the population into several subpopulations or *islands*, which can be allocated to separate processors. IGAs give better solution quality and demand less computation time even when the model is implemented in a serial manner.

Each island evolves independently from the others for a period of time, called the migration interval. At the end of a migration interval, a few individuals are exchanged between islands to avoid premature convergence of individual islands. This scheme helps impart new genetic material to destination islands and increase survival probability of high fitness individuals.

By distributing the islands over the search space, islands collectively participate in exploring the search space; and by maintaining several good solutions at a time, islands individually exploit good solutions and perform limited local exploration. Thus, with the islands helping in maintaining the overall population diversity, the need of high mutation rates is reduced since mutation will be required to maintain diversity within islands (not within the whole population). Reductions in mutation rates reduce the likelihood of destroying individuals of good quality. The interested reader is referred to (Younes *et al.*, 2006) for an application of IGA to dynamic problems.

5.2. Hybridization

Most successful implementations existing today are actually hybrids of more than one technique. Hybridization basically aims to combine and extend strengths of individual techniques and alleviate their weaknesses. For example, GAs are effective at sampling large areas of the search space, whereas local search heuristics are effective at fine-tuning small areas of the search space, and hence the effectiveness of GAs can often be enhanced by hybridizing with some local search techniques. In fact, most hybrid implementations of GAs involve adding some iterative search techniques such as local search. Local search can also be viewed as a means of integrating problem specific knowledge in the GA. Today, these hybrids are very common, known under different names: hybrid genetic algorithms, genetic local search, and Memetic algorithms (MAs).

To ensure an effective and efficient implementation of hybrid GAs, one should balance several factors. They include the timing of the local search within the algorithm (e.g. after each operator or once after the end of the generation), the frequency of embedding local search (e.g. at each generation or once every few generations), the extent of local search (e.g. until a local optimum is found or for a few iterations only), the number of individuals subjected to local search (e.g. to the entire population or to the best individual only). In addressing these issues, one has to evaluate the benefits of hybridization against any additional expenses incurred, since costs of individual techniques in a hybrid contribute to the total computational cost of the resulting algorithm.

A word of caution: using local search till local optima are reached can be computationally expensive when the number of local search iterations to local optimality is not polynomially bounded. As well, applying local search to the entire population can rapidly lead to a profound loss of diversity and thus to premature convergence. Our own experience (Younes *et al.*, 2006) indicates that a good approach is to apply limited iterations of local search to the population's best individual at the end of each generation.

6. Conclusions

Many optimization problems in Chemical Engineering are hard to solve. They are characterized by large solution spaces in mixed types of variables, non-linearity in objective functions and constraints, and large computational costs of evaluating candidate solutions. GAs are well-suited to solving such problems, and hence their popularity in Chemical Engineering.

A hasty application of GAs rarely gives good results. A deep understanding of the problem at hand is key to the success of the implementation, as it guides the user in addressing several aspects in the implementation, including how to select the appropriate encoding scheme and genetic operators, how to integrate problem specific information in the implementation, and how to reduce the computational cost of evaluations.

GAs can lose population diversity quickly, and are thus susceptible to the problem of premature convergence. They are also weak at fine-tuning their solutions. These shortcomings can be alleviated by controlling population diversity and by hybridizing with other algorithms.

Although GAs have been in use for many years, they still provide ample opportunities for further research, particularly in hybridization with other techniques. Another promising direction that has been given little attention in Chemical Engineering is the use of adaptive operator controllers that can determine genetic parameters and identify when to switch between several operators. Such control should keep the total number of function evaluations to a minimum, and hence greatly reduce the run time on expensive objective function evolutions.

References

Agrawal, N., Rangaiah, G., Ray, A.K. and Gupta, S.K. (2006). Multi-objective optimization of the operation of an industrial low-density polyethylene tubular reactor using genetic algorithm and its jumping gene adaptations. *Industrial and Engineering Chemistry Research*, **45**(9), pp. 3182–3199.

Baker, J.E. (1987). Reducing bias and inefficiency in the selection algorithm. *Proceedings of the 2nd International Conference on Genetic Algorithms on*

Genetic Algorithms and their Application, Lawrence Erlbaum Associates, Inc., Mahwah, NJ, USA, ISBN 0-8058-0158-8, pp. 14–21.

Beasley, D., Bull, D.R. and Martin, R.R. (1993a). An overview of genetic algorithms: Part 1, fundamentals. *University Computing*, **15**(2), pp. 58–69.

Beasley, D., Bull, D.R. and Martin, R.R. (1993b). An overview of genetic algorithms: Part 2, research topics. *University Computing*, **15**(4), pp. 170–181.

Cao, K., Feng, X. and Ma, H. (2007). Pinch multi-agent genetic algorithm for optimizing water-using networks. *Computers and Chemical Engineering*, **31**(12), pp. 1565–1575.

Cubillos, F., Acuna, G. and Lima, E. (2007). Real-time process optimization based on grey-box neural models. *Brazilian Journal of Chemical Engineering*, **24**(3), pp. 433–443.

Davis, L. (1985). Applying adaptive algorithms to epistatic domains, Joshi, A. (ed.), *Proc. of the 9th Int. Joint Conf. on Artificial Intelligence*, Morgan Kaufmann, Los Angeles, CA, pp. 162–164.

Elliott, L., Ingham, D., Kyne, A., Mera, N., Pourkashanian, M. and Wilson, C. (2006). A novel approach to mechanism reduction optimization for an aviation fuel/air reaction mechanism using a genetic algorithm. *Journal of Engineering for Gas Turbines and Power*, **128**(2), pp. 255–263.

Eshelman, L.J. and Schaffer, J.D. (1993). Real-coded genetic algorithms and interval schemata. Whitley, L.D. (ed.), *Foundations of Genetic Algorithms 2*, Morgan Kaufmann Publishers, San Mateo, CA.

Fogel, L.J., Owens, A.J. and Walsh, M.J. (1966). *Artificial Intelligence Through Simulated Evolution*, John Wiley and Sons, New York.

Gen, M. and Cheng, R. (1999). *Genetic Algorithms*, John Wiley and Sons, Inc., New York, NY, USA, ISBN 0471315311.

Goldberg, D.E. (1989). *Genetic Algorithms in Search, Optimization and Machine Learning*, Addison-Wesley, Reading, MA.

Goldberg, D.E. and Lingle, Jr., R. (1985). Alleles, loci, and the traveling salesman problem. Grefenstette, J.J. (ed.), *Proc. Int. Conf. on Genetic Algorithms and Their Applications*, Lawrence Erlbaum, Hillsdale, NJ, pp. 154–159.

Haupt, R.L. and Haupt, S.E. (1998). *Practical Genetic Algorithms*, John Wiley & Sons, Inc., New York, NY, USA, ISBN 047-1188735.

He, Y. and Hui, C.-W. (2007). Genetic algorithm for large-size multi-stage batch plant scheduling. *Chemical Engineering Science*, **62**(5), pp. 1504–1523.

Holland, J.H. (1975). *Adaptation in Natural and Artifical Systems*, University of Michigan Press, Ann Arbor, MI.

Istadi, I. and Amin, N. (2007). Modelling and optimization of catalytic-dielectric barrier discharge plasma reactor for methane and carbon dioxide conversion using hybrid artificial neural network — genetic algorithm technique. *Chemical Engineering Science*, **62**(23), pp. 6568–6581.

Jeżowski, J., Bochenek, R. and Poplewski, G. (2007). On application of stochastic optimization techniques to designing heat exchanger- and water networks, *Chemical Engineering and Processing: Process Intensification*, **46**(11), pp. 1160–1174.

Jung, Y.S., Kulshreshtha, C., Kim, J.S., Shin, N. and Sohn, K.-S. (2007). Genetic algorithm-assisted combinatorial search for new blue phosphors in a (ca,sr, ba,mg,eu)_xb_yp_zo_δ system, *Chemistry of Materials*, **19**(22), pp. 5309–5318.

Masoori, M., Boozarjomehry, R. and dlobshiata (2007). Which method is better for the kinetic modeling: Decimal encoded or binary genetic algorithm? *Chemical Engineering Journal*, **130**(1), pp. 29–37.

Michalewicz, Z. (1992). *Genetic Algorithms + Data Structures = Evolution Programs*, Springer-Verlag, Berlin, Heidelberg.

Michalewicz, Z. and Fogel, D.B. (2004). *How to Solve It: Modern Heuristics* (Springer), ISBN 3540224947.

Mitchell, M. (1996). *An Introduction to Genetic Algorithms*, The MIT Press, Cambridge, Massachusetts.

Oliver, I.M., Smith, D.J. and Holland, J.R.C. (1987). A study of permutation crossover operators on the travelling salesman problem, Grefenstette, J.J. (ed.), *Genetic Algorithms and their Applications: Proc. of the Second Int. Conf. on Genetic Algorithms*, lawrence Erlbaum Association, Hillsdale, NJ, pp. 224–230.

Rechenberg, I. (1973). *Evolutionsstrategie: Optimierung technischer Systeme nach Prinzipien der biologischen Evolution*, Frommann-Holzbog, Stuttgart, Germany.

Reeves, C.R. and Rowe, J.E. (2002). *Genetic Algorithms: Principles and Perspectives: A Guide to GA Theory*, Kluwer Academic Publishers, Norwell, MA, USA, ISBN 1402072406.

Spall, J.C. (2003). *Introduction to Stochastic Search and Optimization*, John Wiley and Sons, Inc., New York, NY, USA, ISBN 0471330523.

Tao, J. and Wang, N. (2007). Dna computing based rna genetic algorithm with applications in parameter estimation of chemical engineering processes. *Computers and Chemical Engineering*, **31**(12), pp. 1602–1618.

Till, J., Sand, G., Urselmann, M. and Engell, S. (2007). A hybrid evolutionary algorithm for solving two-stage stochastic integer programs in chemical batch scheduling. *Computers and Chemical Engineering*, **31**(5–6), pp. 630–647.

Whitley, D. (2001). An overview of evolutionary algorithms: practical issues and common pitfalls. *Information and Software Technology*, **43**(14), pp. 817–831.

Wolpert, D.H. and Macready, W.G. (1997). No free lunch theorems for optimization. *IEEE Transactions on Evolutionary Computation*, **1**(1), pp. 67–82.

Wu, L.-L., Chang, W.-X. and Guan, G.-F. (2007). Extractants design based on an improved genetic algorithm. *Industrial and Engineering Chemistry Research*, **46**(4), pp. 1254–1258.

Yang, B., Wang, Z., Chen, C., Yuan, J. and Wang, L. (2007/03). Modeling and optimization for the secondary reaction of fcc gasoline based on the fuzzy neural network and genetic algorithm. *Chemical Engineering and Processing*, **46**(3), pp. 175–180.

Younes, A., Basir, O. and Calamai, P. (2006). A hybrid evolutionary approach for combinatorial problems in dynamic environments. *The Canadian Conference on Electrical and Computer Engineering (IEEE CCECE 2006)* (IEEE, Ottawa, Canada).

Younes, A., Basir, O. and Calamai, P. (2007). Adaptive control of genetic parameters for dynamic combinatorial problems, Doerner, K., Gendreau, M., Greistorfer, P., Gutjahr, W., Hartl, R. and Reimann, M. (eds.), *Metaheuristics — Progress in Complex Systems Optimization*, chapter 11, Operations Research/Computer Scinece Interfaces Series, Springer Verlag, New York, pp. 205–224.

Young, C.-T., Zheng, Y., Yeh, C.-W. and Jang, S.-S. (2007). Information-guided genetic algorithm approach to the solution of minlp problems. *Industrial and Engineering Chemistry Research*, **46**(5), pp. 1527–1537.

Chapter 5

TABU SEARCH FOR GLOBAL OPTIMIZATION OF PROBLEMS HAVING CONTINUOUS VARIABLES

Sim Mong Kai, Gade Pandu Rangaiah* and Mekapati Srinivas

Department of Chemical & Biomolecular Engineering
National University of Singapore, Singapore 117576
**chegpr@nus.edu.sg*

1. Introduction

Glover outlined tabu list, aspiration criterion and other concepts in the form of a meta-heuristic called tabu/taboo search, TS (Glover, 1986 and 1990b). The canonical form of TS (including fundamentals principles and advanced considerations) was introduced by Glover in his two seminal papers (Glover, 1989, 1990a). In brief, TS starts with an initial solution and then systematically searches the neighborhood for the best solution by performing several moves on the current solution to generate a set of solutions in the neighborhood. Based on the tabu memory (which keeps track of the previously visited points in TL), TS checks and allows the newly generated solution only if it is not in TL. It then selects the best among all the allowable solutions as the next solution, and inserts it into TL. The new solution replaces the best solution if it is found to be better, and

*Corresponding author.

is used to generate solutions in the next iteration. These steps are repeated until some termination criterion such as maximum number of iterations is satisfied.

TS had undergone various modifications over the years and gained wide acceptance for combinatorial optimization problems (COPs), which involve discrete solution sets, such as travelling salesman, vehicle routing and flow-shop scheduling problems COPs (e.g. Taillard, 1991; Srivastava and Chen, 1993; Gendreau *et al.*, 1994). However, its use for global optimization of problems having continuous variables was limited. Hu (1992) proposed TS method with random moves for such problems, and his results showed that the proposed method outperforms random search and a composite genetic algorithm (GA) for several test functions and three design problems. Since then, several studies have modified and/or applied TS to optimization problems having continuous variables, which are important in the design and operation of chemical processes.

In the following sections, we first introduce the TS–QN algorithm which was developed and used by our group and apply it to a simple global optimization problem. We then review reported TS methods for continuous problems, discuss existing software for TS and Chemical Engineering applications of TS. Finally, features of TS which can be exploited for global optimization of continuous problems are briefly described before ending the chapter with concluding remarks.

2. Tabu Search with Quasi-Newton (TS–QN) Method

Consider the optimization problem:

$$\begin{aligned} &\text{minimize} \quad f(x_1, x_2, \ldots, x_N) \\ &\text{w.r.t. } x_1, x_2, \ldots, x_N \\ &\text{subject to } x_k^L \le x_k \le x_k^U \quad \text{for } k = 1, 2, \ldots, N. \end{aligned} \tag{1}$$

Here, $x_1, x_2, \ldots, x_N$ are the N decision variables, and superscripts L and U respectively denote the lower and upper bounds on them. Equality and inequality constraints are not considered since many versions of TS including TS–QN were developed and tested for optimization problems with bounds only. If constraints are present, then they have to be taken care of suitably (e.g. penalty function approach).

TS–QN algorithm is described here for several reasons. First, it is easy to understand and implement. Second, it is based upon the Enhanced Continuous TS (ECTS) developed by Chelouah and Siarry (2000). ECTS follows closely to Glover's tenets of TS and represents a significant step in the development of TS for the global optimization of large, continuous problems. Third, use of a quasi-Newton (QN) method speeds up the rate of convergence during the intensification phase, as compared to the use of ECTS or Nelder–Mead simplex search. QN is a modified Newton's method which is well suited for unconstrained problems with bounds as in Eq. (1). Unlike Newton's method, QN does not require the Hessian matrix of second derivatives to be calculated, but makes use of the gradient vector only.

In order to find reliably and accurately the global optimum of continuous problems with multiple minima, TS–QN algorithm consists of two phases: diversification phase to ensure that the entire solution space is well searched, followed by intensification phase near the most promising solution to obtain the global minimum accurately. It employs a modified ECTS algorithm that allows for generation of neighbors within the central hyperrectangle, in the diversification phase of the search. From the Promising List (PL) of solutions found, search intensification is carried out on the most promising solution using QN. The FORTRAN program of TS–QN provided on the CD employs the DBCONF subroutine from the International Mathematics and Statistics Library® (IMSL)[1]. DBCONF is a double-precision subroutine for minimizing a function of n variables (subject to bounds) using QN method and finite-difference approximation for derivatives (Visual Numerics, 2008). If DBCONF is not available, then the user will have to modify the TS–QN program to use another local optimization program for the intensification phase.

Flowchart of TS-QN and pseudo code of TS–QN are shown in Figs. 1 and 2 respectively. Like many other optimization methods, TS–QN involves several parameters: TL size (N_t), PL size (N_p), ball radius for tabu points

---

[1] IMSL is a collection of libraries for numerical analysis that are available in various programming languages such as FORTRAN, C, C# and Java™.

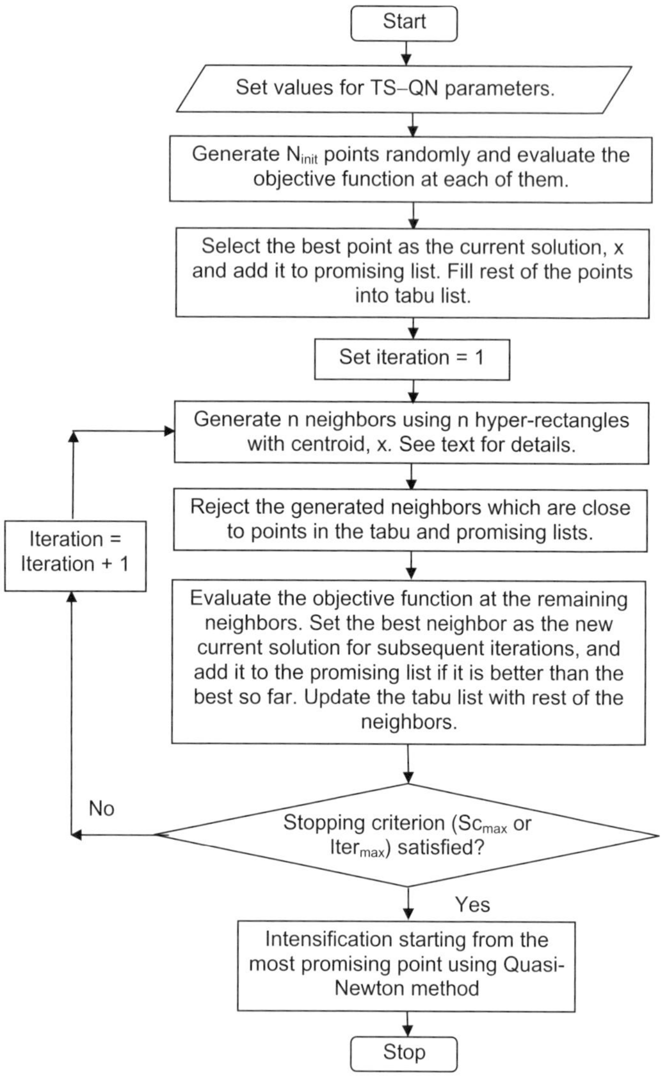

Figure 1. Flowchart of TS–QN (Teh and Rangaiah, 2003).

(ε_t), ball radius for promising points (ε_p), half-length of most outer hyper-rectangle (h_n), number of initial points generated ($N_{initial}$), number of neighbors generated (n), maximum number of successive iterations (Sc_{max}) without improvement in the best function value and maximum

ε_p : Radius from Promising List
ε_t : Radius from Tabu List
$N_{initial}$: Initial no of solutions
n : No of neighbors generated
PL : Promising List
TL : Tabu List
S : Set of neighbors generated
x : Current optimal solution
x_i^* : Current solution

```
TL, PL = Ø                                              // initialize TL, PL to be empty
Generate N_initial solutions randomly within bounds
Select the best solution, let it be x and add it to PL
Add all other solutions to the TL
while termination criterion not satisfied
  s = GenerateNeighbors(x, n)
  for i = 1 to n
    if (ε_t > |s_i − x^TL|) or (ε_p > |s_i − x^PL|)    // Rejects points close to those in TL, PL
      Reject s_i if it falls within ε_t of points in TL
    end if
  end for
  Find the objective function value at all remaining neighbors
  Add all points to the TL, except for the best neighbor
  Add the best neighbor to PL only if it is better than the best so far
end while                                               // end of diversification phase
Select best point in the PL and let it be x.
QuasiNewton(x)                                          // QN for search intensification
```

Figure 2. Pseudo code of TS–QN.

number of iterations ($Iter_{max}$). The first step in the TS–QN algorithm is to provide suitable values for these parameters. Guidelines for this are given in Table 1, and could be used prior to tuning the parameter values for the specific application problem(s).

Next, $N_{initial}$ points ($\mathbf{x}^j$ for j = 1, 2, ..., $N_{initial}$) are randomly generated in the solution space using uniformly distributed random numbers and the bounds on variables, and the objective values at these points are evaluated. The randomly generated point with the lowest objective function value is selected as the current solution and also included in the initially-empty PL, while the rest of the points are included in the initially-empty TL. To facilitate a homogeneous exploration of the neighborhood around the current solution, the neighborhood (bounded by h_n from the current

Table 1. Guidelines for parameters in TS–QN, compared against ECTS.

Parameter	Guidelines for ECTS (Chelouah and Siarry, 2000)	Guidelines for TS–QN (Teh and Rangaiah, 2003)
N_t	7	10
N_p	10	10
$N_{initial}$	—	20N
n	10 or 2N for $N \leq 5$	2N (subject to minimum of 10 and maximum of 30)
Sc_{max}	5N	6N
$Iter_{max}$	50N	50N
ε_t	0.01*	0.01*
ε_p	0.02*	0.01*

*These are for the solution space normalized to 0 to 1 for each decision variable.

solution) is divided into n partitions, with one neighbor randomly generated in each partition. If the random solution generated does not fall within the bounds of the hyper-rectangle, new random solution(s) are generated until the boundary conditions have been satisfied.

Possible methods of partitioning are: geometric, linear and isovolume, along with concentric crowns (Siarry and Berthiau, 1997) or hyper-rectangles (Chelouah and Siarry, 2000). Among these, TS–QN uses geometric partitioning and hyper-rectangles for their simplicity and effectiveness. The partitioning of the neighborhood around the current solution into 4 concentric hyper-rectangles is illustrated in Fig. 3, where s_1, s_2, s_3 and s_4 are the neighbors generated randomly in the hyper-rectangles of half-length h_1, h_2, h_3 and h_4 respectively. Teh and Rangaiah (2003) opted to generate a neighbor in the inner-most hyper-rectangle. Previously, generation of neighbors inside the central hyper-rectangle was prohibited (Chelouah and Siarry, 2000), in order to prevent revisiting of the same area. However, Wang *et al.* (2004) had shown that the generation of multiple neighbors within the central hyper-rectangle is particularly useful during the intensification phase as it improves solution quality and convergence.

The more important thing is to randomly generate only one neighbor in the inner-most hyper-rectangle and between adjacent hyper-rectangles as shown in Fig. 3. The procedure for this is as follows; see also the pseudo

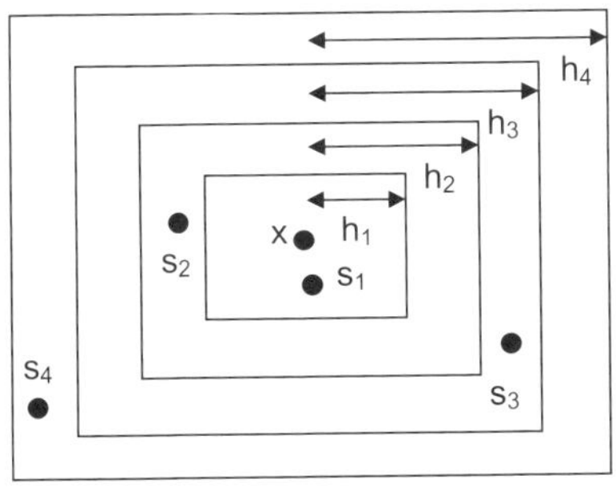

Figure 3. Four concentric hyper-rectangles around the current solution, x with a randomly generated neighbor, s in each hyper-rectangle.

h_n : Half-length of the most outer hyper-rectangle
n : Number of neighbors to be generated
x : Current solution near which neighbors are to be generated

```
1:   Define h_n
2:   for i = n to 2                          // Generation of half-length of hyper-rectangles
3:       h_{i-1} = h_i/2
4:   end for
5:   for i = 1 to n                          // Generation of upper and lower bounds
6:       UB_i = x_i +h_i
7:       LB_i = x_i -h_i
8:   end for
9:   for i = 2 to n                          // Generation of points within hyper-rectangles
10:      Generate a neighbor within UB and LB of a hyper-rectangle
11:      Calculate distance from the generated neighbor to LB
12:        If neighbor generated is not within two adjacent hyper-rectangles
13:          go back to line 10 to regenerate another neighbor
14:        end if
15:  end for
```

Figure 4. Pseudo code for generating neighbors.

code in Fig. 4. First, calculate half-length of the n hyper-rectangles for geometric partitioning by:

$$h_{n-j+1} = h_n/2^{(j-1)} \quad \text{for } j = 1, 2, \ldots, n. \tag{2}$$

To generate a point in the inner-most rectangle, set m = 1, and compute the upper bounds (UB_k^m for k = 1, 2, ..., N) and lower bounds (LB_k^m for k = 1, 2, ..., N) for the rectangle, where N is the dimension of the

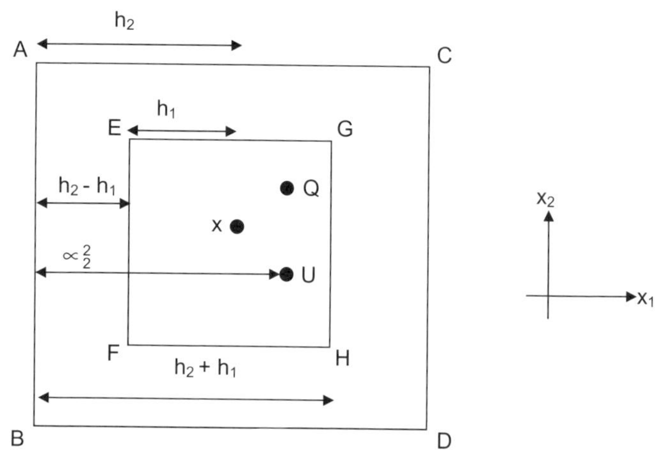

Figure 5. Generation of a neighbor within concentric hyper-rectangles.

problem. If any of these lower (upper) bounds violates the lower (upper bound) of the respective variable, then it is set to the corresponding lower (upper bound) in order to ensure feasibility of the points generated. EF and GH in Fig. 5 represent the LB^1 and UB^1 respectively in x_2-direction. Then, generate a random point, Q $(x_1, x_2^m, \ldots, x_N^m)$ within the inner-most hyper-rectangle by:

$$x_k^m = LB_k^m + R(UB_k^m - LB_k^m) \quad \text{for } k = 1, 2, \ldots, N, \tag{3}$$

where R is a uniformly distributed random number in the interval [0, 1].

To generate a point in the next hyper-rectangle, set $m = m + 1$, compute the upper and lower bounds of the mth hyper-rectangle, check and modify these bounds for any violation of bounds on the variables, and generate a random point using Eq. (2). The random point generated has to be within the region defined by the current hyper-rectangle (ABCD in Fig. 5) and not in the inner, adjacent rectangle (EFGH). This can be verified by calculating the shortest distance from the randomly generated point, U to the lower bound, α_k^m as shown in Fig. 5. If $h_m - h_{m-1} < \alpha_k^m < h_m + h_{m-1}$ for all $k = 1, 2, \ldots, N$, U is in the inner hyper-rectangle EFGH, and so another point has to be generated using Eq. (2); else, the point U is accepted. Repeat this procedure to generate a point in each remaining hyper-rectangles.

Neighbors that fall within ε_t from points in the TL or ε_p from points in the PL are rejected, with those within ε_t being added to the TL, so as to prevent repeated visits to the same regions and ensuring search diversification over the entire solution space. From the remaining neighbors, the one that has the lowest objective function value will be the new current solution, even if it is worse than the preceding iteration (to prevent the search from being entrapped in a local minimum), while the rest will be added to the TL. This solution is included into the PL only if it is better than the best solution so far. The diversification procedure will keep iterating until any of the termination criterion, Sc_{max} or $Iter_{max}$ is satisfied. Then, the most promising solution in the PL is identified and intensification phase is carried out using QN method to locate the global minimum accurately. In the next section, TS–QN is applied to the modified Himmelblau function.

3. Application of TS–QN to the Modified Himmelblau Function

The modified Himmelblau function is (Deb, 2002):

$$\text{minimize } f(x_1,\ x_2) = (x_1^2 + x_2 - 11)^2 + (x_1 + x_2^2 - 7)^2 + 0.1[(x_1 - 3)^2 + (x_2 - 2)^2] \quad (4)$$

w.r.t. x_1 and x_2

subject to $-6 < x_1 < 6$ and $-6 < x_2 < 6$.

It has four minima with a global minimum of 0.0, at $x_1 = 3.0$ and $x_2 = 2.0$. The three local minima are 1.50435 at (3.58149, −1.82080), 3.48713 at (−2.78706, 3.12820) and 7.36735 at (−3.76343, −3.26605).

The TS–QN method with parameters: $N_t = N_p = 10$, $N_{initial} = 40$, number of neighbors = 10, $Sc_{max} = 12$, $Iter_{max} = 100$, $\varepsilon_t = \varepsilon_p = 0.01$ (as per Table 1), is applied to the modified Himmelblau function as follows. Selected results are given in Table 2.

(1) 40 initial solutions were randomly generated, and the solution with the best objective function value was selected as the current solution.
(2) Based on the current solution, 10 neighbors were randomly generated, one for each hyper-rectangle. With the exception of the neighbor in the

Table 2. Values obtained by TS–QN for the modified Himmelblau function.

Iteration	X_1	X_2	Objective value
0 (best initial solution)	−2.735	−3.228	52.024
1	−2.883	−3.365	45.056
2	−2.940	−3.314	39.593
...	...	...	...
5	2.396	2.394	9.532
6	2.726	2.067	2.264
...	...	...	...
15	2.992	1.979	0.013
16	2.976	1.996	0.023
...	...	...	...
29	3.000	1.999	1.709E-5
41 (Termination criterion, $Sc_{max} = 12$ reached)	3.015	2.005	0.010
Most promising point found in the above iterations	3.000	1.999	1.709E-5
Upon QN (Global optimum found)	3.000	2.000	0.000

inner-most hyper-rectangle, each randomly generated neighbor should fall between adjacent hyper-rectangles (as in Fig. 3); if not, a new neighbor is generated until this requirement is met.

(3) The best non-tabu neighbor is selected and used for the next iteration, even if it is worse than the best solution obtained so far. This is illustrated by iteration 16 (Table 2), where the solution obtained is worse than that obtained from iteration 15, but is subsequently used for the generation of neighbors in iteration 17. The TL and PL are updated; if the best non-tabu neighbor is better than the best solution found thus far, it is added to the PL, and all other neighbors are added to the TL. Note the escaping from the local minimum region to the global minimum region in iteration 5.

(4) Steps 2 and 3 are repeated until 12 successive iterations without any improvement in objective value or maximum no of 100 iterations is satisfied. Here, the first condition is satisfied leading to termination of the search (Table 2).

(5) The most promising point obtained from the above iterations is given in Table 2, and it is very near the global minimum. Using this as the initial method, QN method is used to find the minimum accurately.

The above results vary from one trial to another depending on the random number seed, which affects the random numbers generated and used in the iteration.

4. TS Methods for Global Optimization of Continuous Problems

The seminal form of TS was first proposed by Glover (1989, 1990a) as a meta-heuristic for the solving of COPs and was successfully applied to a wide variety of problems, which included design, routing and scheduling. TS was capable of solving NP-hard optimization problems with similar or fewer NFEs as compared to other meta-heuristics, within a reasonable time frame (Gendreau *et al.*, 1994). This had prompted researchers to propose various modifications to TS, adapting it for the optimization of continuous problems.

Hu (1992) was the first to adapt TS for global optimization of continuous problems through generation of random moves in neighbors of varying sizes (defined by a set of steps). However, his implementation only made use of short term memory (STM), and was prone to being stuck at local optima as it only uses the best improving non-tabu moves for the next iteration. This was contrary to the tenets of TS, which allowed for the best non-tabu move, even if it is worse than the existing solution. Even then, the results obtained were promising and demonstrated the potential of TS for continuous problems. This sparked off a series of modifications of TS for global optimization of continuous problems. These methods, summarized in Appendix A in chronological order, proved to be relatively effective against competing methods that were available at that time.

After Hu's (1992) paper, Cvijovic and Klinowski (1995) introduced a TS algorithm for global optimization. In this, the entire solution space was partitioned into disjunct cells, and neighbors were generated in a completely stochastic manner, which runs against the tenet of systematic neighborhood generation. Diversification and intensification measures were introduced through the use of both a TL and an Elite List (equivalent to PL). Rajesh *et al.* (2000) used a similar method to solve continuous problems, which required significantly fewer NFEs compared to GA.

Yang *et al.* (1998) and Machado *et al.* (2001) proposed similar modifications[2] to Hu's algorithm. In their algorithms, known as Universal TS (UTabu) and Common TS respectively, the key modifications were: (1) the best neighbor was selected for the next iteration, even if it was worse than the existing solution, thereby correcting the violation present in Hu's design; (2) restarts from the optimum solution were introduced after every N iterations in order to escape from local optima; (3) TL was not used as it may restrict the search; and (4) interval length for different variables was no longer considered to be the same, which is more realistic and suitable for practical implementations. Compared to UTabu, Common TS includes local minimization from each feasible random move (only if it is not near a stored trajectory in order to avoid unnecessary minimization), before using it as the current solution for the next iteration. Both UTabu and Common TS were able to find the global minimum for all test functions tried, unlike Hu's method which was prone to being trapped in a local minimum. Also, numerical results have validated that the use of TL was counter-effective for UTabu and Common TS, since the global optimum was found only when no TL was used. The lack of a TL clearly does away with adaptive memory, a key hallmark of TS, and hence one may not consider them as TS algorithms.

Methods that adhered closely to Glover's tenets on TS, started to appear for global optimization of continuous problems, shortly following the work of Cvijovic and Klinowski (1995). These include the methods by Battiti and Tecchiolli (1996), Siarry and Berthiau (1997), and Chelouah and Siarry (2000). Battiti and Tecchiolli (1996) proposed Continuous Reactive TS (C–RTS) as a modification to reactive TS (RTS), allowing it to be used for continuous problems. It consists of two phases: RTS is first used to explore the solution space and identify promising areas; and a local minimizer is then used for intensification. The use of a dynamic TL which learns and adapts itself according to the search, allows C–RTS to solve optimization problems efficiently as compared to other methods, including

[2] Yang Shiyou is the common author in both papers, and hence the ideas seem to be carried over from one paper to another leading to the similarity.

other implementations of TS. Though C–RTS was not meant for constrained optimization like most other implementations of TS, it performed favorably for the design of a transformer, with the constraints being handled by the penalty function approach.

Siarry and Berthiau (1997) proposed the Continuous TS (CTS) algorithm, which enabled TS to be applied to continuous problems. Unlike C–RTS which generates neighborhoods based on a tree of boxes and involves relatively complex partitioning, CTS allows neighborhoods to be generated in a simpler fashion, via geometric, linear and isovolume partitioning. However, CTS was only suited for continuous problems of small dimensions (< 5 variables), prompting the development of ECTS by Chelouah and Siarry (2000). ECTS was based on the following modifications to CTS: implementation of diversification and intensification phases, replacement of concentric crowns with hyper-rectangles and the use of a PL. Results on 20 benchmark problems had shown that ECTS was comparable or better than competing methods including C–RTS, though there were cases where ECTS was stuck at the local minimum.

Around the same period, Wang *et al.* (1999) introduced Custom TS (C–TS), which uses a small neighborhood size at the beginning of the search to facilitate convergence, followed by large neighborhood size for thorough search of entire solution space (diversification) towards the end. The different neighborhood sizes used, entails a tradeoff between solution quality and computational requirements. Franze and Speciale (2001) introduced DOPE, which accepted an improving move instead of the best possible non-tabu move in the entire neighborhood. The quality of the solution is maintained through having additional iterations, and NFEs required were substantially less even though more iterations were carried out, as each iteration involved significantly fewer NFEs. Variable move steps, which could facilitate the different requirements during intensification and diversification, were also introduced. However, the performance of DOPE was highly dependent on the starting point used, limiting its later use.

Based on information gathered from past implementations, ECTS was observed to be slow to converge during the intensification phase. Various improvements to ECTS were proposed by researchers to counter this problem. Wang *et al.* (2004) modified the intensification phase of ECTS by allowing the generation of points within the central hyper-rectangle. Their results show that this modification generally reduces NFEs. The proposed algorithm by Wang *et al.* (2004) in the non-revised form generally required more NFEs than ECTS, due to the use of an aspiration criterion. Unlike conventional implementations of TS, Wang *et al.* (2004) checks for the aspiration criterion before TL, which is generally suited for cases where the search is highly restricted with many available moves being tabu.

Other modifications to ECTS involved the introduction of a local minimizer for the intensification phase. Teh and Rangaiah (2003) proposed the use of Nelder–Mead (NM) and Quasi-Newton (QN) for intensification purposes. The algorithms, respectively known as TS–NM and TS–QN, were tested on benchmark functions, Phase Equilibrium Calculations (PECs) and Phase Stability (PS) problems. TS–QN generally required fewer NFEs than TS–NM. When compared to ECTS, TS–QN was more accurate but less reliable in finding the global optimum. The effects of systematic, random and mixed neighborhood generation on TS–QN were subsequently tested by Srinivas and Rangaiah (2007a). TS–S–QN (systematic) was more suited for PECs, while TS–M–QN (mixed) was more suited for PS problems, due to the presence of local minima far away from the global minimum. TS–R–QN (random generation) was found to be the least efficient computationally, since the random generation does not guide the search in any meaningful way. This is in line with the central tenet of TS, which requires a systemic search of the solution neighborhood, where bad strategic moves are favored over good random moves, since real life problems are unlikely to be completely random in nature.

Chelouah and Siarry (2005) introduced Continuous Tabu Simplex Search (CTSS) with normalized variables and two different ways of generating neighbors (varying one variable only or multiple variables from the current solution). More importantly, Nelder–Mead simplex search is nested within the diversification phase and used to intensify the search whenever a promising area is found. This is unlike ECTS where intensification is carried out separately after the termination of the diversification phase and the selection of the most promising area.

Though Chelouah and Siarry (2000) had introduced ECTS as an improvement to CTS, Zheng *et al.* (2005) proposed an alternative method known as Staged Continuous TS (SCTS) based on a three-stage implementation of CTS. The first stage involves the survey of the entire solution space to find prospective points that are likely to lead to global optimum.

Second stage attempts to find a point close to the global minimum, while the last stage involves reduced search space to achieve convergence to the global minimum. The results on the test functions show that SCTS has good reliability but the NFEs required were more than both ECTS and CTSS of Chelouah and Siarry (2000, 2005).

Ji and Tang (2004) proposed Memory TS (MTS) as an improvement of the algorithm of Cvijovic and Klinowski (1995). In this algorithm, an extra variable was introduced to record the optimum solution during the search, and this was mathematically proven to ensure convergence to the global optimum. This is significant as the convergence property of TS is not well understood and meta-heuristics are generally not backed by mathematical proofs.

Hedar and Fukushima (2006) proposed Directed TS (DTS) which is considered as a multi-start method (which uses multiple starting points for search to ensure that the entire solution space is well searched so that the global optimum could be found but is computationally inefficient). DTS consists of three stages: exploration, diversification and intensification. In the exploration phase, two different local search strategies, Nelder–Mead Search (NMS) and Adaptive Pattern Search (APS) are used for the generation of trial points. Unlike common implementations of TS, DTS makes use of a multi-ranked TL, which is ranked according to both recency and objective value. Tabu Regions (TRs) and semi-TRs are constructed around points in the TL to prevent cycling of solution. This is similar to the neighborhood generation of Siarry and Berthiau (1997) where neighbors are generated within the concentric crowns (equivalent to semi-TRs) only and not within the central sphere (equivalent to TR). Long term memory (LTM) is introduced through the Visited Regions List, whereby centers of the visited regions and the frequencies visited are recorded and used to drive the search away from visited regions. Intensification by a local method is carried out to refine the elite solutions visited. For the benchmark problems tested, DTS_{APS} required fewer NFEs and is more reliable than DTS_{NMS}. For problems of small dimensions ($n < 30$), DTS_{APS} consistently obtained higher quality solutions than ECTS, while requiring fewer NFEs. Performance of DTS_{APS} degrades with problem dimension due to the use of a direct search method (APS).

Hedar and Fukushima (2006) noted that the results reported for ECTS by Chelouah and Siarry (2000) contained some inconsistencies. Though the success rate was given as 100% for Rosenbrock and Zakharov functions, the average errors obtained from the successful minimization were actually greater than the criterion for success. Note that the Goldstein Price test function in both ECTS (Chelouah and Siarry, 2000) and DTS papers (Hedar and Fukushima, 2006) contains a wrong coefficient and operator, probably typographical errors since the results obtained using the wrong test function would be vastly different from the global minimum which they stated. The correct coefficient and operator can be found by referring Goldstein and Price (1971). Also, values of two variables which lead to the global minimum, were swopped in the ECTS paper.

Stepanenko and Engels (2007) proposed Gradient TS (GTS) which obtains the global optimum more efficiently through the use of analytical first and second derivatives, followed by QN or Steepest Descent. Numerical results over several test functions show that GTS outperforms other meta-heuristics such as DE, GA, RS and SA. However, GTS is applicable only for functions which are differentiable; hence, Stepanenko and Engels (2008) proposed Gradient Only TS (GOTS) and TS with Powell's Algorithm (TSPA). These methods allow TS to be applied to all functions, as approximate gradient is calculated using a grid of function evaluations, with TSPA using Powell's method instead of QN for finding the next local minimum. GOTS was found to be less efficient than GTS for small problems, but more efficient for large problems, while TSPA does not make use of gradients and hence generally requires more NFEs than GOTS. Stepanenko and Engels (2008) noted that there was an error in the calculation of the NFEs required for 10-variable Rastrigin function, by GTS.

Parallel TS (PTS) has been implemented in various forms over the years, to improve convergence and reduce the dependence on the initial solution. The three main ways in which PTS could be carried out are described by Crainic and Toulouse (1997) and are as follows: (1) parallelization of operations solely for faster computations with no improvement in the solution quality; (2) decomposition of the problem into smaller parts that can be solved separately to yield an improved overall solution; and (3) thorough exploration of solution space through simultaneous searches with synchronization and co-operation between searches. Kalinli and Karaboga (2004) introduced PTS where several basic TS algorithms are run in parallel with exchange of information via the crossover operation of GA. Tests on several benchmark functions have shown that PTS converged more quickly when compared to basic TS, GA and touring ant colony optimization using the same NFEs. This was subsequently applied for digital filter design (Kalinli and Karaboga, 2005).

4.1. Multi-objective TS (MOTS)

Multi-objective optimization (MOO) involves problems with two or more objectives, which are often conflicting. This leads to Pareto-optimal solutions, whereby it is impossible to improve an objective without leading to worsening of at least one other objective. The two approaches for solving MOO problems are: (1) combine the multiple objectives into a composite function through weighting and scaling each objective function based on its relative importance; (2) searching the entirely solution space directly for sets of Pareto-optimal solutions and then deciding on which is the best. Readers are referred to Rangaiah (2008) for more details on MOO methods and their applications in Chemical Engineering.

The first approach is useful if the relative importance of the various objectives are known, with the weights used being properly selected and well defined. Hansen (1997) proposed multi-objective combinatorial optimization (MOCO) where several composite objective function searches by TS are run in parallel, each using different and dynamically updated set of weights. Through comparison, sets of Pareto-optimal solutions can be obtained, and it was subsequently adapted for continuous problems.

Baykasoglu *et al.* (1999) uses the second approach of searching through the entire solution space to obtain Pareto-optimal solutions. This is accomplished through TS with simple neighborhood generation strategy to search the entire Pareto optimal front. On top of TL and Pareto list (of selected Pareto-optimal solutions found) as well as a candidate list (containing all other Pareto-optimal solutions not selected in the current iteration) were used, with the candidate list providing seed solutions for diversification. This method was found to be superior to multi-objective GA when tested on four benchmark problems. Subsequently, Baykasoglu (2006) applied it to four different mechanical component design problems, and it was shown to outperform multi-objective GA and Monte Carlo methods. However, this method allows only improving moves which run against the tenets of TS.

Jaeggi *et al.* (2004) proposed an alternative algorithm which overcomes the problems associated with the previous methods. It uses Hookes & Jeeves local search, coupled with STM and LTM to achieve intensification and diversification. Parallelization is employed to speed up the search through functional decomposition (operations necessary for each iteration, are executed in parallel). Further improvements were introduced by Jaeggi *et al.* (2008), through the novel use of path relinking strategies, known as Path Relinking MOTS (PRMOTS) for continuous optimization problems.

4.2. Use of TL in other stochastic methods

The success of TS in obtaining the global optimum efficiently, has led to incorporating TL in other stochastic methods. Tan *et al.* (2003) introduced an exploratory multi-objective evolutionary algorithm where TL is introduced to prevent search cycling and to maintain search diversity. Salhi and Queen (2004) introduced SA with a descent method and TL, which enables not only the global optimum to be found but also the other local minima which might be useful. Teh and Rangaiah (2003) and Srinivas and Rangaiah (2007a) found that the reliability of TS–QN is rather low as compared to DE–QN. Hence, Srinivas and Rangaiah (2007a, 2007b) incorporated TL into DE, creating a hybrid method known as DETL which has reduced the NFEs required by DE substantially without compromising on the reliability. Altiparmak and Karaoglan (2008) implemented SA with TL, followed by adaptive cooling strategy, which generally leads to better solutions than those found by TS or SA alone.

4.3. Other issues/developments

In a recent paper entitled "Unchartered Domains" (Glover, 2007), Glover reiterated that much remains to be learned about TS, despite the progresses made over the years. Focus on effective use of memory in the search methods, may provide us with a key to more effective use and implementation of adaptive memory. Unlike many other search techniques, which are generally stochastic in nature, TS is deterministic or contains stochastic elements implemented in a systematic manner. Glover (2007) calls into question the relevance of completely stochastic search methods for typical problems. Through having a deep understanding and proper formulation of the problem, one will be able to obtain the optimum solution via non-stochastic means.

5. TS Software

Several software packages having TS are listed in Table 3. They use TS alone or as part of a hybrid method for the optimization of continuous problems. TS only software (namely, 1, 3 and 4 in Table 3) consists of several subprograms for different strategies/features of TS (discussed later in this chapter); these can be called upon to evaluate their effectiveness and/or for solving difficult problems.

6. Chemical Engineering Applications of TS

In recent years, TS has been increasingly applied to global optimization of continuous problems in Chemical Engineering; a chronological summary of these applications is given in Appendix B. Wang *et al.* (1999) proposed a general TS algorithm for optimal design of multi-product batch chemical processes. A dynamic neighborhood size was employed, with a small neighborhood at the beginning to facilitate convergence and a large neighborhood at the end for diversification. Teh and Rangaiah (2003) introduced TS–QN, TS–NM, DE–QN and DE-NM for the solving phase equilibrium, phase stability and parameter estimation problems. TS–QN was a modification to the ECTS algorithm proposed by Chelouah and Siarry (2000).

Linke and Kokossis (2003) proposed a TS algorithm for optimization of general reaction and separation superstructures. This was later modified by Ashley and Linke (2004) who introduced rules and search was guided towards their conformance. The rules were obtained from data mining and were successfully applied for synthesis of complex reactor networks. Montolio-Rodriguez *et al.* (2007) took it a step further by conducting a multi-level search of the solution space using the algorithms proposed by Linke and Kokossis (2003), and Ashley and Linke (2004). The multi-level search enables the relationship between design complexity and performance to be known, which is useful for search refinement.

Rajesh *et al.* (2003) introduced an algorithm based on Cvijovic and Klinowski (1995) to solve bi-level programming (BLP) for Chemical Engineering problems, which was successfully demonstrated for the optimization of

Table 3. Available TS software.

No.	Name (Reference)	Features
1	OpenTS (Harder, 2006)	This is part of Computational Infrastructure for Operations Research (COIN–OR). The framework in Java allows TS implementation in a well-defined, object-oriented manner. Key advantage lies with the use of Java, which is a cross-platform and can be used to build applications that are placed on the web and able to take advantage of multi-processors. It can be used on any problem type with built-in TS functions to avoid repetitive coding.
2	METSlib (Maischberger, 2007)	An open-source object oriented meta-heuristics framework in C++, which includes Random Restart Local Search, Iterated Local Search, SA and TS. For TS, various search strategies are available, with different termination criteria, aspiration criteria and TL available for implementation. It allows one to customize and code new versions as well.
3	Reactive Search — Continuous Optimization (Battiti and Brunato, 2007)	Reactive Search — Continuous Optimization comes in various forms: (1) Microsoft Excel 2007 add-ins; (2) Web Application (Key variables into a Java applet); (3) Stand-alone Program, which includes C-RTS and Reactive Affine Shaker. They allow the programmer to define the system to be optimized with a C++ function, and invoke the powerful reactive search solver.
4	MITS Toolbox (Exler *et al.*, 2008)	Matlab toolbox with TS-based algorithm for solving continuous problems at http://www.iim.csic.es/~gingproc/software.html. See Appendix B for two chemical engineering applications using this software.
5	OptQuest (OptTek Systems, 2008, Frontline Systems, 2008)	OptQuest and OptQuest Engine combine TS, integer programming, scatter search and neural networks into a single, composite search algorithm, which complements the strength of each heuristic, to allow efficient identification and solving of new scenarios. It includes the following products from Frontline Systems: (1) Premium Solver Platform — OptQuest Solver Engine for Excel; (2) Field Installable Solver Engine; and (3) Solver Platform Software Development Kit, which allows one to develop custom applications on a wide range of platforms and languages, for solving large non-smooth optimization problems (up to 5000 variables and 1000 constraints).

a reaction-diffusion equation. Lin and Miller (2004b) introduced a general TS algorithm which can be applied for chemical process optimization. This was tested on Heat Exchanger Network (HEN) synthesis (Lin and Miller, 2004a, 2004b) and pump system configuration (Lin and Miller, 2004b). Based on the algorithm proposed by Lin and Miller (2004b), Chavali *et al.* (2004) and Lin *et al.* (2005) implemented TS for computer-aided molecular design, in particular, for the search of molecules with specified properties for use in transition metal catalysts.

Frewen *et al.* (2005) proposed an algorithm based on the works of Glover and Laguna (1997), and Siarry and Berthiau (1997) for modeling defect and impurity evolution in crystalline silicon processing. This was tested against various stochastic global optimization methods with favorable results. Lin *et al.* (2008) introduced optimal component lumping, which reduces the dimension of the problem through pseudo components with minimal loss of information, and is applicable to Chemical Engineering problems. Exler *et al.* (2008) modified C-RTS for integrated process and control system design, whereby the combinatorial component is used to guide the search, while a local solver is used to improve the efficiency. This was successfully applied to the Tennessee Eastman problem as well as control strategies of wastewater treatment plant.

7. Features of Tabu Search

The hallmark features of TS lies with the implementation of adaptive memory and responsive exploration, which opens up numerous possibilities and allows the solution space to be explored on both strategic and tactical level. These are the very essential qualities which qualify it as an intelligent optimization technique (Glover and Laguna, 1997), unlike traditional search techniques such as branch and bound, which is based upon fixed memory structures, or other meta-heuristics such as SA which is memory-less. Adaptive Memory Programming (AMP) refers to the exploitation of a collection of strategic memory components during the search process, and it involves the attributes of recency, frequency, influence and quality (Glover and Laguna, 1997). Influence generally refers to changes in the solution's quality (special form of influence), structure and feasibility, while quality is based upon how close the solution is to the optimum value.

On the other hand, responsive exploration is based upon the strategic exploration of the solution space. This could be guided through restrictions imposed by TLs, relaxation of tabu conditions by aspiration criteria, as well as intensification, diversification, strategic oscillation and path relinking. One of the main tenets of responsive exploration is that a bad strategic choice would yield useful information as compared to a good random choice, given that most real world problems are not completely random in nature and do have some form of structure (Glover, 1997). Results from the bad strategic choice could provide clues that could be used to modify the existing strategy, directing the search towards the desired direction. This is applicable to both deterministic and probabilistic implementation of TS. For the latter case, stochastic element will be introduced in a structured manner, so that some form of structure remains and the information obtained could be used to guide the search.

Unlike local search which terminates when the best solution in the neighborhood is worse than the existing optimum, TS continues from it based upon the rationale that good moves are more likely to reach the optimal solution. This particular property of TS allows the search to escape from the local optima and increases the success rate of finding the global optimum. When the new solution generated is better than the current solution, the optimum solution will be updated. TS will continue to iterate till the specified termination criterion is satisfied.

Both short term memory (STM) and long term memory (LTM), key components of the adaptive memory component of TS, have their own unique strategies that allow them to modify the current solution neighborhood and the search trajectory. STM based on recency, forms the core of TS and is always implemented to prevent cycling and revisiting, thus reducing NFEs. LTM based on frequency, is usually built upon STM and used for search intensification and diversification, which can also be carried out without the use of LTM. As most problems could usually be solved reasonably well by TS with only STM, the true potential of TS which includes the use of LTM has often been left untapped. Also, many features of STM and LTM which holds great promise are often not incorporated for solving continuous problems. These factors necessitate a review of the key features of STM and LTM, which could be incorporated into the existing algorithms, in order to fully exploit adaptive memory component of TS.

8. Short Term Memory (STM)

Central to all TS algorithms is the implementation of STM, typically in the form of a recency-based TL, with a First-In-First-Out[3] (FIFO) queue structure. Only selected attribute(s) of the move or solution are stored in the TL since storing the complete solution sets would be too memory intensive, especially for large problems, except in the case of Elite Candidate List where complete solution sets are desired to guide the search trajectory in different ways beyond that of objective function evaluation. New moves/solutions generated in the neighborhood of the current solution are compared against attributes stored in the TL and any matching attributes are forbidden. In the case of continuous problems, use of general attributes like move size might cause the search to be severely restricted, leading to reduction in solution quality. Instead, the exact position of the move is usually stored in the TL and a tabu radius is placed around points in the TL to prevent a revisit of the same neighborhood. The tabu status of a move can be overridden through the use of an aspiration criterion which will be discussed later.

As the neighborhood being explored is dynamic and changes with respect to search history, both TL size and the tabu tenure are critical. For basic implementations of TS with a FIFO-based TL, also known as a fixed TL, the tabu tenure of each move is equivalent to TL size and is fixed at an intermediate value, which represents a compromise between cycling (due to small TL size) and deterioration of solution quality (due to large TL size). The simple implementation fails to take advantage of the full capabilities of adaptive memory. Dynamic TL size offers substantial advantage especially for continuous problems of large dimension, since it significantly reduces NFEs required (Battiti and Tecchiolli, 1996).

Dynamic TL can be implemented for STM in a random or systemic fashion. Taillard (1991) implemented random dynamic TL, whereby TL size is randomized and bounded within the optimum range of 0.9 N to 1.1 N. TS implemented with this dynamic TL required 30% less NFEs to find near-optimal solution as compared to fixed TL size. Srivastava and Chen (1993)

[3] TL with a FIFO queue structure means that when the TL is full, the first component which was placed in the TL will be the first to be taken out and replaced.

proposed a dynamic TL that varied systematically in size, increasing each time a specified number of iterations had been carried out. Initially, TL size is kept small so that the entire solution space can be searched freely without any restrictions. As the search progresses, the TL size increases and the search is increasingly restricted and forced to diversify into unexplored regions. This is balanced by the aspiration criterion, which allows promising solutions to override their tabu status. Both methods described here were observed to have comparable performance, and the method to be used would depend on the problem type (Srivastava and Chen, 1993).

9. Aspiration Criterion

The primary purpose of the aspiration criterion is to allow promising moves to override their tabu status so that search is guided towards the global minimum. Generally, TS involves checking TL before aspiration criterion (if the move is tabu), though in certain cases where the search is highly restricted (i.e. most of the moves are tabu), it is more effective to check the aspiration criterion before TL. The different forms of aspiration criterion are outlined below.

Global Aspiration by Objective is one of the most commonly used aspiration criterion, whereby the tabu status is overridden if the objective value is better than the best solution found so far. *Aspiration by Search Direction* allows improving moves to override their tabu status and do not have to be better than the best solution found so far. It is less restrictive than global aspiration by objective, and guides the search along a particular direction.

Moves which lead to large changes in structure and/or solution quality are of high influence. Generally, reversal of a low influence move is permitted only if it is found in between two high influence moves, since the high influence move would have brought the solution far away from the previous solution (result of a low influence move). *Aspiration by Move Influence* allows moves of low influence to be carried out successively until it is unlikely to obtain further gains. Following this, there will be a shift of aspiration criterion to give influential moves more importance. *Aspiration by Default* takes place when all available moves are tabu and not admissible by aspiration criterion. In this case, the least tabu move (i.e. the move which has the shortest remaining tabu tenure) is selected.

10. Candidate List

One of the key distinguishing characteristics of TS is that the neighborhood search is carried out systematically instead of randomly. For complex problems, the neighborhood is large, and hence it is necessary to restrict the number of solutions being examined. Candidate list isolates regions of the neighborhood containing promising moves and examines them during the current iteration. The five common candidate list strategies are: aspiration plus strategy, elite candidate list, successive filter strategy, sequential fan candidate list and bounded change candidate list (Onwubolu, 2002).

TS relies on search history to direct the search trajectory towards productive regions. In *aspiration plus strategy*, certain number of moves is examined to direct the search trajectory. The purpose of "plus" is to allow for additional moves to be considered so that the search is carried out beyond the threshold level, in order to pick up any additional promising moves, before choosing the overall best move. *Elite candidate list* involves the creation of a master list of k best moves out of a large number of moves which have been evaluated; here, k is a parameter to be defined. The moves in the list are generally high-quality local optima, of which the best move will be executed in the subsequent iteration. The process is repeated until the best neighbor generated falls below a certain quality threshold or after performing certain number of iterations (a pre-determined search parameter). The elite candidate list is then regenerated and the search process is repeated until the termination criterion is satisfied. In this way, the best search trajectory is constantly pursued in the search.

Compound moves are often encountered in practical applications. *Successive filter strategy* works by breaking down the compound moves into their individual components. For each component, the moves are filtered such that a certain number of moves that have passed the specified threshold are selected to make up the compound moves. This greatly reduces the number of possible combinations to be considered and generally allows optimal or near-optimal solutions to be found within a reasonable short period of time.

The key advantage of the *sequential fan candidate list* lies with the capability for parallel processing. From each initial solution, a fan of p solution streams is generated based on the p best moves. Subsequent iterations would require p or more best moves (different moves might result in the same solution) to generate another p best solutions. For *Bounded Change Candidate List*, changes to solution components are restricted and kept within strategic bounds to minimize search in unpromising regions. This vastly reduces the computational requirements needed as the solution space to be searched is smaller and is particularly suited for intensification purposes. Problems which can be decomposed are most suited for this strategy, since the bounds for the search are defined by the decomposition process.

11. Long Term Memory (LTM)

The most common implementation of LTM is the *frequency-based memory* that complements the recency-based STM, to refine the direction of the search trajectory. Typically, it would involve measures of both transitional and residential frequency; the former refers to how often solution attribute changes, while the later refers to how often an attribute is found in the solution generated. It is very important for the frequencies to be interpreted based upon their various contexts. For example, a high residential frequency would indicate that a particular attribute is highly attractive if the solution domain is high quality solutions, while the reverse is true if the solution domain is low quality solutions. For solution domain which includes a mix of both high and low quality moves, a high residential frequency could be indicative of an attribute which restricts the search process to the current domain. On the other hand, a high transition frequency for high quality solutions would indicate that a particular attribute is acting as a "crack filler", being varied constantly to fine tune the solution in order to obtain the optimum (Glover and Laguna, 1997).

Based on the information obtained from frequency-based memory, the search can be directed towards promising regions (intensification) or towards unvisited regions (diversification). The penalized frequency method screens out moves that occur frequently through the use of an evaluator function, whereby penalties and incentives are applied to various components of the objective function. On the other hand, constrained frequency method restricts moves to those that contain solution attributes which occur below/above critical frequency. This was implemented by Srivastavas and Chen (1993) for global diversification, whereby critical frequency is set at an arbitrary value of 0.01 and slowly incremented until there is at least one admissible non-improving move.

Frequency based memory can also be used to vary TL size dynamically, as shown by Battiti and Tecchiolli (1994) in Reactive TS (RTS). The solutions visited during the search and their corresponding iteration numbers are stored in LTM, to facilitate the check for repetition after every move, as well as to calculate the interval between repetitions. RTS consists of two basic mechanisms which adapt TL size to the search region: (1) a fast reaction mechanism which increases the size of the TL whenever repetition is encountered, and (2) a slow reduction mechanism to periodically reduce TL size, since certain regions of search space might not require a large TL and having a TL that is larger than required restricts the search severely, leading to deterioration of solution quality. In addition to the two basic mechanisms, there is an additional escape mechanism that kicks in if repetition is occurring on a regular basis beyond a certain threshold. In such cases, the search may be stuck within a local minimum and is diversified by making a number of random moves proportional to the moving average of the cycling length.

12. Intensification

Search intensification is generally carried out in two ways. One is favoring moves with attributes that are common to elite solutions. This commonly involves the examination of elite solutions through LTM and attributes which appear most frequently are considered as desirable. The search trajectory will be guided towards solutions having these desirable attributes, by providing incentives to moves which include them or penalties to moves that remove them during the objective function evaluation. Intensification by decomposition can also be carried out by partial decomposition of the problem or its structure and placing restrictions on parts which will not contain desirable attributes. In this way, the efforts to direct the search trajectory towards desirable attributes can be focused on a particular area.

Second way of intensification is by revisiting attractive regions and conducting a more thorough search. This could involve LTM in the case where there is a high frequency of promising moves being generated in a particular region, which prompts a revisit of the attractive region for more complete exploration, or when the solution quality drops below a threshold and the search is restarted with the best solution from the elite candidate list. In the latter case, intensification does not require the use of LTM. Search intensification can also take place when the diversification phase is completed, and a thorough exploration of the promising list or elite candidate list is made to obtain the global minimum.

12.1. Diversification

The STM largely prevents the cycling of solution through the use of TL, which can be considered as a form of diversification. In the TS–QN algorithm given above, diversification was achieved without the use of LTM; random generation of neighbors in hyper-rectangles as well as the use of STM to create a TL and a PL coupled with tabu balls, prevented cycling and achieved diversification. However, LTM can be used to enhance the existing diversification strategies though the two main methods of modifying choice rules and restarting the search process (Glover and Laguna, 1997). Modifying choice rules could be done through the use of penalized frequency methods and constrained frequency methods outlined earlier. The modification of choice rules can result in *strategic oscillation*, which represents an inter-play of intensification and diversification procedures. Moves are permitted across a feasible boundary to a specified depth, and the process is repeated alternating between the different directions. The use of STM and LTM ensures that the same search trajectory is not revisited.

Restarting the search process takes place when the search appears to be stuck at a minimum, characterized by constant cycling or deterioration of objective function values beyond a certain threshold. In these cases, the entire search will be restarted by using a randomly generated point (which might not be good, if this is far away from promising solutions) or by applying a series of random moves away from the existing optimum (Battiti and Tecchiolli, 1994).

12.2. Path relinking

Similar to the concept of elite candidate list used for intensification, whereby incentives are used to guide the search towards desired attributes, *path relinking* does the same in a more intense manner. In this case, the search will specifically move from the initial solution towards one or more guiding solutions (elite solutions), instead of just favoring moves with the desired attributes, with aspiration criterion allowing for deviation of solutions along the path. Path relinking can be used for both intensification and diversification, similar to that of penalized frequency method.

13. Conclusions

TS had been successfully adopted and applied to the optimization of continuous problems through various modifications proposed by the researchers. It has also found many applications in Chemical Engineering. TS seems to be computationally efficient but its reliability is less than 100% for some continuous problems. Many concepts of TS which had proven to be highly successful for COPs were not incorporated in the optimization of continuous problems, such as the use of LTM and its associated strategies of strategic oscillation and path relinking. Through the incorporation of LTM and its associated strategies, the performance of TS could be enhanced for continuous optimization problems, allowing it to escape from local minima more easily. Further, it is necessary to test promising TS algorithms on large benchmark problems and applications, with and without constraints.

References

Altiparmak, F. and Karaoglan, I. (2008). An adaptive tabu-simulated annealing for concave cost transportation problems. *Journal of the Operational Research Society*, **59**, pp. 331–341.

Ashley, V.M. and Linke, P. (2004). A novel approach for reactor network synthesis using knowledge discovery and optimization techniques. *Chemical Engineering Research and Design*, **82**, pp. 952–960.

Battiti, R. and Tecchiolli, G. (1994). The reactive tabu search. *ORSA Journal on Computing*, **6**(2), pp. 126–140.

Battiti, R. and Tecchiolli, G. (1996). The continuous reactive tabu search: Blending combinatorial optimization and stochastic search for global optimization. *Annals of Operations Research*, **63**, pp. 153–188.

Battiti, R. and Brunato, M. (2007). *Reactive Search — Continuous Optimization*. Retrieved July 7, 2008 from the World Wide Web: http://www.reactive-search.org/.

Baykasoglu, A., Owen, S. and Gindy, N. (1999). A taboo search based approach to find the pareto optimal set in multiple objective optimization. *Engineering Optimization*, **31**, pp. 731–748.

Baykasoglu, A. (2006). Applying multiple objective tabu search to continuous optimization problems with a simple neighborhood strategy. *International Journal for Numerical Methods in Engineering*, **65**, pp. 406–424.

Chavali, S., Lin, B., Miller, D.C. and Camarda, K.V. (2004). Environmentally-benign transition metal catalyst design using optimization techniques. *Computers and Chemical Engineering*, **28**, pp. 605–611.

Chelouah, R. and Siarry, P. (2000). Tabu search applied to global optimization. *European Journal of Operational Research*, **123**, pp. 256–270.

Chelouah, R. and Siarry, P. (2005). A hybrid method combining continuous tabu search and Nelder–Mead simplex algorithms for the global optimization of multi-minima functions. *European Journal of Operational Research*, **161**, pp. 636–654.

Crainic, T.G. and Toulouse, M. (1997). *Parallel Metaheuristics*, Centre de Recherche sur les Transports Universite de Montreal, France.

Cvijovic, D. and Klinowski, J. (1995). Taboo search: an approach to the multiple minima problem. *Science*, **267**, pp. 664–666.

Deb, K. (2002). *Optimization for Engineering Design: Algorithms and Examples*, Prentice-Hall of India, New Delhi.

Exler, O., Antelo, L.T., Egea, J.A., Alonso, A.A. and Banga J.R. (2008). A tabu search-based algorithm for mixed-integer nonlinear problems and its application to integrated process and control system design. *Computers and Chemical Engineering*, **32**(8), pp. 1877–1891.

Franze, F. and Speciale, N. (2001). A tabu-searched-based algorithm for continuous multi-minima problems. *International Journal for Numerical Methods in Engineering*, **50**, pp. 665–680.

Frewen, T.A., Sinno, T., Haeckl, W. and Ammon, W. (2005). A systems-based approach for generating quantitative models of microstructural evolution in silicon materials processing. *Computers and Chemical Engineering*, **29**, pp. 713–730.

Frontline Systems, Inc., Solver technology (2008). nonsmooth optimization. Retrieved July 7, 2008 from the World Wide Web: http://www.solver.com/technology6.htm.

Gendreau, M., Hertz, A. and Laporte, G. (1994). A tabu search heuristic for the vehicle routing problem. *Management Science*, **40**, pp. 1276–1290.

Glover, F. (1986). Future paths for integer programming and links to artificial intelligence. *Computers and Operations Research*, **13**, pp. 533–549.

Glover, F. (1989). Tabu search: part 1. *ORSA Journal on Computing*, **1**, pp. 190–206.

Glover, F. (1990a). Tabu search: part 2. *ORSA Journal on Computing*, **2**, pp. 4–32.

Glover, F. (1990b). Tabu search: a tutorial. *Interfaces*, **20**, pp. 74–94.

Glover, F. (1997). Tabu search and adaptive memory programming — Advances, applications and challenges. *Interfaces in Computer Science and Operations Research*, **7**, pp. 1–76.

Glover, F. (2007). Tabu search — unchartered domains. *Annals of Operational Research*, **149**, pp. 89–98.

Glover, F. and Laguna M. (1997). *Tabu Search*, Kluwer Academic Publishers, Boston.

Goldstein, A.A. and Price, J.F. (1971). On descent from local minima. *Mathematics of Computation*, **115**, pp. 569–574.

Hansen, M.M. (1997). *Tabu Search for Multi-Objective Optimization: MOTS*, Presented at MCDM'97, Cape Town, South Africa, January 6–10.

Harder, R., OpenTS, (2006). Retrieved July 7, 2008 from the World Wide Web: https://projects.coin-or.org/Ots.

Hedar, A.R. and Fukushima, M. (2006). Tabu search directed by direct search methods for non-linear global optimization. *European Journal of Operational Research*, **170**, pp. 329–349.

Hu, N. (1992). Tabu search method with random moves for globally optimal design. *International Journal on Numerical Methods in Engineering*, **35**, pp. 1055–1070.

Jaeggi, D.M., Asselin-Miller, C., Parks, G.T., Kipourus, T., Bell, T. and Clarkson, P.J. (2004). Multi-objective parallel tabu search. *Lecture Notes in Computer Science*, **3242**, pp. 732–741.

Jaeggi, D.M., Parks, G.T., Kipourus, T. and Clarkson, P.J. (2008). The development of a multi-objective tabu search algorithm for continuous optimization problems. *European Journal of Operational Research*, **185**, pp. 1192–1212.

Ji, T. and Tang, H. (2004). Global optimizations and tabu search based on memory. *Applied Mathematics and Computation*, **159**, pp. 449–457.

Kalinli, A. and Karaboga, D. (2004). Training recurrent neural networks by using parallel tabu search algorithm based on crossover operation. *Engineering Applications of Artificial Intelligence*, **17**, pp. 529–542.

Kalinli, A. and Karaboga, N. (2005). A parallel tabu search algorithm for digital filter design. *COMPEL: The International Journal for Computation and Mathematics in Electrical and Electronic Engineering*, **24**, pp. 1284–1298.

Lin, B. and Miller, D.C. (2004a). Solving heat exchanger network synthesis problems with tabu search. *Computers and Chemical Engineering*, **28**, pp. 1451–1464.

Lin, B. and Miller, D.C. (2004b). Tabu search algorithm for chemical process optimization. *Computers and Chemical Engineering*, **28**, pp. 2287–2306.

Lin, B., Chavali, S., Camarda, K. and Miller, D.C. (2005). Computer-aided molecular design using tabu search. *Computers and Chemical Engineering*, **29**, pp. 337–347.

Lin, B., Leibovici, C.F. and Jorgensen, S.B. (2008). Optimal component lumping: Problem formulation and solution techniques. *Computers and Chemical Engineering*, **32**, pp. 1167–1172.

Linke, P. and Kokossis, A. (2003). On the robust application of stochastic optimization technology for the synthesis of reaction/separation systems. *Computers and Chemical Engineering*, **27**, pp. 733–758.

Machado, J.M., Shiyou, Y., Ho, S.L. and Peihong, N. (2001). A common tabu search algorithm for the global optimization of engineering problems. *Computer Methods in Applied Mechanics and Engineering*, **190**, pp. 3501–3510.

Maischberger, M. (2007). METSlib trac. Retrieved July 7, 2008 from the World Wide Web: http://code.100allora.it/metslib/wiki.

Montolio-Rodriguez, D., Linke, D. and Linke, P. (2007). Systematic identification of optimal process designs for the production of acetic acid via ethane oxidation. *Chemical Engineering Science*, **62**, pp. 5602–5608.

Onwubolu, G.C. (2002). *Emerging Optimization Techniques in Production Planning and Control*, Imperial College Press, London.

OptTek Systems, Inc. (2008). Optquest engine. Retrieved July 7, 2008 from the World Wide Web: http://www.optquest.com/optquest.html.

Rajesh, J., Jayaraman, V.K. and Kulkarni, B.D. (2000). Taboo search algorithm for continuous function optimization. *Chemical Engineering Research and Design: Transactions of the Institute of Chemical Engineers*, Part A, **78**, pp. 845–848.

Rajesh, J., Gupta, K., Kusumakar, H.S., Jayaraman, V.K. and Kulkarni, B.D. (2003). A tabu search based approach for solving a class of bilevel programming problems in chemical engineering. *Journal of Heuristics*, **9**, pp. 307–319.

Rangaiah, G.P. (editor). (2009). *Multi-Objective Optimization: Techniques and Applications in Chemical Engineering*, World Scientific, Singapore.

Salhi, S. and Queen, N.M. (2004). A hybrid algorithm for identifying global and local minima when optimizing functions with many minima. *European Journal of Operational Research*, **155**, pp. 51–67.

Siarry, P. and Berthiau, G. (1997). Fitting of tabu search to optimize functions of continuous variables. *International Journal for Numerical Methods in Engineering*, **40**, pp. 2449–2457.

Srinivas, M. and Rangaiah, G.P. (2007a). A study of differential evolution and tabu search for benchmark, phase equilibrium and phase stability problems. *Computers and Chemical Engineering*, **31**, pp. 760–772.

Srinivas, M. and Rangaiah, G.P. (2007b). Differential evolution with tabu list for global optimization and its application to phase equilibrium and parameter estimation problems. *Industrial and Engineering Chemistry Research*, **46**, pp. 3410–3421.

Srivastava, B. and Chen, W.H. (1993). Part type selection problem in flexible manufacturing systems: tabu search algorithms. *Annals of Operations Research*, **41**, pp. 279–297.

Stepanenko, S. and Engels, B. (2007). Gradient tabu search. *Journal of Computational Chemistry*, **28**, pp. 601–611.

Stepanenko, S. and Engels, B. (2008). New tabu search based global optimization methods. Outline of algorithm and study of efficiency. *Journal of Computational Chemistry*, **29**, pp. 768–780.

Taillard, E. (1991). Robust taboo search for the quadratic assignment problem. *Parallel Computing*, **17**, pp. 443–455.

Tan, K.C., Khor, E.F., Lee, T.H. and Yang Y.J. (2003). A tabu-based exploratory evolutionary algorithm for multi-objective optimization. *Artificial Intelligence Review*, **19**, pp. 231–260.

Teh, Y.S. and Rangaiah, G.P. (2003). Tabu search for global optimization of continuous functions with application to phase equilibrium calculations. *Computers and Chemical Engineering*, **27**, pp. 1665–1679.

Visual Numerics (2008). IMSL numerical libraries family of products. Retrieved July 7, 2008 from the World Wide Web: http://www.vni.com/products/imsl/fortran/overview.php.

Wang, C., Quan, H. and Xu, X. (1999). Optimal design of multi-product batch chemical processes using tabu search. *Computers and Chemical Engineering*, **23**, pp. 427–437.

Wang, M., Chen, X. and Qian, J. (2004). An improvement of continuous tabu search for global optimization. *Proceedings of the 5th World Congress on Intelligent Control and Automation*, June 15–19.

Yang, S., Ni, G., Li, Y., Tian, B. and Li, R. (1998). An universal tabu search algorithm for global optimization of multi-modal functions with continuous variables in electromagnetics. *IEEE Transactions on Magnetics*, **34**(5), pp. 2901–2904.

Zheng, R.T., Ngo, N.Q., Shum, P., Tjin, S.C. and Binh, L.N. (2005). A staged continuous tabu search algorithm for the global optimization and its applications to the design of fiber bragg gratings. *Computational Optimization and Applications*, **30**, pp. 319–335.

Appendix A. Different implementations of TS for global optimization of continuous problems.

No.	Methods (References)	Features and merits	Testing and comparison	Remarks
1	TS Method with Random Moves (Hu,1992)	Set of step sizes is defined, and feasible random moves are generated for each neighborhood defined by step size. Different neighborhood sizes guide the search in a systematic manner (instead of blind sampling) and also reduce the likelihood of search being trapped at a local optimum. N-dimensional problems will require 2^N random moves, which is expensive for larger problems.	Tested on 1–2 variable problems of Dixon and Szego, and a few engineering problems. The proposed TS is better than or comparable to GA, and requires significantly fewer NFEs than random search.	Differs from key tenets of TS set out by Glover, as only moves which involve decrease in value of objective function are considered; this may lead to search being stuck at a local optimum. Efficiency of the algorithm is highly dependent on step size.
2	Taboo Search (Cvijovic and Klinowski, 1995)	For N variables, neighborhood structure consists of disjunct cells formed by partitioning each of the N axes into a specified number of intervals.	Tested on several benchmark problems: Branin RCOS (2 variables), Goldstein Price (2 variables), Shubert (2 variables), Hartmann (3 and 6 variables), Rastrigin (2 variables).	Neighborhood generation strategy is against the tenets of TS, being completely stochastic instead of systematic. This lowers the probability of finding the global optimum. Address of cells is added to the TL, instead

(Continued)

Appendix A. (*Continued*)

No.	Methods (References)	Features and merits	Testing and comparison	Remarks
		In each iteration, n_S sample points are drawn from n_C randomly chosen cells. Two separate lists are maintained: (1) TL that stores addresses of visited moves (for diversification), and (2) Elite list that stores addresses of best solutions (for intensification). The algorithm can be vectorized and run in parallel.	Compared with pure random search, multi-start and two variants of SA, NFEs required for TS were consistently the lowest. Rajesh *et al.* (2000) used a similar method for solving NLP, MILP and MINLP problems; their results show that TS requires significantly less NFE compared to GA.	of move attributes, which might be memory intensive for large problems.
3	Continuous Reactive TS, C-RTS (Battiti and Tecchiolli, 1996)	C-RTS uses RTS (Battiti and Tecchiolli, 1994) for combinatorial optimization problems to identify promising areas, followed by an adaptive random search (affine shaker) algorithm to find the local minimum.	NFEs required by C-RTS for Dixon-Szego test functions, were comparable to or better than other methods, whereas CPU time is competitive for most cases. C-RTS was also tested for discontinuous	C-RTS is meant for unconstrained global optimization of problems, which do not have to be continuous and differentiable. Affine shaker, a general purpose local minimizer, can be substituted with efficient

(*Continued*)

Appendix A. *(Continued)*

No.	Methods (References)	Features and merits	Testing and comparison	Remarks
		In RTS, TL size adapts during the search based on (1) fast operating mechanism which increases TL size when cycling occurs, and (2) slow operating mechanism which decreases TL size when it is greater than the moving average of repetition interval. It also has additional diversification mechanism via random moves, when cycling occurs beyond a threshold, which is indicative of the search being trapped within a small space.	functions with constraints, through the use of penalty methods. In this case, C-RTS$_{ave}$ requires much less NFEs then C-RTS$_{min}$. Note that C-RTS$_{min}$ and C-RTS$_{ave}$, use, respectively, minimum and average value of points for evaluation of boxes.	methods such as steepest descent, when dealing with differentiable functions.
4	Continuous TS, CTS (Siarry and Berthiau, 1997)	CTS is an adaptation of TS for continuous problems, while adhering to Glover's tenets. The paper describes generation of neighbors (using linear, geometric and isovolume partitioning) and the management of tabu list.	CTS was tested for Goldstein-Price (2 variables), Hartmann (3 variables) and Rosenbrock (2, 5 and 10 variables) functions. It was comparable to SA for functions with 2 variables. For 5 and 10 variable Rosenbrock	Linear and geometric partitioning are better than isovolume partitioning. A single neighbor is generated within each partition. Isovolume partitioning however, might allow a better exploration of the entire search space, especially

(Continued)

Appendix A. (*Continued*)

No.	Methods (References)	Features and merits	Testing and comparison	Remarks
			functions, NFEs for CTS was respectively 10 and 20 times more than that for SA.	during search diversification. CTS is suitable for small problems.
5	Universal TS, UTabu (Yang *et al.*, 1998)	UTabu is an Improvement of TS of Hu (1992). It accepts the best neighbor for the next iteration even if it is worse than the existing best solution. UTabu does not have TL as it could restrict search and trap the search at a local minimum for continuous problems. It restarts from the optimum solution after every N iterations, so as to escape from a local optimum.	UTabu was compared with Hu's Tabu, SA and UTabu with TL, for a benchmark problem with 15^5 local minima. Only UTabu and SA were able to reach the global minimum. NFEs for UTabu was less than that required for SA.	Without TL, UTabu may not be considered as TS.
6	Custom TS, C–TS (Wang *et al.*, 1999)	C–TS has dynamic neighborhood size: small in the beginning of the search for faster convergence (may get stuck at local optimum), and large towards the end (may reduce the accuracy) for thorough	C–TS was tested on four applications in multi-product batch chemical processes. Compared with Mathematical Programming (MP), standard TS (i.e. without dynamic	Computational experience with respect to initial solution, dynamic neighborhood, double tabu list, dynamic tabu size, diversification and intensification, step size of

(*Continued*)

Appendix A. (*Continued*)

No.	Methods (References)	Features and merits	Testing and comparison	Remarks
		search of the entire solution space. For the application to multi-product batch processes, effect of the number of units and unit size on objective differs greatly; this leads to unbalanced changes of the optimization variables. Hence, two separate TL were employed, one for number of units and another for unit size. Further, step size of continuous variables is varied dynamically and depending on the variable value since a large step size may result in a local optimum while a small step size will increase computational requirements.	neighborhood sizes) and SA, C–TS found comparable or better objective function values. Also, CPU time required for C–TS was 4–20% of MP, 10–30% of standard TS and 20–25% of SA.	continuous variables and termination criterion, was discussed in detail.

(*Continued*)

Appendix A. (*Continued*)

No.	Methods (References)	Features and merits	Testing and comparison	Remarks
7	Enhanced Continuous TS, ECTS (Chelouah and Siarry, 2000)	ECTS is the improved CTS through the following: (1) use of hyper-rectangles instead of crown balls for neighborhood generation, so as to simplify selection of points in the neighborhood, (2) using both TL & PL with tabu balls to prevent revisit of the solution space, and (3) diversification and intensification steps in sequence. The intensification step involves periodic halving of radius of tabu balls and size of hyper-rectangles, for an accurate solution.	ECTS was tested on 20 benchmark problems (Branin RCOS, Easom, Goldstein Price, Shubert, De Joung, Hartmann, Shekel, Rosenbrock and Zakharov problems) involving 2 to 100 variables. NFEs required by ECTS were less than or comparable to those for CTS, C-RTS, TS of Cvijovic and Klinowski (1995) and enhanced SA.	Slow convergence of intensification procedure in the ECTS was subsequently improved by using a local minimizer. ECTS is shown to be effective for large, continuous problems (more than 10).
8	Common TS (Machado *et al.*, 2001)	Common TS follows the procedure of UTabu (Yang *et al.*, 1998) along with local minimization from the random feasible move which is not near a stored move trajectory, to prevent unnecessary minimization.	Five test functions including one used to test UTabu, were employed. Common TS was able to find global minimum for all of them, and required 3 to 7 times less NFEs than SA.	Common TS was applied to an electrical engineering problem.

(*Continued*)

Appendix A. (*Continued*)

No.	Methods (References)	Features and merits	Testing and comparison	Remarks
9	DOPE (Franze and Speciale, 2001)	DOPE algorithm is based on pattern search and TS to solve continuous problems by discretizing the continuous space so that it can be treated as a COP (allows algorithm to work in a deterministic manner). It includes variable move step to facilitate different requirements during intensification and diversification. An improving move (instead of searching for the best possible in the neighborhood) is accepted to reduce NFEs, while maintaining solution quality through having more iterations. Two TLs, one for moves and one for states, each with different tabu tenures are employed.	DOPE was tested on smooth functions: De Joung and Zakharov, Rosenbrock function with narrow valley, Goldstein-Price function with far local minima, and Shubert function with near local minima. NFEs required by DOPE were comparable or less than those by ECTS, and were consistently less than NFEs required by enhanced SA.	DOPE algorithm is deterministic in nature, with random component triggered when search is stuck. Its performance in finding the global optimum is highly dependent on starting point. Application of DOPE to a device parameter extraction problem is described in the paper.

(*Continued*)

Appendix A. (*Continued*)

No.	Methods (References)	Features and merits	Testing and comparison	Remarks
10	TS–NM & TS–QN (Teh and Rangaiah, 2003; Srinivas and Rangaiah, 2007a)	Modification of ECTS through the use of ECTS for diversification, followed by local minimization (by Nelder–Mead, NM or quasi-Newton, QN method) for intensification. Srinivas and Rangaiah (2007a) tried various ways of generation of neighbors: TS–S–QN (systematic, using hyper-rectangles), TS–R–QN (random) and TS–M–QN (mixed).	TS–QN, TS–NM, GA–QN and GA–NM were tested on more than 10 benchmark problems with 2 to 20 variables. Compared to ECTS, TS–QN is more accurate, slightly less reliable in finding global minimum, requires less NFEs for some problems and more NFEs for other problems.	TS–QN was found to be more efficient and reliable than TS–NM because NM is a direct search method whereas QN is a gradient search method. Further, TS–QN was found to require fewer NFEs compared to GA–QN for the problems tested. TS–R–QN was less efficient computationally, as the random generation does not guide the search in a meaningful way.
11	Improvement to ECTS (Wang *et al.*, 2004)	The improvements proposed to ECTS are: (1) intensification phase allows for generation of points within the central hyper-rectangle as well; and (2) aspiration criterion was checked first before the tabu condition.	The proposed algorithm was tested on several benchmark problems with 2 to 5 variables. Use of intensification strategy reduces NFEs (generally by 10%) and increases accuracy.	Checking aspiration criterion before the tabu condition is likely to be inefficient in engineering applications.

(*Continued*)

Appendix A. *(Continued)*

No.	Methods (References)	Features and merits	Testing and comparison	Remarks
12	Parallel TS, PTS (Kalinli and Karaboga, 2004)	Several independent basic TS are executed from different initial solutions. Asynchronous exchange of information takes place between the algorithms based on crossover operator of GA.	PTS was tested for several benchmark problems with 2 to 20 variables, and shown to have better convergence than the basic TS, GA and touring ant colony optimization using the same NFEs.	PTS was applied to training of artificial neural networks for linear and nonlinear systems.
13	Memory TS, MTS (Ji and Tang, 2004)	MTS ensures convergence to the global optimum through saving the best solution found during the search. Results can be refined further through the use of a local optimizer.	MTS was tested on several benchmark problems with 2 to 6 variables. Compared to pure random search, multi-start, SA and TS (of Cvijovic and Klinowski, 1995), it requires fewer NFEs.	Convergence of MTS was proved but it is not tested on benchmark problems with large number of variables.
14	Continuous Tabu Simplex Search, CTSS (Chelouah and Siarry, 2005)	CTSS uses ECTS for diversification phase with the following modifications: (1) normalization of variables to (0,1), and (2) two ways of generating neighbors - disturb single variable to obtain	CTSS was compared with ECTS, GA, continuous hybrid algorithm (CHA), enhanced SA, C-RTS, TS and INTEROPT, for test functions with 2 to 6 variables.	CTSS was applied to the design of an eddy current sensor for non-destructive control.

(Continued)

Appendix A. (*Continued*)

No.	Methods (References)	Features and merits	Testing and comparison	Remarks
		neighborhood (useful for less than 5 variables) and disturb multiple variables to obtain hypercube neighborhood (usually 2 variables, or up to 1/3 of all variables if more than 5 variables). Unlike ECTS, which does intensification at the end of the diversification phase, Nelder–Mead simplex search is used for intensification whenever a promising area is located during the diversification phase. Best point in the promising list is the final solution.	For functions with less than 5 variables, results show that CTSS and ECTS are better than GA and CHA, and CTSS requires about half of NFEs compared to ECTS. Compared to C-RTS, CTSS requires 10–50% fewer NFEs for problems with more than 2 variables but it requires more NFEs for 2-variable problems.	

(*Continued*)

Appendix A. (*Continued*)

No.	Methods (References)	Features and merits	Testing and comparison	Remarks
15	Staged Continuous TS, SCTS (Zheng *et al.*, 2005)	SCTS is a 3-stage implementation of CTS: (1) search of the entire solution space to find a prospective point which is likely to lead to a global minimum, (2) finding a point close to the global minimum, and (3) convergence to the global minimum from the solution of stage 2	SCTS and an improved GA (IGA) were compared for 13 test functions with 1 to 20 variables. The results show that SCTS was better than IG for reliably finding the global minimum with fewer NFEs.	For benchmark functions of 2–3 variables, NFEs required for SCTS were more than those for ECTS and CTSS reported by Chelouah and Siarry (2000, 2005). SCTS was applied to an engineering problem (design of fiber Bragg gratings).
16	Directed TS, DTS (Hedar and Fukushima, 2006)	DTS employs three search procedures: exploration, diversification and intensification, and can be considered as a multi-start method. A local method: either Nelder–Mead (NM) or Adaptive Pattern Search (APS), is used to generate trial points for exploration search. Semi-tabu regions, multi-ranked	DTS was tested on many benchmark problems with 2 to 30 variables. DTS based on APS requires fewer NFEs and is more reliable compared to that based on NM, which is unsuitable for large problems. DTS-APS requires fewer NFEs compared to ECTS for small problems. However, its	DTS based on APS was shown to perform better than GA and scatter search, for NFEs more than 1000, i.e. if the algorithm is allowed to run for a substantially long period of time, DTS would yield better solutions. The paper notes absence of some details in the published results

(*Continued*)

Appendix A. *(Continued)*

No.	Methods (References)	Features and merits	Testing and comparison	Remarks
		TL and visited region list are introduced in the diversification search. Exploration and diversification searches are repeated before intensification search involving a local method starting from some of the best points visited.	performance degenerates with increasing problem dimension due to the use of a direct search method. Also DTS-APS is more accurate than ECTS and DOPE.	for fair comparison, and also some inconsistencies in ECTS results reported in Chelouah and Siarry (2000).
17	Gradient TS, GTS (Stepanenko and Engels, 2007)	GTS combines gradient methods (steepest descent and quasi-Newton) with tabu concepts (list, region and direction) and multi-start for diversification.	GTS was tested on 7 benchmark problems including Rastrigin, Ackley and Griewangk, with 2 to 50 variables. Requires substantially fewer NFEs compared to random search, DE, GA, SA and two other versions of TS.	GTS converges quickly with no oscillations unlike DE (in the Mathematica software). It is limited to objective functions which are differentiable.

(Continued)

Appendix A. (*Continued*)

No.	Methods (References)	Features and merits	Testing and comparison	Remarks
18	Gradient Only TS, GOTS & TS with Powell's Algorithm, TSPA (Stepanenko and Engels, 2008)	GOTS and TSPA are modifications of GTS for problems where gradients are unavailable. GOTS approximates the gradient using a grid of function evaluations. TSPA uses Powell's algorithm for steepest descent followed by a grid of function evaluations for modest descent.	Tested on the following problems: Rastrigin and Ackley (2 to 50 variables), Griewangk (2 to 20 variables), Levy (4, 7 variables), Branin RCOS (2 variables), Goldstein Price (2 variables) and Hansen (2 variables). Generally, GOTS requires more NFEs than GTS for small problems but is more efficient for larger problems. TSPA generally requires more NFEs than GOTS, with a much larger increase in NFEs when the dimension increases.	Larger step size allows local minimum to be overcome with fewer steps; however, NFEs might increase as the search becomes less accurate.

Appendix B. Chemical engineering applications of TS.

No.	Application(s) (References)	Motivation/Scope	Features of the method and problem	Remarks
1	Multi-product batch chemical processes (Wang *et al.*, 1999)	General TS algorithm for optimal design of multi-product batch chemical processes	See No. 6, Custom TS in Appendix A.	See No. 6, Custom TS in Appendix A.
2	Phase stability, phase equilibrium and parameter estimation (Teh and Rangaiah, 2003; Srinivas and Rangaiah, 2007a and b)	To explore potential of TS for these problems in chemical engineering, and compare its performance with other SGO methods	Teh and Rangaiah (2003) used ECTS for diversification phase, followed by a local minimization method (either Quasi-Newton, QN or Nelder–Mead, NM) for intensification phase. Subsequently, Srinivas and Rangaiah (2007a) tried different methods of neighborhood generation in TS, whereas Srinivas and Rangaiah (2007b) proposed Differential Evolution with Tabu List (DETL) and compared its performance with other SGO methods.	TS–QN was found to be more efficient and reliable than TS–NM because NM is a direct search method whereas QN is a gradient search method. Differential Evolution (DE) was more reliable but computationally less efficient compared to TS. Overall, the performance of DETL was better than DE and TS.

(Continued)

Appendix B. (*Continued*)

No.	Application(s) (References)	Motivation/Scope	Features of the method and problem	Remarks
3	Synthesis of Reaction and Separation Systems (Linke and Kokossis, 2003)	To expand the use of TS for chemical process synthesis, optimization of general reaction and separation superstructures by TS and SA was presented and compared.	Tried both dynamic neighborhood (Wang *et al.*, 1999) and fixed neighborhood in TS. Used a two-stage simulation, which offered 5–15% savings of CPU time. In TS, two simulation bounds are introduced, one for the best non-tabu state and the other for the best tabu-state. Simulation is carried out with low accuracy and its performance is evaluated. If the performance is close to the lower bound, then high accuracy simulation will be carried out. This is unlike common implementation of TS, where high accuracy simulation is carried out throughout.	TS was able to obtain high quality solutions more efficiently (about 40% less simulations) when compared to SA. Dynamic neighborhood was found to have little or no effect on the search efficiency for this application.
4	Bi-level programming (BLP) problems in Chemical Engineering (Rajesh *et al.*, 2003)	Use of TS to solve BLP problems	TS based on Cvijovic and Klinowski (1995) was applied to both inner and outer loop of BLP. For the given values of variables in the outer loop,	TS-based BLP successfully solved a reaction-diffusion equation (formulated as an optimization problem)

(*Continued*)

Appendix B. (*Continued*)

No.	Application(s) (References)	Motivation/Scope	Features of the method and problem	Remarks
			inner loop optimizes the inner-loop objective. The resulting solution is then used as the domain for outer-loop optimization. In TS, partition parameter for each variable has to be carefully chosen; it should be large for an accurate solution, while a small value limits the exploration of the optimum.	besides several mathematical problems.
5	Superstructure involving CSTR and PFTR (Ashley and Linke, 2004)	To develop robust techniques for synthesis and optimization of complex reactor networks.	TS for optimization of reaction processes, based on Linke and Kokossis (2003). Initial structure simulated is assessed for rule conformance. If rules are violated, moves are directed towards conformance.	Rules are obtained from data mining, providing a pre-analysis of the solution space, reducing the number of poor solutions being tried. Rule-based TS consistently outperforms TS itself

(*Continued*)

Appendix B. (*Continued*)

No.	Application(s) (References)	Motivation/Scope	Features of the method and problem	Remarks
6	Heat Exchanger Network (HEN) design (Lin and Miller, 2004a)	To implement and test TS for HEN synthesis problems, and thus demonstrate potential of TS for solving continuous problems.	TS is similar to that in Lin and Miller (2004b) outlined below, except for the generation of neighbors by the following HEN-specific strategies. (1) Single variable strategy: new solutions are generated by varying one variable at a time (i.e. adding or dropping a single heat exchanger). (2) Multiple dimension strategy: new solutions are generated by replacing all the variables with random binary values. (3) Random subset strategy: a subset of variables is varied randomly, while keeping the rest unchanged. This represents a tradeoff between the previous two strategies.	TS was tested on 3 HEN synthesis problems in the literature. It found the global optimum most of the time. For one problem TS required less computational time compared to a deterministic approach reported in the literature.

(*Continued*)

Appendix B. (*Continued*)

No.	Application(s) (References)	Motivation/Scope	Features of the method and problem	Remarks
7	Heat Exchanger Network (HEN) synthesis and pump system configuration (Lin and Miller, 2004b)	To present general TS algorithm for chemical process optimization problems, guidelines for TS parameters and constraint handling techniques.	Neighbor solutions are generated randomly. When best neighbor generated is not better than the existing solution used for the iteration, it is classified as tabu and the second best neighbor is then used for generating new neighbors. For continuous problems, it is unlikely to revisit exactly the same solution; therefore, 20% of the area (centered at the solution) surrounding each tabu solution is classified as tabu. Long term memory involves frequency based tabu regions, which determines if search has been stuck, directing it to unvisited or less visited areas. Restart operation is based upon search history. Aspiration criterion is the best objective function value or sigmoid function	The reliability of TS to find the global minimum was shown on nonlinear programming test problems from the literature. Then, a HEN synthesis problem (with 2 hot and 2 cold streams) and a 3-level pump system problem were solved by TS, which found the global optimum with nearly 100% reliability. Finally, TS was tested on the classic 10sp1 (5 hot streams, 5 cold streams) HEN problem. It found the optimum solution of 10 heat exchangers

(*Continued*)

Appendix B. (*Continued*)

No.	Application(s) (References)	Motivation/Scope	Features of the method and problem	Remarks
			(designed to strike a balance between intensification and diversification). For intensification, elite candidate list is used, together with a sine function that generates neighbors close to the current solution and gradually reduce the search space dynamically, such that the final solution can be found with high precision. Both the maximum time and the termination-on–convergence (i.e. negligible improvement over several iterations) criteria are used concurrently. Equality constraints can be solved and substituted directly, to reduce the dimension of the problem. Inequality constraints can also be solved as equality constraints by adding slack variables.	consistently, and the annual cost was lower (by 0.28%) than the best value reported previously.

(*Continued*)

Appendix B. (*Continued*)

No.	Application(s) (References)	Motivation/Scope	Features of the method and problem	Remarks
8	Computer-aided molecular design (Chavali *et al.*, 2004; Lin *et al.*, 2005)	Search for molecules that meet certain specified properties, for use in transition metal catalysts	Implementation of TS, based on Lin and Miller (2004b), for computer-aided molecular design is described in detail in Lin *et al.* (2005).	TS provided a list of suitable candidate molecules (near optimal solutions) while using smaller computation time as compared to with DICOPT solver in GAMS.
9	Modeling of impurity and defect evolution in silicon materials processing (Frewen *et al.*, 2005)	Several SGO methods were tested for robust modeling of defect and impurity evolution in crystalline silicon processing.	TS was based on Glover and Laguna (1997) for adaptive memory and responsive exploration. The neighborhood generation follows Siarry and Berthiau (1997). In addition, TS was compared with simulated annealing, genetic algorithm (GA), hybrid GA (GA combined with NM after a specified number of generations) and particle swarm optimization.	Hybrid GA was found to be superior (less sensitive to initial guess and faster convergence) compared to other methods tested, for the problems considered.

(*Continued*)

Appendix B. (*Continued*)

No.	Application(s) (References)	Motivation/Scope	Features of the method and problem	Remarks
10	Production of acetic acid via ethane oxidation (Montolio-Rodriguez *et al.*, 2007)	Optimal process design of heterogeneous gas phase reaction and relationship between design complexity and performance.	Superstructure optimization based on Linke and Kokossis (2003) and Ashley and Linke (2004). Instead of a general search of the entire superstructure, a multi-level search is conducted to obtain relationship between design complexity and performance, thus guiding the search.	First level search identifies performance of conventional and base case designs. Second level search optimizes entire structure to identify absolute performance limit, which provides insights into major process features to refine the search.
11	Optimal component lumping, OCL (Lin *et al.*, 2008)	Component lumping (to reduce the problem dimension when a process involves numerous components) leads to information loss.	TS is based on Lin and Miller (2004b). Both discrete and continuous formulation of the optimization problem for OCL, are described. TS allows reformulation of the problem such that the number of independent variables is greatly reduced.	Solutions obtained by TS were as good as or better than those obtained by BARON. Discrete formulation is preferable for the OCL problem.

(*Continued*)

Appendix B. (*Continued*)

No.	Application(s) (References)	Motivation/Scope	Features of the method and problem	Remarks
		OCL minimizes such loss of information.	The OCL problem has two main approximations: mixture property is calculated by a linear weighting scheme, and objective function is the deviation between the property of each component in the lump and the average of the lump.	
12	Wastewater treatment plant and Tennessee Eastman process (Exler *et al.*, 2008)	Integrated process and control system design.	Enhancement of C–RTS by Battiti and Tecchiolli (1996) known as mixed integer TS (MITS), whereby combinatorial component is used to guide the search into promising areas. Involves the generation of new decision vectors and the use of a novel local solver, Mixed Integers Sequential Quadratic Programming to improve the efficiency.	Formulated as a mixed integer dynamic optimization (MIDO) problem. Better performance compared to two modern MINLP solvers, OQNLP (based on scatter search) and MINLPbb.

Chapter 6

DIFFERENTIAL EVOLUTION: METHOD, DEVELOPMENTS AND CHEMICAL ENGINEERING APPLICATIONS

Chen Shaoqiang, Gade Pandu Rangaiah* and Mekapati Srinivas

Department of Chemical & Biomolecular Engineering
National University of Singapore, Singapore 117576
**chegpr@nus.edu.sg*

1. Introduction

In global optimization problems where little knowledge of the problem is available (i.e. black-box problems), stochastic optimization methods are preferred due to their ability to escape local optima. One of these techniques, pioneered by Price and Storn since 1995, is the method of differential evolution (DE), which has received much attention over the years. DE was first developed as a tool to solve the Chebychev polynomial fitting problem (Price *et al.*, 2005). With parameters (i.e. decision variables) of grossly different magnitudes and multiple local minima, this problem is challenging for most general purpose optimizers.

DE, however, was able to tackle the Chebychev polynomial fitting problem with ease. This finding generated much interest in this method, and

*Corresponding author.

numerous developments have been made to DE over the years. Like other stochastic methods, it does not require *a priori* information about the system to be solved, and can be deployed for solving most optimization problems. DE is simple to understand, and the parameters involved are easy to manipulate; it has relatively fast convergence and its success in finding the global optima for many known problems has been more than encouraging (Price *et al.*, 2005).

In this chapter, DE is described in the next section, along with a brief discussion on tuning its parameters; both a flowchart and pseudo code for DE are presented for the benefit of readers new to DE. Then, modifications and enhancements made to DE by several researchers are reviewed; this section is useful to researchers considering further enhancements to DE. Finally, reported applications of DE in Chemical Engineering are summarized for the benefit of researchers and practitioners to solve Chemical Engineering problems. The chapter ends with concluding remarks on DE, its developments and applications.

2. Description of DE

An optimization problem with a single objective, f is as follows:

$$\begin{aligned} &\text{minimize } f(x_1, x_2, \ldots, x_N) \\ &\text{with respect to } x_1, x_2, \ldots, x_N \\ &\text{subject to } l_n \leq x_n \leq u_n \quad \text{for } n = 1, 2, \ldots, N \end{aligned} \tag{1}$$

Here, N is the number of decision variables; l_n and u_n are respectively the lower and upper bound on x_n, and they take large negative and positive value respectively for unbounded problems. In addition, an optimization problem may contain inequality constraints (namely, g_k for $k = 1, 2, \ldots, K$) and equality constraints (namely, h_m for $m = 1, 2, \ldots, M$). The bounds on decision variables and constraints define the feasible space for optimization search.

For simplicity, only unconstrained and unbounded problems are considered in the subsequent description of DE. If there are bounds and/or constraints, they have to be handled by a suitable technique; for example, constraints can be handled by the penalty function technique. Note that decision variables (x_n) are referred as "parameters" in parameter estimation problems. Price *et al.* (2005) employ parameters instead of variables in their book devoted to DE, perhaps, in view of the origin of DE for parameter estimation in Chebyshev polynomials. However, for consistency throughout this book and with many other books on optimization, we use decision variables (or, simply, variables) instead of "parameters" in this chapter.

DE is both simple (see the flowchart and pseudo code in Figs. 1 and 2 respectively) and natural (Price *et al.*, 2005). It has its basis in Darwin's natural selection theory, and is similar to genetic algorithms (GAs) except for one important factor: several GAs, particularly earlier versions, encode decision variables as bit strings whereas DE encodes them as floating-point numbers. Hence, GAs are more suited for combinatorial problems, while DE is more adapted to continuous problems. A flowchart of the classic DE algorithm is shown in Fig. 1, and the main steps of this algorithm are as follows.

Initial population: DE uses a population of N_p individuals or points in the search space. Randomly initialize N_p N-dimensional vectors over the specified search space. Each of these vectors, whose elements are the values of decision variables, represents a point in the search space.

Mutation: This is done by selecting two vectors randomly from the current population, finding the difference vector between them, multiplying the difference vector by the mutation factor (F in the range 0 to 2) to obtain the scaled, difference vector, and finally adding the scaled, difference vector to a third randomly chosen population vector to find the mutant vector. Note that the three population vectors involved in mutation must be distinct, and they must also be different from the target vector used in the next step.

Crossover: Another vector is chosen from the population as the target vector. A trial vector is created by getting each variable value from either the mutant vector or the target vector according to the crossover probability (Cr in the range 0 to 1).

Selection: The trial vector is selected for the new population if it has a lower objective function value (for the case of minimization problems); otherwise, the target vector is selected for the new population. This greedy criterion is the selection force for the classic DE. Mutation, crossover and

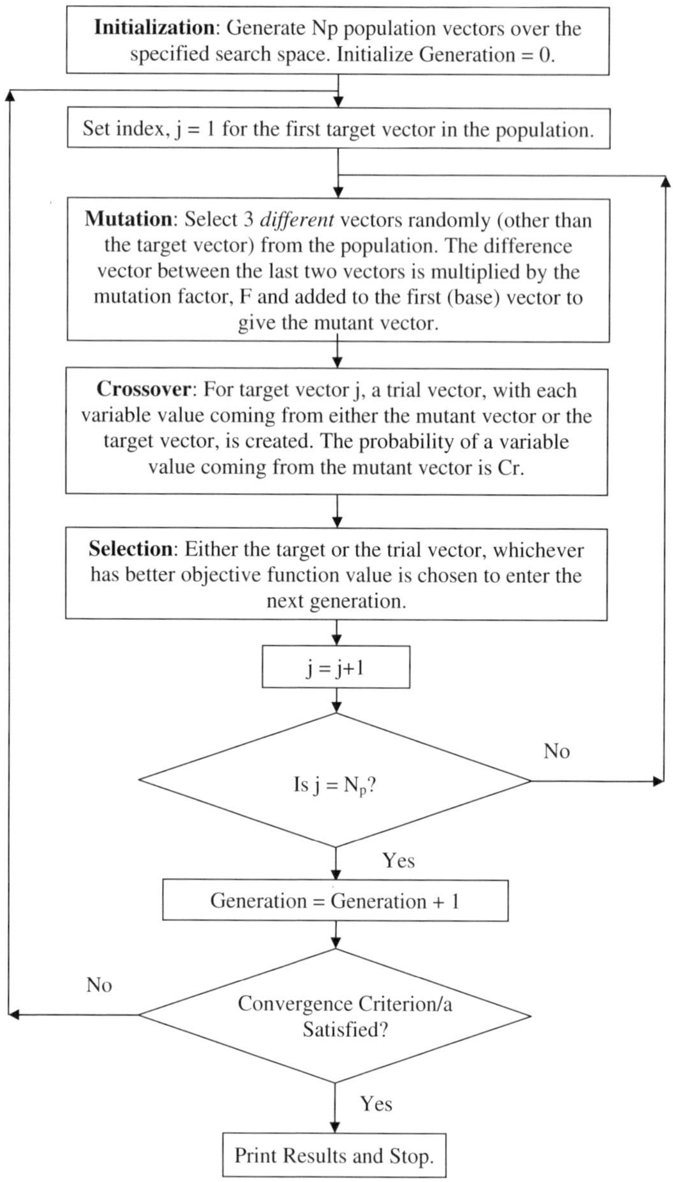

Figure 1. Flow chart for differential evolution; See the pseudo code in Fig. 2 for equations for mutation, crossover and selection operations.

```
// Initialization of population
for j = 1 to Np
for n = 1 to N
  Xj,n = rand(un, ln);   // Selects a random value between upper and lower bounds of nth
                         // variable in the search space specified
end
end
// end of initialization

do
{
for j = 1 to Np
{
  r0 = rand (1:Np); // Select a random integer from 1 to Np for base vector
  r1 = rand (1:Np); // Select a random integer from 1 to Np for difference vector
  r2 = rand (1:Np); // Select a random integer from 1 to Np for difference vector
// Note that r0, r1 and r2 must be distinct, and they also must be different from j.

// Mutation and Crossover
for n = 1 to N
  if rand(0,1) <= Cr
     Un = Xr0,n + F * (Xr1,n − Xr2,n ) ; // Mutates and scales by F followed by crossover
  else
     Un = Xj,n; // U is the trial vector
  end
end
// End of mutation and crossover

// Selection for the next population, Y
if f(U) < f(Xj)
  Yj = U;
else
  Yj = Xj;
end
// End of selection
}
end

X = Y; // Next generation, replace original population with new one
Generation = Generation + 1;
}while (convergence criteria not met)
end
```

Figure 2. Pseudo code for differential evolution.

selection steps are carried out for each vector in the current population as the target vector. This completes one generation in the DE algorithm.

Convergence check and iteration: The above steps of mutation, crossover and selection in a generation are repeated until the specified convergence criteria are satisfied. The usual convergence criteria are the specified number of generations, variance in the new population and error from the known minimum.

DE algorithm has only three parameters: N_p, F and Cr besides initialization of the population, selection of three random vectors in the mutation operation, and the convergence criteria. Price *et al.* (2005) refer these as "control parameters", partly to distinguish "parameters" used for decision variables. Effect of N_p, F and Cr on the performance of DE and recommended values for them, are summarized in Table 1. For understanding the effect of these parameters on DE, consider the modified Himmelblau

Table 1. Parameters in differential evolution and guidelines for their tuning.

Parameter	Effect and guidelines
Population Size, N_p	A higher N_p increases the reliability of finding the global minimum but slows convergence. An increase in N_p can be accompanied by a decrease in F for an improvement in robustness. A minimal N_p is generally needed for good convergence properties but excessively high N_p results in unnecessarily slow convergence without significant decrease in failure rates (Onwubolu and Babu, 2004). A number between 5 to 20 times N is generally recommended for N_p, and a starting value of 10 N can be used (http://www.icsi.berkeley.edu/~storn/code.html).
Mutation Factor, F	F should typically be between 0 and 2. DE is sensitive to F compared to Cr, and so care should be taken in tuning this parameter. A value of around 0.8 is good for most problems (http://www.icsi.berkeley.edu/~storn/code.html), and increasing F increases the reliability of finding the global optimum at the expense of computational effort. Some problems require low F values but such problems are atypical in general (Price *et al.*, 2005).
Crossover Probability, Cr	Cr is a fine-tuning parameter between 0 and 1. Generally, any problem can be solved successfully with either high Cr (0.9 to 1) or low Cr (0 to 0.1). Performance of DE is not very sensitive to either high or low Cr. A starting value of 0.9 is recommended (http://www.icsi.berkeley.edu/~ storn/code.html).

function (Deb, 2002):

$$\text{minimize } f(x) = (x_1^2 + x_2 - 11)^2 + (x_1 + x_2^2 - 7)^2 + 0.1[(x_1 - 3)^2 + (x_2 - 2)^2] \quad (2)$$

with respect to x_1 and x_2

subject to $-6 < x_1 < 6$ and $-6 < x_2 < 6$.

It has four minima with a global minimum of 0.0 at $x_1 = 3.0$ and $x_2 = 2.0$. The three local minima are 1.50435 at (3.58149, −1.82080), 3.48713 at (−2.78706, 3.12820) and 7.36735 at (−3.76343, −3.26605). A computer code to solve this problem by DE is developed based on the DeMAT software of Price *et al.* (2005), and given in the folder of this chapter on the CD accompanying the book. Several exercises are provided at the end of this chapter for readers to try and understand the effect of DE parameters.

The recommended values of $N_p = 10\text{N}$, $F = 0.8$ and $\text{Cr} = 0.9$ (Table 1) should work for most cases. In the event that they do not give a satisfactory result, the following can be tried. If computational efficiency is to be improved, N_p and/or F can be reduced. The parameters in DE are related; so, if one value is being tweaked, the rest should be kept constant as the combined effect of the parameters is often unknown. For this reason, Cr should be best left untouched as DE is more sensitive to F than Cr; moreover, all problems can be solved with a high Cr (0.9) or a low Cr (0.1) (Price *et al.*, 2005). To confirm convergence to the global minimum, N_p and/or F can be increased but at expense of some computational performance. If the parameter tuning fails to solve the problem satisfactorily, alternative strategies such as including jitter and dither (i.e. changing of F values randomly) can be tried; these modifications are discussed later in the chapter.

DE algorithm can be customized, which is important and useful because the classic DE, although simple to understand and implement, lacks the flexibility to deal with various problems. The classic DE (Figs. 1 and 2) just consists of initialization by uniform random distribution, selection of vectors for mutation, crossover by uniform binomial distribution and selection by greedy criterion. It is adequate for general problems but, if the algorithm fails to perform satisfactorily (either in terms of reliability or speed) and the tuning does not help, it becomes imperative to modify the algorithm to enhance performance. Some of these modifications are discussed, together with their uses and limitations, extensively by Price *et al.* (2005), and they are summarized in the subsequent section on 'Developments of DE', together with the newer approaches that many researchers have experimented with in the recent years.

DE algorithm has been studied and coded in many languages including C, MATLAB and java etc. A list of available software can be found on the internet (http://www.icsi.berkeley.edu/~storn/code.html) and on the CD accompanying the book devoted to DE (Price *et al.*, 2005). Readers interested in using DE are advised to look at these comprehensive sources of computer codes.

3. Developments of DE

DE allows certain extent of customizability to permit users to fully utilize it for efficient and reliable solution of their optimization problem if *a priori* information on the problem is available. The important features of DE and their customizability, which highlight the variations of the algorithm proposed over the years, are described in this section. The focus here is on journal papers. Developments up to 2002 can also be found at http://www2.lut.fi/~jlampine/debiblio.htm.

3.1. Initialization

Recall that initialization refers to selection of N_p vectors for the initial population. Table 2 shows the modifications proposed for this initialization since 2002. These can be divided into two groups: (1) distribution, and (2) constraints and formulation.

Distribution: A normal initialization selects N_p vectors randomly over the initial bounds for variables or over an arbitrarily chosen space if no bounds for variables are defined. The choice of vectors normally follows a uniform random distribution but Gaussian or Halton distributions (Price *et al.*, 2005) can be used, and advanced strategies include seeding individuals by opposition based learning (Rahnamayan *et al.*, 2006). Some techniques use parallel distributions/populations and clustering (Wang *et al.*, 2007).

Table 2. Modifications proposed for the initialization step.

Modified part (Reference)	Modification(s)	Remarks
Distribution (Price *et al.*, 2005)	Discussed uniform random, Gaussian and Halton distributions	Uniform random distribution is preferred for their ease of use and generally better performance
Distribution (Wang *et al.*, 2007)	Introduction of concept of clustering. Population is grouped into many subsets initially, and the number of subsets changes with generations for faster convergence	Performance comparison on 28 benchmark functions shows that the algorithm is efficient, especially for high dimension problems
Distribution (Rahnamayan *et al.*, 2008)	Initialization using opposition based learning to obtain fitter starting candidates	Tests on 34 benchmark functions (with 2 to 100 variables) indicate that the proposed algorithm speeds up convergence
Constraints and Formulation (Huang and Wang, 2002)	Introduction of rounding off into the mutation operation, to make DE applicable to integers. Constraints are handled by the augmented Lagrange function	Solved fuzzy decision-making mixed-integer nonlinear programming (MINLP) problems for optimizing multi-product batch chemical plants for multiple objectives
Constraints and Formulation (Lin *et al.*, 2003)	Extends DE to mixed integer problems, using the augmented Lagrange method to deal with constraints	Method better than the conventional penalty method, according to the reported test results
Constraints and Formulation (Munawar and Gudi, 2005)	Converts MINLP into NLP using the "binary condition" so DE can be applied; constraints are handled using penalty function	A deterministic method was used after DE to improve convergence. The proposed hybrid algorithm was tested on 7 case studies, and concluded to be better than integer-DE and NLP-DE

(*Continued*)

Table 2. (*Continued*)

Modified part (Reference)	Modification(s)	Remarks
Constraints and Formulation (Balamurugan and Subramanian, 2008)	Introduction of rounding off to make DE applicable to integers	Implemented with gene swap operator to determine the optimal fuels options for all power generating units
Constraints and Formulation (Wu *et al.*, 2008)	Modifies search space dynamically and cuts down feasible regions to accelerate search	Search space modification was combined with a vaccination operator to develop an improved DE (see Table 5). Test results show that the improved DE has better performance. See Table 7 for an application

Constraints and Formulation: DE was originally proposed for unconstrained continuous variables, but it has since been extended for other problems such as mixed integer problems and handling constraints (Price *et al.*, 2005; Huang and Wang, 2002; Lin *et al.*, 2003; Munawar and Gudi, 2005; Balamurugan and Subramanian, 2008). Initial bounds are usually fixed, but techniques have been developed to modify the search space dynamically (Wu *et al.*, 2008).

3.2. Mutation operation

The mutation operation allows flexibility in choosing the base vector, scaling factor (F) and the mutation algorithm. Improvements and modifications to the mutation process over the years are outlined in Table 3, and briefly summarized thereafter.

Base Vector Selection: There are a few ways to choose the base vector to which the scaled difference vector is added in the mutation operation. One strategy is by a totally random algorithm (Price *et al.*, 2005); although this is relatively simple to code, there are chances that some vectors are picked more frequently than others, which undoubtedly reduces variation. The standard way is to choose by stochastic universal sampling, which

Table 3. Modifications proposed for the mutation process.

Modified part (Reference)	Modification(s)	Remarks
Base Vector Selection, Mutation Operator (Storn, 1996; Storn and Price, 1997)	Strategy for choosing best-so-far vectors and rand-to-best vectors, and concept of adding two vector differences instead of one	These variants of DE are useful under different circumstances, and allow the user to have greater selection of algorithms to use
Base Vector Selection, (Price *et al.*, 2005)	Stochastic universal sampling to provide a more representative population sample	Stochastic universal sampling is now the standard for the random seeding in DE
Base Vector Selection, (Price *et al.*, 2005)	Alternative strategies such as choosing best or random vector as base vector	Discussed extensively on methods to select random vectors, and degeneracy
Base Vector Selection, (Bergey and Ragsdale, 2005)	Selection pressure to develop offspring that have higher chance to survive	For De Jong test suite, modified DE was found to require less number for function evaluations
Base Vector Selection, (Kaelo and Ali, 2006)	Tournament selection among 3 random vectors to select the base vector, and also randomly chosen F	Results on 50 test problems show that the modified DE is considerably faster than the classic DE.
Tuning F (Zaharie, 2002; Price *et al.*, 2005)	Introduction of a minimal F needed for convergence, jittering and dithering (which cause F to be a random variable), and controlling F based on population diversity	Experience with changing F shows that it is beneficial for certain problems
Tuning F (Ali and Torn, 2004)	Two population sets, pre-calculated differentials (which are updated every few generations) and F value based on objective function values	Local search was also implemented in a few modified algorithms. Modifications decrease number of function evaluations by 10–50%.
Tuning F (Liu and Lampinen, 2005)	Fuzzy logic to tune F	Easy to tune F than the normal trial and error approach

(*Continued*)

Table 3. (*Continued*)

Modified part (Reference)	Modification(s)	Remarks
Tuning F (Sarimveis and Nikolakopoulos, 2005)	Line-up DE in which solutions are ranked, and probability of mutation and crossover decreases with fitness values. Only two parameters: N_p and termination criteria	Augmented Lagrangian for handling constraints. Tested on both unconstrained and constrained problems
Tuning F (Qin and Suganthan, 2005; Huang *et al.*, 2006)	Self adapt Cr and F	Dynamically changing F takes out the trouble in the normal tuning process
Tuning F (Omran *et al.*, 2005)	Concept of self adaptive DE, where F is automatically tuned and stored separately for each individual	Outperforms normal DE in all benchmark functions tested
Tuning F (Becerra and Coello, 2006)	Cultured DE: situation knowledge provides the best vector, normative knowledge helps to scale F, and topographical and historical knowledge helps to record local minima etc., to aid in evolution	Cultural algorithms extract domain knowledge during the search, to fine tune evolutionary strategies. For benchmark constrained problems, cultured DE is very efficient compared to another DE and other methods in the literature
Tuning F (Coelho and Mariani, 2007)	Adjusts F adaptively using a sinusoidal function or a Gaussian function or a combination of both	Solution found is better than that found by other methods
Tuning F (Liu *et al.*, 2007)	Immunology algorithms to change F, and modified DE with three sub-populations, each using a different mutation operator	Modified is DE superior to other optimizers such as particle swarm, genetic algorithms and DE, for the two applications studied
Tuning F (Chiou, 2007)	A scaling factor on F based on the 1/5 success rule, and hence changing F dynamically in the hybrid DE of Chiou and Wang (1999)	Modified hybrid DE is better than simulated annealing, genetic algorithm and hybrid DE, for the two applications studied

(*Continued*)

Table 3. (*Continued*)

Modified part (Reference)	Modification(s)	Remarks
Tuning F (Coelho, 2008)	Uses chaotic sequence (Lozi's map) to change the value of F during the optimization	F is dynamic, increasing variance and success rate in finding the global solution.
Tuning F (Nobakhti and Wang, 2008)	Randomized adaptive DE in which a basket of random values of F is first generated and then updated periodically	F is self-adaptive, taking away the pains of selecting a single value. Performance on benchmark problems is promising
Tuning F (Yan, 2008)	Adaptive mutation operator, in which F is scaled by an exponential parameter	Modification overcomes premature convergence. See Table 8 for a ChE application
Tuning F (Yuan *et al.*, 2008)	Chaos theory (logistic map) to tune F and also Cr	DE with self-adaptive parameters found a better solution for the application studied
Mutation Operator (Fan and Lampinen, 2003)	Uses 3 points to create a "difference" vector instead of 2 points. This is called trigonometric DE	Trigonometric DE is found to be faster in most cases
Mutation Operator (Pan *et al.*, 2008)	Discrete DE, which uses a perturbation operator instead of a mutation operator, to adapt it for the discrete domain	Discrete DE without and with local search to further improve performance, was applied to 120 benchmark instances of flowshop scheduling problems

ensures that no base vector gets selected more than once (Price *et al.*, 2005). Other ways to choose the base vector include choosing the best vector (Storn, 1996), using selection pressure to rank and choose the base vector (Bergey and Ragsdale, 2005), and tournament ranking among the vectors to see which one is the base and which ones are the vectors for creating the difference vector (Kaelo and Ali, 2006).

Mutation Factor F: Empirically, the mutation factor F is stated to be a value between 0 and 2 but convergence may be slow if F is more than 1.

Also, there is a minimum F for the search to converge even though selection pressure is absent (Zaharie, 2002; Price *et al.*, 2005). This critical F has been determined to be a function of both N_p and Cr. The norm is to use a constant F; however, recent approaches have used dynamically changing scale factors successfully. It can be changed randomly through dither (different random F for each vector) or jitter (different random F for each variable of each vector) (Zaharie, 2002), according to fuzzy logic (Liu and Lampinen, 2005), according to objective/fitness values (Ali and Torn, 2004; Sarimveis and Nikolakopoulos, 2005), or it can be self adaptive (Qin and Suganthan, 2005; Omran *et al.*, 2005; Becerra and Coello, 2006; Coelho and Mariani, 2007; Nobakhti and Wang, 2008; Yan, 2008). F can be tuned by other concepts such as immunology algorithms (Liu *et al.*, 2007), one fifth rule (Chiou, 2007), or chaos theory (Coelho, 2008; Yuan *et al.*, 2008).

Mutation Algorithm: Instead of taking a difference vector as the form of mutation, other techniques have been suggested. One way is to add two difference vectors together as the mutation operation (Storn, 1996), and another way is to use three points and perform a trigonometric mutation operation (Fan and Lampinen, 2003). To do mutation for discrete domain, the perturbation operator has been suggested instead of the mutation operator (Pan *et al.*, 2008). These variants have success for some cases but the classic DE does not seem to be inferior to them; in fact, it is often more successful on black-box problems due to its ability to map out the topography of the decision space (Price *et al.*, 2005).

3.3. Crossover

Table 4 presents the variations in the crossover step with respect to tuning Cr and operator itself. There have been various attempts to tune Cr automatically and dynamically (Abass, 2002; Qin and Suganthan, 2005; Sarimveis and Nikolakopoulos, 2005; Yuan *et al.*, 2008) even though DE is more sensitive to F than to Cr. Like mutation, crossover is a significant aspect in DE, and researchers have been looking for ways to optimize this operator itself. Classic DE uses a uniform binomial crossover; but one point, N-point and exponential selection have been tried (Price *et al.*, 2005). Other novel strategies include using two populations for preferential

Table 4. Modifications proposed for the crossover process.

Modified part (Reference)	Modification(s)	Remarks
Tuning Cr (Abass, 2002)	Introduction of self adaptive DE, where Cr is automatically tuned and stored separately for each individual	The self adaptive DE was presented for multi-objective optimization problems
Tuning Cr (Qin and Suganthan, 2005)	Self adapt Cr and F	Dynamically changing Cr simplifies the tuning process.
Tuning Cr (Sarimveis and Nikolakopoulos, 2005)	Line-up DE in which solutions are ranked, and probability of mutation and crossover decreases with fitness values. Only two parameters: N_p and termination criteria	Augmented Lagrangian for handling constraints. Tested on both unconstrained and constrained problems
Tuning Cr (Yuan *et al.*, 2008)	Chaos theory (logistic map) to tune Cr and F	DE with self-adaptive parameters found a better solution for the application studied
Crossover Operator (Sun *et al.*, 2005)	Crossover is replaced by the estimation of distribution algorithm (EDA), where a trial solution is generated using the probability model constructed based on the best M solutions	Results on benchmark problems tested show that DE/EDA outperforms DE and EDA
Crossover Operator (Price *et al.*, 2005)	One-point, N-point, exponential and uniform binomial distribution, and also arithmetic recombination in which combination of values is used	Discussed the various ways of selecting the variable to be crossovered, and explained their pros and cons
Crossover Operator (Ali, 2007)	Uses 2 population sets, one auxiliary set to store discarded trial points and to make use of them in preferential crossover	Results on test problems show the algorithm to be more effective than the original DE in terms of number of function evaluations and success rate
Crossover Operator (Balamurugan and Subramanian, 2008)	Introduction of gene swap (i.e. randomly swapping two genes in a chromosome at a given probability) in addition to crossover	Used along with rounding off for integer problems in the optimization of power outputs for all power generating units

crossover (Ali, 2007), using a gene swap algorithm (Balamurugan and Subramanian, 2008), an estimation of distribution algorithm (Sun *et al.*, 2005) and arithmetic recombination (Price *et al.*, 2005).

3.4. Selection

Classic DE selects between the target vector and trial vector using a greedy criterion. Alternatives such as selection by age and tournament selection have been studied (Price *et al.*, 2005). To deal with special problems, other selection strategies have been discussed; multi-objective problems can be handled by Pareto dominance (Babu *et al.*, 2005); and selection by considering constraints and infeasible regions was explored (Lampinen, 2002; Huang *et al.*, 2007; Zhang *et al.*, 2008). Selection procedures based on improving diversity of the solution set have been invented through the use of tabu lists (Srinivas and Rangaiah, 2007a), vaccinations (Wu *et al.*, 2008), selection through norms (Zhu *et al.*, 2005) and simulated annealing (Yan *et al.*, 2006). Table 5 summarizes the proposed modifications to the selection process.

Table 5. Proposed strategies for the selection process.

Reference	Modification(s)	Remarks
Lampinen (2002)	Selection criteria based on concept of Pareto dominance for multi-constrained problems	Extends the use of DE to multi-constrained problems
Price *et al.* (2005)	Discussed selection strategies (by age and/or objective function value) and ways to select (one-to-one or tournament)	Introductory description to selection strategies for optimizers
Babu *et al.* (2005)	DE for multi-objective problems; dominated solutions are removed in the initial population, and Pareto dominance concept is used in the selection step	Multi-objective DE was applied to a styrene reactor (see Table 8)
Zhu *et al.* (2005)	Selection criterion modified to include norm of variables (weights in neural networks) besides objective values	Motivation is to develop neural networks with smaller weights, which have better generalization performance

(Continued)

Table 5. (*Continued*)

Reference	Modification(s)	Remarks
Yan *et al.* (2006)	Introduction of the concept of acceptance based on simulated annealing	Results on standard test suits show that the modified method can escape from local minima better
Huang *et al.* (2007)	A method to handle constraints: two populations, one representing solutions and one representing penalty factors, are used and the two evolve and interact together. So, penalty factors are self-adjusted	A special penalty function is proposed. Results on 6 benchmark problems and 3 engineering problems (all constrained) show that the algorithm is both efficient and robust
Srinivas and Rangaiah (2007a)	Uses a tabu list in the selection step, to improve diversity of accepted solutions	Tests on benchmark, phase equilibrium and parameter estimation problems show that DE with tabu list (DETL) outperforms DE and tabu search. DETL was evaluated on many NLP and MINLP applications by Srinivas and Rangaiah (2007c)
Wu *et al.* (2008)	Vaccination operator, which takes characteristics of excellent individuals and imparts them to others based on probability	By using vaccines for selection, the degenerate phenomenon is reduced. See Table 7 for application to a parameter estimation problem
Zhang *et al.* (2008)	Introduction of dynamic stochastic selection (DSS), where stochastic ranking balances search between feasible and infeasible regions based on probabilities	Effectiveness of DE with DSS for constrained optimization is shown on CEC'06 test functions and four engineering problems

3.5. Hybrid methods and new generation

Several studies have proposed hybrid methods by combining DE with other methods as well as modifications to the generation step. These novel methods are summarized in Table 6.

Hybrid Methods: Several hybrid methods combine DE with other optimizers in sequence. The most common are those that make use of

Table 6. Hybrid methods and changes to continuation of generations.

Modified Part (Reference)	Modification(s)	Remarks
Hybrid Method (Hendtlass, 2001)	Combines DE and particle swarm optimization, with each of them in alternate steps	Hybrid algorithm outperforms its components on most cases
Hybrid Method (Chiou *et al.*, 2004)	Combines ant colony search with DE, doing the former in conjunction with DE operators	Results show that the modified algorithm is better in terms of solution precision
Hybrid Method (Schmidt and Thierauf, 2005)	Combines DE with threshold acceptance algorithm (which is similar to simulated annealing)	Tackles nonlinear and non-convex problems well
Hybrid Method (Yang *et al.*, 2007)	Introduction of a neighborhood search (like local search) to DE	Searches using a random Gaussian number
Hybrid Method (Kaelo and Ali, 2006)	Localization involving reflection and contraction during the selection step, to intensify search	Numerical results indicate that the proposed change improves DE substantially
Hybrid Method (Bhat *et al.*, 2006)	Simplex search at the end of DE, to improve convergence	Hybrid method was twice as fast as DE for a ChE application (Table 7)
Hybrid Method (Chang and Low, 2008)	Ant colony search is used to select one of the five mutation operations in the hybrid DE (with acceleration and migration steps). Also, an orthogonal array technique is used to find a good initial point for search	The proposed algorithm was applied to an electrical engineering problem
Hybrid Method (Srinivas and Rangaiah, 2007b)	Includes a local search (Quasi Newton method) after DE	Improves computational efficiency of the traditional DE
Hybrid Method (Omran *et al.*, 2008)	Combines DE with bare bones particle swarm optimizer; attractor points are explored using the difference vector operator. It has the advantage of requiring very little parameters tuning	The proposed method performed well compared to other approaches for benchmark functions, noise in objective function and 100-variable problems. A clustering algorithm based on the hybrid method was proposed and applied to three types of imagery data

(*Continued*)

Table 6. (*Continued*)

Modified Part (Reference)	Modification(s)	Remarks
Hybrid Method (He *et al.*, 2008)	GA is the main algorithm here, while DE and sequential quadratic programming are used for fine tuning the solution obtained by GA	Effectiveness of the proposed algorithm was illustrated on two applications, and compared with the results by other methods
Hybrid Method (Yang *et al.*, 2007)	Neighborhood search in the mutation operation of DE, and self adapting parameters	The proposed method was used within a new cooperative co-evolution framework to solve benchmark functions with up to 1000 variables
New Generation (Chiou and Wang, 1999)	Hybrid DE with additional acceleration phase (for faster convergence) and migration phases (for escaping from local minima)	Hybrid DE was evaluated on several test problems, and then used to solve a fed-batch fermentation process (see Table 8)
New Generation (Babu and Angira, 2006)	Maintains one array of population instead of two, and the population is updated as soon as a better solution is found	Tested on test functions and application problems (see Table 8). See Srinivas and Rangaiah (2007a, c) for comprehensive evaluation of this modified DE

neighborhood search or deterministic searches (Kaelo and Ali, 2006; Bhat *et al.*, 2006; Srinivas and Rangaiah, 2007b; Yang *et al.*, 2007, 2008). In addition, DE was used in conjunction with particle swarm optimization (Hendtlass, 2001; Omran *et al.*, 2008), ant colony search (Chiou *et al.*, 2004; Chang and Low, 2008), threshold acceptance algorithm (Schmidt and Thierauf, 2005), genetic algorithms and sequential quadratic programming (He *et al.*, 2008).

New Generation: Two main changes proposed to the classic DE are: one approach suggests updating the new generation simultaneously and dynamically (Babu and Angira, 2006), and another approach suggests obtaining the new generation by migration and/or acceleration phases depending on whether speed or reliability is desired (Chiou and Wang, 1999). Both these approaches have been evaluated and used on many applications.

4. Chemical Engineering Applications

With its ease of use, DE is a good choice for a general purpose optimizer. It has been used in many Chemical Engineering applications, which can be classified into two types:

(1) Model building and parameter estimation, where the aim is to reduce model error and then to control a process or to determine unknown coefficients. This includes neural network training, regression analysis and other modeling methods.
(2) Process optimization, where the goal is to maximize yield or maximize profit by determining the best operating conditions. Decision variables are mainly Chemical Engineering variables (flow rate, temperature etc.)

Tables 7 and 8 summarize the Chemical Engineering problems where DE has been applied successfully. In addition, Srinivas and Rangaiah (2007c) evaluated DE with tabu list (DETL) on many application problems taken from the literature, which are either NLPs or MINLPs. Their results show that DETL requires, on average, about 40% fewer objective function evaluations compared to both DE and modified DE of Babu and Angira (2006).

5. Conclusions

Since its inception in 1995, DE has undergone many changes and developments, and is now one of the effective global optimization techniques. This direct search algorithm, based on nature's evolution, has the benefit of being simple and instinctive for beginners to understand, and yet powerful and flexible for researchers to customize for their applications. DE code and concepts are readily available on the internet (http://www.icsi.berkeley.edu/~storn/code.html), and it is good as a general, global optimizer with a few parameters. With these advantages, it is not surprising that DE has been used in many applications in Chemical Engineering, where the bulk of the problems involve continuous variables, which is what DE is adept at. Furthermore, developments in DE, summarized in this chapter, have improved its capabilities to tackle a variety of problems (namely, constrained, mixed integer, multi-objective and large dimensioned), provided modifications which can be customized to suit the problem, enhanced the speed and reliability of the method, and made tuning of its parameters easier. Given

Table 7. Applications of DE in modeling and parameter estimation.

Application (Reference)	Objective	Selected/Independent decision variables	Remarks
Estimation of heat transfer parameters in a trickle bed reactor (Babu and Sastry, 1999)	Sum of squared errors	Thermal conductivity of bed, wall heat transfer coefficient, effective diameter of packing, liquid and gas flow rates	DE is faster and more reliable compared to radial temperature profile method
Parameter estimation problems in fermentation using high ethanol-resistant yeast (Wang and Sheu, 2000)	Mean squared errors in a weighted min-max problem	Eleven parameters which include maximum specific growth rate, maximum specific ethanol production rate, saturation, inhibition and yield coefficients	Uses hybrid DE which includes acceleration and migration operations
Modeling (1) true boiling point curve, and (2) effect of pressure on entropy, of crude oil (Chen *et al.*, 2002)	Sum of squared errors	Weights in the feedforward neural network models	Uses an improved DE which includes Levenberg-Marquardt and random perturbation strategies
Modeling of process control problems by fuzzy cognitive maps (Papargeorgiou and Groumpos, 2005)	Sum of the exceeded bounds	Weights of fuzzy cognitive maps	DE is found to be robust, effective and efficient for the problems studied
Determination of kinetic parameters in fixed-film bio-reactors (Kiranmai *et al.*, 2005)	Errors between predicted and actual values	Five kinetic and other parameters	Model simplifications could lead to erroneous results. DE is faster than GA for this application

(*Continued*)

Table 7. *(Continued)*

Application (Reference)	Objective	Selected/Independent decision variables	Remarks
Modeling recombinant *E. Coli* growth (Ko *et al.*, 2006)	Time-weighted mean squared errors	Several kinetic and other parameters in S-system and Monod models	Uses hybrid DE (with acceleration and migration operations). Additional experiments were performed to validate the models.
Biofilm modeling — biodegradation of phenol case study (Bhat *et al.*, 2006)	Errors between predicted and actual ones	Biofilm thickness, kinetic and other parameters in the biofilm and kinetic models	Simplex method at the end, improves the efficiency of DE
Macro-kinetic model for oxidation of p-xylene (Yan, 2007)	Mean of squared, relative errors	Weights and thresholds in the neural network model	Four rate constants are modeled by an artificial neural network
First order liquid phase reaction and catalytic gas oil reaction (Angira and Santosh, 2007)	Fractional difference variation	A few rate constants	Trigonometric DE was found to be faster than DE
Estimation of biomass and intracellular protein in *E. Coli* (Ko and Wang, 2007)	Mean squared errors	Tunable parameters in the model	Hybrid DE of Chiou and Wang (1999) was used
Fuzzy inference systems for modeling four-effect evaporator and continuous stirred tank reactor (Eftekhari *et al.*, 2008)	Mean squared errors	Membership function width and position	DE was superior to ANFIS, a well-known technique for fuzzy technique

(Continued)

Table 7. (*Continued*)

Application (Reference)	Objective	Selected/Independent decision variables	Remarks
Estimation of apparent thermal diffusivity during drying of banana (Mariani *et al.*, 2008)	Differences between predicted and actual temperatures	Three to four parameters in the expression for thermal diffusivity as a function of temperature and moisture content	DE was successful in this application
Controller design for the ALSTOM gasifier (Nobakhti and Wang, 2008)	Integral of absolute error	20 tuning parameters in the proportional-integral controllers	DE with an adaptive mutation factor (F) was used
Rate expression for sulfur dioxide oxidation, and diesel catalytic cracking reaction (Wu *et al.*, 2008)	Error between predicted and actual values	Six parameters in sulfur dioxide rate expression and three kinetic parameters in cracking reaction	Immune DE for these problems was faster and more reliable than the classic DE
Soft sensor for naphtha 95% cut point in crude distillation unit (Yan, 2008)	Mean of squared errors	Parameters in the ridge regression estimator	DE with modified mutation operator was employed

Table 8. Applications of DE in process optimization.

Application (Reference)	Objective	Selected/Independent decision variables	Remarks
Fed-batch fermentation process (Chiou and Wang, 1999)	Maximize ethanol production	Glucose feed concentration and flow rate, initial glucose concentration, initial volume and fermentation time	Using hybrid DE (with migration and acceleration operations) and multiplier updating method for constraints
Multi-product batch chemical plant — two examples (Huang and Wang, 2002)	Maximize revenue, and minimize investment, operation cost and total production time, simultaneously	Number and size of parallel equipment at each stage, and location and size of storage	The multi-objective optimization (MOO) problem was converted into a fuzzy goal optimization problem, which was solved using mixed integer hybrid DE
Optimal control problems in (1) continuous stirred tank reactor and (2) bi-functional catalyst blending (Cruz *et al.*, 2003)	(1) Minimize squared deviation from steady-state temperature and concentration and control effort; (2) maximize concentration in the reactor	Trajectory of cooling fluid flow rate in problem 1, and trajectory of mass fraction of the hydrogenation catalyst in problem 2	DE was concluded to be reliable and relatively efficient compared to breeder genetic algorithms and iterative dynamic programming
Two-stage fermentor with cell recycling and extractor after the first stage, for lactic acid production (Wang and Lin, 2003)	Maximize lactic acid productivity and glucose conversion simultaneously	Biomass, glucose and lactic acid concentrations, feed glucose concentration, dilution rate, flow rate and bleed ratio and volume ratios	The MOO problem was converted into a fuzzy goal optimization problem, which was solved using hybrid DE. Inclusion of extractor improves the process

(*Continued*)

Table 8. *(Continued)*

Application (Reference)	Objective	Selected/Independent decision variables	Remarks
Beer fermentation process (Wang, 2003)	Maximize ethanol production and minimize fermentation time simultaneously	Trajectory of cooling rate, concentrations of three sugars (glucose, maltose and maltotriose) and fermentation time	The MOO problem was converted into a fuzzy goal optimization problem, which was solved using hybrid DE
Fermentation process for ethanol, and simultaneous saccharification and fermentation (Kapadi and Gudi, 2004)	Maximize ethanol production rate, and maximize lactic acid production rate	Trajectory of feed rate (in both applications), feed concentration of glucose, initial concentration of glucose and initial working volume (in ethanol process only)	Non-uniform control vector parameterization was used. Case of two feeds was also considered
Adiabatic styrene reactor (Babu *et al.*, 2005)	Maximize yield, selectivity, and productivity of styrene	Ethyl benzene feed flow rate and temperature, inlet pressure and steam to reactant ratio	Multi-objective DE was proposed and used for solving the MOO problem
Seven process synthesis and design problems in the literature (Angira and Babu, 2006)	Maximize product yield or maximize profit or minimize costs	Reactor volumes, number of units, in flow, out flow, temperature and composition	MINLP problems solved by the modified DE
Alkylation process operation (Babu and Angira, 2006)	Maximize profit	Olefin feed rate, acid addition rate, alkylate yield, acid strength, motor octane number, isobutane to olefin ratio	Solved by the modified DE

(Continued)

Table 8. (*Continued*)

Application (Reference)	Objective	Selected/Independent decision variables	Remarks
Heat exchanger network design (Babu and Angira, 2006)	Minimize the total heat exchange area	Area of each of the 3 heat exchangers, and 2 exit temperatures	Solved by the modified DE
Two CSTRs in series for reaction A→B→C (Babu and Angira, 2006)	Maximize product concentration	Volumes of the two CSTRs	Two local minima comparable to the global minimum value
(1) Isothermal CSTR, and (2) batch reactor for A→B→C (Babu and Angira, 2006)	Maximize (1) product yield, and (2) intermediate product	(1) Holding time and temperature, and (2) trajectory of reaction temperature	Solved by the modified DE
Fed-batch fermentor for protein production using *E. Coli* (Ko and Wang, 2006)	Maximize protein production rate	Feed rate trajectory, feed glucose concentration and fermentation time	Hybrid DE was used, and two additional experiments were conducted for validation
(1) Plug flow reactor for A→B and A→C, and (2) Batch reactor for A→B→C (Angira and Santosh, 2007)	Maximize yield of product B	Trajectory of (1) the control variable, and (2) reaction temperature	Used trigonometric mutation in DE
Plug flow reactor catalyst blend (Angira and Santosh, 2007)	Maximize yield	Composition of catalyst, mole fraction of components	Used trigonometric mutation in DE
Bi-functional catalyst for hydroisomerization of methyl cyclo-pentane (Angira and Santosh, 2007)	Maximize yield	Composition of catalyst, mole fraction of components	Multi-modal problem, solved by trigonometric mutation in DE

(*Continued*)

Table 8. (*Continued*)

Application (Reference)	Objective	Selected/Independent decision variables	Remarks
Fed-batch bioreactor for secreted protein (Angira and Santosh, 2007)	Maximize a function of product yield	Substrate feed rate profile, concentration of components	Used trigonometric mutation in DE
Phase equilibrium (PE) and phase stability (PS) problems (Srinivas and Rangaiah, 2007a, b)	Minimize Gibbs free energy (for PE), and tangent plane distance function (for PS)	Moles of each component in each phase (for PE), and mole fraction of each component (for PS)	DE was shown to be more reliable but computationally less efficient compared to tabu search
Shell and tube heat exchanger design (Babu and Munawar, 2007)	Minimize cost of exchanger	Tube diameter, length, pitch, passes, shell head type, baffle spacing and cut	DE faster than GA for the design problem studied
Biocompatible solvent design for extractive ethanol fermentation process (Cheng and Wang, 2007)	Maximize extraction of ethanol	Solvent mass flow rate and binary variables for appearance and position of a structural group in the molecule	The mixed integer nonlinear programming problem was solved by mixed integer hybrid DE
Lactic acid recovery by esterification in a CSTR and hydrolysis in a reactive distillation (Rahman *et al.*, 2008)	Minimize total annual cost	CSTR temperature, number of stages, reflux ratio, feed location, number and position of reactive stages, catalyst mass	The problem involves many nonlinear, equality constraints solved by Newton-Raphson method
Heat exchanger network (HEN) (Yerramsetty and Murty, 2008)	Minimize total annual cost	String vector (for arrangement of exchangers), heat load, heat capacity flow rates of split streams	Solved five case studies taken from the literature, and found better HENs in some cases

these developments and merits, DE is attractive and useful both as a simple general optimizer and as a sophisticated tool to solve Chemical Engineering problems.

References

Abass, H. (2002). The self adaptive Pareto differential evolution algorithm. *Proceedings of IEEE Congress on Evolutionary Computation*, pp. 831–836. Honolulu, USA, IEEE Press.

Ali, M.M. (2007). Differential evolution with preferential crossover. *European Journal of Operational Research*, **181**, pp. 1137–1147.

Ali, M.M. and Torn, A. (2004). Population set-based global optimization algorithms: some modifications and numerical studies. *Computers and Operations Research*, **31**, pp. 1703–1725.

Angira, R. and Babu, B.V. (2006). Optimization of process synthesis and design problems: A modified differential evolution approach. *Chemical Engineering Science*, **61**, pp. 4707–4721.

Angira, R. and Santosh, A. (2007). Optimization of dynamic systems: A trigonometric differential evolution approach. *Computers and Chemical Engineering*, **31**, pp. 1055–1063.

Babu, B.V. and Sastry, K.K.N. (1999). Estimation of heat transfer parameters in a trickle bed reactor using differential evolution and orthogonal collocation. *Computers and Chemical Engineering*, **23**, pp. 327–339.

Babu, B.V. and Angira, R. (2006). Modified differential evolution (MDE) for optimization of nonlinear chemical processes. *Computers and Chemical Engineering*. **30**, pp. 989–1002.

Babu, B.V. and Munawar, S.A. (2007). Differential evolution strategies for optimal design of shell and tube heat exchangers. *Chemical Engineering Science*, **62**, pp. 3720–3739.

Babu, B.V., Chakole, P.G. and Mubeen, J.H.S. (2005). Multi-objective differential evolution (MODE) for optimization of adiabatic styrene reactor. *Chemical Engineering Science*, **60**, pp. 4822–4837.

Balamurugan, R. and Subramanian, S. (2008). Hybrid integer coded differential evolution — dynamic programming approach for economic load dispatch with multiple fuel options. *Energy Conversion and Management*, **49**, pp. 608–614.

Becerra, R.L. and Coello, C.A.C. (2006). Cultured differential evolution for constrained optimization. *Computational Methods in Applied Mechanical Engineering*, **195**, pp. 4303–4322.

Bergey, P.K. and Ragsdale, C. (2005). Modified differential evolution: A greedy random strategy for genetic recombination, *Omega 33*, pp. 255–265.

Bhat, T.R., Venkataramani, D., Ravi, V. and Murty, C.V.S. (2006). An improved differential evolution method for efficient parameter estimation in biofilter modeling. *Biochemical Engineering Journal*, **28**, pp. 167–176.

Chang, Y.P. and Low, C.Y. (2008). An ant direction hybrid differential evolution heuristic for the large-scale passive harmonic filters planning problem. *Expert Systems with Applications*, **35**, pp. 894–904.

Chen, C.W., Chen, D.Z. and Cao, G.Z. (2002). An improved differential evolution algorithm in training and encoding prior knowledge into feedforward networks with application in chemistry. *Chemometrics and Intelligent Laboratory Systems*, **64**, pp. 27–43.

Cheng, H.C. and Wang, F.S. (2007). Trade-off optimal design of a biocompatible solvent for an extractive fermentation process. *Chemical Engineering Science*, **62**, pp. 4316–4324.

Chiou, J.P. and Wang, F.S. (1999). Hybrid method of evolutionary algorithms for static and dynamic optimization problems with application to a fed batch fermentation process. *Computers and Chemical Engineering*, **23**, pp. 1277–1291.

Chiou, J.P., Chang, C.F. and Su, C.T. (2004). Ant direction hybrid differential evolution for solving large capacitor placement problems. *IEEE Transactions on Power Systems*, **19**(4), pp. 1794–1800.

Chiou, J.P. (2007). Variable scaling hybrid differential evolution for large scale economic dispatch problems. *Electric Power Systems Research*, **77**, pp. 212–218.

Coelho, L.D.S. (2008). Reliability — Redundancy optimization by means of a chaotic differential evolution approach. *Chaos, Solitons and Fractals*, **41**, pp. 594–602.

Coelho, L.D.S. and Mariani, V.C. (2007). Improved differential evolution algorithms for handling economic dispatch optimization with generator constraints. *Energy Conversion and Management*, **48**, pp. 1631–1639.

Cruz, I.L.L., Willigenburg, L.G.V. and Straten, G.V. (2003). Efficient differential evolution algorithms for multi-modal optimal control problems. *Applied Soft Computing*, **3**, pp. 97–122.

Deb, K. (2002). *Optimization for Engineering Design: Algorithms and Examples*, Prentice-Hall of India, New Delhi.

Eftekhari, M., Katebi, S.D., Karimi, M. and Jahanmiri, A.H. (2008). Eliciting transparent fuzzy model using differential evolution. *Applied Soft Computing*, **8**, pp. 466–476.

Fan, H.Y. and Lampinen, J. (2003). A trigonometric mutation operation to differential evolution. *Journal of Global Optimization*, **27**, pp. 105–129.

He, D.K., Wang, F.L. and Mao, Z.Z. (2008). A hybrid genetic algorithm approach based on differential evolution for economic dispatch with valve-point effect. *Electrical Power and Energy Systems*, **30**, pp. 31–38.

Hendtlass, T. (2001). A combined swarm differential evolution algorithm for optimization problems. *Lecture Notes in Computer Science, Proceedings of the 14th International Conference on Industrial and Engineering Applications of Artificial Intelligence and Expert Systems*, **2070** pp. 11–18.

Huang, H.J. and Wang, F.S. (2002). Fuzzy decision-making design of chemical plant using mixed integer hybrid differential evolution. *Computers and Chemical Engineering*, **26**, pp. 1649–1660.

Huang, V.L., Qin, A.K. and Suganthan, P.N. (2006). Self adaptive differential evolution algorithm for constrained real parameter optimization. *IEEE Congress on Evolutionary Computation*, pp. 17–24.

Huang, F.Z., Wang, L. and He, Q. (2007). An effective co-evolutionary differential evolution for constrained optimization. *Applied Mathematics and Computation*, **186**, pp. 340–356.

Kaelo, P. and Ali, M.M. (2006). A numerical study of some modified differential evolution algorithms. *European Journal of Operational Research*, **189**, pp. 1176–1184.

Kapadi, M.D. and Gudi, R.D. (2004). Optimal control of fed batch fermentation involving multiple feeds using differential evolution. *Process Biochemistry*, **39**, pp. 1709–1721.

Kiranmai, D., Jyothirmai, A. and Murty, C.V.S. (2005). Determination of kinetic parameters in fixed film bio reactors: an inverse problem approach. *Biochemical Engineering Journal*, **23**, pp. 73–83.

Ko, C.L. and Wang, F.S. (2006). Run-to-run fed-batch optimization for protein production using recombinant *Escherichia coli*. *Biochemical Engineering Journal*, **30**, pp. 279–285.

Ko, C.L. and Wang, F.S. (2007). On-line estimation of biomass and intracellular protein for recombinant *Escherichia coli* cultivated in batch and fed-batch modes. *Journal of the Chinese Institute of Chemical Engineers*, **38**, pp. 197–203.

Ko, C.L., Wang, F.S., Chao, Y.P. and Chen, T.W. (2006). S-system approach to modeling recombinant *Escherichia coli* growth by hybrid differential evolution with data collocation. *Biochemical Engineering Journal*, **28**, pp. 10–16.

Lampinen, J. (2002). Multi-constrained nonlinear optimization by the differential evolution algorithm, *6th Online World Conference on Soft Computing in Industrial Applications*.

Lin, Y.C., Huang, K.S. and Wang, F.S. (2003). An evolutionary Lagrange method for mixed integer constrained optimization problems. *Engineering Optimization*, **35**(3), pp. 267–284.

Liu, J.H. and Lampinen, J. (2005). A fuzzy adaptive differential evolution algorithm, *Soft Computing*, **9**(6), pp. 428–462.

Liu, B., Lu, J., Wang, Y. and Tang, Y. (2007). An effective parameter extraction method based on memetic differential evolution algorithm. *Microelectronics Journal*, **39**, pp. 1761–1769.

Mariani, V.C., Lima, A.G.B.D. and Coelho, L.D.S. (2008). Apparent thermal diffusivity estimation of the banana during drying using the inverse method. *Journal of Food Engineering*, **85**, pp. 569–579.

Munawar, S.A. and Gudi, R.D. (2005). A nonlinear transformation based hybrid evolutionary method for MINLP solution. *Chemical Engineering Research and Design*, **83**(A10), pp. 1218–1236.

Nobakhti, A. and Wang, H. (2008). A simple self-adaptive differential evolution algorithm with application on the ALSTOM gasifier, *Applied Soft Computing*, **8**, pp. 350–370.

Omran, M.G.H., Salman, A. and Engelbrecht, A.P. (2005). Self adaptive Differential Evolution. *Computational Intelligence and Security* 2005, Part I, LNAI 3801, pp. 192–199, Springer, Berlin.

Omran, M.G.H., Engelbrecht, A.P. and Salman, A. (2008). Bare bones differential evolution. *European Journal of Operational Research*, **196**, pp. 128–139.

Onwubolu, G.C. and Babu, B.V. (2004). *New Optimization Techniques in Engineering*, Heidelberg, Germany, Springer-Verlag.

Pan, Q.K., Tasgetiren, M.F. and Liang, Y.C. (2008). A discrete differential evolution algorithm for the permutation flowshop scheduling problem. *Computers and Industrial Engineering*, **55**, pp. 795–816.

Papageorgiou, E.I. and Groumpos, P.P. (2005). A new hybrid method using evolutionary algorithms to train fuzzy cognitive maps. *Applied Soft Computing*, **5**, pp. 409–431.

Price, K.V., Storn, R.M. and Lampinen, J.A. (2005). *Differential Evolution — A Practical Approach to Global Optimization*, Springer, Germany.

Qin, A.K. and Suganthan, P.N. (2005). Self-adaptive differential evolution algorithm for numerical optimization. *IEEE Congress on Evolutionary Computation*, **2**, pp. 1785–1791.

Rahman, I., Ahmad, A., Kumar, P. and Kulkarni, B.D. (2008). Optimization of a continuous process for the recovery of lactic acid using differential evolution algorithm. *Chemical Product and Process Modeling*, **3**(1), Article 6.

Rahnamayan, S., Tizhoosh, H.R. and Salama, M.M.A. (2008). Opposition versus randomness in soft computing techniques. *Applied Soft Computing*, **8**, pp. 906–918.

Sarimveis, H. and Nikolakopoulos, A. (2005). A line up evolutionary algorithm for solving nonlinear constrained optimization problems. *Computers and Operations Research*, **32**, pp. 1499–1514.

Schmidt, H. and Thierauf, G. (2005). A combined heuristic optimization technique. *Advances in Engineering Software*, **36**, pp. 11–19.

Srinivas, M. and Rangaiah, G.P. (2007a). Differential Evolution with tabu list for global optimization and its application to phase equilibrium and parameter estimation problems. *Industrial and Engineering Chemistry Research*, **46**, pp. 3410–3421.

Srinivas, M. and Rangaiah, G.P. (2007b). A study of differential evolution and tabu search for benchmark, phase stability and phase equilibrium problems. *Computers and Chemical Engineering*, **31**, pp. 760–772.

Srinivas, M. and Rangaiah, G.P. (2007c). Differential evolution with tabu list for solving nonlinear and mixed-integer nonlinear programming problems. *Industrial and Engineering Chemistry Research*, **46**, pp. 7126–7135.

Storn, R. (1996). On the usage of differential evolution for function optimization, *Proceedings of the 1996 Biennial Conference of the North American Fuzzy Information Processing Society*, June 19–22, Berkeley, CA, USA, pp. 519–523, IEEE Press, New York.

Storn, R. and Price, K.V. (1997). Differential evolution — a simple and efficient heuristic for global optimization over continuous spaces. *Journal of Global Optimization*, **11**, pp. 341–359.

Sun, J.Y., Zhang, Q.F. and Tsang, E.P.K. (2005). DE/EDA: A new evolutionary algorithm for global optimization. *Information Sciences*, **169**, pp. 249–262.

Wang, F.S. (2003). Fuzzy goal attainment of a beer fermentation process using hybrid differential evolution. *European Symposium on Computer Aided Chemical Engineering*, **13**, pp. 1139–1144, Elsevier Science.

Wang, F.S. and Sheu, J.W. (2000). Multi-objective parameter estimation problems of fermentation processes using a high ethanol tolerance yeast. *Chemical Engineering Science*, **55**, pp. 3685–3695.

Wang, F.S. and Lin, K.J. (2003). Performance analysis and fuzzy optimization of a two-stage fermentation process with cell recycling including an extractor for lactic acid production. *Chemical Engineering Science*, **58**, pp. 3753–3763.

Wang, Y.J., Zhang, J.S. and Zhang, G.Y. (2007). A dynamic clustering based differential evolution algorithm for global optimization *European Journal of Operational Research*, **183**, pp. 56–73.

Wu, Y.L., Lu, J.G. and Sun, Y.X. (2008). An improved differential evolution for optimization of chemical process. *Chinese Journal of Chemical Engineering*, **16**(2), pp. 228–234.

Yan, X.F. (2007). Data mining micro-kinetic approach based on ANN and its application to model industrial oxidation of p-xylene to terephthalic acid. *Chemical Engineering Science*, **62**, pp. 2641–2651.

Yan, X.F. (2008). Modified nonlinear generalized ridge expression and its application to develop naphtha cut point soft sensor. *Computers and Chemical Engineering*, **32**, pp. 608–621.

Yan, J.Y., Ling, Q. and Sun, D.M. (2006). A differential evolution with simulated annealing updating method. *International Conference on Machine Learning and Cybernetics*, pp. 2103–2106.

Yang, Z., He, J. and Yao, X. (2007). Making a difference to differential evolution, chapter in Michalewicz, Z. and Siarry, P. (eds.), *Advances in Metaheuristics for Hard Optimization*, pp. 397–414, Springer.

Yang, Z., Tang, K. and Yao, X. (2008). Large scale evolutionary optimization using cooperative coevolution. *Information Sciences*, **178**, pp. 2985–2999.

Yerramsetty, K.M. and Murty, C.V.S. (2008). Synthesis of cost-optimal heat exchanger networks using differential evolution. *Computers and Chemical Engineering*, **32**, pp. 1861–1876.

Yuan, X.H., Zhang, Y.C., Wang, L. and Yuan, Y.B. (2008). An enhanced differential evolution algorithm for daily optimal hydro generation scheduling. *Computers and Mathematics with Applications*, **55**, pp. 2458–2468.

Zaharie, D. (2002). Critical values for the control parameters of differential evolution algorithms, *Proceedings of MENDEL 2002, 8th International Conference on Soft Computing*, pp. 62–67, June 5–7, Brno, Czech Republic.

Zhang, M., Luo, W.J. and Wang, X.F. (2008). Differential evolution with dynamic stochastic selection for constrained optimization, *Information Sciences*, **178**, pp. 3043–3074.

Zhu, Q.Y., Qin, A.K., Suganthan, P.N. and Huang, G.B. (2005). Evolutionary extreme learning machine. *Pattern Recognition*, **38**, pp. 1759–1763.

6. Exercises

For the following exercises, use the code *Rundeopt.m* in the folder of this chapter on the CD accompanying the book for solving the modified Himmelblau function (Eq. 2). This and other m-files in the folder are based on the DeMAT software of Price *et al.* (2005). Matlab software is required for running these m-files. Suggested changes in the tuning parameters will have to be made in *Rundeopt.m* file and saved before running. The animated plot produced by this code shows how the population in each generation varies after the mutation and crossover operations. The red circles represent the population vectors. Run the code several times and observe how DE searches the terrain. Record the success rate of the algorithm in finding the global minimum and the average number of iterations it takes.

(1) Try $N_p = 5, 10, 20, 25$ and 50, and run the code several times for each setting. What happens to the success rate and the speed of convergence? Does the search gets trapped in a local minimum? What is the best setting which achieves a balance between the success rate and speed of convergence? Is the recommended tuning guideline of $N_p = 2N$ reasonable?

(2) To understand the effect of F, try a very low value of F (e.g. 0.1). Use the best setting for N_p found in the previous exercise. Does convergence happens? Also, try very large values of F (e.g. 1.5 and 2). Is convergence fast or slow? Is the recommended F ≈ 0.8 a good choice?

(3) Choose a low value of Cr, say 0.1 along with $N_p = 20$ and F $= 0.8$. Note the movement of the points (red circles) in the search domain. Do they move a lot, or is it much less than the case when Cr is high? Is convergence fast or painfully slow? What does this tell about the use of Cr in increasing variation?

(4) Try initializing in an area which does not include the global minimum, for example from 0 to -6 for both variables. Is it still possible for convergence to the global minimum or does it happen at a local minimum? Is DE able to search outside its initial space? What is the success rate of locating the global minimum as compared to the original case where initialization is done over the whole feasible space?

(5) Change the value to reach (F_VTR in the code) to a bigger number such as 0.01 instead of 0.000001, and solve the problem several times. Does DE algorithm converge significantly faster? What does this show about DE? Is the algorithm built to find an accurate answer? If not, suggest ways to find accurate answer efficiently.

Chapter 7

ANT COLONY OPTIMIZATION: DETAILS OF ALGORITHMS SUITABLE FOR PROCESS ENGINEERING

V. K. Jayaraman

Evolutionary Computing Group
Center for Development of Advance Computing
Pune University Campus, Pune-411008, India
jayaramanv@cdac.in

P. S. Shelokar, P. Shingade, V. Pote, R. Baskar and B. D. Kulkarni[†]

Chemical Engineering & Process Development Division
National Chemical Laboratory, Pune 411008, India
[†]*bd.kulkarni@ncl.res.in*

1. Introduction

Process modeling and control applications have significantly benefited by adopting artificial intelligence methods, machine learning algorithms and soft computing paradigms (Wilson and Martinez, 1997). Recently, optimization strategies that do not require derivative information have gained wider acceptance (Biegler and Grossmann, 2004). These methods are simple, accurate, have the advantage of easy implementation and require little prior knowledge of the optimization problem. Associated with these methods are numerous studies with successful results on a wide range of process problems (e.g. Banga and Seider, 1996; Jayaraman *et al.*, 2000; Rouhiainen and Tade, 2003; Jeżowski, Bochenek and Ziomek, 2005; Luus, 2006). Most of these methods are derived from heuristics that naturally generate numerous variations and inherently prevent premature termination. Meta-heuristics are advanced heuristics derived from observation of nature; for instance, simulated annealing (Kirkpatrick, Gellatt and Vecchi, 1983) mimics the motion of molecules during cooling of metals, and genetic algorithm (Michalewicz, 1996) is based on the evolutionary ideas of natural selection and genetics.

Ant colony optimization (ACO) is a relatively recent meta-heuristic (Dorigo and Gambardella, 1997) that mimics foraging behavior of real life ant colonies. During food search process, the ants exchange information by depositing pheromone on the ground. The ants sense the pheromone and get attracted to pheromone rich paths. It is the capability of depositing and getting attracted to pheromone rich paths that makes this food search process autocatalytic. This cooperative behavior results in the establishment of the shortest route between food source and their nest as shown in Fig. 1(b). Consider two ants randomly taking different directions from

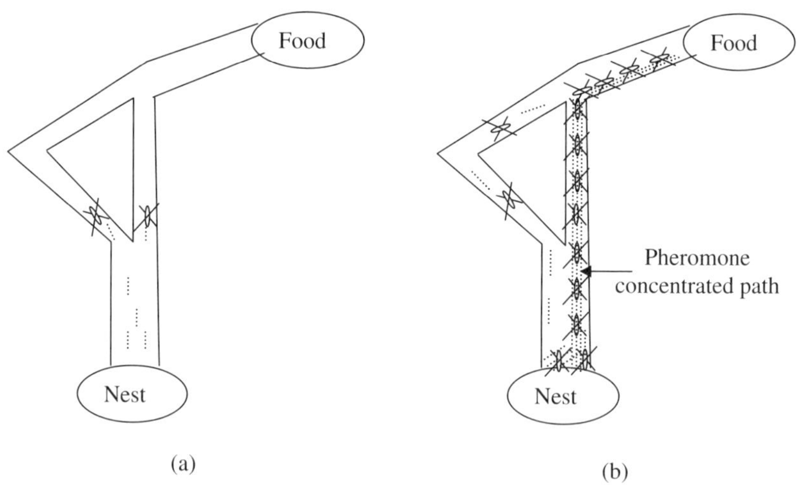

Figure 1. Foraging behavior of ants: (a) two ants randomly start exploring different paths in search of the food; (b) establishment of the optimal route on the shortest path.

their nest in search of food such that a path chosen by one of them turns out to be the shorter than the one taken by the other (Fig. 1(a)). The ant following the shorter path has less distance to travel from the food source to the nest. Hence, it will get attracted to this pheromone rich shorter path and will return to the nest by this path while simultaneously enhancing the pheromone concentration of the path. The other ant after arriving at the food source will choose the pheromone rich shorter path (with a very high probability), and soon all other ants in the vicinity will follow these ants. In this way, the shorter path will be established as the route from the nest to the food source and vice versa. Such a pheromone-meditated cooperative search process leads to the intelligent swarm behavior (Fig. 1(b)). ACO has been successfully explored to solve difficult continuous and combinatorial process optimization problems (Jayaraman *et al.*, 2000; Shelokar, Jayaraman and Kulkarni, 2003, 2004a,b).

This chapter provides a brief overview of different approaches based on ACO algorithm for solving process optimization problems such as single objective unconstrained/constrained continuous problems (Mathur *et al.*, 2000), combinatorial single and multiobjective optimization (Jayaraman *et al.*, 2000), rule based classification (Shelokar, Jayaraman and Kulkarni, 2004a; Parpinelli, Lopes and Freitas, 2002) and data clustering (Shelokar, Jayaraman and Kulkarni, 2004b). In all these approaches, software ants are deputed to iteratively obtain the best solutions. We illustrate the usefulness of ACO paradigm by solving some benchmark problems in the literature.

2. ACO for Continuous Function Optimization

The first applications of artificial ant colonies for continuous functions were developed by Parmee and co-workers (Bilchev and Parmee, 1995; Bilchev, Parmee and Darlow, 1996). In these papers, only local search procedures were employed. These approaches were further extended and modified (Wodrich and Bilchev, 1999; Mathur *et al.*, 2000) by using the concept of local and global ants to make it an effective global optimization procedure for continuous problems. Both local and global ants were deputed to effectively search and explore the problem space. They iteratively move to solutions of improved fitness by repeatedly searching the space locally and globally. The global ants search the problem space employing the processes of random walk and pheromone trail diffusion. The local ants can deposit and sense pheromone. Thus, they use this capability to search the problem space in the neighborhood of solutions. We refer this algorithm as continuous ACO (CACO), and is described in detail in the next section. Another ant-related approach to continuous optimization is the API algorithm (Monmarché, Venturini and Slimane, 2000). In API, ants perform their search independently, but starting from the same nest which is shifted periodically.

Dréo and Siarry (2004) proposed continuous interacting ant colony (CIAC) approach for solving continuous problems. In CIAC, ants communicate between them using stigmergic information and direct communication. Ants move through the search space, being attracted by pheromone deposited in spots, and guided by some direct communication between individuals. Recently, Socha and Dorigo (2008) proposed another ant algorithm for continuous domains. This approach, at any given level, maintains an archive of solutions generated in the previous iterations. The solutions in the archive are sorted based on the fitness and subsequently they are ranked (rank one is associated with the best solution in the archive). This rank is used to assign weights to the solutions. Ants use these weights as pheromone information to generate new solutions. Each ant probabilistically selects a solution from the archive to create an element of new solution. After all ants have completed creation of the new solutions the archive is updated and the process is repeated again until the stopping criterion is satisfied.

3. Continuous Ant Colony Optimization Algorithm (CACO)

The CACO algorithm initially generates a given number of R solutions randomly in the decision variable space. Every solution is equivalent to an individual of a population in the standard genetic algorithm. The fitness values of these trial solutions are first evaluated and then sorted in the ascending order of fitness value. (We consider minimization of the objective function. For unconstrained optimization problems, usually fitness is equivalent to the value of the objective function.) The solutions at the top are stronger (better fitness) than the solutions at the bottom, which are weaker (poor fitness).

Table 1. Unconstrained test problems.

Function	Equation
Goldstein & Price (GP)	$GP(x_1, x_2) = (1+(x_1+x_2+1)^2(19-14(x_1+x_2)+3(x_1^2+x_2^2)+6x_1x_2)) \times(30+(2x_1-3x_2)^2(18-32x_1+12x_1^2 +48x_2-36x_1x_2+27x_2^2))$ • search domain: $-2 \leq x_1, x_2 \leq 2$; • global minimum: $(x_1, x_2) = (0, -1)$, $GP(x_1, x_2) = 3$;
Easom (ES)	$ES(x_1, x_2) = -\cos(x_1)\cos(x_2)\exp(-\lfloor(x_1-\pi)^2+(x_2-\pi)^2\rfloor)$ • search domain: $-100 \leq x_j \leq 100$, $j = 1, \ldots, n$; • global minimum: $(x_1, x_2) = (\pi, \pi)$; $ES(x_1, x_2) = -1$;
Shubert (SH)	$SH(x_1, x_2) = \left(\sum_{i=1}^{5} i\cos((i+1)x_1+i)\right)\left(\sum_{i=1}^{5} i\cos((i+1)x_2+i)\right)$ • search domain: $-10 \leq x_1, x_2 \leq 10$; • global minima with $SH(x_1, x_2) = 0$;
Rosenbrock (RS)	$RS_n(x) = \sum_{j=1}^{n-1}[100(x_j^2 - x_{j+1})^2 + (x_j - 1)^2]$ • search domain: $-5 \leq x_j \leq 10$, $j = 1, \ldots, n$; • global minimum: $x = (1, \ldots, 1)$; $RS_n(x) = 0$;
Zakharov (ZA)	$ZA_n(x) = \sum_{j=1}^{n} x_j^2 + \left(\sum_{j=1}^{n} 0.5jx_j\right)^2 + \left(\sum_{j=1}^{n} 0.5jx_j\right)^4$ • search domain: $-5 \leq x_j \leq 10$, $j = 1, \ldots, n$; • global minimum: $x = (0, \ldots, 0)$; $ZA_n(x) = 0$;
Hartman (HR)	$HR_{3,4} = \sum_{i=1}^{4} c_i \exp[-\sum_{j=1}^{n} \alpha_{ij}(x_j - p_{ij})^2]$ • search domain: $0 \leq x_j \leq 1$, $j = 1, \ldots, 3$; • global minimum: $x = (0.11, 0.555, 0.855)$; $HR_{3,4}(x) = -3.86278$;

To illustrate different steps in CACO algorithm, we consider Hartman function given in Table 1 as:

$$HR_{3,4} = \sum_{i=1}^{4} c_i \exp\left[-\sum_{j=1}^{n} \alpha_{ij}(x_j - p_{ij})^2\right]$$

$$\text{search domain}: \ 0 \leq x_j \leq 1, \quad j = 1, \ldots, 3.$$

The solution of Hartman function is a vector of three decision variables. The decision variables have lower and upper bounds as zero and one respectively. To initiate the algorithm, suppose we randomly generate

$R = 10$ trial solutions. The fitness f of these solutions is first evaluated, and the ten solutions are sorted in the decreasing order as given:

S. No.	x_1	x_2	x_3	$f(x)$	Age	τ
1	0.9134	0.4854	0.9340	−2.8385	0	1
2	0.2785	0.4218	0.7431	−2.5892	0	1
3	0.6324	0.8003	0.6787	−1.4950	0	1
4	0.1270	0.9572	0.8491	−0.9727	0	1
5	0.9575	0.7922	0.6555	−0.7895	0	1
6	0.0975	0.1419	0.7577	−0.7784	0	1
7	0.5469	0.9157	0.3922	−0.7535	0	1
8	0.8147	0.1576	0.6557	−0.4980	0	1
9	0.9649	0.9595	0.1712	−0.0040	0	1
10	0.9058	0.9706	0.0357	−0.0002	0	1

To explore new solutions, the CACO algorithm deploys a problem-specific number of ants (A) out of which a percentage work as global ants (G) and the remaining work as local ants (L). At any given iteration, the global ants perform global search while local ants carry out local search. The ratio of global to local ants is the algorithm parameter and need to be tuned. In our simulations 80% of total ants (A) work as global ants and remaining ants operate as local ants.

3.1. Local search

In CACO, each trial solution has an associated age and pheromone value, which is initialized to zero and one respectively at the beginning of iteration process. The local ants can deposit pheromone at these trial solutions. To generate L new solutions, local ants select solutions with a probability proportional to the current pheromone values of these solutions. The probability of selecting a solution i out of R solutions is given by

$$P_i(t) = \frac{\tau_i(t)}{\sum_k \tau_k(t)}, \tag{1}$$

where i is the solution index and $\tau_i(t)$ is the current pheromone value of solution i at iteration t. After selecting the solution, the ant calculates a step size using the current age of the solution. The ant employs this step size to locally find a new solution in the neighborhood of the current solution. To do this, the ant perturbs each variable of the current solution in either direction with equal probability. If the new solution has an improved fitness then this will replace the current one and the pheromone of the solution is updated proportional to the improvement in the fitness. The size of the ant's move in the local search depends on the age of a solution. The search radius is highest for age equal to zero and minimum for maximum age, with the variation in radius set to follow a linear relationship. Age of the solution is decremented by one if the ant's move is successful (found improved solution) or it is incremented by one otherwise.

3.2. Global search

During global search, G weak solutions are replaced by new solutions created out of appropriate combinations of the stronger solutions. For this reason the stronger solutions are termed as parent solutions and the newly created solutions are called as offspring solutions. The global search procedure essentially consists of random walk (RW) and trail diffusion (TD) processes. Some percentage of G, usually 80–90% of global ants implement RW and remaining carry out TD process (Mathur *et al.*, 2000). In our simulations 80% of global ants perform RW operation.

The random walk consists of the following operations (Wodrich and Bilchev, 1999; Mathur *et al.*, 2000): (i) the crossover operation (which combine existing parents to produce offsprings) and (ii) the mutation operation. The crossover operation is carried out as follows: select a parent randomly and set the first element of the offspring's position vector to be the same as that of the first element of the parent's position vector. The second element of the offspring's position vector is set to the corresponding element of a randomly chosen another parent, with a probability equal to the crossover probability, CP. Subsequent elements of the offspring's position vector are chosen by the same procedure. Thus, for $CP = 1$, each element of the offspring's position vector is obtained from a different parent. For $CP = 0$, the offspring is identical to the randomly chosen single parent.

To show the random walk operation, consider the Hartman function. Suppose the algorithm deputed $G = 4$ global ants and 80% of G equal to three ants perform random walk operation. In this case corresponding to G

ants, we can use top six $(R - G)$ points as parents. Suppose the first ant has arbitrarily chosen a parent six and set the first element of offspring to the corresponding element of this parent. To set second and third elements of the offspring, the ant employs selection probability proportional to crossover probability. In our simulation study, a value of crossover probability was set to $CP = 0.65$. If the value of crossover probability is greater than a selection probability value, then the chosen parent gives its element value to the corresponding element of the offspring. Selection probability is generated as a random number uniformly distributed in the range (0, 1). Suppose the ant randomly chosen parents one and four with selection probabilities say, 0.35 and 0.21 respectively. The offspring generated by the ant employing crossover operation is given as:

offspring by crossover operation 0.0975 0.4854 0.8491

Note that the first element of offspring is contributed by parent six, second element by parent one and the last element by parent four.

The algorithm now applies mutation operation on this offspring. Every element of this offspring undergoes mutation operation with a probability proportional to the mutation probability, *MP*. In our simulations a value of mutation probability was set to $MP = 0.5$. The mutation step size is calculated as per the relation:

$$\Delta(I, S) = S\left(1 - r^{(1-t/I)^b}\right), \tag{2}$$

where r is a random number uniformly distributed in the range (0, 1) and is generated at every iteration at the beginning of the mutation operation, S is the maximum step size to perturb the value of decision variable in the offspring and is calculated as the difference between upper and lower bounds for the decision variable, I is a given number of maximum iterations and usually employed as the main criterion to terminate the simulation, t is the current iteration number, and b is a positive parameter controlling the degree of non-linearity. We have chosen value of $b = 10$ as mentioned in Wodrich and Bilchev (1999) and Mathur *et al.* (2000). Equation (2) shows that mutation step $\Delta(I, S)$ decreases non-linearly as the algorithm progresses. Thus, the algorithm tries to explore concentrated search towards the end of iterations. A value of $I = 1{,}000$ was set for the simulation study on the unconstrained test functions (Table 1).

To illustrate the mutation operation, consider the offspring above generated by crossover operation. Suppose at the current iteration level $t = 1$ and at the beginning of mutation process the ant obtains $r = 0.25$, a random number uniformly distributed in the range (0, 1). To carry out mutation on the first element (0.0975) of the offspring, the ant generates a probability value, say 0.15, a random number uniformly distributed in the range (0, 1). Since this probability value is less than the mutation probability ($0.15 < MP$), the first element of the offspring undergoes mutation operation. To calculate the mutation step $\Delta(I, S)$, the ant obtains the step size S which is equal to one for all the decision variables of Hartman function. Using Eq. (2), the mutation step is calculated as $\Delta(I, S) = 0.253$. The algorithm uses this value to perturb the current value of the element in either direction with equal probability. The value of first element after perturbation can be given as ($0.0975 + 0.253 = 0.3505$). Similarly, the ant applies mutation process for the remaining two elements of the offspring. The offspring after mutation operation can be given as:

offspring after mutation operation	0.3505	0.4854	0.5956

The remaining two global ants similarly apply crossover and mutation processes to generate their offspring for the illustrative Hartman function.

The second global search procedure viz., trail diffusion (TD) is quite similar to arithmetic crossover in genetic algorithm (Michalewicz, 1996). In our simulations, 20% of G global ants perform trail diffusion operation. In TD, an ant generates an offspring employing arbitrarily selected two parents. The elements of the offspring's position vector are generated by using one of the following three options selected with selection probabilities described next:

Option 1: using both the parents given by equation

$$x_i(\mathit{offspring}) = \alpha x_i(\mathit{parent}_1) + (1 - \alpha)x_i(\mathit{parent}_2)$$

Option 2: using the first parent given as:

$$x_i(\mathit{offspring}) = \alpha x_i(\mathit{parent}_1) \quad , \qquad (3)$$

Option 3: using the second parent given as:

$$x_i(\mathit{offspring}) = (1 - \alpha)x_i(\mathit{parent}_2)$$

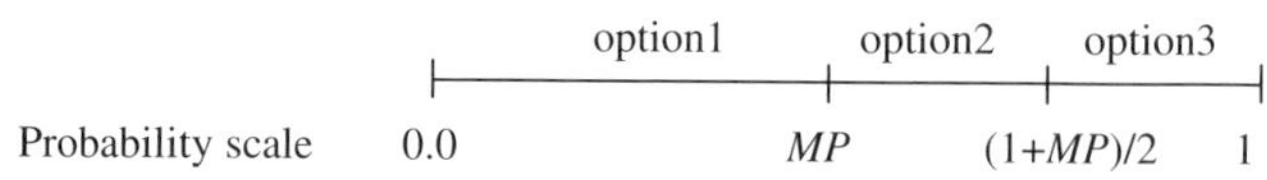

Figure 2. Selection probabilities for three alternatives in trail diffusion process of CACO algorithm.

where α is the weight factor and its value is drawn from a random number uniformly distributed in the range (0, 1). The selection probability values corresponding to three options are shown in Fig. 2.

The selection probability for the first option is set equal to *MP* (which is mainly done to avoid introducing another algorithm parameter). While the selection probabilities for option 2 and option 3 are set to $(1 + MP)/2$ and one respectively. In our simulation study, 20% of global ants G undertake TD process. For illustrative Hartman function, we have considered $G = 4$ global ants. Out of four global ants, three ants (80% of G) already performed RW process. The remaining global ant undertakes TD process. Suppose the ant arbitrarily selects two parents say two and five from the $(R - G)$ parents and also generates $\alpha = 0.025$, a random number uniformly distributed in the range (0, 1). To apply one of the three alternatives to generate an offspring, suppose the ant generates a selection probability say, 0.15, a random number uniformly distributed in the range (0, 1). Since this probability value is less than the selection probability of option 1 ($0.15 < MP$), the ant applies option 1 to generate the offspring given as:

offspring by TD process	0.9404875	0.78294	0.65769

The G number of solutions (offsprings) generated by global ants replace weaker solutions.

3.3. Pheromone evaporation

In real ant systems, a process of food resource exhaustion (which causes the ants to lose interest in the search) is known to occur. This feature is incorporated in the CACO algorithm by reducing the pheromone value of all solutions by a certain value after each iteration step. This process of evaporation introduces a bias for fitter solutions by low probability for selection of non-improving solutions during the algorithm progress. The pheromone evaporation is carried out by

$$\tau_i(t+1) = \rho \cdot \tau_i(t), \tag{4}$$

where ρ is the persistence of pheromone value normally lies in the range (0.9, 0.95).

3.4. Steps in CACO algorithm

Recall L is the number of local ants. G is the number of global ants. Total number of ants, $A = L + G$.

(1) At iteration counter $t = 1$, create randomly R solutions and set the age and pheromone values of all solutions equal to 0 and 1 respectively.
(2) Local search: an ant selects a solution using probability based on pheromone values; finds a new solution in the neighborhood of current one using the age of the current solution; if fitness is improved, then replace the current solution by the new solution, decrease the age by one and deposit the pheromone value proportional to change in the fitness; otherwise, increase the age by one.
(3) Global search: G new solutions are generated by random walk and trail diffusion operations.
(4) Update pheromone values using Eq. (4).
(5) Update $t = t + 1$, repeat steps 2 to 4 till the iteration counter t reaches the given maximum number of iterations.

4. ACO Algorithm for Combinatorial Optimization

Optimization methods based on foraging behavior of real life ant colonies have successfully been applied for solving different combinatorial optimization problems such as traveling salesman (Dorigo and Gambardella, 1997), quadratic assignment (Gambardella, Taillard and Dorigo, 1999), job shop scheduling (Colorni *et al.*, 1994), vehicle routing (Gambardella, Taillard and Agazzi, 1999), sequential ordering (Gambardella and Dorigo, 1997), graph coloring (Costa and Hertz, 1997), and knapsack problem (Kong, Tian and Kao, 2008). In this section, we show application of ACO for solving scheduling problem of serial multiproduct batch plants (Jayaraman *et al.*, 2000).

5. Scheduling of Serial Multiproduct Batch Plant

Batch processes have always been of considerable importance in chemical manufacturing due to their flexibility in processing multiple products and ability to accommodate diverse operating conditions. A serial multiproduct process operation consists of a series of M batch units joined together by semi-continuous units. In this process operation, all products N pass through all units M in the same sequence (Kim, Jung and Lee, 1996). The processing time, pc_{ij} of product i in unit j is known *a priori*. A batch of N products with minimum makespan has been of particular interest and has been thoroughly analyzed in chemical process engineering applications (Kim, Jung and Lee, 1996). This problem comprises of two tasks: (1) determination of completion times dealing with the details of operation schedule for a given sequence, and (2) determination of the optimal sequence.

This batch scheduling problem comprises of several different subcategories depending on the way of handling the intermediate products between successive stages. In this chapter, we specifically consider that each of N ($i = 1, \ldots, N$) jobs follow the same job order on M ($j = 1, \ldots, M$) machines and further make the following assumptions:

(1) processing time pc_{ij} for each job on each machine is known *a priori*
(2) any suitable intermediate policy can be considered
(3) each job can be processed on one machine at a time
(4) jobs are not pre-emptable

For batch chemical processes, different intermediate storage policies are considered in the literature, namely, the unlimited intermediate storage (UIS), finite intermediate storage (FIS), no intermediate storage (NIS), and zero wait processing (ZW). Heuristic and MILP formulations for makespan minimization for all different storage policies are available in the literature (e.g. Kim, Jung and Lee, 1996). Here, we have considered UIS and ZW policies to illustrate the application of ACO.

5.1. UIS policy

Under UIS policy, when the immediate processing unit is not available for the product coming out from the current processing unit, a temporary storage facility is always accessible for the processed product. The expression for the completion time C_{ij} using processing time pc_{ij} as (Kim, Jung and Lee, 1996):

$$C_{ij} = \max\lfloor C_{(i-1)j}, C_{i(j-1)}\rfloor + pc_{ij}. \tag{5}$$

Kim, Jung and Lee (1996) have developed the following expression for the completion time C_{ij} using the transfer times tt_{ij} and setup times st_{ikj}

$$C_{ij} = \max\lfloor C_{i(j-1)}, C_{(i-1)j} + st_{(i-1)ij} + tt_{i(j-1)}\rfloor + pc_{ij} + tt_{ij}. \tag{6}$$

Here, C_{ij} is the completion time of product i on batch unit j where the product is finished to transfer out of the unit j and to fill in the unit $j+1$, tt_{ij} is the transfer time for product i out of batch unit j, st_{ikj} is the setup time required for processing product k after it has processed product i on batch unit j. We have applied Eq. (6) to show the application of ACO for scheduling of batch process.

5.2. ZW policy

Under ZW policy, it is assumed that an immediate processing unit is always available when the product comes out from the current processing unit. This ZW policy is applied in chemical process industry when batches of products are usually unstable reaction intermediates. The expression for the completion time C_{ij} using processing time pc_{ij} is given as (Kim, Jung and Lee, 1996):

A relationship between C_{ij} and C_{iM} is given as:

$$C_{ij} = C_{iM} - \sum_{k=j+1}^{M} pc_{ik}. \tag{7}$$

C_{iM} for the ZW policy is represented as:

$$C_{iM} = \max\left(C_{(i-1)1} + \sum_{k=1}^{M} pc_{ik}, C_{(i-1)2} + \sum_{k=2}^{M} pc_{ik}, \ldots, C_{(i-1)M} + \sum_{k=M}^{M} pc_{ik}\right). \tag{8}$$

The expression for the completion time C_{ij} using transfer times, tt_{ij} and setup times, st_{ikj} is given as (Kim, Jung and Lee, 1996):

For the first product in the sequence:

$$C_{1M} = \sum_{k=1}^{M} pc_{1k} + \sum_{k=0}^{M} tt_{1k}. \tag{9}$$

For the second to Nth product in the sequence:

$$C_{iM} = \max \left[\begin{array}{l} \left(C_{(i-1)1} + st_{(i-1)i1} + \sum_{k=1}^{M} pc_{ik} + \sum_{k=0}^{M} tt_{ik} \right), \ldots, \\ \left(C_{(i-1)(M-1)} + st_{(i-1)i(M-1)} \right. \\ \quad \left. + \sum_{k=(M-1)}^{M} pc_{ik} + \sum_{k=(M-2)}^{M} tt_{ik} \right), \\ \left(C_{(i-1)M} + st_{(i-1)iM} + pc_{iM} + \sum_{k=(M-1)}^{M} tt_{ik} \right) \end{array} \right]. \tag{10}$$

We have applied Eq. (10) to show the application of ACO for scheduling of batch process.

Equations (6) and (10) suggest that setup time is product order dependent and transfer time is independent of the order of products in a sequence. The objective of scheduling of serial multiproduct batch plant is to find the optimal order of N jobs in a batch that produces the minimum value of the total completion time C_{NM}. For N jobs scheduling problem, there are $N!$ possible solutions. If D is the set of solutions then the objective function can be given as:

$$F(d) = \min_{d \in D}(C_{NM}). \tag{11}$$

6. ACO Algorithm for Scheduling of Serial Multiproduct Batch Plant

In the scheduling problem of multiproduct batch process, products are processed in the order they follow in a given sequence. Thus, relative position of products is important in a sequence. To apply ACO algorithm for finding an optimal sequence of N jobs, we have used position index of products as solution attributes. For example, suppose we have $N = 5$ products, one of the solutions from a total of 125 possible solutions could be given as:

Jobs	1	2	3	4	5
Position	4	1	5	2	3

In this solution (schedule) product 1 is at position 4 while product 2 is at position 1 and so on. We depute certain number of ants A and each ant produces a solution. In our simulations, we used the number of ants $A = 50$ for problems unto 8 jobs and 10 units. For scheduling problems with more than 8 jobs and 10 units, the number of ants were kept $A = 100$. Since each job can take any of the N possible positions in the order, pheromone τ is defined by $N \times N$ matrix where τ_{ij} indicates the affinity of product i towards position j. At the start of the algorithm, the elements of pheromone matrix are initialized by $\tau_{ij} = 1/N$. Each ant starts with an empty solution vector of size N and builds the complete solution by positioning all the products at N different locations.

At any iteration number t, an ant constructs a solution using pheromone. To generate a partial schedule d, the ant assigns position j to product i by either choosing the position with maximum pheromone using certain probability q_0 as:

$$j = \arg \max_{l \in \xi} (\tau_{il}), \tag{12}$$

or choosing probabilistically one of the N positions as:

$$p_{ij} = \frac{\tau_{ij}}{\sum_{l \in \xi} \tau_{il}}, \quad l \in \xi, \tag{13}$$

where ξ is a set of locations in a sequence not occupied by products. This keeps the eye on available positions for products not positioned in the current partial solution. Thus, it acts like a memory to the ant. q_0 is *a priori* defined number in the range (0, 1). In our simulations, we found a value of $q_0 = 0.65$ has produced good algorithm performance for scheduling batch plants. According to Eq. (12), an ant selects position j for product i with maximum value of pheromone. In Eq. (13), we get the normalized probabilities of possible available positions ($l \in \xi$) for product i. The ant then calculates cumulative probabilities for those positions and generates a random number in the range (0, 1). The ant selects position j for product i whose value of cumulative probability is close to and greater than the value of generated random number. The ant repeats this process until all the products are ordered in the sequence.

Once all the ants build their solutions, algorithm updates pheromone information as:

$$\tau_{ij} = (1-\rho)\tau_{ij} + \rho \sum \Delta\tau_{ij}^{k}, \tag{14}$$

$$\Delta\tau_{ij}^{k} = \begin{cases} F_k(d) & \text{if job } i \text{ is at position } j \text{ in schedule } d \\ 0 & \text{otherwise} \end{cases}. \tag{15}$$

Here, $F_k(d)$ is the objective function (fitness) defined in Eq. (11) for solution d generated by ant k. At the end of an iteration, the best solution is compared with the global best solution so far found by the algorithm. If the former is better than the latter, then update the global best solution and continued to the next iteration. The termination criterion for ACO algorithm is *a priori* defined number of maximum iterations.

7. ACO for Multiobjective Scheduling of Serial Multiproduct Batch Plant

ACO has been applied to solve multiobjective combinatorial problems. T'kindt *et al.* (2002) solved the 2-machine flow shop scheduling problem with the objective of minimizing both the makespan and the total completion time criteria. In their algorithm, the former objective is assumed to be optimized prior to the latter. The ACO algorithm by T'kindt *et al.* (2002) uses intensification and diversification processes to generate solutions. Intensification is emphasized using a local search procedure: adjacent pair-wise interchange (API) neighborhood operator and Johnson's algorithm whilst diversification is emphasized using a feature from simulated annealing. Doerner *et al.* (2006) proposed Pareto ACO (PACO) algorithm for project portfolio selection. In PACO, for each objective k pheromone information is stored in a vector τ_j^k, $j = 1, \ldots, N$, where τ_j^k indicates whether or not adding a project j promises favorable effects on a portfolio's outcome.

In this chapter, ACO is applied to solve multiobjective flow shop scheduling problem with the objective of minimizing makespan and average total residence time of products. The proposed ACO generates a Pareto set of optimal schedules for the batch plant.

7.1. Multiobjective optimization (MOO) problem definition

A general multiobjective problem can be described as a vector function f that maps a tuple of m parameters (decision variables) to a tuple of n objectives (Fonseca and Fleming, 1994).

$$\begin{aligned} \min/\max \quad \bar{y} &= f(\bar{x}) = (f_1(\bar{x}), f_2(\bar{x}), \ldots, f_n(\bar{x})) \\ \text{subject to} \quad \bar{x} &= (x_1, x_2, \ldots, x_m) \in X \\ \bar{y} &= (y_1, y_2, \ldots, y_n) \in Y \end{aligned}, \tag{16}$$

where $\bar{x}$ is the decision vector, X is the parameter space, $\bar{y}$ is the objective vector, and Y is the objective space. In MOO, we have a set of optimal solutions called Pareto set, which consists of all decision vectors for which the corresponding objective vectors cannot be improved in a given dimension without worsening in another. This can be explained as follows: assume a minimization problem and two decision vectors $\bar{x}_1$ and $\bar{x}_2 \in X$. Then $\bar{x}_1$ is said to dominate $\bar{x}_2$ if and only if:

$$\begin{aligned} & f_i(\bar{x}_1) \leq f_i(\bar{x}_2) \quad \text{for all objectives } f_i,\ i \in \{1, 2, \ldots, n\} \\ & \text{and} \\ & f_i(\bar{x}_1) < f_i(\bar{x}_2) \quad \text{for at least one objective } f_i,\ i \in \{1, 2, \ldots, n\} \end{aligned}. \tag{17}$$

All decision vectors which are not dominated by any other decision vectors of a given set are called non-dominated or Pareto set. For multiobjective scheduling problem, we applied two objectives makespan and average residence (completion) time, *ART*. The average residence time is calculated over completion time of each product at the end of M unit as:

$$ART = \frac{\sum_i C_{iM}}{N}. \tag{18}$$

Solution construction procedure for multiobjective scheduling is similar to that for single objective scheduling problem described earlier. The main difference lies in the procedure of depositing pheromones. In the ACO approach for multiobjective scheduling problem, only one objective is optimized at a time using a scheme similar to lexicographic ordering. In lexicographic ordering, the user is asked to rank the objectives in the order of importance. The optimum solution is then obtained by minimizing the objective functions, starting with the most important one and proceeding according to the assigned order of importance (Fonseca and Fleming, 1994).

In the present ACO, once all the ants generate their solutions, pheromone is deposited using an objective (makespan or average residence time) selected by an approach similar to lexicographic ordering. In our approach, at each iteration, the selection of primary objective was done with equal probability. The algorithm first sorts the solutions based on the primary objective and then selects the first half of the sorted solutions. Subsequently, the pheromone concentrations are updated using secondary objective of these solutions. In this ACO approach for MOO, the non-dominated solutions obtained are externally stored. At the end of an iteration, solutions generated by ants are checked for their non-dominance and the non-dominated solutions are added to the external set. The solutions of this external set further undergo non-dominance check and are updated by removing dominated solutions. The final output of ACO for multiobjective scheduling is the external set of non-dominated solutions.

8. ACO for Knowledge Discovery in Process Data

Fault diagnosis methods based on historic process knowledge and data have gained considerable importance. For example, Venkatasubramanian, Vaidyanathan and Yamamoto (1990) introduced neural networks for detecting and diagnosing process failures in steady-state processes. van Kampen *et al.* (1997) proposed genetic algorithm based learning classification rules from an ion chromatography database. Mo *et al.* (1998) applied *k*-means clustering method for qualitative interpretation and compression of process data. Shelokar, Jayaraman and Kulkarni (2004a) applied ACO for rule based classification of multiple process faults; Shelokar, Jayaraman and Kulkarni (2004b) proposed ACO based clustering analysis of wines. Yoo, Lee and Vanrolleghem (2005) applied fuzzy clustering for interpreting patterns and analysis of acute leukemia gene expression data. A comprehensive review of process history knowledge and data based methods can be found elsewhere (Venkatasubramanian, Rengaswamy and Kavuri, 2003). This chapter briefly describes application of ACO for data clustering (Shelokar, Jayaraman and Kulkarni, 2004b).

9. ACO for Data Clustering

Clustering is a technique to cluster objects into more than one group so as to maximize separability between these groups. Clustering algorithms specify the number of groups and maximize an objective function that is a measure of separability of these groups. In this manner, clustering becomes a well-defined optimization problem. A good introduction to contemporary data clustering algorithms can be found in Han and Kamber (2001). Most of the clustering criteria are non-convex and non-linear so that the problem may have local minima, which are not necessarily optimal (Selim and Ismail, 1984). Moreover, they possess exponential complexity in terms of number of clusters and become an NP-hard problem when number of clusters exceeds three (Welch, 1983). Recently meta-heuristics like tabu search (Sung and Jin, 2000), genetic algorithms (Murthy and Chowdhury, 1996) and simulated annealing (Selim and Al-sultan, 1991) have been successfully employed for clustering.

In this chapter, we have applied ACO for data clustering (Shelokar, Jayaraman and Kulkarni, 2004b) called as *AntzClust*. Consider a problem of partitioning N objects (instances) into a given number of clusters K. The quality of partitioning generated by an ant is measured using Euclidean distance metric. The objective is to minimize the value of this dissimilarity metric given as (Sung and Jin, 2000):

$$\operatorname*{Min}_{d\in D} F(d) = \sum_{j=1}^{K}\sum_{i=1}^{N}\sum_{v=1}^{n} w_{ij}\|x_{iv} - m_{jv}\|^2, \tag{19}$$

$$\text{s.t.} \sum_{j=1}^{K} w_{ij} = 1, \quad i = 1, \ldots, N, \tag{20}$$

$$\sum_{i=1}^{N} w_{ij} \geq 1, \quad j = 1, \ldots, K, \tag{21}$$

where x_{iv} is a value of vth attribute of ith object, m_j is a jth cluster center calculated as an average of objects assigned to cluster j and w_{ij} indicates allocation of object i to cluster j as:

$$w_{ij} = \begin{cases} 1 & \text{if object } i \text{ is in cluster } j \\ 0 & \text{otherwise} \end{cases}. \tag{22}$$

Each ant generates a possible clustering solution where all objects N are assigned to K clusters. Suppose a dataset consists of $N = 8$ objects that need to grouped into $K = 3$ clusters. A possible solution an ant can produce is shown below as:

Objects	1	2	3	4	5	6	7	8
Cluster label	3	1	1	2	3	2	2	1

In this solution object 1 is assigned to cluster 3, object 2 is grouped in cluster 1 and so on. To generate a solution d, ant starts with an empty solution and incrementally builds a solution by assigning each object to one of K clusters by either choosing the cluster label with maximum pheromone using certain (exploitation) probability q_0 as:

$$j = \arg\max(\tau_{ki}), \tag{23}$$

or choosing probabilistically one of the K cluster labels as:

$$p_{ji} = \frac{\tau_{ji}}{\sum_{k=1}^{K} \tau_{ki}}, \tag{24}$$

where q_0 is *a priori* defined number in the range (0, 1) and τ_{ji} is a pheromone deposited at label j for object i. In our simulations, we found $q_0 = 0.65$ has produced good performance of ACO for data clustering. The pheromone τ_{ji} is of size $K \times N$ for problem of clustering N objects into K clusters. According to Eq. (23), an ant selects cluster label j for object i with maximum value of pheromone. In Eq. (24) we get the normalized probabilities of possible cluster labels for object i. The ant then calculates cumulative probabilities for those labels and generates a random number in the range (0, 1). The ant selects label j for object i whose value of cumulative probability is close to and greater than the value of generated random number. The ant repeats this process until cluster labels are assigned to all the objects in the solution.

Once all the ants build their solutions, the algorithm updates pheromone information as:

$$\tau_{ji} = (1-\rho)\tau_{ji} + \rho \sum \Delta\tau_{ji}^{l}, \tag{25}$$

$$\Delta\tau_{ji}^{l} = \begin{cases} 1/F_l(d) & \text{if object } i \text{ is assigned to cluster } j \text{ in solution } d \\ 0 & \text{otherwise} \end{cases}. \tag{26}$$

Here, $F_l(d)$ is the objective function defined in Eq. (19) for solution d generated by ant l. At the end of an iteration, the best solution is compared with the global best solution so far found by the algorithm. If the former is better than the latter, then update the global best solution and proceed to the next iteration. The termination criterion for ACO algorithm is *a priori* defined number of maximum iterations.

10. ACO Algorithm for Rule Based Classification

ACO has been applied to discover classification rules from the dataset (Shelokar, Jayaraman and Kulkarni, 2004a; Parpinelli, Lopes and Freitas, 2002) where the classification task is formulated as a complex search optimization problem. The solutions (rules) discovered represent hidden relationships between attributes and class. The application of ACO for discovery of classification rules is described in detail in the chapter entitled "Ant Colony Optimization for Classification and Feature Selection" in this book.

11. Results and Discussion

Different applications of ACO algorithm are illustrated by employing test examples from the process engineering literature. Different approaches and illustrative examples discussed in this chapter are provided on the CD supplied with the book.

11.1. CACO for unconstrained continuous functions

Table 1 shows a set of six unconstrained multimodal test functions. This test set was previously employed by Chelouah and Siarry (2003, 2005).

In this simulation study CACO parameters were set as: number of solutions $R = 100$, number of global ants $G = 40$, percent distribution of G ants for RW and TD operation = 80:20, crossover probability $CP = 0.65$, mutation probability $MP = 0.5$, number of local ants $L = 10$, evaporation rate $\rho = 0.9$, maximum age value $age = 10$, maximum number of iterations $I = 1{,}000$. Simulations were performed 100 times for each test example. The results are reported in Table 2. The performance is reported using success rate (SR) and number of function evaluations (NFE). The success rate is a measure of probability of reaching the global optimal solution and is mentioned as a percentage. The number of function evaluations is calculated based on all 100 trials (irrespective of their success). Table 2 shows the results by CACO and also by continuous hybrid algorithm (CHA) and continuous tabu simplex search (CTSS) algorithm proposed by Chelouah and Siarry (2003 and 2005). CHA algorithm is a hybrid method combining genetic and Nelder–Mead simplex algorithms (Chelouah and Siarry, 2003). From the results shown in Table 2, CTSS took the least number of average function evaluations (233) followed by CACO and CHA with 320 and 454 respectively with 100% success rate. CTSS employs sophisticated simplex search with certain number of moves to improve the quality of current solution. In the present CACO, a simple local search is employed to find improved solution in the neighborhood of current solution. Thus, hybridization of CACO with standard local search can enhance the performance.

Table 2. Results using CACO for unconstrained test functions.

Function	Required accuracy	% Success rate			Average number of function evaluations		
		CACO	CTSS	CHA	CACO	CTSS	CHA
ES	0.001	100	100	100	338	325	952
GP	0.001	100	100	100	164	119	259
SH	0.01	100	100	100	450	283	345
RS_2	0.001	100	100	100	662	369	459
ZA_2	0.0001	100	100	100	163	78	215
$H_{3,4}$	0.0001	100	100	100	141	225	492
Overall	—	100	100	100	320	233	454

Table 3. Constrained MINLP test problems.

Example	Variable (binary + real)	Equation	Constraints (equality + inequality)	Global optimum
Kocis and Grossmann (1988)	2(1 + 1)	$f = 2x + y$	$g_1 = 1.25$ $-x^2 - y \leq 0$	$f = 2$ $(1, 0.5)$
Kocis and Grossmann (1987)	2(1 + 1)	$\min f = -y + 2x_1$ $- \ln(x_1/2)$	$g_1 = -x_1 - \ln(x_1/2)$ $+y \leq 0$ $0.5 \leq x_1 \leq 1.4$ $y = \{0, 1\}$	$f = 2.124$ $(1, 1.375)$
Floudas (1995)	2(1 + 2)	$\min f = -0.7y$ $+5(x_1 0.5)^2$ $+0.8$	$g_1 = -\exp(x_1 - 0.2)$ $-x_2 \leq 0$ $g_2 = x_2 + 1.1y \leq 0$ $0.2 \leq x_1 \leq 1$ $2.22554 \leq x_2 \leq 1$ $y = \{0, 1\}$	$f = 1.07654$ $(1, 0.94194, 2.1)$

12. CACO for Constrained Continuous Functions

Table 3 shows three benchmark constrained MINLP problems employed. These test problems were applied by Costa and Oliveira (2001) and Srinivas and Rangaiah (2007). For each test problem 100 simulations were performed. Table 4 shows the results in terms of SR and NFE. Table 4 shows the results reported by six different methods. The results obtained by GA and $(\mu + \lambda)$ ES were studied by Costa and Oliviera (2001). The results from M-SIMPSA and M-SIMPSA-pen were studied by Cardoso *et al.* (1997) and DETL was recently studied by Srinivas and Rangaiah (2007). To solve constrained optimization problems using CACO, we applied the commonly employed penalty function approach.

Example 1 is a process synthesis problem with non-convexity in the constraint. It has also been solved by other authors (Kocis and Grossmann 1988; Cardoso *et al.*, 1997; Costa and Oliviera, 2001; Srinivas and Rangaiah, 2007). As can be seen from Table 4, M-SIMPSA required the least NFE (607) among the results reported by six methods. All the methods reported results with 100% reliability. The average NFE (6283) required by CACO is somewhat less than that reported (6787) using GA. Example 2 with a non-linear constraint has been proposed by Kocis and Grossmann (1987)

Table 4. Performance of CACO on constrained MINLP test problems.

Example	CACO		GA		$(\mu + \lambda)$ES		M-SIMPSA		M-SIMPSA–pen		DETL	
	NFE	SR	NFE	SR	NFE	SR	NFE	SR	NFE	SR	NFE	SR
Kocis and Grossmann (1988)	6283	100	6787	100	1518	100	607	99	16282	100	$627 + 8$	100
Kocis and Grossmann (1987)	10120	100	13939	100	2255	100	10582	83	14440	100	$489 + 6$	94
Floudas (1995)	42320	95	107046	90	1749	100	**	0	38042	100	$1350 + 9$	86

and also studied by (Cardoso *et al.*, 1997; Costa and Oliviera, 2001; Srinivas and Rangaiah, 2007). For this problem, CACO, GA, $(\mu + \lambda)$ ES and M-SIMPSA-pen gave success rate of 100%. While M-SIMPSA and DETL provided success rates of 83 and 94 respectively. Example 3 was firstly studied by Floudas (1995) and presents a non-linear constraint. It has also been solved by (Cardoso *et al.*, 1997; Costa and Oliviera, 2001; Srinivas and Rangaiah, 2007). For this problem $(\mu + \lambda)$ ES and M-SIMPSA-pen produced 100% convergence rate. While CACO, GA, DETL produced success rate of 95, 90 and 86. Although SR rate of CACO is higher than that of DETL, the number of function evaluations required (1359) by DETL is significantly less than that required (42320) by CACO. DETL employed deterministic method in local search operation at the end of global search. Inclusion of a deterministic method for a local search in CACO could be interesting.

12.1. Scheduling of batch processes

Application of ACO for scheduling of batch processes is illustrated using a benchmark problem studied by Kim, Jung and Lee (1996). The objective is to find an optimal schedule with minimum makespan for a batch of 8 products to be processed on 4 machines. The problem formulations consist of UIS and ZW policies with transfer and set up times. In UIS formulation, ACO generated two different schedules as: P5-P7-P1-P2-P6-P8-P4-P3 and P5-P4-P1-P7-P2-P6-P3-P8 for the optimal makespan value of 173. In case of ZW formulation, ACO produced the optimal schedule as: P5-P1-P6-P3-P7-P4-P2-P8 with makespan value of 195. We tested the performance of ant algorithm on test examples with 4, 6, 8, 10 and 20 products, and units varying between 3 and 10. We generated processing time with uniform distribution in the range of (1, 30) and for set-up and transfer times the range was (1, 10). For smaller sized problems unto 8 products, we used the mean deviation from the optimum as the criterion for evaluation purposes; while for large problems, the mean deviation from the best solution obtained during 25 trials was used. The results in the form of success rate are shown in Table 5. For this simulation study, parameters for ACO algorithm were set as: number of ants $A = 30$ (and 100 for large problcms), exploitation

Table 5. Performance of ACO for the multiproduct batch plant.

Example(s)	% Success rate
$4 \times 3, 4 \times 6, 4 \times 8, 4 \times 10$	100
$6 \times 3, 6 \times 6, 6 \times 8, 6 \times 10$	100
$8 \times 3, 8 \times 6, 8 \times 8, 8 \times 10$	100
$10 \times 3, 10 \times 6$	100
10×8	99
10×10	98
20×3	98
20×6	96
20×8	94
20×10	92

probability $q_0 = 0.65$, evaporation rate $\rho = 0.9$, maximum number of iterations $I = 1{,}000$.

In this study to show the application of ACO for generation of Pareto solutions for MOO of batch processes, we consider two competing objectives minimization of makespan and average residence time (ART). To illustrate the application of ACO for generation of Pareto set of schedules, we employed the problem of 8 products to be processed on 4 machines under UIS policy. This problem was earlier solved by ACO for minimization of the single objective makespan. The algorithm for MOO generated a complete non-dominated set of schedules shown in Table 6. The algorithm generated 5 schedules of which schedule P5-P7-P1-P2-P6-P8-P4-P3 with makespan of 173 is the optimal schedule for single objective makespan problem. The schedule P5-P7-P2-P6-P8-P3-P4-P1 is the optimal schedule with the best ART value of 115.375. The optimal values in Table 6 show that the objectives: makespan and ART are conflicting in nature.

12.2. ACO for data clustering

To show the application of ACO for data clustering, we applied two well-known benchmark datasets: iris and wine, which are publicly available at UCI Machine Learning Repository (www.ics.uci.edu/~mlearn/MLRepository.html). Iris dataset consists of 150 samples defined by

Table 6. Pareto set generated by ACO for multiproduct chemical batch plant.

Makespan	ART	Corresponding sequences
205.0	115.375	P5-P7-P2-P6-P8-P3-P4-P1
183.0	116.000	P5-P7-P1-P6-P8-P3-P4-P2
182.0	116.375	P5-P7-P1-P3-P6-P8-P4-P2
175.0	116.875	P5-P7-P1-P6-P2-P8-P4-P3
173.0	118.375	P5-P7-P1-P2-P6-P8-P4-P3

four attributes, sepal-length, sepal-width, petal-length and petal-width. Using these attributes values, plant type iris-versicolor, iris-setosa and iris-virginica can be recognized. Thus, this dataset consists of samples belonging to three groups. The objective is to group these samples into three clusters, each belonging to a particular iris plant type. Wine dataset has a total of 178 samples defined by the following 13 attributes derived from chemical analysis: alcohol, malic acid, ash, alkalinity of ash, magnesium, total phenols, flavanoids, nonflavanoids phenols, proanthocyaninsm, color intensity, hue, OD280/OD315 of diluted wines and proline. This dataset consists of samples from three different cultivars. Thus, this dataset consists of samples belonging to three groups. The objective is to group these samples into three clusters, each belonging to a particular cultivar.

For this study, ACO parameters were set as: number of ants $A = 30$, exploitation probability $q_0 = 0.65$, evaporation rate $\rho = 0.9$, maximum number of iterations $I = 1{,}000$. We also applied simulated annealing (SA) for data clustering proposed by Selim and Al-sultan (1991). SA is easy to implement and known to be successful for solving different combinatorial optimization problems. The termination criterion for SA is kept as the maximum number of function evaluations. It is equivalent to the number of times the objective function (Eq. (19)) is executed during ACO simulation. With each method, 10 trials were carried out on each dataset. The results are reported as the average NFE over 10 different trials. Both ACO and SA converged to the optimal solutions in all the 10 trials. For clustering Iris data, ACO took 10,998 function evaluations while SA required 29,103 function evaluations to generate the optimal solution. For winc dataset,

ACO algorithm took somewhat more number of average function evaluations (9306) as compared to that of (7907) SA. Standard genetic algorithm was also tried for both iris and wine data clustering problems; it required 38,128 and 33,551 average function evaluations over 10 different trials respectively.

13. Conclusions

In this chapter, we have applied ACO for solving continuous, combinatorial and multiobjective process optimization problems. The illustrated examples include process synthesis problems belonging to MINLP type and multiproduct scheduling of batch plants. Application of ACO is also highlighted for solving data clustering problems, which occur in process engineering applications. In this study, we have employed simple local search process. One area of study is to develop different pheromone models in the local search operation. Recently, Socha and Dorigo (2008) have proposed ACO to continuous and mixed-variable problems. Research in this respect is ongoing and should result in new efficient ACO implementations for continuous and mixed-variable problems. ACO is particularly well suited for parallel implementations, since ants operate in an independent and asynchronous way. An overview of the trends and strategies in designing parallel meta-heuristics can be found in Randall and Lewis (2002). A parallel implementation of ACO for data clustering is interesting for speedy grouping of a large number of instances in a database. Similarly a parallel implementation of ACO for discovery of classification rules is more practical when the algorithm learns rules on large database. Dorigo and co-workers (e.g. Dorigo and Gambardella, 1997; Gambardella, Taillard and Dorigo, 1999; Gambardella and Dorigo, 2000) have studied ACO for solving combinatorial optimization problems. In MOO using ACO, an important research direction is how effectively the pheromone updating can be implemented for handling different objectives. Different pheromone updating methods have been proposed in the literature such as pheromone deposition independently by objectives, weighted contribution based on importance of objectives, Pareto solutions based etc. A comprehensive study comparing these approaches is warranted for MOO by ACO.

Abbreviations

ACO	Ant Colony Optimization
CACO	Continuous Function Ant Colony Optimization
CHA	Continuous Hybrid Algorithm combining Genetic and Nelder–Mead Simplex Algorithms
CTSS	Continuous Tabu Simplex Search Algorithm
DETL	Differential Evolution with Tabu List
$(\mu + \lambda)$ES	μ(parents) + λ(children) Evolutionary Strategy
GA	Genetic Algorithm
MINLP	Mixed Integer Non-linear Programming
M-SIMPSA	Minlp Simplex Simulated Annealing Algorithm
M-SIMPSA-pen	Minlp Simplex Simulated Annealing Algorithm with Penalty Function
NFE	Number of Function Evaluations
RW	Random Walk operation in CACO Algorithm
SA	Simulated Annealing
SR	Success Rate
TD	Trail Diffusion operation in CACO Algorithm
UIS	Unlimited Intermediate Storage Policy
ZW	Zero Wait Policy

References

Banga, J.R. and Seider, W.D. (1996). Global optimization of chemical processes using stochastic algorithms. In Floudas, C.A. and Pardalos, P.M. (eds.), *State of the Art in Global Optimization*. Dordrecht, The Netherlands, Kluwer Academic.

Biegler, L.T. and Grossmann, I.E. (2004). Retrospective on optimization. *Computers and Chemical Engineering*, **28**, pp. 1169–1192.

Bilchev, G. and Parmee, I. (1995). The ant colony metaphor for searching continuous design spaces. *Lecture Notes in Computer Science*; Fogarty, T. (ed.), Springer-Verlag: Berlin, **993**, p. 25.

Bilchev, G., Parmee, I. and Darlow, A. (1996). The inductive genetic algorithm with applications to the fault coverage test code generation problem. *4th European Congress on Intelligent Techniques and Soft Computing*; *Proceedings, EUFIT '96*, Verlag Mainz: Mainz, Germany, **1**, p. 452.

Cardoso, M.F., Salcedo, R.L., Azevedo, S.F. and Barbosa, D. (1997). A simulated annealing approach to the solution of minlp problems. *Computers and Chemical Engineering*, **21**, pp. 1349–1364.

Chelouah, R. and Siarry, P. (2003). Genetic and Nelder–Mead algorithms hybridized for a more accurate global optimization of continuous multiminima functions. *European Journal of Operational Research*, **148**, pp. 335–348.

Chelouah, R. and Siarry, P. (2005). A hybrid method combining continuous tabu search and Nelder–Mead simplex algorithms for the global optimization of multiminima functions. *European Journal of Operational Research*, **161**, pp. 636–654.

Colorni, A., Dorigo, M., Maniezzo, V. and Trubian, M. (1994). Ant System for job-shop scheduling. *Belgian Journal of Operations Research, Statistics and Computer Science*, **34**, pp. 39–53.

Costa, D. and Hertz, A. (1997). Ants can color graphs. *Journal of the Operational Research Society*, **48**, pp. 295–305.

Costa, L. and Oliveira, P. (2001). Evolutionary algorithms approach to the solution of mixed integer non-linear programming problems. *Comput. Chem. Eng.*, **25**, pp. 257–266.

Doerner, K.F., Gutjahr, W.J., Hartl, R.F., Strauss, C. and Stummer, C. (2006). Pareto ant colony optimization with ILP preprocessing in multiobjective project portfolio selection. *European Journal of Operational Research*, **171**, pp. 830–841.

Dorigo, M. and Gambardella, L.M. (1997). Ant colony system: A cooperative learning approach to the traveling salesman problem. *IEEE Trans. Evolut. Comput.*, **1**, pp. 53–66.

Dréo, J. and Siarry, P. (2004). Continuous interacting ant colony algorithm based on dense heterarchy, *Future Generation Computer Systems*, **20**, pp. 841–856.

Floudas, C.A. (1995). *Non-Linear and Mixed-integer Optimizations Fundamentals and Applications*; Oxford University Press: Oxford, UK, 1995.

Fonseca, C.M. and Fleming, P.J. (1994). An overview of evolutionary algorithms in multiobjective optimization. *Technical Report*, Department of Automatic Control and Systems Engineering, University of Sheffield, Sheffield, UK.

Gambardella, L.M. and Dorigo, M. (1997). *HAS-SOP: An Hybrid Ant System for the Sequential Ordering Problem.* (Tech. Rep. 11–97). Lugano, Switzerland: IDSIA.

Gambardella, L.M., Taillard, E.D. and Agazzi, G. (1999). MACS-VRPTW: A multiple Ant Colony System for vehicle routing problems with time windows. Corne, D., Dorigo, M. and Glover, F. (eds.), *New Ideas in Optimization*. Maidenhead, UK: McGraw-Hill.

Gambardella, L.M., Taillard, E.D. and Dorigo, M. (1999). Ant colonies for the quadratic assignment problem. *Journal of the Operational Research Society*, **50**(2), pp. 167–176.

Gambardella, L.M. and Dorigo, M. (2000). Ant Colony System hybridized with a new local search for the sequential ordering problem, *INFORMS J. Comput.*, **12**, p. 237.

Han, J.W. and Kamber, M. (2001). *Data mining: Concepts and Techniques*. Morgan Kaufmann, San Francisco, CA.

Jayaraman, V.K., Kulkarni, B.D., Karale, S. and Shelokar, P. (2000). Ant colony framework for optimal design and scheduling of batch plants. *Computers and Chemical Engineering*, **24**(8), pp. 1901–1912.

Jeżowski, J., Bochenek, R. and Ziomek, G. (2005). Random search optimization approach for highly multi-modal non-linear problems. *Advances in Engng. Software*, **36**(8), pp. 504–517.

Kim, M., Jung, J.H. and Lee, I. (1996). Optimal scheduling of multiproduct batch processes for various intermediate storage policies. *Ind. Eng. Chem. Res.*, **35**, pp. 4058–4066.

Kocis, G.R. and Grossmann, I.E. (1987). Relaxation strategy for the structural optimization of process flow sheets. *Ind. Eng. Chem. Res.*, **26**, pp. 1869–1880.

Kocis, G.R. and Grossmann, I.E. (1988). Global optimization of non-convex mixed-integer non-linear programming (MINLP) problems in process synthesis. *Ind. Eng. Chem. Res.*, **27**, pp. 1407–1421.

Kong, M., Tian, P. and Kao, Y. (2008). A new ant colony optimization algorithm for the multidimensional Knapsack problem. *Computers and Operations Research*, **35**, pp. 2672–2683.

Luus, R. (2006). Parametrization in non-linear optimal control problems. *Optimization*, **55**, pp. 65–89.

Mathur, M., Karale, S.B., Priye, S., Jayaraman, V.K. and Kulkarni, B.D. (2000). Ant Colony Approach to Continuous Function Optimization. *Ind. Eng. Chem. Res.*, **39**, pp. 3814–3822.

Michalewicz, Z. (1996). *Genetic Algorithms + Data Structures = Evolution Programs*, Springer, New York.

Mo, K.J., Eo, S., Shin, D. and Yoon, E.S. (1998). Qualitative interpretation and compression of process data using clustering method. *Computers and Chemical Engineering*, **22**, pp. S555–S562.

Monmarché, N., Venturini, G. and Slimane, M. (2000). On how *Pachycondyla apicalis* ants suggest a new search algorithm. *Future Generation Computer Systems*, **16**, pp. 937–946.

Murthy, C.A. and Chowdhury, N. (1996). In search of optimal clusters using genetic algorithms. *Pattern Recognition Letters*, **17**, pp. 825–832.

Kirkpatrick, S., Gellatt, C.D. and Vecchi, M.P. (1983). Optimization by simulated annealing, *Science* **220**, pp. 671–680.

Parpinelli, R.S., Lopes, H.S., and Freitas, A.A. (2002). An ant colony algorithm for classification rule discovery. Abbas, H., Sarker, R. and Newton, C. (eds.), Data mining: A heuristic approach (pp. 191–208), Idea group publishing, London, UK.

Randall, M. and Lewis, A. (2002). A parallel implementation of ant colony optimization. *J. Parallel Distributed Comput.*, **62**, p. 1421.

Rouhiainen, C. and Tade M.O. (2003). Genetic algorithms for optimal scheduling of chlorine dosing in water distribution systems, *20th Convention of the Australian Water Association*, Perth, Australia, 6–10 April 2003.

Selim, S.Z. and Ismail, M.A. (1984). K-means type algorithms: a generalized convergence theorem and characterization of local optimality. *IEEE Trans. Pattern Anal. Mach. Inteli.*, **6**, pp. 81–87.

Selim, S.Z. and Al-sultan, K.S. (1991). A simulated annealing algorithm for the clustering problem, *Pattern Recognition*, **24**, pp. 1003–1008.

Shelokar, P.S., Jayaraman, V.K. and Kulkarni, B.D. (2003). Multiobjective optimization of reactor-regenerator system using ant algorithm. *Petroleum Science Technology*, **21**, pp. 1167–1184.

Shelokar, P.S., Jayaraman, V.K. and Kulkarni, B.D. (2004a). An ant colony classifier system: application to some process engineering problems. *Computers and Chemical Engineering*, **28**, pp. 1577–1584.

Shelokar, P.S., Jayaraman, V.K., and Kulkarni, B.D. (2004b). An ant colony approach for clustering. *Analytica Chimica Acta*, **509**, pp. 187–195.

Socha, K. and Dorigo, M. (2008). Ant colony optimization for continuous domains. *European J. Oper. Res.*, **185**, pp. 1155–1173.

Srinivas, M. and Rangaiah, G.P. (2007). Differential evolution with tabu list for solving non-linear and mixed-integer non-linear programming problems. *Ind. Eng. Chem. Res.*, **46**, pp. 7126–7135.

Sung, C.S. and Jin, H.W. (2000). A tabu-search-based heuristic for clustering, *Pattern Recognition*, **33**, pp. 849–858.

T'kindt, V., Monmarchë, N., Tercinet, F. and Laügt, D. (2002). An ant colony optimization algorithm to solve a 2-machine bicriteria flowshop scheduling problem. *European Journal of Operational Research*, **142**, pp. 250–257.

van Kampen, A.H.C., Ramadan, Z., Mulholland, M., Hibbert, D.B. and Buydens, L.M.C. (1997). Learning classification rules from an ion chromatography database using a genetic based classifier system. *Analytica Chimica Acta*, **344**, pp. 1–15.

Venkatasubramanian, V., Vaidyanathan, R. and Yamamoto, Y. (1990). Process fault detection and diagnosis using neural networks-I. Steady-state processes. *Computers and Chemical Engineering*, **14**(7), pp. 699–712.

Venkatasubramanian,V., Rengaswamy, R. and Kavuri, S.N. (2003). A review of process fault detection and diagnosis: Part II: Qualitative models and search strategies. *Computers and Chemical Engineering*, **27**, pp. 313–326.

Welch, J.W. (1983). Algorithmic Complexity: Three NP-hard Problems in Computational. Statistics, *J. Stat. Comput. Sim.*, **15**, pp. 17–25.

Wilson, J.A. and Martinez, E.C. (1997). Neuro-fuzzy modeling and control of a batch process involving simultaneous reaction and distillation. *Computers and Chemical Engineering*, **21**(1), pp. S1233–S1238.

Wodrich, M. and Bilchev, G. (1997). Cooperative distributed search: the ants' way. *Control Cybernetics*, **26**, pp. 413–446.

Yoo, C.K., Lee, I.-B. and Vanrolleghem, P.A. (2005). Interpreting patterns and analysis of acute leukemia gene expression data by multivariate fuzzy statistical analysis. *Computers and Chemical Engineering*, **29**, pp. 1345–1356.

Acknowledgement

Author VKJ acknowledges financial support received under grant no: SR/S3/CE/050/2009 from the DST, New Delhi, India. Author PSS acknowledges a partial financial support received under Fast Track Scheme for Young Scientist from the DST, New Delhi, India.

Chapter 8

PARTICLE SWARM OPTIMIZATION FOR SOLVING NLP AND MINLP IN CHEMICAL ENGINEERING

Bassem Jarboui, Houda Derbel and Mansour Eddaly

FSEGS, route de l'aéroport km 4
Sfax 3018, Tunisie

Patrick Siarry*

LiSSi, Université de Paris 12
61 avenue du Général de Gaulle
94010 Créteil, France
siarry@univ-paris12.fr

1. Introduction

Social insects consist of many interacting individuals that live in colonies and can act together more efficiently than an individual insect. A large number of optimization problems are solved by insect colonies, especially social ones. Among those problems are the real world optimization short path problem, when foraging for food, and the assignment problem, when allocating tasks between workers. Interaction between swarm insects is performed through direct or indirect communication, depending on the nature of information to be communicated. For instance, a social bee can

*Corresponding author.

communicate the location of a food source directly by making a characteristic dance in presence of other bees, by physical contact, by exchange of food or liquid or any local information throughout the group. Other forms of indirect communication require that an individual should modify some characteristics of the environment to guide the behavior of other members at a later time. Consequently, swarm intelligence is based on systems of natural behaviors concerning more than one agent.

The idea of particle swarm optimization was originated from the observation of scientists about the social behavior of birds and fishes. There is an impressive choreography when fishes and birds move in a synchronous way without colliding. This process was simulated by Reynolds (1987). He assumed that the flocks of birds, which are called boids, are guided by three objectives: to avoid collision with a neighbor, to match the speed of a neighbor and to join the center of the flock. Socially, the three concepts are still present with humans, but in a much more complex way. For example, flock centering is analogous to the convergence of two minds to the same point of the cognitive space, which is called an agreement. From an evolutionary computation point of view, a swarm is analogous to a population and a particle to an individual.

Particle Swarm Optimization (PSO) is an important tool of swarm intelligence used for solving Global Optimization (GO) problems. This optimization technique started from simulation of the social behavior by Kennedy and Eberhart (1995). PSO is an effective technique for GO problems due to its easy implementation, inexpensive computation and low memory requirements (Eberhart and Kennedy, 1995). We begin this chapter by defining particle swarms and looking at rules that control their evolution, to end with some characteristics and applications of PSO.

A typical PSO algorithm can be described as follows: a set of entities, called particles, move in the search space of an objective function. Each particle has to decide its movement through an evaluation of the function and its current location. This is based on the previous situation related to the current particle and the other members of the swarm.

Although PSO and Genetic Algorithm (GA) have some common characteristics, they are quite different. On one hand, both algorithms start with a swarm randomly generated and use random methods to update the population. On the other hand, PSO does not include genetic operators like mutation and crossover, in contrast with GA. Salman *et al.* (2002) carried out a comparative study between PSO and GA on the randomly generated mapping problem instances. The results showed that not only the solution given by PSO has a better quality but it is less time consuming. Some researchers considered hybrid approaches of PSO with GA. For example, Shi *et al.* (2005) hybridized PSO with a variable size population GA. In this work, each of GA and PSO is executed separately. After that, the members of their respective solutions are mixed according to their fitness.

The remaining parts of this chapter are organized as follows. The initial algorithm is detailed in Sec. 2. Section 3 presents simple modifications added to the original PSO. Section 4 deals with some features of the PSO algorithm; to begin with, we discuss modifications to the basic PSO to ensure convergence as well as the particle swarm structure; then we describe two important basic PSO algorithms, which are *global best* and *local best PSO*. Both binary and discrete PSO are presented in Sec. 5. Sections 6 and 7 deal with general properties of a multi-start algorithm and the gravity center technique. The application of PSO to Mixed-Integer Nonlinear Programming (MINLP) and in particular to Chemical Engineering problems are detailed in Secs. 8 and 9. Section 10 summarizes the computational results obtained by applying different variants of PSO in the setting of MINLP in Chemical Engineering. Finally, we conclude with some ideas of future directions.

2. The Initial Version of the PSO Algorithm

The initial program was a simulation of a bird flock (Reynolds, 1987; Heppner and Grenander, 1990). The simulation includes birds that look for food by learning the behavior of nearby birds to the source. Let minimize $f(\vec{x})$ subject to $\vec{x}^L \le \vec{x} \le \vec{x}^U$ be a global optimization problem to solve. PSO randomly assigns particles to space positions and velocities. Each particle i consists of three vectors with dimension D (D being the dimension of the search space). These are the current position $\vec{x}_i$, the previous best position $\vec{p}_i$ and the velocity $\vec{v}_i$ in the search space. The current position $\vec{x}_i$ represents the problem solution and its coordinates. If the current position is better than the previous best one, then its coordinates are used to update the vector $\vec{p}_i$. The best value of best function $f(\vec{x}_i)$ is stored in a variable called $pbest_i$. Each particle will act according to the behavior of its neighbor and is influenced by the best point found by the particles in the neighborhood of the particle i, denoted by $\vec{p}_g$. The velocity is the distance traveled by a particle to reach a new position. A neighborhood is simply represented by a graph where vertexes are particles: if an edge connects two of them, then they can interact together. $\vec{p}_i$ and $\vec{p}_g$ will represent the range of variation of velocity of one particle.

The *original algorithm* for implementing PSO is described below. It includes local search algorithms by self experience and global search ones by neighboring experience. This feature allows intensification, as well as diversification.

(1) Generate random positions and velocities for a population of particles. Initialize $\vec{p}_i$ to the starting position and $pbest_i$ to the fitness.
(2) Evaluate the fitness/objective function of the position $\vec{x}_i$ for each particle.
(3) Update $pbest_i$ and $\vec{p}_i$ with the best values.
(4) Store the index of the best particle in the neighborhood in the variable g.
(5) Update velocity and position of the particle at iteration t as follows:

$$\vec{v}_i(t+1) = \vec{v}_i(t) + \varphi_1 \vec{r}_1(\vec{p}_i(t) - \vec{x}_i(t)) + \varphi_2 \vec{r}_2(\vec{p}_g(t) - \vec{x}_i(t)), \quad (1)$$

$$\vec{x}_i(t+1) = \vec{x}_i(t) + \vec{v}_i(t). \quad (2)$$

Here, $\vec{r}_1$ and $\vec{r}_2$ are two random vectors variables following the uniform distribution between 0 and 1, φ_1 and φ_2 are weights of the stochastic acceleration to adjust the oscillation of one particle around $\vec{p}_i$ and $\vec{p}_g$. Coefficients φ_1 and φ_2 are used to strengthen the attraction of the particle to the best position. They control the maximum step size of the particle, and are called "learning factors".
(6) Return to the second step until a stopping criterion (such as maximum number of iterations and/or desired fitness) is reached.

Equations (1) and (2) require that each particle is pulled to the best location depending on the neighborhoods in the problem space. Hence, Eq. (1) is made up of three components as follows.

- The previous velocity $\vec{v}_i$ which stores the last movement of the particle. This memory helps the particle to save the flight direction and change in the current direction.

- The cognitive component $\varphi_1 \vec{r}_1(\vec{p}_i(t) - \vec{x}_i(t))$ is the process by which a particle acquires information on its environment and about best positions. This component helps the particles to find the equilibrium by reaching their suitable positions. Kennedy and Eberhart (1995) describe this part as the "nostalgia" of the particle.
- The social component $\varphi_2 \vec{r}_2(\vec{p}_g(t) - \vec{x}_i(t))$ measures the performance of a particle i in comparison with other particles in the neighborhood.

3. Simple Modifications to the Original PSO

Although basic PSO is applied to solve different problems such as permutation problems (Salerno, 1997) and multilayer neural networks (Eberhart and Kennedy, 1995; Eberhart and Shi, 1998), the convergence towards good solutions remains a problem. For this reason, some concepts, namely, inertia weight and velocity constriction, are developed to improve the convergence of the basic PSO. These are discussed as follows.

3.1. Maximum velocity

In the original version of PSO, each component of the vector $\vec{v}_i$ is within the range $[-V_{\max}, +V_{\max}]$ so that the balance between exploration and exploitation is ensured. If the velocity of any particle exceeds $V_{\max}$, then it will be set to the maximum allowed. If $V_{\max}$ is large enough, then particles could reach good solutions, otherwise particles explore the space of solutions too slowly. Thus, $V_{\max}$ is an important parameter that may help to override local optima.

3.2. Inertia weight

The original Eq. (1) is modified by including a coefficient w for $\vec{v}_i$ (Shi and Eberhart, 1998), which eliminates the need for $V_{\max}$ that can lead to a non-converging trajectory of a particle. The modified Eqs. (1) and (2) are:

$$\vec{v}_i(t+1) = w\vec{v}_i(t) + \varphi_1 \vec{r}_1(\vec{p}_i(t) - \vec{x}_i(t)) + \varphi_2 \vec{r}_2(\vec{p}_g(t) - \vec{x}_i(t)), \quad (3)$$

$$\vec{x}_i(t+1) = \vec{x}_i(t) + \vec{v}_i(t+1). \quad (4)$$

The coefficient w is a tuning parameter of the velocity, and is called the "inertia weight". Larger the inertia weight is, larger the exploration of the current search will be. The three parameters: w, φ_1 and φ_2 have great influence on the performance of the algorithm. Experiments showed that, when setting inertia weight, the maximum velocity strategy may be avoided (Engelbrecht, 2007). In fact, $V_{\max}$ could be simply set to the value of the dynamic range of each variable. This allows a particle to adequately explore the search subspaces.

If we interpret the two last terms in Eq. (3) as an exogenous force $\vec{f_i}$ that attracts the particle, then the rate of change in particle's velocity is $\Delta\vec{v}_i = \vec{f_i} - (1 - w)\vec{v}_i$. The term $(1 - w)$ can be interpreted as a disagreement factor. Consequently, the inertia weight can be interpreted as the fluidity of the medium in which a particle moves, as mentioned in Poli *et al.* (2007). Different ways are proposed to adjust the inertia weight. For instance, w was adapted with a fuzzy system in Eberhart and Shi (2000) and it was effectively adjusted using a random component in Eberhart and Shi (2001).

The use of the inertia weight controls the development of the search and improves the performance of PSO algorithm in various applications. When $w \geq 1$, velocities are increasing and particles deviate from good directions, whereas when $w < 1$, particles slow down to a velocity equal to 0, which makes the local exploitation easier. A method that is usually used to vary the inertia factor is weighting with temporal decrease. This strategy consists in decreasing this coefficient over the time. Each iteration represents a time increment. The main idea is to slow the progress of the particle in order to save the optimum and hope to converge when the number of iterations increases (Van Den Bergh, 1999). However, this method may increase computational time for convergence.

3.3. Constriction factor

Researchers noticed that if the PSO algorithm does not encompass velocity, the system "explodes" within a few iterations. Kennedy (1998) showed that the trajectories of non-stochastic one-dimensional particles contained interesting regularities when $0 \leq \varphi_1 + \varphi_2 \leq 4.0$. The first attempt to look at trajectories of a particle was reported in Ozcan and Mohan (1999).

Their analysis showed that particles were "surfing the waves" of underlying sinusoidal curves. After that, Clerc and Kennedy (2002) demonstrated that the results given in Ozcan and Mohan (1999) were the signature of a five-dimensional attractor. Hence, they developed constriction factor in Eq. (1) to simultaneously achieve three important targets: to control the convergence of the particle, to prevent explosion and to eliminate the arbitrary setting of $V_{\max}$.

Among different ways to implement the constriction factor, we present one simple strategy to perform this task, mentioned in Clerc and Kennedy (2002):

$$\vec{v}_i(t+1) = \chi(\vec{v}_i + \varphi_1 \vec{r}_1(\vec{p}_i(t) - \vec{x}_i(t)) + \varphi_2 \vec{r}_2(\vec{p}_g(t) - \vec{x}_i(t))), \quad (5)$$

$$\vec{x}_i(t+1) = \vec{x}_i(t) + \vec{v}_i(t+1), \quad (6)$$

where $\varphi = \varphi_1 + \varphi_2 > 4$, $k \in [0, 1]$ and the constriction factor is $\chi = \frac{2k}{\left|2-\varphi-\sqrt{\varphi^2-4\varphi}\right|}$. In this approach, φ often is taken equal to 4.1 ($\varphi_1 = \varphi_2 = 2.05$), $k = 1$ and $\chi \approx 0.7298$. This is equivalent to an inertial approach such that $w = 0.729$ and $\varphi_1 = \varphi_2 = 1.49445$.

3.4. Stopping criteria

A stopping criterion must be chosen arbitrarily while taking into consideration the convergence of the population (Engelbrecht, 2007). The convergence is detected when the population of the candidate solutions is stagnant. Various termination criteria are proposed in the literature.

- Maximum number of iterations: This condition is usually used in the case of a time restriction to evaluate the best solution.
- An acceptable solution has been found: Based on this criterion, the algorithm will stop when a particle $\vec{x}_i$ is close to the known or desired optimum $\vec{x}^*$ i.e. $|f(\vec{x}_i) - f(\vec{x}^*)| \leq \varepsilon, \varepsilon \geq 0$ and f is the fitness function.
- No improvement is observed during a given number of iterations: an improvement could be observed at the level of particle position or velocity. In fact, the algorithm is stopped as soon as the average change in particle position or velocity is small enough to guarantee the convergence.

4. Neighborhood Selection Strategies

Although a particle is independent from other particles in the same neighborhood, neighborhoods are founded on spatial similarity. However, the development of neighborhood in a basic PSO is always based on particle indices, which helps not only to avoid expensive computation of spatial ordering of particles but also to transmit information about good solutions to all particles independent of their location. In a PSO algorithm, there is a communication between neighborhoods, which pushes particles towards a single position represented by the global best particle.

4.1. Global best / Local best PSO

There exist two basic PSO algorithms, named *global best* (*gbest*) and *local best* (*lbest*) *PSO*. These algorithms are distinguished by the size of their neighborhood. As indicated by its name, the neighborhood of a *gbest PSO* is the global swarm. Hence, the velocity of a particle i in Eq. (1) refers to the best position $\vec{p}_g(t)$ obtained by the whole swarm. Let n be the total number of particles in the swarm; then, for a minimization problem, the global best position $\vec{p}_g(t)$ is equal to the vector defined by: $f(\vec{p}_g(t)) = \min_{i=1,\ldots,n}\{f(\vec{p}_i(t))\}$. Algorithm 1 details the different steps of the *gbest* PSO.

In the *local best PSO*, smaller neighborhoods are considered. Particles exchange information locally, according to partial knowledge of the solution space. In the *lbest PSO* case, the velocity of a particle depends on the best position in the neighborhood of particles. The local best position $\vec{p}_{g_i}(t)$ of particle i is the vector defined by: $f(\vec{p}_{g_i}(t)) = \min_{j\in\Omega_i}\{f(\vec{p}_j(t))\}$ where Ω_i denotes the neighborhood of particle i. Algorithm 2 below details the different steps of *lbest PSO*.

Note that *gbest PSO* is a particular case of *lbest PSO,* where the neighborhood corresponds to all particles. Both *gbest* and *lbest PSO* are seeking the global best particle. Because of the limited diversity, *lbest PSO* is less likely to fall into a local optimum. On the other hand, *gbest PSO* converges more rapidly than *lbest PSO* due to larger particle inter-connectivity (Eberhart *et al.*, 1996; Krink *et al.*, 2002).

Algorithm 1: *gbest* PSO

```
Initialize the particle positions and their velocities
loop
  for i = 1, ..., n
    //Update the personal best position
    if (f(x⃗_i) < f(p⃗_i)) then
      p⃗_i = x⃗_i;
    end if
    //Update the global best position
    g = 1;
    for j = 2, ..., n
      if (f(p⃗_j) < f(p⃗_g)) then
        g = j;
      end if
    end for
    for j = 1, ..., D
      //Update the particle velocity and position
      v_ij(t + 1)=w v_ij(t) + φ_1 r_1j(p_ij − x_ij(t)) + φ_2 r_2j(p_gj − x_ij(t))
      x_ij(t + 1) = x_ij(t) + v_ij(t + 1)
    end for
  end for
until stopping criterion is reached
```

Algorithm 2: *lbest* PSO

```
Initialize the particle positions and their velocities
loop
  for i = 1, ..., n
    //Update the personal best position
    if (f(x⃗_i) < f(p⃗_i)) then
      p⃗_i = x⃗_i;
    end if
  //Update the local best position
  g_i = i;
  for j ∈ Ω_i
```

if $(f(\vec{p}_j) < f(\vec{p}_{g_i}))$ **then**
$g_i = j;$
end if
end for
for $j = 1, \ldots, D$
//Update the particle velocity and position
$v_{ij}(t+1) = wv_{ij}(t) + \varphi_1 r_{1j}(p_{ij} - x_{ij}(t)) + \varphi_2 r_{2j}(p_{g_i j} - x_{ij}(t))$
$x_{ij}(t+1) = x_{ij}(t) + v_{ij}(t+1)$
end for
end for
until stopping criterion is reached

5. Extensions of Basic PSO

Various modifications of basic PSO have been proposed to perform a search in binary or discrete spaces. PSO was an effective technique for optimizing continuous variables as well as discrete variables. For the latter case, values of velocities and positions are discretized after updating according to Eq. (1). Many redefinitions of discrete PSO algorithms are proposed in the literature and the way of discretization depends on the nature of the studied problem. Kennedy and Eberhart (1997) have presented a binary version of PSO. The component values of $\vec{x}_i$ must be taken in the set $\{0, 1\}$ and the velocity gives the probability that a bit will take the value 1. Laskari *et al.* (2002) have proposed a method based on truncation of real values to their nearest integer. Liao *et al.* (2007) have proposed an extension to the binary PSO algorithm of Kennedy and Eberhart (1997) for solving a flowshop scheduling problem. They have shown that the new proposed algorithm is competitive in comparison both with continuous PSO of Tasgetiren *et al.* (2004), developed for the same problem, and genetic algorithms. Miranda and Fonseca (2002) have proposed to use a probabilistic rounding rather than the direct rounding to the next integer. Salman *et al.* (2002) have adopted a continuous version of PSO for the task assignment problem, which is classified as a discrete combinatorial optimization problem. They simply converted real values to positive integers by dropping the sign and the fraction part. Yin (2004) has presented a discrete particle swarm algorithm for optimal polygonal approximation of digital curves. The works of Jarboui *et al.* (2007 and 2008) deal with combinatorial PSO for solving interesting problems such as partitional clustering and permutation flowshop. Pan *et al.* (2008) have developed a discrete particle swarm optimization algorithm for the no-wait flowshop scheduling problem. In their study, the particles were coded as discrete job permutations and a new position update method was proposed based on the discrete domain.

Binary PSO is included in discrete PSO, which was first developed by (Kennedy and Eberhart, 1997). A particle position is coded by a binary vector. It takes the value of 1 if it changes. In this case, according to Eq. (1), the number of bits that are changed from one iteration to another represents the velocity. Therefore, if the particle does not move, then the velocity is equal to zero. On the other hand, for binary PSO, the velocity component of a particle can be interpreted as the probability threshold to be 1 or 0. For example, $v_{ij}(t) = 0.6$ means that the jth component of the velocity of particle i is 60% of 1. These terms could be normalized in the interval [0, 1] by simply dividing them by $V_{\max}$. This kind of normalization may lead to bad solutions rapidly if $V_{\max j} \gg v_{ij}$ (Engelbrecht, 2007). A better way of generating probabilities is the use of the sigmoid function $s(x_{ij}(t)) = \frac{1}{1+e^{-x_{ij}(t)}}$ which evaluates positions as follows:

$$x_{ij}(t) = \begin{cases} 1 & \text{if } r_{ij} < s(x_{ij}(t)) \\ 0 & \text{otherwise} \end{cases}, \tag{7}$$

with r_{ij} is a variable uniformly distributed in the range [0, 1].

6. Multi-Start PSO Algorithms

Multi-start algorithms refer to strategies that increase the diversity at the time of convergence of the swarm towards the same point, by introducing a continual randomness in the swarm. Kennedy and Eberhart (1995) were the pioneers in considering the randomness in the PSO process, called craziness. Since then many questions such as how to evaluate craziness and what are the consequences on the swarm, have been studied.

Position vectors and velocity vectors can be randomly initialized, either simultaneously or separately (Løvberg, 2002; Schutte and Groenuold,

2005). If position vectors are kept constant and velocities are randomized, then particles can save their current and previous best positions and the solution space is diversified. If the search has not provided a new best solution, then the particle continues its moves towards its personal best position. If the position vectors are randomly initialized, particles acquire new personal positions because of the loss of the memory (Løvberg, 2002; Løvberg and Krink, 2002). So, the particle is prevented from being stuck towards its previously best found position. In order to reinitialize the velocities, each velocity component must be set to a random value while taking into account the maximum allowed value. This procedure is frequently used in the literature. On other hand, new particle positions are randomly reinitialized in the entire search space using uniformly distributed random numbers (Engelbrecht, 2007). Reinitializing either particle's position or velocity is deeply related with the decision of when this will be done. Several alternatives in the literature address this question (e.g. Clerc, 1995 and 2004; Van Den Bergh, 2002).

7. Gravity Center Technique

As was seen in the previous section, randomization process can lead to loss of memory in some cases. To overcome this drawback, it is possible to reorganize memories so that they communicate with one another. This idea is achieved by either mixing memories or developing the center of gravity technique (Clerc, 2005). The basic idea behind the gravity center technique is that when the memory swarm starts to converge, the particle which has more chance to be the best is the center of gravity. In the work of Clerc (1999), this point is called the queen and calculated as a new temporary particle. In fact, the gravity center is a good alternative for piloting the search in particle swarm techniques.

8. Application of PSO to Mixed-Integer Nonlinear Programming (MINLP)

8.1. Applications to constrained problems

In general, a constrained optimization problem specifies restrictions on the search space. These restrictions can be classified as boundary constraints, equality constraints and inequality constraints, which can be linear or nonlinear. There exist various handling approaches applied in evolutionary algorithms such as ignoring infeasible solutions, transforming the problem to an unconstrained problem, ranking solutions according to the degree of violation. For solving constrained problems with PSO, the main idea is to deal with infeasible particles. If the number of infeasible particles is too small, then a simple way to deal with constraints is to exclude infeasible particles from being the personal best or the neighborhood global best solutions. Another approach to get rid of infeasible particles is to replace them with new positions from the feasible space, which are randomly generated. Other ways of PSO applied to constrained problems use penalty function approach (Parsopoulas and Vrahatis, 2002; Tandom *et al.*, 2002).

We are interested in penalty function approaches, which add a penalty to the objective function to penalize infeasible solutions. Let the MINLP problem be in the following form:

$$\begin{aligned}
&\text{minimize} \;\; f(\vec{x}, \vec{y}) \\
&\text{subject to} \\
&\qquad g_i(\vec{x}, \vec{y}) \le 0, \quad i = 1, \ldots, m_g \\
&\qquad h_j(\vec{x}, \vec{y}) = 0, \quad j = 1, \ldots, m_h, \\
&\qquad \vec{x}^L \le \vec{x} \le \vec{x}^U, \\
&\qquad \vec{y}^L \le \vec{y} \le \vec{y}^U,
\end{aligned} \tag{8}$$

where $\vec{x}$ and $\vec{y}$ are vectors of continuous and discrete variables respectively. When applying a penalty method, the objective function takes the new form: $f(\vec{x}, \vec{y}) + p(\vec{x}, \vec{y})$ where $p(\vec{x}, \vec{y})$ is the penalty function on violated constraints.

PSO was applied for various applications and schemes. It is a successful tool not only in swarm intelligence but also for solving nonlinear optimization problems with constraints. Nonlinear programming (NLP) refers to optimizing problems in the presence of equality and inequality constraints. If we have continuous and discrete variables, the problem is called a mixed-integer nonlinear programming (MINLP). Several examples in engineering disciplines can be formulated as a nonlinear program due to the nonlinearity of the objective function or constraints. Different solution techniques are proposed to solve MINLP problems. In general, those approaches are suitable for convex problems. When applied to non-convex problems, these algorithms could move away from the global optimum. Yiqing *et al.* (2007) have developed an improved PSO algorithm for solving non-convex NLP/MINLP problem with equality and/or inequality constraints. They showed that their improved PSO is more advantageous than numerous stochastic optimization techniques from the viewpoint of excluding local optima. Because engineering problems are formulated as combinatorial optimization problems, the discrete binary version of PSO is adopted to solve those problems.

8.2. Applications in chemical engineering

Chemical engineering problems often require optimization algorithms that are able to use nonlinear equality constraints. Indeed, problems involving constraints are always treated by eliminating infeasible solutions, as mentioned in Sec. 10. This is essentially provided by penalty and barrier function methods for solving nonlinear problems. However, when the number of constraints is large, performance of these approaches can be unsatisfactory. Yiqing *et al.* (2007) pay attention to such cases. They propose a transformation of the classic MINLP to facilitate the handling of penalty functions with PSO algorithm. They test their approach on eight test problems studied by different authors in the literature (Costa and Olivera, 2001; Wang and Smith, 1994). These problems illustrate different examples in Chemical Engineering such as planning, production control and network design in water utilization process.

9. Application of PSO to the NLP and MINLP Problems

Several aspects involved in the PSO application to the NLP and MINLP problems are outlined in this section.

9.1. Initialization

We begin with generating an initial population by using uniformly distributed random numbers to cover the whole search space. This is performed using the equation: $x_{ij} = x_j^L + r_j \times (x_j^U - x_j^L) \forall i = 1, \ldots, n, j = 1, \ldots, D$

where r_{ij} is a random value following the uniform distribution between 0 and 1.

9.2. Handling discrete variables and bound checking

Integer and binary variables in the MINLP present in our application are simply treated as continuous variables and rounded to the nearest integer in the vector solution. Usually, PSO research is guided to find the optimal value of the fitness function inside a valid hyperspace. This is conducted by enforcing particles satisfy bounds. If any of lower or upper bounds is violated, then the particle is set exactly to the violated bound. This approach gives good computational efficiency when the global optimum is located at the bounds of the decision variables. In other cases, it can be detrimental to the convergence of the algorithm.

9.3. Handling constraints

In this study, constraints are handled using the penalty function strategy. Each infeasible solution is penalized in the objective function by adding the following sum:

$$K + R\left\{\sum_{i=1}^{m_g} \max\left\{0, g_i(\vec{x}, \vec{y})\right\} + \sum_{j=1}^{m_h} \left|h_j(\vec{x}, \vec{y})\right|\right\}, \tag{9}$$

where,

$$K = \begin{cases} M & \text{if } \sum_{i=1}^{m_g} \max\{0, g_i(\vec{x}, \vec{y})\} + \sum_{j=1}^{m_h} \left|h_j(\vec{x}, \vec{y})\right| > 0 \\ 0 & \text{otherwise} \end{cases}, \tag{10}$$

with $M > \max_{i=1,\ldots,n}(f_i(\vec{x}, \vec{y}))$. The penalty function consists of a constant factor K and the sum of constraint violations multiplied by a high value R. The value of this function is added to the objective function for a minimization problem and subtracted from it in the case of maximization. We introduce the constant factor K in order to maximize the efficiency of the algorithm. Even if the parameter R is a large value, we do not have any guarantee to check an infinitely small violation of constraints. So, the use of R only is not enough to force solutions to the feasible space.

10. Numerical Results

Gbest origin, gbest CF, gbest WTD, lbest origin, lbest CF and lbest WTD were coded in C++ programming language. The source codes of the two programs (*gbest* and *lbest*) are available on the CD accompanying the book, and the way of using these programs are described in the source codes. All experiments were run in Windows XP on a desktop PC with Intel Pentium IV, 3.2 GHz processor and 512 MB memory.

The performances of the six PSO algorithms were tested on a set of 24 examples that can be classified into two classes, according to the nature of their variables. The first one consists of 16 examples belonging to the NLP where all the variables are continuous, whereas the second is composed of 8 examples belonging to the MINLP, where some variables are binary or integer. Reference and number of variables in these problems are given in Tables 1 and 2. These 24 examples were recently used for testing differential evolution with tabu list by Srinivas and Rangaiah (2007), who provided their mathematical formulations as an Excel file available at

Table 1. NLP benchmark problems.

Examples	References	Inequality constraints	Real variables
NLP-1	Deb (2000)	2	2
NLP-2	Himmelblau (1972)	38	5
NLP-3	Deb (2000)	9	13
NLP-4	Deb (2000)	6	8
NLP-5	Deb (2000)	4	7
NLP-6	Deb (2000)	3	5
NLP-7	Deb (2000)	8	10
NLP-8	Ryoo and Sahinidis (1995)	9	2
NLP-9	Umeda and Ichikawa (1971)	0	2
NLP-10	Sahinidis and Grossmann (1991)	1	2
NLP-11	Ryoo and Sahinidis (1995)	4	2
NLP-12	Visweswaran and Floudas (1990)	12	6
NLP-13	Ryoo and Sahinidis (1995)	4	1
NLP-14	Ryoo and Sahinidis (1995)	3	2
NLP-15	Ryoo and Sahinidis (1995)	2	2
NLP-16	Ryoo and Sahinidis (1995)	8	3

Table 2. MINLP benchmark problems.

Examples	References	Inequality constraints	Real variables	Binary variables	Integer variables
MINLP-1	Ryoo and Sahinidis (1995)	2	1	1	0
MINLP-2	Kocis and Grossmann (1987)	1	1	1	0
MINLP-3	Floudas (1995)	3	2	1	0
MINLP-4	Kocis and Grossmann (1989)	4	2	1	0
MINLP-5	Floudas (1995)	9	3	4	0
MINLP-6	Cardoso *et al.* (1997)	3	3	0	2
MINLP-7	Salcedo (1992)	13	7	0	3
MINLP-8	Salcedo (1992)	61	16	0	6

Table 3. Parameters of algorithms.

Algorithms	Parameters
gbest origin and lbest origin	Three parameters are fixed in this variant: number of initial particles $n = 100$, the weights of the stochastic acceleration φ_1 and φ_2 are equal to 1.5.
gbest CF and lbest CF	Number of initial particles $n = 100$ as in *origin* variant, $\varphi_1 = \varphi_2 = 2.05$ and $\chi = 0.7298$. So, referring to the inertial approach, $w = 0.729$ and $\varphi_1 = \varphi_2 = 1.49445$.
gbest WTD and lbest WTD	For these algorithms, $n = 100$, $\varphi_1 = \varphi_2 = 1.5$ and w varies decreasing from 1 to 0.75 by a factor of 0.9995 after each generation.

http://pubs.acs.org/doi/suppl/10.1021/ie070007q. All constraints in these examples are formulated as inequality constraints.

All parameters of each algorithm are fixed experimentally following the nature of *origin, CF or WTD* variant (i.e. *gbest origin* and *lbest origin* have the same parameters and similarly for *gbest CF* and *lbest CF* and *gbest WTD* and *lbest WTD*). The values of the parameters of the algorithms are given in Table 3. These parameters are set taking into account the solution quality, the CPU time and previous fixed parameters in the literature of PSO. We have set 1,000 generations as a stopping criterion for each replication.

The performance measure employed in our numerical study is the Global Optimum Found Rate (GOFR), which is equal to the number of times that each algorithm reaches the global optimum divided by the number of replications. In our experiments, we have taken 100 replications. A replication is qualified as successful if the global optimum is obtained with an absolute error of 10^{-6}.

Table 4 presents the GOFR results for NLP. For NLP-1, *gbest WTD*, *gbest CF* and *lbest CF* have a GOFR of 100% in finding the global optimum and *lbest WTD* is classified in the second range with a GOFR of 99%. For NLP-2, *gbest WTD* is the best one in terms of GOFR. Among 100 replications, *lbest origin* is unable to reach the optimal value. However, *gbest CF*, *gbest WTD* and *lbest CF* succeeded in finding the global optimum 84 times, 83 times and 82 times, respectively. For NLP-6, *gbest CF* and *gbest WTD* outperform other algorithms. In fact, these two variants have a GOFR of 100%. Also, it's seen that *lbest WTD* and *lbest CF* are better than *gbest origin*, and they are able to obtain the global optimum 96 times and 93 times among 100 replications, respectively.

Table 4. GOFR of the PSO algorithms for NLP problems.

	gbest origin	*gbest CF*	*gbest WTD*	*lbest origin*	*lbest CF*	*lbest WTD*
NLP-1	87	**100**	**100**	27	**100**	99
NLP-2	69	**84**	83	0	82	72
NLP-3	0	0	0	0	0	0
NLP-4	0	0	0	0	0	0
NLP-5	0	0	0	0	0	0
NLP-6	0	**100**	**100**	0	96	93
NLP-7	0	0	0	0	0	0
NLP-8	22	**50**	**50**	8	14	14
NLP-9	**100**	**100**	**100**	6	**100**	**100**
NLP-10	**100**	**100**	**100**	91	**100**	**100**
NLP-11	99	94	98	99	**100**	99
NLP-12	0	**93**	92	0	54	38
NLP-13	**100**	86	87	99	61	76
NLP-14	11	43	**51**	11	20	14
NLP-15	13	**41**	40	1	11	5
NLP-16	43	**99**	98	6	59	48

For NLP-8, *gbest CF* appears to be the best one; it has a GOFR of 50% similar to *gbest WTD*. The algorithms, *lbest WTD* and *lbest CF* are better than *lbest origin* in terms of GOFR (14% for *lbest CF* and *lbest WTD* and 8% for *lbest origin*). However, *gbest* versions remain better than *lbest* versions. For NLP-9 and NLP-10, all approaches, except *lbest origin*, have converged to the global optimum value with GOFR = 100%. For NLP-11, although the results of different algorithms are very close regarding the GOFR, *lbest CF* is the best one with GOFR of 100%. *Gbest origin*, *lbest origin* and *lbest WTD* dominate the remaining approaches. For NLP-12, *gbest CF* is the best one with GOFR of 93% and *gbest WTD* is the second one with GOFR of 92%. For NLP-13, *gbest origin* is the best and *lbest origin* is second best with GOFR of 100% and 99% respectively. For NLP-14, *gbest WTD* outperforms all other approaches. It is able to reach the optimal value 51 times among 100 replications. For NLP-15, *gbest CF* slightly dominates *gbest WTD,* the first has GOFR of 41% and the second has GOFR of 40%. Among the *lbest* version algorithms, *lbest CF* is the best one for GOFR. For NLP-16, *gbest CF* and *gbest WTD* are the best algorithms, with 99% and 98 % as GOFR, respectively. We note that, for NLP-3, NLP-4, NLP-5 and NLP-7, all algorithms have not reached the global optimum for any replication.

Table 5 refers to the obtained results for the MINLP. For MINLP-1, *gbest origin* and *gbest CF* outperform in terms of GOFR (100% for both algorithms). Moreover, *gbest* version performs better than *lbest* version. For MINLP-2 and MINLP-4, *gbest origin* seems better than other algorithms with GOFR of 88% for the first problem and 98% for the second. For MINLP-3, *gbest CF* is the best one in terms of GOFR (73%). For MINLP-5, *gbest WTD* is slightly better than other approaches, with GOFR of 26%. For MINLP-6, all approaches, except *lbest origin*, have reached the global optimum value, with GOFR of 100%. For MINLP-7 and MINLP-8, the GOFR is around of 0%.

Figures 1–12 present the average value of the objective function obtained, for each proposed approach, out of 100 replications, according to the number of generations. Figure 1 presents the obtained results for NLP-1 and shows that *gbest WTD* and *gbest CF* appear better than all other approaches, in terms of solution's quality and speed of convergence. Regarding *lbest* variants, *lbest CF* is better than *lbest WTD* and *lbest origin*.

Table 5. GOFR of the PSO algorithms for MINLP problems.

	gbest origin	*gbest CF*	*gbest WTD*	*lbest origin*	*lbest CF*	*lbest WTD*
MINLP-1	**100**	**100**	99	98	88	83
MINLP-2	**88**	83	83	50	55	55
MINLP-3	64	**73**	67	2	28	28
MINLP-4	**98**	87	92	72	71	66
MINLP-5	24	21	**26**	15	15	24
MINLP-6	**100**	**100**	**100**	92	**100**	**100**
MINLP-7	0	0	0	0	0	0
MINLP-8	0	0	0	0	0	0

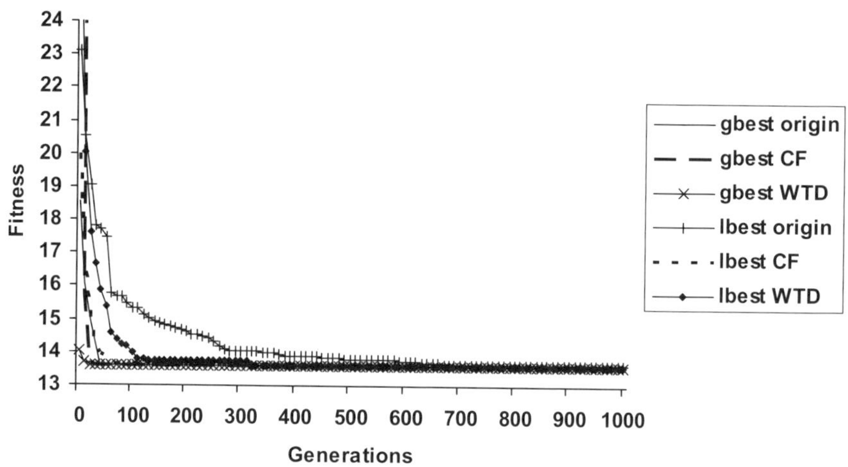

Figure 1. Results for NLP-1.

The latter algorithm needs a lot of iterations to converge. In Fig. 2, it is clear that the results found for NLP-2 by *lbest origin* are far away from other algorithms results. *gbest WTD* and *gbest CF* are the best ones. The results of NLP-5 are illustrated in Fig. 3. We can see that, except for the *lbest origin* approach, all others are very close.

The obtained results for NLP-6 are presented in Fig. 4 and show that the curves of *gbest CF* and *gbest WTD* are superimposed and these two

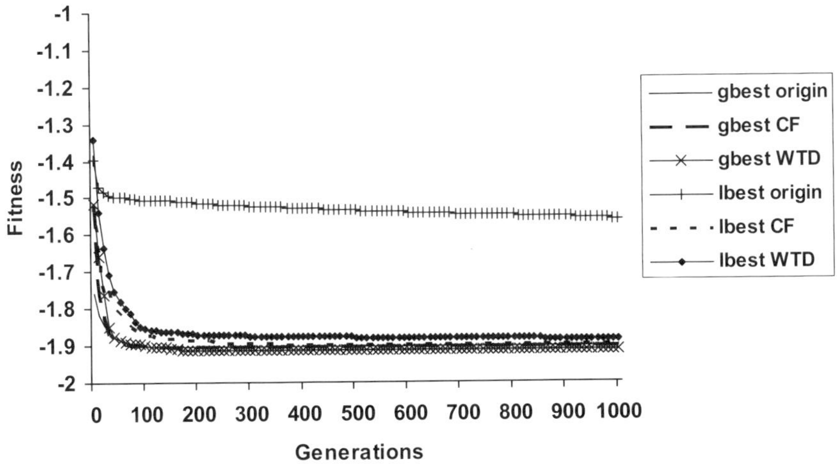

Figure 2. Results for NLP-2.

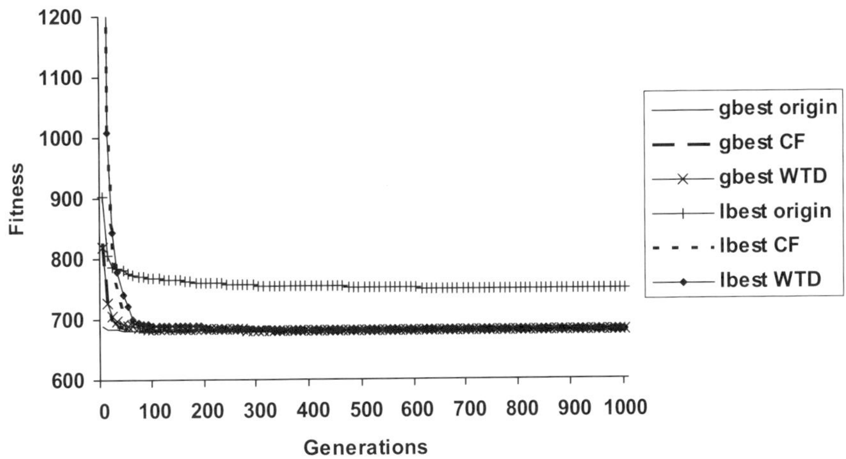

Figure 3. Results for NLP-5.

algorithms outperform the remaining algorithms. For *lbest* version, *lbest CF* is better than *lbest WTD* and *lbest origin*. Figure 5 shows that all algorithms, except for *lbest origin*, have converged to the global optimum of NLP-10. Furthermore, *gbest* versions are very similar and are faster than

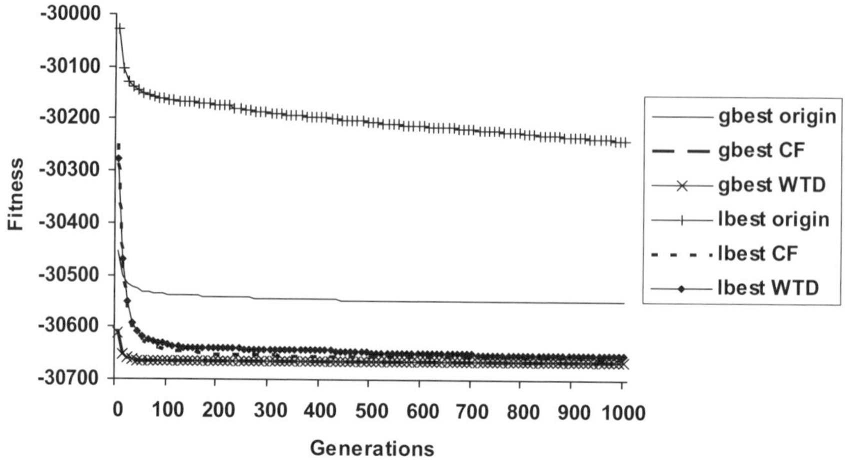

Figure 4. Results for NLP-6.

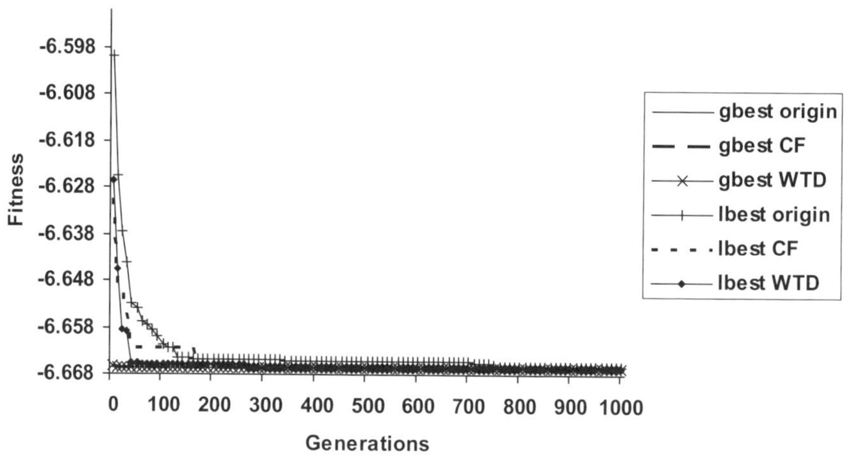

Figure 5. Results for NLP-10.

lbest versions in term of speed of convergence. *lbest WTD*, *lbest CF* and *lbest origin* are ranked in such order regarding their speed of convergence over 1,000 generations. In Fig. 6, *lbest* version is better than *gbest* version for solving NLP-11. In addition, *lbest CF* is better than other approaches on average. From Fig. 7, we can observe that, on average, *gbest CF*, *gbest WTD*,

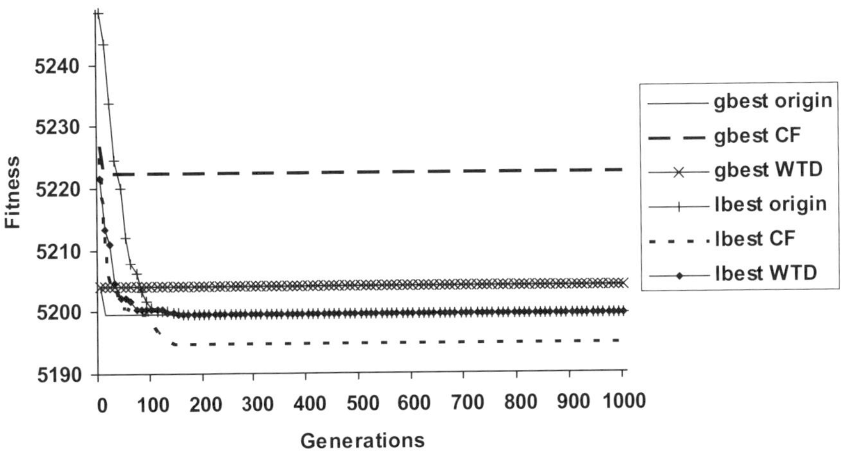

Figure 6. Results for NLP-11.

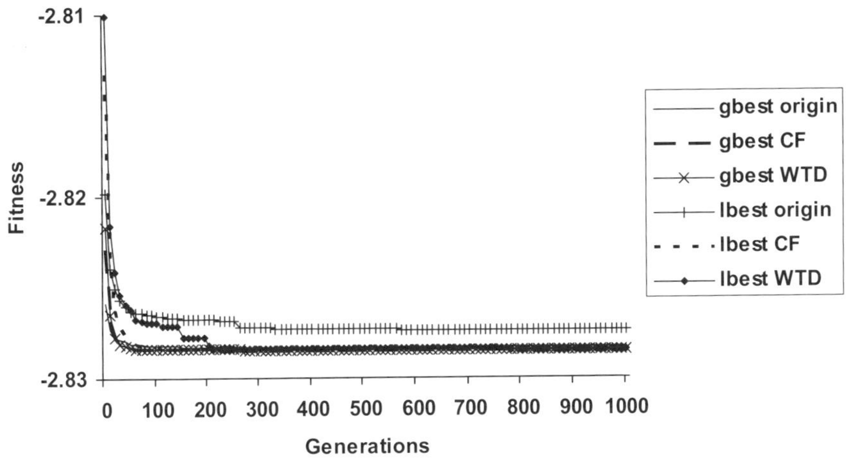

Figure 7. Results for NLP-14.

lbest CF and *lbest WTD* have converged to the same value for NLP-14. Figure 8 presents the obtained results for NLP-16 and shows that the curves of *gbest* variant approaches are below the *lbest* approaches curves. Consequently, *gbest* performs better than *lbest* in average.

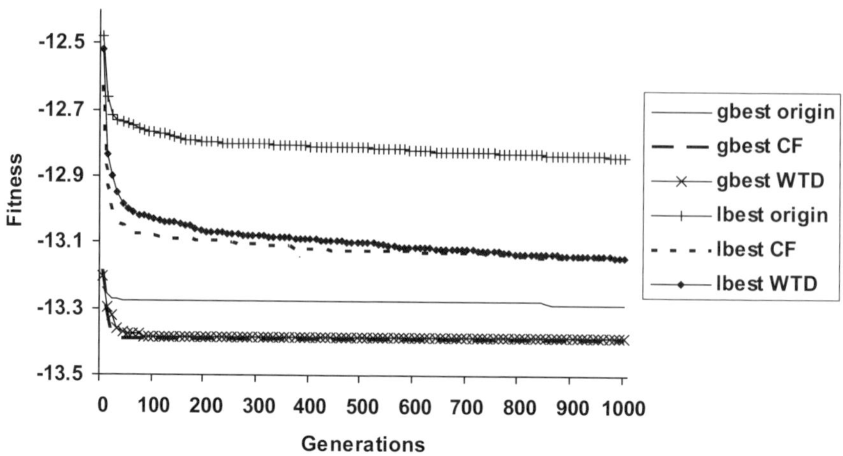

Figure 8. Results for NLP-16.

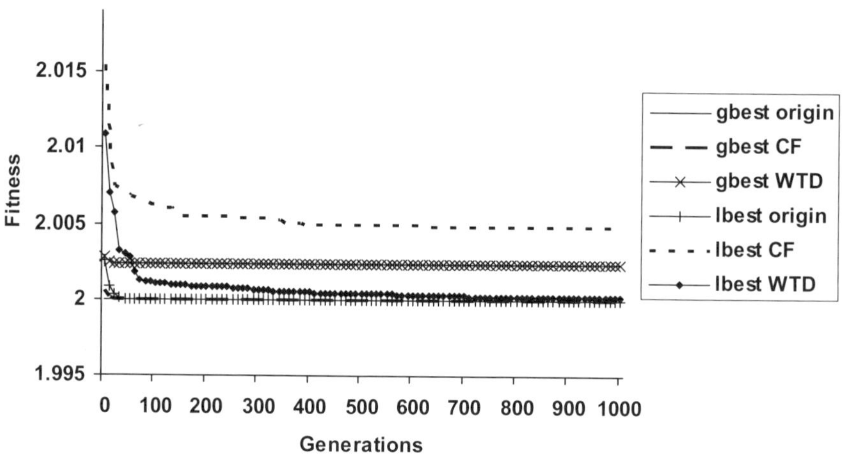

Figure 9. Results for MINLP-1.

Figures 9–12 show the results provided by the different algorithms for solving some MINLP problems. Figure 9 shows that *gbest origin* and *gbest CF* outperform the others, but *gbest CF* requires more iterations to reach the global minimum of MINLP-1 than *gbest origin*. Besides, *lbest origin* performs better than *lbest CF* and *lbest WTD*, on average. Figure 10 shows that *gbest* algorithms are better than *lbest* algorithms, and that *gbest CF* is

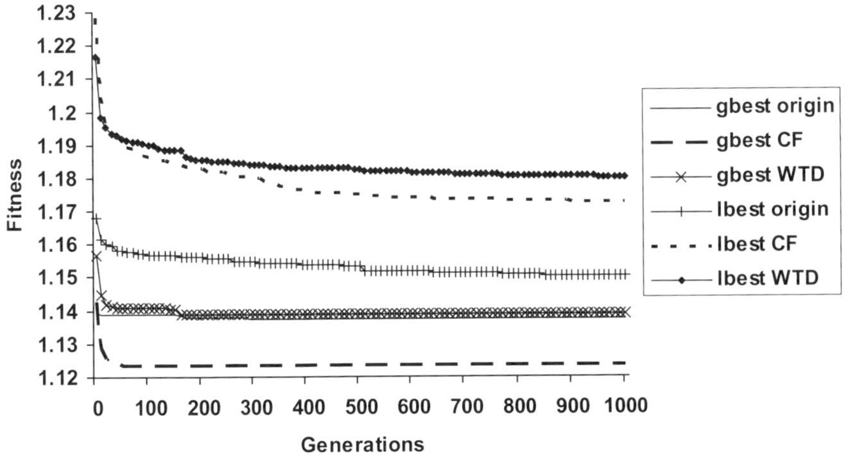

Figure 10. Results for MINLP-3.

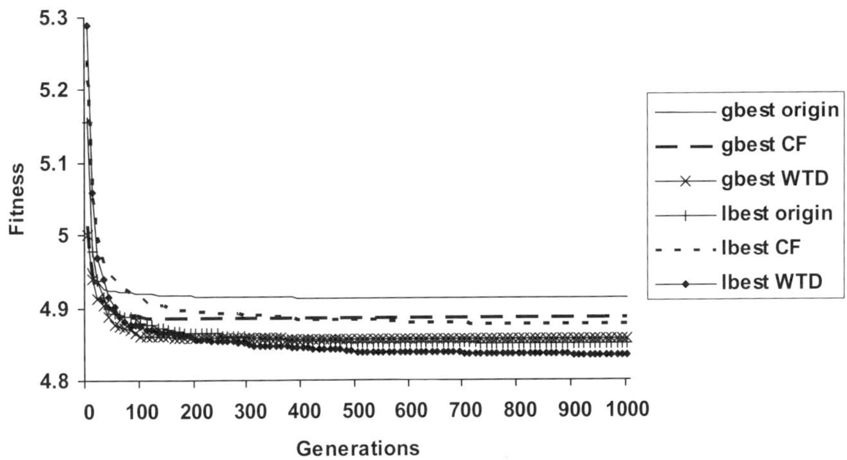

Figure 11. Results for MINLP-5.

the best algorithm. Figure 11 shows that, in term of the fitness function, on average, *lbest WTD* dominates other approaches as the number of generations increases, and that, among *gbest* versions, *gbest WTD* is the best one. Figure 12 shows that *gbest CF* and *gbest WTD* outperform other algorithms and that they have almost the same shape of curves.

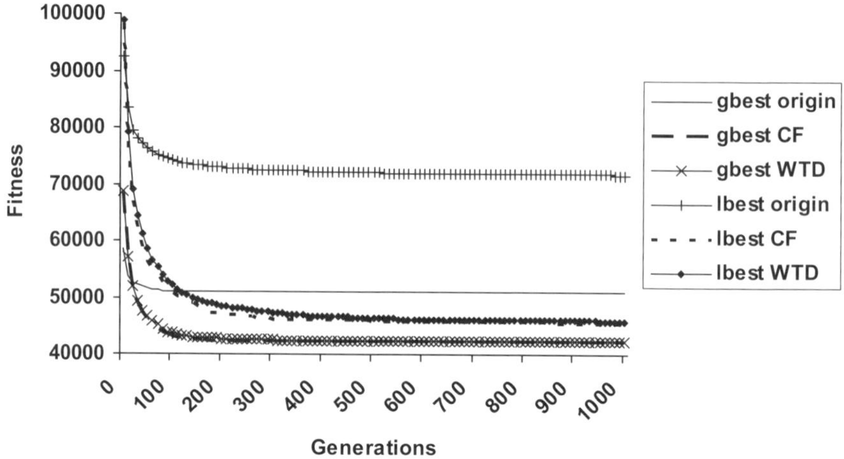

Figure 12. Results for MINLP-7.

To conclude, *gbest* versions have a higher speed of convergence. Therefore, these algorithms can provide good solutions in short time but, sometimes, this rapidity can be detrimental to the solution's quality, as shown in Figure 6, where *gbest CF* has provided bad solutions although it has converged in the first generations. On the other hand, *lbest* versions seem unable to find the global optimum in reasonable time. In addition, we find that the constriction factor variant and the weighted timing decreasing variant are almost similar, both in the *gbest* and *lbest* versions. Concerning the *gbest origin* and *lbest origin* versions, the former is able to find good results some times (Figs. 4, 11 and 12), whereas the latter remains the worst variant. Finally, *gbest CF* and *gbest WTD* are the most stable variants, which can provide good results in reasonable time.

Concerning the CPU times, we have observed that all algorithms appear almost similar and require 0.1 sec on average. However, *gbest* seems slightly faster than the others with 0.07 sec. In general, most proposed approaches provided good results for the problems with small number of variables and constraints. But, for large scale problems, some approaches are unable to reach the global optimum consistently and converge rapidly to a local optimum. This is caused either by small size of initial population or by insufficient diversification.

11. Conclusion and Future Directions

In this chapter, we have reviewed the different variants of PSO algorithm for solving NLP and MINLP formulated in Chemical Engineering. The proposed algorithms can be classified into two versions, called *gbest* and *lbest* version. In each version, three variants are considered: original PSO, constriction factor PSO and weighted timing decreasing PSO variant. So, in total six algorithms were developed. We have performed computational experiments based on test problems available in the Chemical Engineering literature. The results show that most proposed approaches are efficient in finding the global optimum for the problems with small size. But, for large scale problems, when the number of variables increases, these approaches encounter some difficulties and converge in a local optimum. As future direction, we suggest to introduce a diversification process by applying the multi-start strategies for re-initializing either the velocity components or/and the position components. Besides, also adding an improvement phase to the PSO algorithm can be beneficial in terms of solution quality and computational time. This can be done through hybridization with another evolutionary algorithm.

References

Cardoso, M.F., Salcedo, R.L., Feyo de Azevedo, S. and Barbosa, D. (1997). A simulated annealing approach to the solution of MINLP problems. *Computers and Chemical Engineering*, **21**, pp. 1349–1364.

Clerc, M. (1999). The swarm and the queen: towards a deterministic and adaptive particle swarm optimization. *Proceedings of the IEEE Congress on Evolutionary Computation*, **3**, pp. 1951–1957.

Clerc, M. and Kennedy, J. (2002). The particle swarm explosion, stability and convergence in a multidimensional complex space. *IEEE Transactions on Evolutionary Computation*, **6**(1), pp. 58–73.

Clerc, M. (2004). Discrete particle swarm optimization. *New Optimization Techniques in Engineering*, *Lecture Notes in Computer Science*, Springer-Verlag, Vol. **3612**.

Clerc, M. (2005). *Particle Swarm Optimization*, Hermes-Lavoisier.

Costa, L. and Olivera, P. (2001). Evolutionary algorithms approach to the solution of mixed integer nonlinear programming problems. *Computers and Chemical Engineering*, **25**, pp. 257–266.

Deb, K. (2000). An efficient constraint handling method for genetic algorithms. *Computer Methods in Applied Mechanics and Engineering*, **186**, pp. 311–318.

Eberhart, R. and Kennedy, J. (1995). A new optimizer using particle swarm theory. *Proceedings of the Search International Symposium on Micro Machine and Human Science*, pp. 39–43.

Eberhart, R.C., Simpson, P.K. and Dobbins, R.W. (1996). *Computational Intelligence PC Tools*, Academic Press Professional, 1st edition.

Eberhart, R.C. and Shi, Y. (2000). Comparing inertia weights and constriction factors in particle swarm optimization. *Proceedings of the IEEE International Conference on Evolutionary Computation (CEC)*, pp. 84–88.

Eberhart, R.C. and Shi, Y. (2001). Teaching and optimizing dynamic systems with particle swarms. *Proceedings of the IEEE International Conference on Evolutionary Computation (CEC)*, pp. 94–100.

Engelbrecht, A.P. (2007). *Computational Intelligence: An Introduction*, John Wiley and Sons.

Floudas, C.A. (1995). *Nonlinear and Mixed-Integer Optimization, Fundamentals and Applications*, Oxford University Press, Oxford.

Heppner, F. and Grenander, U. (1990). *A Stochastic Non-linear Model for Coordinated Bird Flocks*. AAAS Publications, Washington, DC.

Himmelblau, D.M. (1972). *Applied Nonlinear Programming*, McGraw-Hill, New York.

Jarboui, B., Cheikh, M., Siarry, P. and Rebai, A. (2007). Combinatorial particle swarm optimization (CPSO) for partitional clustering problem. *Applied Mathematics and Computation*, **192**, pp. 337–345.

Jarboui, B., Ibrahim, S., Siarry, P. and Rebai, A. (2008). A combinatorial particle swarm optimization for solving permutation flowshop problems. *Computers and Industrial Engineering*, **54**, pp. 526–538.

Kennedy, J. and Eberhart, R.C. (1995). Particle swarm optimization. *Proceedings of IEEE International Joint Conference on Neural Networks*, pp. 1942–1948.

Kennedy, J. and Eberhart, R.C. (1997). A discrete binary version of the particle swarm algorithm. *Proceedings of the World Multiconference on Systemics, Cybernetics and Informatics*, pp. 4104–4109.

Kennedy, J. (1998). The behavior of particles. *Evolutionary Programming VII, Lecture Notes in Computer Science*, Vol. **1447**, pp. 581–589.

Kocis, G.R. and Grossmann, I.E. (1987). Relaxation strategy for the structural optimization of process flow sheets. *Industrial and Engineering Chemistry Research*, **26**, pp. 1869–1880.

Krink, T., Vesterstrom, J.S. and Riget, J. (2002). Particle swarm optimization with spatial particle extension. *Proceedings of the 4th Congress on Evolutionary Computation*, **2**, pp. 1474–1479.

Laskari, E.C., Parsopoulas, K.E. and Vrahatis, M.N. (2002). Particle swarm optimization for integer programming. *Proceedings of the IEEE 2002 Congress on Evolutionary Computation*, Honolulu (HI), pp. 1582–1587.

Løvberg, M. and Krink, T. (2002). Extending particle swarm optimizers with self-organized criticality. *Proceedings of the IEEE Congress on Evolutionary Computation*, **2**, pp. 1588–1593.

Løvberg, M. (2002). *Improving Particle Swarm Optimization by Hybridization of Stochastic Search Heuristics and Self-Organized Criticality*, Master's Thesis, Department of Computer Science, University of Aarhus, Denmark.

Liao, C.J., Tseng, C.T. and Luarn, P. (2007). A discrete version of particle swarm optimization for flowshop scheduling problems. *Computers and Operations Research*, **34**, pp. 3099–3111.

Miranda, V. and Fonseca, N. (2002). New evolutionary particle swarm algorithm applied to voltage/var control. *14th Power Systems Computation Conference*.

Ozcan, E. and Mohan, C.K. (1999). Particle swarm optimization: surfing the waves. *Proceedings of the Congress of Evolutionary Computation*, **3**, pp. 1939–1944.

Pan, Q.K., Tasgetiren, M.F. and Liang, Y.C. (2008). A discrete particle swarm optimization algorithm for the no-wait flowshop scheduling problem. *Computers and Operations Research*, **35**, pp. 2807–2839.

Parsopoulas, K.E. and Vrahatis, M.N. (2002). Particle swarm optimization method for constrained optimization problems. *Intelligent Technologies , Theory and Applications: New trends in Intelligent Technologies*, Sincak, P., Vascak, J., Kvasnicka, V. and Pospichal, J. (eds.), pp. 214–220.

Poli, R., Kennedy, J. and Blackwell, T. (2007). Particle Swarm optimization: An overview. *Swarm Intelligence*.

Reynolds, C.W. (1987). Flocks, herds and schools: a distributed behavioral model. *Computer Graphics*, **21**(4), pp. 25–34.

Ryoo, H.S. and Sahinidis, N.V. (1995). Global optimization of non-convex NLPs and MINPs with applications in process design. *Computers and Chemical Engineering*, **19**, pp. 551–566.

Sahinidis, N.V. and Grossmann, I.E. (1991). Convergence properties of generalized benders decomposition. *Computers and Chemical Engineering*, **15**, pp. 481–491.

Salcedo, R.L. (1992). Solving non-convex nonlinear programming and mixed-integer nonlinear programming problems with adaptive random search. *Industrial and Engineering Chemistry Research*, **31**, pp. 262–273.

Salman, A., Ahmad, I. and Almadani, S. (2002). Particle swarm optimization for task assignment problem. *Microprocessors and Microsystems*, **26**, pp. 363–371.

Salerno, J. (1997). Using the particle swarm optimization technique to train a recurrent neural model. *Proceedings of the IEEE International Conference on Tools with Artificial Intelligence*, pp. 45–49.

Schutte, J.F. and Groenwold, A.A. (2005). A study of global optimization using particle swarms. *Journal of Global Optimization*, **31**, pp. 93–108.

Shi, Y. and Eberhart, R.C. (1998). A modified particle swarm optimizer. *Proceedings of the IEEE International Conference on Evolutionary Computation (CEC)*, pp. 69–73.

Shi, X.H., Liang, Y.C., Lee, H.P., Lu, C. and Wang, L.M. (2005). An improved GA and a novel PSO-GA-based hybrid algorithm. *Processing Letters*, **93**, pp. 255–261.

Srinivas, M. and Rangaiah, G.P. (2007). Differential evolution with Tabu list for solving nonlinear and mixed-integer nonlinear programming problems. *Industrial and Engineering Chemistry Research*, **46**, pp. 7126–7135.

Tandom, V., El-Mounayri, H. and Kishawy, H. (2002). Nc end milling optimization using evolutionary computation. *International Journal of Machine Tools and Manufacture*, **42**(5), pp. 595–605.

Tasgetiren, M.F., Liang, Y.C., Sevkli, M. and Gencyilmaz, G. (2004). Particle swarm optimization algorithm for makespan and maximum lateness minimization in permutation flowshop sequencing problem. *Proceedings of the 4th International Symposium on Intelligent Manufacturing Systems*, Turkey, Sakarya.

Umeda, T. and Ichikawa, A.A. (1971). A modified complex method for optimization. *Industrial and Engineering Chemistry Process Design and Development*, **10**, pp. 229–236.

Van Den Bergh, F. (1999). Particle swarm weight initialization in multi-layer perceptron artificial neural network. *Development and Practice of Artificial Intelligence Techniques* (Durban, South Africa), pp. 41–45.

Van Den Bergh, F. (2002). *An Analysis of Particle Swarm Optimizers*, PhD Thesis, Department of Computer Science, University of Computer Science, University of Pretoria, South Africa.

Visweswaran, V. and Floudas, C.A. (1990). A global optimization algorithm (GOP) for certain classes of non-convex NLPs — II. Application of theory and test problems. *Computers and Chemical Engineering*, **14**, pp. 1419–1434.

Wang, Y.P. and Smith, R. (1994). Wastewater minimization. *Chemical Engineering Science*, **49**(7), pp. 981–1006.

Yin, P.Y. (2004). A discrete particle swarm algorithm for optimal polygonal approximation of digital curves. *Journal of Visual Communication and Image Representation*, **15**, pp. 241–260.

Yiqing, L., Xigang, Y. and Yongjian, L. (2007). An improved PSO algorithm for solving non-convex NLP/MINLP problems with equality constraints. *Computers and Chemical Engineering*, **31**, pp. 153–162.

Chapter 9

AN INTRODUCTION TO THE HARMONY SEARCH ALGORITHM

Gordon Ingram* and Tonghua Zhang

Centre for Process Systems Computations
Department of Chemical Engineering
Curtin University of Technology
GPO Box U1987, Perth WA 6845, Australia
**g.ingram@curtin.edu.au*

1. Introduction

The Harmony Search (HS) draws its inspiration not from a biological or physical process like most other meta-heuristic optimization techniques, but from an artistic one — the improvisation process of musicians. Geem *et al.* (2001) explain the analogy between optimization and musical performance that embodies the spirit and language of HS:

Global optimization	Improvised musical performance
• Seeks a global optimum by testing values for each decision variable, which has a set of feasible values	• Seeks a wonderful harmony by playing notes on each instrument, which has a range of playable pitches
• Is judged by the objective function	• Is judged by an aesthetic standard
• Requires a number of iterations	• Involves a number of improvisations

As Yang (2008) remarks, when musicians improvise they have several choices: playing an existing score from memory, performing variations on an existing piece, or creating an entirely new composition. Like other stochastic global optimization (SGO) methods, HS 'combines rules and randomness' to imitate the optimizing process that inspired it (Lee and Geem, 2005). The basic HS includes three operations drawn from the observations above: (a) playing a note from a collection of favored notes stored in memory, (b) playing a note close in pitch to one that is in memory, and (c) randomly playing a note from the entire range of the instrument. The operations are loosely depicted in Fig. 1. These operations allow a mixture of global exploration of the variable space and local solution refinement that have analogs in other SGO methods. HS handles discrete and continuous variables with similar ease, and can incorporate constraints as needed.

The principal creator and champion of HS is Zong Woo Geem, who developed the algorithm through combined interests in civil engineering and music. Geem *et al.* (2001) is most often cited as introducing HS. Since 2001, interest in HS methods has been increasing as suggested in Fig. 2. Work has proceeded into applying HS to various problems and on enhancing the algorithm itself. Around half the studies published to date use some modification of the basic method. Yang (2008) regards HS as one of the major modern techniques for meta-heuristic optimization.

The primary aim of this chapter is to introduce the Chemical Engineering community to the HS method and its current applications. The algorithm has been reported to be a competitive alternative to other SGO techniques in related engineering disciplines, but has received little attention for process applications. Being a relatively young algorithm, there are good opportunities for exploring variations on the basic method to tailor it to Chemical

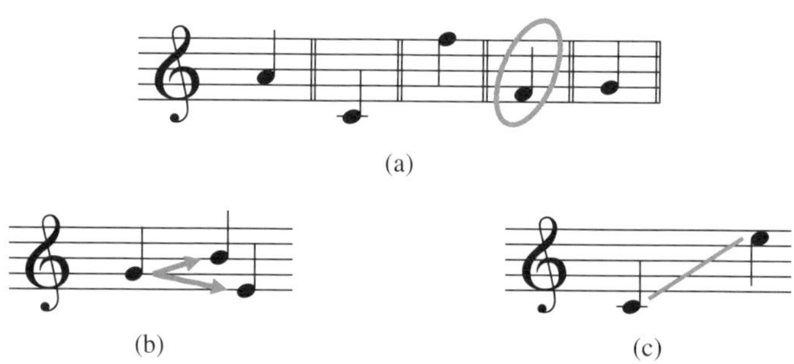

Figure 1. Operations of the basic Harmony Search: (a) playing from memory, (b) pitch adjusting, and (c) random playing.

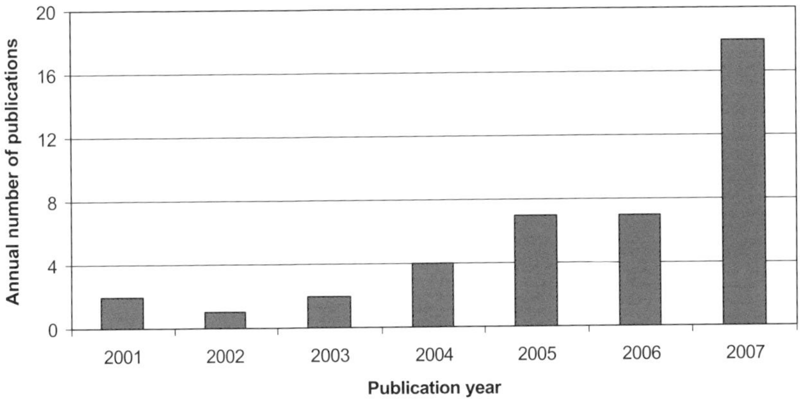

Figure 2. The growing interest in Harmony Search techniques. (Source: Compendex + Inspec database search, topic: "harmony search" in all fields, duplicates removed, 18 July 2008.)

Engineering problems. With this in mind, the secondary aim of the chapter is to provide a convenient summary of the many modifications to the basic algorithm. This is the first time that a comprehensive summary of HS modifications has been presented in the literature.

The chapter is arranged as follows. Section 2 outlines the applications of the HS, including the few in Chemical Engineering. The basic HS algorithm is described in detail in Sec. 3, while Sec. 4 summarizes many modifications on the basic method. Section 5 covers the availability of HS software. In Sec. 6, an illustrative example is briefly presented. Conclusions and future research directions are indicated in Sec. 7. The CD-ROM provides supplementary material on HS applications in Sec. 2, Matlab codes from Sec. 6, and sample solutions to the Exercises at the end of the chapter.

2. Applications of the Harmony Search

A range of optimization applications have used HS methods as presented in Fig. 3. The most common type of problem tackled using HS is design, comprising about a third of published applications to date (Fig. 3a). 'General optimization', which refers principally to testing HS on benchmark problems, is the second most frequent application at ~20%. Other kinds of optimization problems that use HS — parameter estimation;

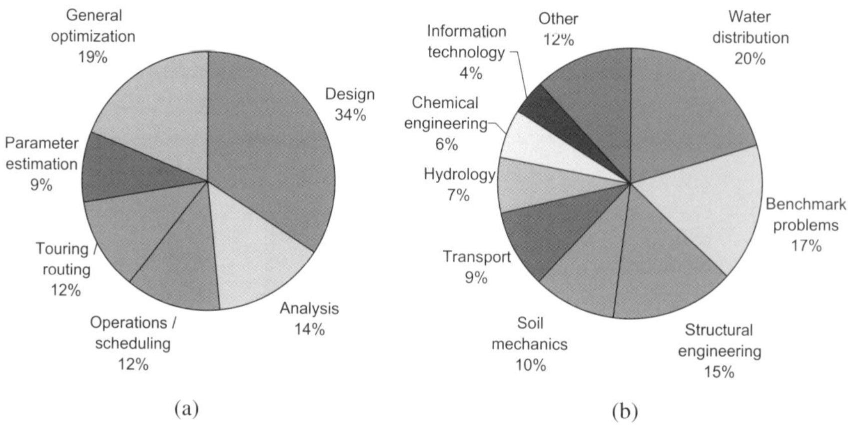

Figure 3. Approximate categorization of published applications of the Harmony Search: (a) by type of optimization problem, and (b) by discipline area.

touring/route finding problems, such as the traveling salesman problem; operations/scheduling; and analysis — all have approximately the same number of published applications. 'Analysis' in this context describes problems such as clustering web documents (Mahdavi *et al.*, 2008), detection of abnormalities in medical images (Dong *et al.*, 2008) and performing slope stability analysis (Cheng *et al.*, 2008).

Currently there are few applications related to Chemical Engineering (Fig. 3b). Most HS applications are associated with water distribution, structural engineering and benchmark problems, perhaps reflecting Geem's background in civil engineering and the relative youth of the algorithm. The 'other' category in Fig. 3(b) includes solving Sudoku puzzles (Geem, 2007a), musical composition (Geem and Choi, 2007), ecological conservation (Geem and Williams, 2007), medical imaging (Dong *et al.*, 2008), manufacturing (Zarei *et al.*, 2008), and utilization of combined heat and power systems (Vasebi *et al.*, 2007).

Geem (2008b) provides an overview of industrial HS applications in nine categories. A summary of around forty HS applications is also presented on the CD-ROM accompanying this text. The application area, optimization objective, number of variables (integer and real) and number of constraints (equality and inequality) are tabulated. The type of HS algorithm used, whether the basic algorithm or some modification of it, is also noted.

The largest problem tackled to date appears to be the design of a water distribution network by Geem (2007c) involving 454 discrete variables using the basic HS method. Geem and Williams (2007) applied a specially modified HS algorithm to solve a maximal species covering problem with 441 binary variables. Omran and Mahdavi (2008) used basic and modified HS to solve nine different unconstrained benchmark problems with 100 continuous variables each.

2.1. Selected applications broadly related to Chemical Engineering

The design of water distribution networks (Geem *et al.*, 2001, 2002; Geem, 2006a,b, 2007c; Geem and Park, 2006) and the structural design of 2D and 3D trusses, domes and grillage structures (Lee and Geem, 2004; Lee *et al.*, 2005; Ryu *et al.*, 2007; Saka, 2007; Degertekin, 2008; Erdal and Saka, 2008) are the two most common real-world HS applications. As noted in Table 1,

Table 1. Typical features of the two most common real-world HS applications.

	Design of water distribution networks	Structural design
System	Water distribution network of given connectivity usually containing loops, with at least one source and several sinks	Structure made up of fixed-length beams with given connectivity and known applied loads
Objective	Minimize capital cost of pipes	Minimize weight of structure
Decision variables	Pipe diameters (discrete standard sizes)	Beam designations (discrete) or cross-sectional areas (continuous)
Physical model	Continuity and pressure loss equations, often solved by calling external software (EPANET, KYPIPE)	Force calculations that predict the stresses in each beam and deflections at each joint, for example by using the Finite Element Method (FEM)
Constraints	Minimum required flowrate and minimum pressure at each sink	Maximum stress in each beam, maximum deflection at selected joints
Number of variables	8–454	2–27

generally these applications are discrete-variable, constrained optimization problems that require the solution of models of the physical phenomena involved. In this way, they are similar to some Chemical Engineering optimization problems. Note that the HS algorithm is able to solve discrete, continuous and mixed optimization problems. Oddly, few mixed-variable problems have been reported, but one such is the lowest-cost, constrained design of a pressure vessel (Lee and Geem, 2005).

Also of interest are the parameter estimation applications for fitting a model to a set of data. These include hydrograph fitting for flood routing (Kim *et al.*, 2001), rainfall-runoff modeling (Paik *et al.*, 2005), heterogeneous aquifer characterization for groundwater modeling (Ayvaz, 2007), and energy modeling in the transport sector (Ceylan *et al.*, 2008). These are typically unconstrained, continuous-valued optimization problems; the smallest problem estimated two model parameters and the largest forty.

2.2. Chemical Engineering applications

Although many published HS applications are peripherally related to Chemical Engineering, we will focus on three that are essentially Chemical Engineering applications.

Tian *et al.* (2005b) used HS for nonlinear model predictive control (NMPC) of two discrete-time, single-input single-output systems. The control objective was to drive the predicted closed loop system output $y_p(k)$ at each time k along a reference trajectory $y_r(k)$ with increment constraints on the manipulated variable $u(k)$:

$$\min_{[u(k),\ldots,u(k+C-1)]}(J)$$

$$= \frac{1}{2}\left\{\sum_{j=1}^{P}(y_p(k+j)-y_r(k+j))^2+\sum_{i=1}^{C}\lambda_i\Delta u^2(k+i-1)\right\},$$

where P is the prediction horizon, C is the control horizon, λ_i is a weighting factor and $\Delta u(k) = u(k) - u(k-1)$. They used $C = 2$, meaning that the HS solves for two decision variables at each time step. Favorable results — fast computations and good set-point tracking, both with and without process noise — were achieved using the Harmony Annealing Algorithm developed by Tian *et al.* (2004).

Geem and Hwangbo (2006) considered the design of a heat pipe on a satellite in which the two objectives were to minimize the mass of the heat pipe while maximizing conductance across the heat pipe's thermal join. This was formulated as an unconstrained, single-objective optimization problem having five continuous-valued design variables:

$$\min_{[L_f, L_c, t_f, t_b, T_{op}]} (Z) = \left| \frac{G(L_f, L_c, t_f, t_b, T_{op}) - G^*}{G^*} \right| + \left| \frac{M(L_f, L_c, t_f, t_b) - M^*}{M^*} \right|,$$

where G is the thermal conductance; M is the mass; L_f, L_c, t_f and t_b are geometrical design variables; T_{op} is the operating temperature, which is also a design variable; and G^* and M^* are the optimal values of G and M determined by optimizing each function separately. The basic HS found a better solution, with 0.6% lower mass and 1.6% higher conductance, than reported previously.

Fesanghary *et al.* (2008a) investigated the optimal design of shell and tube heat exchangers. The objective function was to minimize total annual cost, comprising a capital component related to installed heat transfer area and material of construction, and an operating component associated with the pumping energy needed to overcome tube side and shell side pressure drops. Heat transfer coefficients and pressure drops were calculated as a function of operating conditions, for example the Bell-Delaware method was used for the shell side heat transfer coefficient, and constant material properties were assumed. After performing a sensitivity study, nine design variables were selected: eight related to the exchanger geometry (tube diameter, tube pitch ratio, baffle cut, ...) and one for the material of construction (carbon steel, admiralty or copper-brass). The Improved Harmony Search of Mahdavi *et al.* (2007) obtained a solution within 0.2% of the global optimum, compared to 0.7% for the Genetic Algorithm (GA). While both HS and GA took a few seconds to find near-optimal solutions, around 164 hours were required to find the global optimum by exhaustive enumeration.

3. Basic Harmony Search Algorithm

The HS algorithm seeks to solve the single objective optimization problem:

$$\text{Minimize} \quad f(\mathbf{x}) \quad \text{by varying decision, or design, variables } \mathbf{x}, \tag{1a}$$

$$\text{subject to} \quad \mathbf{g}(\mathbf{x}) \leq \mathbf{0}, \tag{1b}$$

$$\mathbf{h}(\mathbf{x}) = \mathbf{0}, \tag{1c}$$

$$\text{and} \quad x_i^L \leq x_i \leq x_i^U, \ \text{if } x_i \text{ is continuous, or,} \tag{1d}$$

$$x_i \in \mathbf{X}_i = \{X_{i,1}, X_{i,2}, \ldots, X_{i,K_i}\}, \ \text{if } x_i \text{ is discrete.} \tag{1e}$$

3.1. Basic harmony search for continuous decision variables

The basic HS method (Geem *et al.*, 2001; Lee and Geem, 2005; Yang, 2008) is presented in Algorithm 1 for the case in which all decision variables are continuous, that is, $x_i \in [x_i^L, x_i^U]$, $i = 1, \ldots, N$, where x_i^L and x_i^U are lower and upper bounds on x_i. Note that in the HS there is no need to discretize the decision variables x_i. The HS algorithm consists of five steps: (1) specifying the algorithm parameters, (2) initializing harmony memory, (3) improvising a new harmony, (4) updating harmony memory if required, and (5) repeating steps 3 and 4 until a termination criterion is satisfied.

Algorithm 1. Basic Harmony Search algorithm for continuous decision variables.

Input HS algorithm parameters: $f(\mathbf{x})$, $\mathbf{x^L}$, $\mathbf{x^U}$, HMS, MaxImp, HMCR, PAR, $\mathbf{b}$
Randomly initialize HM with vectors satisfying lower and upper bounds, then sort HM
Repeat
    Improvise a new harmony, $\mathbf{x}'$ (Algorithm 2)
    If $\mathbf{x}'$ is better than the worst harmony in HM, $\mathbf{x_{worst}}$, **then**
        Replace $\mathbf{x_{worst}}$ with $\mathbf{x}'$ in HM, then sort HM
    EndIf
Until termination criterion is satisfied
Return the best harmony in HM, $\mathbf{x_{best}}$

Specifying the algorithm parameters

Along with the objective function $f(\mathbf{x})$, and lower and upper bounds on the decision variables, $\mathbf{x^L}$ and $\mathbf{x^U}$, the algorithm requires several parameters:

HMS	Harmony Memory Size, HMS ≥ 1
MaxImp	Maximum number of Improvisations, MaxImp > 1
HMCR	Harmony Memory Considering Rate, $0 \leq$ HMCR ≤ 1
PAR	Pitch Adjusting Rate, $0 \leq$ PAR ≤ 1
b	Bandwidth vector used in pitch adjusting, $\mathbf{b} \in \mathbb{R}^N > \mathbf{0}$,

where N is the number of decision variables. Alternatively, in place of **b**, the number of pitch intervals $\mathbf{p} \in (\mathbb{Z}^+)^N$ may be specified such that $b_i = (x_i^U - x_i^L)/p_i$ (Yang, 2008).

Initializing harmony memory

The algorithm maintains a store of solution vectors, known as Harmony Memory HM, that is updated during the optimization process. Harmony Memory is the HMS $\times$ $(N + 1)$ augmented matrix:

$$\text{HM} = \left[\begin{array}{cccc|c} x_1^1 & x_2^1 & \dots & x_N^1 & f(\mathbf{x^1}) \\ x_1^2 & x_2^2 & \dots & x_N^2 & f(\mathbf{x^2}) \\ \vdots & \vdots & \ddots & \vdots & \vdots \\ x_1^{\text{HMS}} & x_2^{\text{HMS}} & \dots & x_N^{\text{HMS}} & f(\mathbf{x^{HMS}}) \end{array}\right],$$

in which elements x_i^j represent the pitch (value) of the ith instrument (decision variable) in the jth harmony (solution vector). Thus, each row of HM represents a solution vector $\mathbf{x} = (x_1, \dots, x_N)$ along with its fitness $f(\mathbf{x})$. In some representations (e.g. Lee and Geem, 2005), HM consists of the decision variables only, with the objective function values stored separately.

To initialize HM, each decision variable is randomly[1] assigned a value within it bounds: $x_i^j = \text{rnd}([x_i^L, x_i^U])$, $i = 1, \dots, N$; $j = 1, \dots, \text{HMS}$,

[1] The function rnd(X) returns a number uniformly randomly drawn from the set X, which may be a continuous range or a set of discrete values.

and the corresponding objective function values are evaluated. Harmony memory is then sorted so that the best harmony with minimum $f(\mathbf{x})$ is in row 1 and the worst is in row HMS.

Improvising a new harmony

The basic HS algorithm improvises (generates) a new harmony, $\mathbf{x}' = (x'_1, \ldots, x'_N)$, by considering each decision variable separately. As shown

---

Algorithm 2. Improvising a new harmony in the basic Harmony Search for continuous variables.

Input HM, N, $\mathbf{x^L}$, $\mathbf{x^U}$, HMS, HMCR, PAR, $\mathbf{b}$
For each decision variable, $i = 1, \ldots, N$
 If rnd([0,1]) $\leq$ HMCR **then**
  Harmony memory considering: randomly select any variable-i pitch in HM
  $x'_i \leftarrow x^j_i$ where $j = \text{rnd}(\{1, \ldots, \text{HMS}\})$
  If rnd([0,1]) $\leq$ PAR **then**
   Pitch adjusting: randomly adjust x'_i within a small bandwidth, $\pm b_i$
   $x'_i \leftarrow x'_i + \text{rnd}([-b_i, b_i])$
   $x'_i \leftarrow \min(\max(x'_i, x^L_i),\ x^U_i)$ *Ensure bounds remain satisfied*
  EndIf
 Else
  Random playing: randomly select any pitch within upper and lower bounds
  $x'_i \leftarrow \text{rnd}([x^L_i, x^U_i])$
 EndIf
EndFor
Return the new harmony $\mathbf{x}' = (x'_1, \ldots, x'_N)$

---

in Algorithm 2, for each variable x'_i, one of three HS operations is randomly selected:

Harmony memory considering is chosen with probability	HMCR × (1 − PAR)
Pitch adjusting is chosen with probability	HMCR × PAR
Random playing is chosen with probability	(1 − HMCR)

These operations were illustrated in Fig. 1.

Harmony memory considering means selecting at random from values of x_i^j already stored in HM (Fig. 1a). *Pitch adjusting* consists of harmony memory considering to find an x_i^j followed by the adjustment of its pitch by a small random amount specified by the bandwidth vector **b** (Fig. 1b). Note that bounds checking is enforced here. *Random playing* is simply randomly setting x'_i to any value between the lower and upper bounds for variable i (Fig. 1c).

The global solution space is explored by random playing and by harmony memory considering during early improvisations. Local solution refinement is undertaken by pitch adjusting and by memory considering in later improvisations. For two typical sets of HS parameters, HMCR = 0.8, PAR = 0.3 (Lee and Geem, 2004) and HMCR = 0.9, PAR = 0.35 (Lee and Geem, 2005), the probabilities for performing each operation are:

Operation	Lee and Geem (2004)	Lee and Geem (2005)
Harmony memory considering	0.56	0.585
Pitch adjusting	0.24	0.315
Random playing	0.20	0.100

Updating harmony memory

If the new harmony is better than the worst harmony currently stored in HM as judged by the value of $f(\mathbf{x}')$, then the worst harmony in HM is replaced by the new harmony, and HM is re-sorted.

Termination criterion

The algorithm concludes when the number of improvisations (iterations) reaches MaxImp. The best solution vector in HM, $\mathbf{x_{best}}$, is returned. Optionally, all the other vectors stored in HM may also be returned — these represent (HMS–1) alternative near-optimal solutions that may be considered alongside $\mathbf{x_{best}}$.

3.2. The harmony search approach compared with other global optimization methods

As Geem *et al.* (2001) point out, HS and other meta-heuristic optimization methods aim to find a good (near-optimal) solution without compromising any nonlinear characteristics of the problem within a reasonable computational time using reasonable memory resources. In contrast to traditional optimization techniques, HS and similar algorithms require neither careful initial guesses nor expensive gradient calculations.

HS calls upon information from past solution vectors stored in HM, which is similar to the genetic pool in GA, pheromone trails in Ant Colony Optimization (ACO) and the tabu list in the Tabu Search (TS) (Geem *et al.*, 2001; Mahdavi *et al.*, 2008). The adaptation rate, for pitch adjusting for instance, can vary during the optimization, similar to the transition probability changing with temperature in Simulated Annealing (SA) (Geem *et al.*, 2001), although this does appear only in variations on the basic HS (Kim *et al.*, 2004; Mahdavi *et al.*, 2007; Dong *et al.*, 2008).

The differences between GA and HS are discussed by Geem *et al.* (2001), Lee and Geem (2005), Geem (2006b), and Degertekin (2008):

(i) HS generates a new solution from all vectors in HM, not just two (the parents) as in GA;
(ii) HS considers each decision variable independently, and does not need to preserve the structure of the gene, although some inter-variable dependence may be imposed if desired (Geem, 2006a);
(iii) In contrast to binary-coded GA, HS does not need to encode and decode the decision variables into binary strings, and it treats

continuous variables without any loss of precision. In this way, it is similar to real-coded GA (Michalewicz, 1996).

Degertekin (2008) outlines the differences between ant colony optimization and HS:

(i) HS can select values outside its memory with a certain probability compared to ACO, which develops new solutions only from stored values;
(ii) HS may perform local searches on any solution vector, while ACO uses local searching only for selected elite vectors;
(iii) HS updates HM after each new solution vector is formed so that the most recent information is always used unlike ACO, which updates its memory only after n new vectors are generated, where n is the number of ants.

Clearly, HS incorporates features similar to other meta-heuristic optimization methods, although it does seem to have inherently high flexibility. Lee and Geem (2005), Degertekin (2008) and others argue that this flexibility allows HS to produce better solutions. Section 3.7 discusses the numerical performance of HS relative to other SGOs.

3.3. Basic harmony search for discrete and mixed variable problems

The basic HS algorithm easily accommodates discrete decision variables. If variable x_i is discrete with K_i distinct values then $x_i \in \mathbf{X}_i = \{X_{i,1}, X_{i,2}, \ldots, X_{i,K_i}\}$ where $X_{i,1} < X_{i,2} < \cdots < X_{i,K_i}$. For a problem involving only discrete decision variables, several changes are needed to Algorithm 1, which describes the basic HS method for continuous variables:

(i) On line 1, the continuous-variable bounds vectors $\mathbf{x}^{\mathbf{L}}$ and $\mathbf{x}^{\mathbf{U}}$ are replaced by the sets of allowable discrete values $\mathbf{X}_i, i = 1, \ldots, N$, and the bandwidth vector $\mathbf{b}$ is replaced by the neighboring index vector $\mathbf{m} \in \mathbb{Z}^N > \mathbf{0}$;
(ii) On line 2, HM is randomly initialized with allowable discrete values;
(iii) On line 3, the new harmony is improvised using Algorithm 3.

Algorithm 3. Improvising a new harmony in the basic Harmony Search for discrete variables.

1 **Input** HM, N, $\mathbf{X}_i$ $(i = 1, \ldots, N)$, HMS, HMCR, PAR, $\mathbf{m}$
2 **For** each decision variable, $i = 1, \ldots, N$
3   **If** rnd([0,1]) $\leq$ HMCR **then**
4   *Harmony memory considering: randomly select any variable-i pitch in HM*
5         $x_i' \leftarrow x_i^j$ where $j = \text{rnd}(\{1, \ldots, \text{HMS}\})$
6   **If** rnd([0,1]) $\leq$ PAR **then**
7   *Pitch adjusting: randomly select a value neighboring* x_i^j, *as defined by* m_i
8   Find L, the index of x_i^j in $\mathbf{X}_i$ from HM considering, that is $L : x_i^j = X_{i,L}$
9             $l = \text{rnd}(\{(L - m_i), (L - m_i + 1), \ldots, (L - 1), (L + 1), \ldots, (L + m_i - 1), (L + m_i)\})$
10             $l \leftarrow \min(\max(l, 1),\ K_i)$ *Ensure index remains valid*: $1 \leq l \leq K_i$
11             $x_i' \leftarrow X_{i,l}$
12   **EndIf**
13   **Else**
14   *Random playing: randomly select any pitch from the set of allowable values*
15         $x_i' \leftarrow X_{i,k}$ where $k = \text{rnd}(\{1, \ldots, K_i\})$
16   **EndIf**
17 **EndFor**
18 **Return** the new harmony $\mathbf{x}' = (x_1', \ldots, x_N')$

Pitch adjusting, lines 7–11 in Algorithm 3, is now a discrete operation defined by the neighboring index vector $\mathbf{m}$, which contains small positive integers. As an example, say x_i can take on discrete values from the set $\{10, 20, 30, 40, 50, 60\}$ and that the value 30 was chosen during harmony memory considering. If $m_i = 1$ then the new pitch-adjusted x_i' would be randomly selected from values neighboring 30, that is, $\{20, 40\}$; if $m_i = 2$, then x_i' would be selected from both first and second neighbors, namely $\{10, 20, 40, 50\}$. Note that the definition of **m** used here is slightly different from that commonly used elsewhere (Lee and Geem, 2004).

Optimization problems having both discrete and continuous variables can be handled in a straightforward manner by combining the methods of Secs. 3.1 and 3.3.

3.4. Stochastic partial derivative for discrete variables

Geem (2008a) introduced a quantity termed the 'stochastic partial derivative' for discrete decision variables:

$$\left.\frac{\partial f}{\partial x_i}\right|_{x_i=X_{i,k}} = \underbrace{\frac{1}{K_i}(1-\mathrm{HMCR})}_{\textit{Random playing}} + \underbrace{\frac{n_i(X_{i,k})}{\mathrm{HMS}}\mathrm{HMCR}\cdot(1-\mathrm{PAR})}_{\textit{Harmony memory considering}}$$

$$+ \underbrace{\frac{\frac{1}{|\mathrm{K_p}|}\sum_{l\in\mathrm{K_p}} n_i(X_{i,l})}{\mathrm{HMS}}\mathrm{HMCR}\cdot\mathrm{PAR}}_{\textit{Pitch adjusting}},$$

where $n_i(X_{i,k})$ is the number of times that discrete variable x_i has the value $X_{i,k}$ in HM, the set $\mathrm{K_p} = \{\max(k-m_i,1),\ldots,k-1,k+1,\ldots,\min(k+m_i,K_i)\}$, and $|\mathrm{K_p}|$ denotes the number of elements in $\mathrm{K_p}$. Calling $(\partial f/\partial x_i)|_{x_i=X_{i,k}}$ a 'derivative' is misleading, however, because it is actually the *probability* that the value $X_{i,k}$ will be chosen for x_i' when generating a new harmony. Clearly, the quantity $\partial f/\partial x_i$ depends on the values of the HS parameters and the current contents of HM. Geem (2008a) uses the so-called derivative to show how the contents of HM evolve to favor optimal values of the decision variables in several test cases.

3.5. Selection of algorithm parameter values

As defined in Sec. 3.1, the parameters of the basic HS algorithm are HMS, HMCR, PAR, MaxImp, and either **m** or **b** for discrete or continuous variables, respectively. Table 2 summarizes values for the parameters that were either selected or recommended for a range of applications. Sensitivity analyzes have considered ranges larger than those shown.

Table 2. Parameter values used or recommended in applications of the basic harmony search.

Application	*N*	HMS	HMCR	PAR	MaxImp	Notes and reference
Benchmark problems	2–10	20	0.9	0.35	15,000–230,000	Constrained and unconstrained problems; Lee and Geem (2005)
	2–5, 15, 30, 100	5	0.9	0.3	50,000	Continuous (**b** = **0.01**) and integer problems; Omran and Mahdavi (2008)
Structural design	10–29	20	0.8	0.3	50,000	Lee and Geem (2004)
	8–27	20–40	0.8–0.9	0.3–0.45	30,000–80,000	**m** = **1**; Lee *et al.* (2005)
	2–13	30	0.9	0.45	4,000	**m** = **1**; Saka (2007)
	2–20	50	0.8	0.4	5,000	**m** = **3**; Degertekin (2008)
	2	50	0.9	0.5	?	**m** = **1**; Erdal and Saka (2008)
Water distribution network design	8	100	0.95	0.05	5,000	Geem (2006b)
	34	50	0.93	0.18	200,000	
	21	50	0.9	0.1	20,000	
	30	30–100	0.7–0.95	0.3–0.7	10,000	
	9	30–100	0.7–0.95	0.3–0.7	5,000	
Hydrological parameter estimation	3	100	0.95	0.05	5,000	$\mathbf{b} = (\mathbf{x}^{\mathbf{U}} - \mathbf{x}^{\mathbf{L}})/1{,}000$; Kim *et al.* (2001); Geem (2006b)
Transport energy model parameter estimation	4, 7	10	0.8	0.4	100,000	Ceylan *et al.* (2008)
Aquifer modeling parameter estimation	2–10	10	0.9	0.45	50,000	Ayvaz (2007)
Scheduling multiple dam operation	48	30	0.95	0.05	35,000	Geem (2007b)
Multiple pump operation	40	19	0.97	0	3,500	Binary variables; Geem (2005a)
School bus routing	10	20	0.9–0.95	0	1,000	Discrete variables with 4 values; Geem (2005b)
Music composition	22, 48	10	0.9	0.3	3,000	Geem and Choi (2007)

Several studies have interpreted the effect of the algorithm parameters (Mahdavi *et al.*, 2007; Vasebi *et al.*, 2007; Degertekin, 2008; Fesanghary *et al.*, 2008b). Small values of the Harmony Memory Size can lead to the HS stalling in local minima. Increasing HMS generally provides better solutions, but more iterations are required to achieve them. So, if MaxImp is limited and rapid solution is important, having a moderate HMS will usually provide the best solutions. When the Harmony Memory Considering Rate is close to zero, HS behaves like a purely random search, rarely using the favored vectors accumulated in HM. On the other hand, HMCR very near one impedes the algorithm's exploration of variable space outside the range currently stored in HM, potentially causing the algorithm to become caught in local minima. A low Pitch Adjusting Rate with small **b** or **m** can slow HS's ability to home in on nearby minima, but large values of these parameters may cause the algorithm to skitter around minima, behaving more like a random search. In contrast to some other studies, Vasebi *et al.* (2007) found that the HS was relatively insensitive to PAR.

Table 2 may be used as a guide for selecting appropriate parameter values, but it is also helpful to provide some recommendations. Vasebi *et al.* (2007) state that selecting HMS between N and $2N$ is reasonable for most problems, and this range is also favored by Cheng *et al.* (2008). Geem has recommended setting HMS between 10 and 50, HMCR between 0.7 and 0.95, PAR between 0.2 and 0.5 (Lee *et al.*, 2005), and has also suggested HMCR $= 0.9$, PAR $= 0.3$ and $\mathbf{b} = \mathbf{0.01}$ (Omran and Mahdavi, 2008). Often $\mathbf{m} = \mathbf{1}$ for discrete variables (Geem *et al.*, 2005a); however, Degertekin (2008) found that $\mathbf{m} = \mathbf{3}$ was better in a structural engineering application. For binary variables pitch adjusting may be inappropriate, so PAR is set to zero (Geem, 2005a; Geem and Williams, 2007). Sensitivity analyzes are sometimes used to select parameter values for particular applications (Geem, 2007a; Ceylan *et al.*, 2008; Fesanghary *et al.*, 2008b).

3.6. Handling equality and inequality constraints

Upper and lower bounds on continuous variables and valid values for discrete variables, Eqs. (1d) and (1e), are automatically handled by the HS algorithm. Equality and inequality constraints, Eqs. (1c) and (1b), often appear in real-world problems as model equations and performance criteria, respectively, as indicated in Table 1. To date, constraints in HS applications have been handled in several ways: (i) penalty functions, (ii) a rejection strategy, (iii) external solvers for equality constraints, or (iv) special procedures used in particular problems.

Penalty functions have been used most often for constraint handling (Kim *et al.*, 2004; Geem, 2005b; Geem *et al.*, 2005a,c; Geem, 2006b; Geem, 2007a,b; Degertekin, 2008). A constrained, or augmented, version of the objective function $f_{con}(\mathbf{x})$ is formed by adding a penalty term or penalty factor to $f(\mathbf{x})$, then $f_{con}(\mathbf{x})$ is minimized. For example, to ensure a set of inequality constraints, $H_j(\mathbf{x}) \geq H_j^{\min}$, $j = 1, \ldots, C$, were satisfied, Geem (2006b) used a step-quadratic penalty term:

$$f_{con}(\mathbf{x}) = f(\mathbf{x}) + \sum_{j=1}^{C} (a\text{sgn}(\varepsilon_j) + b\varepsilon_j^2),$$

where $\varepsilon_j = \max(H_j^{\min} - H_j(\mathbf{x}), 0)$, and a and b are positive penalty coefficients.

Less commonly used is a rejection strategy, in which new harmonies that do not satisfy the constraints, Eqs. (1b) and (1c), are simply discarded (Lee and Geem, 2004; Lee *et al.*, 2005). The discarded solutions are counted towards the number of improvisations tried. Some studies combine penalty and rejection methods along with an adaptive error strategy that penalizes constraint violations more severely during later iterations (Erdal and Saka, 2008; Forsati *et al.*, 2008).

When the equality constraints are actually the equations of a mathematical model representing a real-world system, an external software package has sometimes been used to solve model equations at each iteration. Examples include the use of EPANET or KYPIPE to solve mass and energy balances for flow in a pipe network (Geem *et al.*, 2002; Kim *et al.*, 2004; Geem, 2006b) and SPMsim to analyze the behavior of a mooring system (Ryu *et al.*, 2007).

Special procedures are used to ensure the feasibility of new solutions in some graph-based optimization problems. Geem *et al.* (2000) and Geem and Park (2006) develop feasible new solutions by choosing only among adjacent nodes and preventing loop formation in a rectilinear pipe layout problem. Forsati *et al.* (2008) use various methods — rejecting edges that fail bandwidth constraints, generating only valid tree structures, a solution repair strategy — to optimize multi-cast routing in a computer network. In a different type of application, the maximal covering species problem, Geem and Williams (2007) keep track of the sum of digits when forming new binary solution vectors to ensure the target number of land parcels is chosen.

3.7. Performance of the harmony search compared to other SGO methods

The basic HS is reported to be competitive with other meta-heuristic optimization methods. Lee and Geem (2004) showed that HS generally found slightly (0.01–4.6%) lower structure weights in eight structural design problems than previously published solutions using mathematical or other SGO approaches. Lee and Geem (2005) studied six unconstrained and six constrained benchmark problems, five structural design problems and a parameter estimation problem. HS found solutions near the global optima for the benchmark problems, and similar or better solutions for the other problems compared to previous results. Geem (2006b) applied HS to the optimal design of five water distribution networks. Compared to previous solutions using GA, SA or TS, HS found solutions of the same or lower cost, and generally with fewer function evaluations. Geem (2008b) lists many comparisons between HS and other meta-heuristic and mathematical optimization methods for a range of real-world applications.

HS iterations are said to be fast. Geem *et al.* (2005a) suggested that HS iterations were slightly faster GA ones in a route finding application. Paik *et al.* (2005) reported HS was much faster per iteration than GA in their parameter estimation problem, taking only about 5% of the computing time for the same number of iterations. Cheng *et al.* (2008) remark that HS is the fastest global optimization algorithm for small-scale ($N \leq 25$) problems.

4. Variations on the Basic Harmony Search Algorithm

Despite the reported success of the basic algorithm, there is a growing diversity of modified HS algorithms that seek improved performance. These efforts may be classified into (i) minor variations, for example regarding initialization; (ii) methods for changing the HS parameters during iteration; and (iii) new or modified HS operators. Sometimes several individual innovations are combined. In this section, the individual changes have been teased apart to emphasize the diversity of approaches. All modified algorithms are claimed to be superior to the basic HS.

4.1. Minor variations on the basic harmony search

Harmony memory initialization

Most applications use simple random HM initialization as presented in Secs. 3.1 and 3.3, but several schemes aimed at improving initialization have been proposed. Geem and Williams (2007), Degertekin (2008) and Geem (2008b) suggest generating more random vectors than needed, for example 2–5 $\times$ HMS, and placing only the best HMS of these into HM. In addition, Geem and Williams (2007) reduce duplication in the initial HM by selective pruning so that a vector can appear at most twice. For constrained optimization problems, several authors (e.g. Lee *et al.*, 2005; Degertekin, 2008) include only strictly feasible vectors in the initial HM, while others allow near-feasible (Erdal and Saka, 2008) or partially-feasible (Forsati *et al.*, 2008) vectors.

For particular situations, special procedures may be appropriate. In an application similar to parameter estimation, Mahdavi *et al.* (2008) used statistics derived from the data points to seed HM. Geem and Williams (2007) initialized HM with sparse binary vectors using a procedure that guaranteed a specified number of 1s and was biased towards producing favorable values of the objective function.

Sorting harmony memory

Some interpretations of the basic HS (Ayvaz, 2007; Vasebi *et al.*, 2007) do not sort harmony memory during initialization or updating. The relative advantages of sorting versus not sorting have not been discussed.

Termination criteria

In the majority of HS implementations, the algorithm is terminated when the number of improvisations reaches a predetermined number, MaxImp. In an attempt to reduce computing time, Ayvaz (2007) and Degertekin (2008) halt the algorithm before MaxImp, if $f(\mathbf{x}_{\mathbf{best}})$ does not change for a certain number of successive improvisations. Both used 1,000 improvisations as the limit; however, this corresponds to 20% of MaxImp for the structural engineering case studied by Degertekin, but only 2% for Ayvaz's hydrogeological application. Mahdavi *et al.* (2008) terminate the algorithm at MaxImp, or earlier if the parameter ACF, Average Change in Fitness, falls below a threshold value. Cheng *et al.* (2008) monitor the difference between the best and worst harmonies in HM, $\Delta f = |f(\mathbf{x}_{\mathbf{best}}) - f(\mathbf{x}_{\mathbf{worst}})|$. When Δf becomes less than a critical amount, 10^{-4} in their application, the algorithm is said to have reached quasi-convergence. The parameters HMCR and PAR are changed slightly, and then the algorithm run is continued for a fixed number of improvisations before terminating.

In some applications and benchmark problems, the objective function may have a minimum that is known a priori and can be used to terminate the algorithm early.

4.2. Techniques involving parameter adaptation

Variable HMCR and PAR

Dong *et al.* (2008) increased HMCR linearly with *Imp*, the current improvisation number, reasoning that more random playing early on helps global searching, but later, more frequent consultation of HM is needed as it becomes filled with better solutions. PAR was decreased linearly with *Imp*. The new algorithm, which also used a modified pitch adjusting operator, was named the Adaptive Harmony Search (AHS). Kim *et al.* (2004), in their ReHS algorithm, also varied HMCR and PAR during solution, but give no details. Cheng *et al.* (2008) check when convergence has slowed, then decrease HMCR by 0.05, increase PAR by 0.05, and continue for a fixed number of iterations.

Variable PAR and bandwidth

Mahdavi *et al.* (2007) argued that PAR should be increased and the bandwidth b_i decreased as optimization proceeds to make local searching more frequent and more focused during later iterations. They proposed a linear increase in PAR and an exponential decay for b_i:

$$\mathrm{PAR}(Imp) = \mathrm{PAR}_{\min} + (\mathrm{PAR}_{\max} - \mathrm{PAR}_{\min}) \cdot \frac{Imp}{\mathrm{MaxImp}}$$

$$b_i(Imp) = b_{\max i} \cdot \exp\left[\ln\left(\frac{b_{\min i}}{b_{\max i}}\right) \cdot \frac{Imp}{\mathrm{MaxImp}}\right].$$

Typically, $\mathrm{PAR}_{\min} = 0.45$ and $\mathrm{PAR}_{\max} = 0.9$. This Improved Harmony Search (IHS) algorithm has been applied or further modified in subsequent studies (Fesanghary *et al.*, 2008a,b; Forsati *et al.*, 2008; Omran and Mahdavi, 2008; Zarei *et al.*, 2008).

Tightening of variable bounds

To reduce the range of random playing during later iterations, the upper and lower bounds on the variables may be tightened: $x_i^L \to x_i^{L'} > x_i^L$, $x_i^U \to x_i^{U'} < x_i^U$. Geem *et al.* (2001) restricted the bounds to the current range stored in HM at MaxImp/2. Cheng *et al.* (2008) placed modified bounds of $\pm\eta(x_i^U - x_i^L)$ with $\eta = 0.1$ centered about the best solution in HM near the end of iterations.

4.3. Techniques that modify the HS operators

Biased selection

In HM considering, Li and Chi (2007) and Cheng *et al.* (2008) used biased random selection, so that vectors with better fitness had a greater chance of selection. HM was sorted, with the best vector stored in row 1. Row j was chosen to supply the new value for x_i' with probability

$$\frac{\delta(1-\delta)^{j-1}}{\sum_{k=1}^{\mathrm{HMS}} \delta(1-\delta)^{k-1}},$$

where $\delta \in (0, 1]$ is the bias parameter. Note that the chance of choosing row $j+1$ is $(1-\delta)$ times the chance of choosing row j, in contrast to the basic HS, where every row has the same chance, 1/HMS. During iterations, δ was varied in a stepped manner, for example $0.1 \to 0.5 \to 0.8$. Paik *et al.* (2005) found that proportionate selection, as used in simple GA, degraded performance. Geem and Williams (2007) used a different biased selection process for binary variables when random playing to ensure vectors contained few 1s.

Pitch adjusting replaced by global best solution

Omran and Mahdavi (2008) replaced basic pitch adjusting for variable i with the selection of element i from the best solution vector currently in HM, that is, lines 7–9 in Algorithm 2 are replaced by $x_i' \leftarrow x_{\text{best}\,i}$. This is the global best concept from Particle Swarm Optimization (PSO). The Global-best Harmony Search (GHS) algorithm combines the IHS of Mahdavi *et al.* (2007) with this global best modification.

Pitch adjusting replaced by mutation

Li *et al.* (2007b) used a non-uniform mutation operator from GA to replace the standard HS pitch adjusting operation, lines 7–9 in Algorithm 2:

If rnd([0, 1]) ≤ 0.5 **then**

$$x_i' \leftarrow x_i' + \Delta(Imp, x_i^U - x_i')$$

Else

$$x_i' \leftarrow x_i' - \Delta(Imp, x_i' - x_i^L)$$

EndIf

where $\Delta(Imp, y) = y \cdot \text{rnd}([0, 1]) \cdot (1 - Imp/\text{MaxImp})^b$ and b is a constant.

Biased pitch adjusting

When pitch adjusting with discrete variables for $m_i > 1$, several studies have favored selecting immediately adjacent pitches over those further removed. Most often three pitch adjusting rates are defined, PAR1 for selecting nearest neighbors, PAR2 for second nearest and PAR3 for third nearest, with PAR1 > PAR2 > PAR3 (Geem *et al.*, 2005c; Dong *et al.*, 2008; Forsati *et al.*, 2008).

Random playing replaced by simulated annealing

Tian *et al.* (2004) replaced random playing in the basic HS with very fast simulated annealing: the Metropolis algorithm is applied to the *worst* solution in HM to get the new solution x_i'. The new algorithm, called the Harmony Annealing Algorithm (HAA), has been used by Tian *et al.* (2005a,b).

Ensemble consideration for new harmonies

Geem (2006a) introduced a new operator, ensemble consideration, that allows strongly correlated variables to influence each other when generating a new harmony. Ensemble consideration is performed with a certain probability, ECR, after a new harmony vector has been generated (line 15, Algorithm 2). Typically, ECR ≈ 0.01. The new value for the ith variable is

$$x_i' \leftarrow f_s(x_j', \mathbf{x}_i, \mathbf{x}_j), \quad \max_{j:j\neq i}(\mathrm{corr}^2(\mathbf{x}_i, \mathbf{x}_j)),$$

where $\mathbf{x}_i = (x_i^1, \ldots, x_i^{\mathrm{HMS}})$, $\mathbf{x}_j = (x_j^1, \ldots, x_j^{\mathrm{HMS}})$, $\mathrm{corr}(\mathbf{u}, \mathbf{v})$ is the correlation function, and $f_s(v^*, \mathbf{u}, \mathbf{v})$ is a selection function that returns the value u^* from $\mathbf{u}$ that is most often associated with the value v^* in $\mathbf{v}$ when considering the pairs (u_k, v_k). It is unclear how this would be implemented for continuous variables.

Sequential Quadratic Programming (SQP) refinement of new harmonies

Fesanghary *et al.* (2008b) modified IHS by applying SQP to polish new solution vectors (Algorithm 2, line 16) with a low probability, for example 10%. In addition, at the end of iterations, SQP is used to refine all vectors in HM. These modifications make up the Hybrid HS Algorithm (HHSA).

Special training of elite harmonies

Paik *et al.* (2005) detected when the rate of solution improvement becomes unacceptably low then applied a special search, elite training, to stimulate further improvement. In elite training each variable is discretized and then adjusted until no improvement occurs, while other variables remain constant. This special search is performed when $|(f(\mathbf{x_{best}}) - f(\mathbf{x_{worst}}))/f(\mathbf{x_{worst}})| \times 100\% < C$, or when $Imp \approx \mathrm{MaxImp}$. They used $C = 2\%$ initially, then reduced C by 0.1% each time the special search was called. Paik *et al.* (2005) made special searching demand-driven, in contrast to Fesanghary *et al.* (2008b) who used SQP at any time with a small probability.

Generation of multiple harmonies

Instead of a single new harmony per improvisation, n new harmonies may be generated in parallel. The best HMS out of the (HMS $+ n$) available harmonies are retained in HM. Cheng *et al.* (2008) used $n = 0.1 \times$ HMS. Li *et al.* (2006a) generated one new harmony using the basic HS and p new harmonies using a chaos harmony method, so $n = p + 1$.

Chaotic harmonies

Li *et al.* (2006a) generated multiple chaotic harmonies in parallel with a single HS harmony in each improvisation. The chaos variables $\chi_i^0 \in (0, 1)$ are randomly initialized. At each improvisation, they vary according to the logistic map $\chi_i^{Imp} = 4\chi_i^{Imp-1}(1-\chi_i^{Imp-1})$ and new harmonies are generated according to $x_i' = l_i + (u_i - l_i)\chi_i^{Imp}$, where l_i and u_i are modified bounds. Three variations on the chaotic harmony method are available: simple, fixed and dynamic. In the simple method, $l_i = x_i^L$ and $u_i = x_i^U$. In the fixed method, the variable bounds are divided into q equal subintervals, and so q new harmonies are generated for each chaos variable. In the dynamic method, the bounds vary with the current contents of HM.

Prevention of overlapping harmonies

To encourage diversity, several studies have limited the number of identical vectors appearing in HM (Paik *et al.*, 2005; Geem and Park, 2006; Geem and Williams, 2007).

4.4. Specialized applications and hybrid algorithms

For route finding problems, such as the Traveling Salesman Problem (TSP), the basic HS does not work well (Geem *et al.*, 2001). Improved performance with TSPs has been reported by replacing simple pitch adjusting with two new operators: the neighboring-city-going operator and the city-inverting operator (Geem *et al.*, 2001). They provide some dependence between successive decision variables, x_{i-1} and x_i, representing the previous and current cities in the TSP, that is absent in the basic HS.

Broadly related to TSPs is the bandwidth-and-delay-constrained multicast routing problem studied by Forsati *et al.* (2008). They developed two algorithms: HSPR, which uses a repair strategy to fix infeasible new solutions generated by HS, and HSNPI, which generates new solutions using a method similar to Geem *et al.* (2005c).

Two other specialized applications are the document clustering problem studied by Mahdavi *et al.* (2008) and the maximal covering species problem of Geem and Williams (2007). In the former application, two algorithms that combined HS with the K-means partitioning algorithm performed better than either HS or K-means alone. In the latter problem, several techniques were used to assist the solution process: modified HM initialization, biased random selection, constraint checking during new solution generation, duplicate solution removal and periodic non-random selection.

Several hybrid SGO algorithms have been developed that use HS to supplement the principal algorithm, by using some HS concept or calling HS as a subroutine. HS hybrids that are mostly GA include Li *et al.* (2005), Li and Li (2007), Li *et al.* (2007a); while Li *et al.* (2007c,d) and Li *et al.* (2008) are mostly PSO methods. Li *et al.* (2006c) combined four techniques: GA, HS, TS and the Nelder–Mead simplex method.

4.5. Reflection on the HS variations

Sections 4.1–4.4 illustrate the diversity of modifications to the basic HS method. Often several changes are packaged together as a single modified algorithm, but the effect of the individual changes and the reason for the particular combination chosen is usually not spelt out. Further improvements in the HS method may result from a systematic, comprehensive evaluation of the many different innovations, both singly and in combination.

5. Harmony Search Software

Currently, there is little publicly-available HS software. Geem maintains a few program examples, based on Visual Basic, on the site http://www.

hydroteq.com/. Yang (2008) presents a Matlab/Octave program to optimize the Rosenbrock function. The CD-ROM accompanying this text also includes some Matlab programs that demonstrate HS optimization.

6. Illustrative Example

The modified Himmelblau function

$$f(x_1, x_2) = (x_1^2 + x_2 - 11)^2 + (x_1 + x_2^2 - 7)^2 \\ + 0.1[(x_1 - 3)^2 + (x_2 - 2)^2],$$

with $x_1, x_2 \in [-6, 6]$ is a benchmark minimization problem that is regarded as difficult (Srinivas and Rangaiah, 2007). The global minimum is $f(3, 2) = 0$. Figure 4 shows one instance of the basic HS solution of the problem using the parameter values: HMS = 10, MaxImp = 10,000, HMCR = 0.8, PAR = 0.4 and $\mathbf{b} = (\mathbf{x}^{\mathbf{U}} - \mathbf{x}^{\mathbf{L}})/1{,}000$, which are within the usual range given in Sec. 3.5.

Based on 1,000 runs of the HS algorithm using different random starting values, the basic HS method with given parameter values found the global minimum with a success rate of 77%, as judged by finding the global minimum function value within an absolute error of 10^{-6}. The average number of function evaluations, based on only the successful runs, was 2643, requiring on average 0.06 seconds computing time (Matlab 7.5.0, Pentium D 3.4 GHz, 2 GB RAM PC).

Figure 4 shows the ability of HS to escape from local minima. Early iterations are attracted towards the local minimum at ~(3.6, −1.8), but at about *Imp* ≈ 2900, the region near the global minimum is found, and subsequent local searching leads to a successful optimization. Note that Fig. 4(a) also illustrates that variables are treated independently in the basic HS, with strong search bands located around $x_1 \sim 3$, $x_1 \sim 3.6$, $x_2 \sim 2$ and $x_2 \sim -1.8$ in this example. Matlab files for the basic HS that can reproduce these results are supplied on the CD-ROM.

The exercises at the end of this chapter explore different objective functions, the effect of HS parameters on algorithm performance, the ability to perform local versus global searching, modified HS initialization, and constrained optimization.

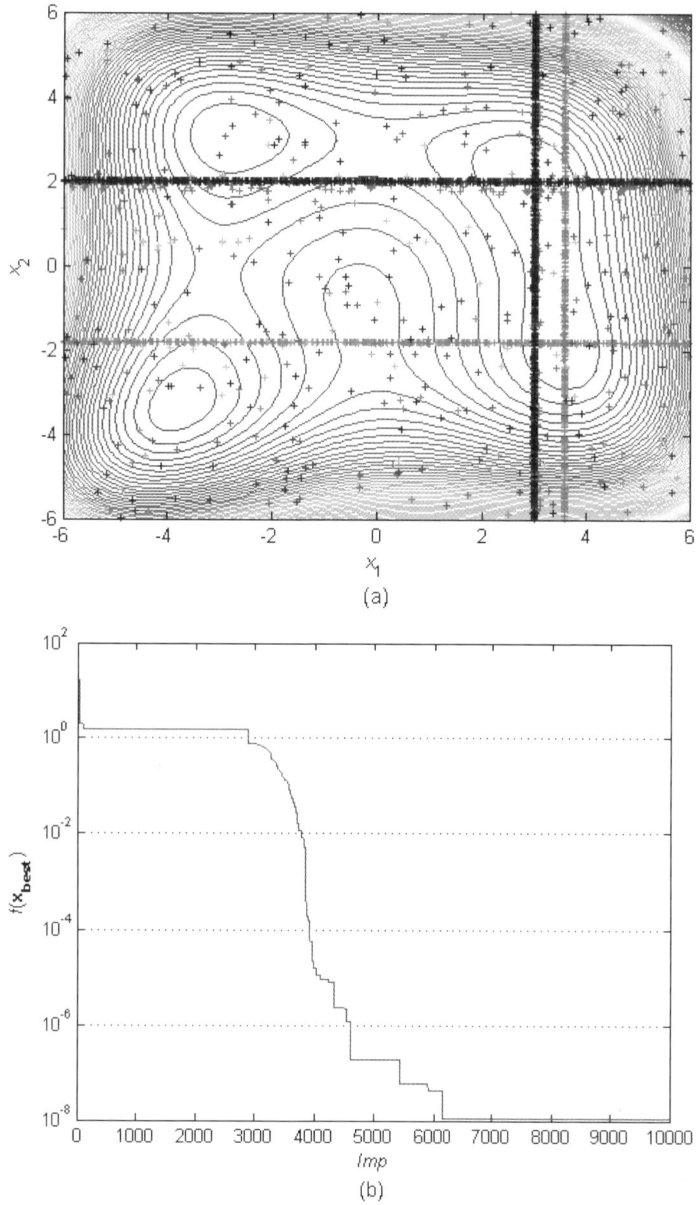

Figure 4. (a) Contours of the modified Himmelblau function showing new harmonies generated by the basic HS algorithm, with early iterations shown by light markers and later iterations by dark markers; and (b) the corresponding convergence history.

7. Conclusions and Opportunities for Further Research

HS is a relatively recent SGO method that draws its inspiration from improvised musical performance. It has operations similar to other meta-heuristic optimization routines — the use of memory, processes for local and global searching — yet it remains conceptually simple and has substantial flexibility. HS has proven to be competitive with other global optimization techniques. It has been applied to benchmark problems and a range of real-world applications, including the design of water distribution networks, optimization of truss structures, and stability analysis of soil slopes. However, HS techniques appear to be little known in the Chemical Engineering community.

Two broad directions for future HS research may be identified: enhancements to the HS algorithm and further applications in Chemical Engineering. Among potential improvements to the HS approach are: (i) methods for automatic parameter setting; (ii) routines for selecting among the various HS operators and parameter adaptation methods; (iii) techniques for dealing with noise in the objective function; and (iv) guidelines for choosing a constraint handling method. Since there are so few Chemical Engineering applications, it would be beneficial to compare HS with other techniques on a range of typical Chemical Engineering problems. There is good scope for exploring the usefulness of HS for mixed discrete-continuous variable problems, including superstructure optimization; adaptive control; scheduling of batch plant operation; and multi-objective optimization.

Acknowledgment

We thank Professor Moses Tadé for his support and feedback during the writing of this chapter. Tonghua Zhang acknowledges support from the ARC Discovery Project DP0770420.

References

Ayvaz, M.T. (2007). Simultaneous determination of aquifer parameters and zone structures with fuzzy c-means clustering and meta-heuristic harmony search algorithm. *Advances in Water Resources*, **30**, pp. 2326–2338.

Ceylan, H., Ceylan, H., Haldenbilen, S. and Baskan, O. (2008). Transport energy modeling with meta-heuristic harmony search algorithm, an application to Turkey, *Energy Policy*, **36**, pp. 2527–2535.

Cheng, Y.M., Li, L., Lansivaara, T., Chi, S.C. and Sun, Y.J. (2008). An improved harmony search minimization algorithm using different slip surface generation methods for slope stability analysis. *Engineering Optimization*, **40**, pp. 95–115.

Degertekin, S.O. (2008). Optimum design of steel frames using harmony search algorithm. *Structural and Multidisciplinary Optimization*, **36**, pp. 393–401.

Dong, H., Bo, Y., Gao, M. and Zhu, T. (2008). Improved harmony search for the detection with photon density wave. *International Symposium on Photoelectronic Detection and Imaging 2007: Related Technologies and Applications, Beijing, China, Proceedings of SPIE*, **6625**, 662523.

Erdal, F. and Saka, M.P. (2008). Effect of beam spacing in the harmony search based optimum design of grillages. *Asian Journal of Civil Engineering* (*Building and Housing*), **9**, pp. 215–228.

Fesanghary, M., Damangir, E. and Soleimani, I. (2008a). Design optimization of shell and tube heat exchangers using global sensitivity analysis and harmony search algorithm. *Applied Thermal Engineering*, doi:10.1016/j.applthermaleng.2008.05.018.

Fesanghary, M., Mahdavi, M., Minary-Jolandan, M. and Alizadeh, Y. (2008b). Hybridizing harmony search algorithm with sequential quadratic programming for engineering optimization problems. *Computer Methods in Applied Mechanics and Engineering*, **197**, pp. 3080–3091.

Forsati, R., Haghighat, A.T. and Mahdavi, M. (2008). Harmony search based algorithms for bandwidth-delay-constrained least-cost multicast routing. *Computer Communications*, **31**, pp. 2505–2519.

Geem, Z.W. (2005a). Harmony search in water pump switching problem. *Advances in Natural Computation, Lecture Notes in Computer Science*, Springer-Verlag, Berlin, **3612**, pp. 751–760.

Geem, Z.W. (2005b). School bus routing using harmony search. *Late Breaking Paper at 2005 Genetic and Evolutionary Computation Conference (GECCO 2005), June 25–29, 2005, Washington, DC, USA*, Association for Computing Machinery, New York (CD-ROM).

Geem, Z.W. (2006a). Improved harmony search from ensemble of music players. *Knowledge-Based Intelligent Information and Engineering Systems, Lecture Notes in Computer Science*, Springer-Verlag, Berlin, **4251**, pp. 86–93.

Geem, Z.W. (2006b). Optimal cost design of water distribution networks using harmony search. *Engineering Optimization*, **38**, pp. 259–280.

Geem, Z.W. (2007a). Harmony search algorithm for solving Sudoku. *Knowledge-Based Intelligent Information and Engineering Systems, Lecture Notes in Computer Science*, **4692**, pp. 371–378.

Geem, Z.W. (2007b). Optimal scheduling of multiple dam system using harmony search algorithm. *Computational and Ambient Intelligence, Lecture Notes in Computer Science*, **4507**, pp. 316–323.

Geem, Z.W. (2007c). Harmony search algorithm for the optimal design of large-scale water distribution network. *Proceedings of the 7th International IWA Symposium on Systems Analysis and Integrated Assessment in Water Management (Watermatex 2007), May 7–9, 2007, Washington, DC, USA*, International Water Association, London (CD-ROM).

Geem, Z.W. (2008a). Novel derivative of harmony search algorithm for discrete design variables. *Applied Mathematics and Computation*, **199**, pp. 223–230.

Geem, Z.W. (2008b). Harmony search applications in industry. *Soft Computing Applications in Industry, Studies in Fuzziness and Soft Computing*, **226**, pp. 117–134.

Geem, Z.W. and Choi, J.-Y. (2007). Music composition using harmony search algorithm. *Applications of Evolutionary Computing, Lecture Notes in Computer Science*, Springer-Verlag, Berlin, **4448**, pp. 593–600.

Geem, Z.W. and Hwangbo, H. (2006). Application of harmony search to multi-objective optimization for satellite heat pipe design. *Proceedings of the US-Korea Conference on Science, Technology, and Entrepreneurship (UKC 2006), August 10–13, 2006, Teaneck, NJ, USA*, Korean-American Scientists and Engineers Association, Vienna, VA, USA (CD-ROM).

Geem, Z.W., Kim, J.H. and Loganathan, G.V. (2001). A new heuristic optimization algorithm: Harmony search. *Simulation*, **76**, pp. 60–68.

Geem, Z.W., Kim, J.H. and Loganathan, G.V. (2002). Harmony search optimization: Application to pipe network design. *International Journal of Modelling and Simulation*, **22**, pp. 125–133.

Geem, Z.W., Kim, T.G. and Kim, J.H. (2000). Optimal layout of pipe networks using harmony search. *Proceedings of the 4th International Conference on Hydro-Science and -Engineering, September 26–29, 2000, Seoul, Korea*, Korea Water Resources Association, Seoul (CD-ROM).

Geem, Z.W., Lee, K.S. and Park, Y. (2005a). Application of harmony search to vehicle routing. *American Journal of Applied Sciences*, **2**, pp. 1552–1557.

Geem, Z.W., Lee, K.S. and Tseng, C.-L. (2005b) Harmony search for structural design. *Proceedings of 2005 Genetic and Evolutionary Computation Conference (GECCO 2005), June 25–29, 2005, Washington, DC, USA*, Association for Computing Machinery, New York, pp. 651–652.

Geem, Z.W. and Park, Y. (2006). Harmony search for layout of rectilinear branched networks. *WSEAS Transactions on Systems*, **5**, pp. 1349–1354.

Geem, Z.W. and Tseng, C.-L. (2002a). Engineering applications of harmony search. *Late-Breaking Papers of 2002 Genetic and Evolutionary Computation Conference (GECCO 2002), July 9–13, 2002, New York, USA*, pp. 169–173.

Geem, Z.W. and Tseng, C.-L. (2002b). New methodology, harmony search and its robustness. *Late-Breaking Papers of 2002 Genetic and Evolutionary Computation Conference (GECCO 2002), July 9–13, 2002, New York, USA*, pp. 174–178.

Geem, Z.W., Tseng, C.-L. and Park, Y. (2005c). Harmony search for generalized orienteering problem: Best touring in China. *Advances in Natural Computation, Lecture Notes in Computer Science*, Springer-Verlag, Berlin, **3612**, pp. 741–750.

Geem, Z.W. and Williams, J.C. (2007). Harmony search and ecological optimization. *International Journal of Energy and Environment*, **1**, pp. 150–154.

Kim, J.H., Baek, C.W., Jo, D.J., Kim, E.S. and Park, M.J. (2004). Optimal planning model for rehabilitation of water networks. *Water Science and Technology: Water Supply*, **4**, pp. 133–147.

Kim, J.H., Geem, Z.W. and Kim, E.S. (2001). Parameter estimation of the nonlinear Muskingum model using harmony search. *Journal of the American Water Resources Association*, **37**, pp. 1131–1138.

Lee, K.S. and Geem, Z.W. (2004). A new structural optimization method based on the harmony search algorithm. *Computers and Structures*, **82**, pp. 781–798.

Lee, K.S. and Geem, Z.W. (2005). A new meta-heuristic algorithm for continuous engineering optimization: Harmony search theory and practice. *Computer Methods in Applied Mechanics and Engineering*, **194**, pp. 3902–3933.

Lee, K.S., Geem, Z.W., Lee, S.H. and Bae, K.W. (2005). The harmony search heuristic algorithm for discrete structural optimization. *Engineering Optimization*, **37**, pp. 663–684.

Li, H.-Q. and Li, L. (2007). A novel hybrid real-valued genetic algorithm for optimization problems. *Proceedings of 2007 International Conference on Computational Intelligence and Security (CIS 2007), December 15–19, 2007, Harbin, China*, IEEE, Los Alamitos, California, pp. 91–95.

Li, L., Chen, Z.-Y., Chi, S.-C., Cheng, Y.-M. and Wang, Y.-J. (2008). Three-dimensional slope stability analysis based on NURBS simulation. *Yantu Gongcheng Xuebao/Chinese Journal of Geotechnical Engineering*, **30**, pp. 212–218 (in Chinese).

Li, L., Chi, S. and Lin, G. (2006a). Chaos harmony search method and its application to local factor of safety method for soil slopes. *Yanshilixue Yu Gongcheng Xuebao/Chinese Journal of Rock Mechanics and Engineering*, **25**(Supp. 1), pp. 2763–2769 (in Chinese).

Li, L. and Chi, S.-C. (2007). Application of new version of harmony search algorithm in slope stability analysis. *Journal of Water Resources and Architectural Engineering*, **5**, pp. 1–6 + 24 (in Chinese).

Li, L., Chi, S.-C., Cheng, Y.-M. and Lin, G. (2007a). Improved genetic algorithm and its application to determination of critical slip surface with arbitrary shape in soil slope. *Shuili Xuebao/Journal of Hydraulic Engineering*, **38**, pp. 157–162 (in Chinese).

Li, L., Chi, S.-C. and Chu, X.-S. (2006b). Location of non-circular slip surface using the modified harmony search method based on correcting strategy. *Yantu Lixue/Rock and Soil Mechanics*, **27**, pp. 1714–1718 (in Chinese).

Li, L., Chi, S.-C. and Lin, G. (2005). Genetic algorithm incorporated with harmony procedure and its application to searching of non-circular critical slip surface in soil slopes. *Shuili Xuebao/Journal of Hydraulic Engineering*, **36**, pp. 913–918 + 924 (in Chinese).

Li, L., Chi, S.-C., Lin, G. and Chu, X.-S. (2007b). Slope stability analysis using extremum principle by Pan Jiazheng and harmony search method. *Yantu Lixue/Rock and Soil Mechanics*, **28**, pp. 157–162 (in Chinese).

Li, L., Huang, Z. and Liu, F. (2007c). An improved particle swarm optimizer for truss structure optimization. *Computational Intelligence and Security, Lecture Notes in Artificial Intelligence*, Springer-Verlag, Berlin, **4465**, pp. 1–10.

Li, L.J., Huang, Z.B., Liu, F. and Wu, Q.H. (2007d). A heuristic particle swarm optimizer for optimization of pin connected structures. *Computers and Structures*, **85**, pp. 340–349.

Li, Q., Yang, S. and Ruan, Y. (2006c). A hybrid algorithm for optimizing multimodal functions. *Wuhan University Journal of Natural Sciences*, **11**, pp. 551–554.

Mahdavi, M., Chehreghani, M.H., Abolhassani, H. and Forsati, R. (2008). Novel meta-heuristic algorithms for clustering web documents. *Applied Mathematics and Computation*, **201**, pp. 441–451.

Mahdavi, M., Fesanghary, M. and Damangir, E. (2007). An improved harmony search algorithm for solving optimization problems. *Applied Mathematics and Computation*, **188**, pp. 1567–1579.

Michalewicz, Z. (1996). *Genetic Algorithms + Data Structures = Evolution Programs*, Springer-Verlag, Berlin, pp. 97–106.

Omran, M.G.H. and Mahdavi, M. (2008). Global-best harmony search. *Applied Mathematics and Computation*, **198**, pp. 643–656.

Paik, K., Kim, J.H., Kim, H.S. and Lee, D.R. (2005). A conceptual rainfall-runoff model considering seasonal variation. *Hydrological Processes*, **19**, pp. 3837–3850.

Ryu, S., Duggal, A.S., Caspar, N.H. and Geem, Z.W. (2007). Mooring cost optimization via harmony search. *Proceedings of OMAE07, 26th International Conference on Offshore Mechanics and Arctic Engineering, June 10–15, 2007, San Diego, California, USA*, American Society of Mechanical Engineers, San Diego.

Sahinidis, N.V. and Grossmann, I.E. (1991). Convergence properties of generalized Benders decomposition. *Computers and Chemical Engineering*, **15**, pp. 481–491.

Saka, M.P. (2007). Optimum geometry design of geodesic domes using harmony search algorithm. *Advances in Structural Engineering*, **10**, pp. 595–606.

Srinivas, M. and Rangaiah, G.P. (2007). Differential evolution with tabu list for global optimization and its application to phase equilibrium and parameter

estimation problems. *Industrial and Engineering Chemistry Research*, **46**, pp. 3410–3421.
Tian, Y.-H., Bo, Y.-M. and Gao, M.-F. (2004). Harmony annealing algorithm for multi-dimensional function optimization. *Computer Simulation*, **21**, pp. 79–82 (in Chinese).
Tian, Y.-H., Bo, Y.-M. and Gao, M.-F. (2005a). Parameters choice criteria in harmony annealing for function optimization. *Computer Simulation*, **22**, pp. 70–74+89 (in Chinese).
Tian, Y.-H., Bo, Y.-M. and Gao, M.-F. (2005b). The application of harmony annealing algorithm for predictive control of nonlinear systems. *Automation in Petro-Chemical Industry*, **2**, pp. 39–42 (in Chinese).
Vasebi, A., Fesanghary, M. and Bathaee, S.M.T. (2007). Combined heat and power economic dispatch by harmony search algorithm. *International Journal of Electrical Power and Energy Systems*, **29**, pp. 713–719.
Yang, X.-S. (2008). *Nature-Inspired Metaheuristic Algorithms*, Luniver Press, Frome, UK.
Zarei, O., Fesanghary, M., Farshi, B., Saffar, R.J. and Razfar, M.R. (2008). Optimization of multi-pass face-milling via harmony search algorithm. *Journal of Materials Processing Technology*, doi:10.1016/j.jmatprotec.2008.05.029.

Exercises

(1) Use basic HS to minimize the two-variable Rastrigin function (Srinivas and Rangaiah, 2007):

$$f(x_1, x_2) = 20 + x_1^2 + x_2^2 - 10\cos(2\pi x_1) - 10\cos(2\pi x_2)$$

for $x_1, x_2 \in [-600, 600]$, which has the global minimum $f(0,0) = 0$. Use the same HS parameters as in Sec. 6. Comment on how the success rate of the algorithm is affected by the convergence tolerance *tol*, the absolute difference in the objective function value, for $tol = 10^{-1}$, 10^{-2}, 10^{-3}, 10^{-4}, 10^{-5} and 10^{-6}.

(2) Using the modified Himmelblau function and basic HS algorithm as in Sec. 6, explore the effect of the HS parameters, HMS, HMCR, PAR and bandwidth, on the solution. Consider the success rate, and for successful runs, the number of function evaluations and computing time.

(3) Investigate the effect of initialization on the performance of the basic HS algorithm. During HM initialization, generate $N_{init} \times$ HMS random vectors, then place only the best HMS of these into the initial HM. Consider $N_{init} = 1$ (the default for basic HS), 2, 3, 4, 5, and

10, 20, ..., 100. Use the modified Himmelblau function (Sec. 6) and the two-variable Rastrigin function (Exercise 1) with the HS parameter values given in Sec. 6. Evaluate the performance of the algorithm in terms of SR, NFE and the average final value of the objective function.

(4) Consider the nonconvex NLP by Sahinidis and Grossmann (1991):

$$\begin{array}{ll} \min & f(x_1, x_2) = -x_1 - x_2 \\ \text{s.t.} & x_1 x_2 \leq 4 \\ & 0 \leq x_1 \leq 6 \\ & 0 \leq x_2 \leq 4 \end{array},$$

which has the global minimum $f(6, 2/3) = -20/3$. For the basic HS algorithm, look into the effect of the penalty function parameters on the solution. Minimize a function of the form:

$$f_{con}(\mathbf{x}) = f(\mathbf{x}) + \alpha \left(\frac{Imp}{\text{MaxImp}} \right)^{\beta} \sum_{j=1}^{m} (f_{pen,j}(\mathbf{x}))^{\gamma},$$

where

$$f_{pen,j}(\mathbf{x}) = \begin{cases} \max\{0, g_j(\mathbf{x})\}, & 1 \leq j \leq q \\ |h_j(\mathbf{x})|, & q + 1 \leq j \leq m \end{cases};$$

q is the number of inequality constraints; m is the total number of constraints; and α, β and γ are penalty function parameters. Use the HS parameters given in Sec. 6, and consider $\alpha = \{10^3, 10^4, 10^5, 10^6\}$, $\beta = \{0, 1, 2\}$ and $\gamma = \{1, 2\}$. Note that $\beta = 0$ corresponds to a static penalty strategy, while $\beta > 0$ indicates a dynamic strategy.

Sample solutions to the exercises can be found on the accompanying CD-ROM.

Chapter 10

META-HEURISTICS: EVALUATION AND REPORTING TECHNIQUES

Abdunnaser Younes and Ali Elkamel

Department of Chemical Engineering, University of Waterloo Waterloo, Ontario, Canda

Shawki Areibi

School of Engineering, University of Guelph Guelph, Ontario, Canada

1. Introduction

Many decision problems are too difficult to be solved exactly within a reasonable amount of time. For such problems, heuristics become the only methods of choice. A heuristic method (also called an approximation algorithm or an inexact procedure) is a procedure consisting of well-defined set of steps that quickly provides a good solution for a given problem. With heuristics, the guarantee of finding optimal solutions is traded with the efficiency of getting good solutions. Thus, heuristic methods are useful in many real-world applications, where running time takes priority over solution quality or simply where exact methods are inapplicable.

Meta-heuristics are a relatively new class of approximate methods that are designed to solve hard optimization problems where classical heuristic methods have failed to be effective, robust or efficient. Unlike heuristic methods, meta-heuristics are general guidance strategies that are not aimed at particular problems, and hence do not have unique sets of steps. Meta-heuristics use concepts from artificial intelligence, mathematical, physical or even biological science to guide classical heuristics to improve their performance by avoiding getting trapped in local optima.

Meta-heuristic search methods tend to find promising solutions by searching the solution space using two strategies: exploration and exploitation. Exploration pressures cause the meta-heuristic to evaluate solutions that have not been visited earlier during the search, while exploitation pressures help the meta-heuristic to utilize knowledge acquired during the search to zoom down on promising regions to further search for better solutions. In the past two decades several meta-heuristic approaches have emerged for handling complex optimization problems [Osman and Kelly (1996)], such as Tabu search, simulated annealing, genetic algorithms, particle swarm optimization to name just a few.

The use of meta-heuristic methods is motivated by their effectiveness, efficiency, robustness, and ease of implementation. However, as these methods do not guarantee optimality of their solutions, any desirable characteristic that motivates their use has to be demonstrated by a careful design of experiments and choice of test problems, coupled with a clear reporting of the results. These aspects are frequently overlooked, even though the literature contains several publications that point to their importance.

This chapter describes experimentation setup and result reporting that take into consideration effects of stochastic sampling, tradeoff between solution quality and cost, and fair algorithm comparison. We do not attempt to investigate all aspects of report writing but merely focus on those points that are particular to meta-heuristics and are the cause of frequent misunderstanding.

2. General Considerations

This section highlights some issues that have to be addressed during experimentation and reporting of meta-heuristic methods.

2.1. Performance measures

Of the merits attributed to meta-heuristics, only solution quality and efficiency can be measured quantitatively. Hence, measure of performance are generally based on solution quality, running time, or both.

2.1.1. Solution quality

The first measures of performance of a stochastic algorithm should cluster around the quality of the solution found by the algorithm, as this is the goal of the optimization process. These measures include the value of the best found solution and the value of first acceptable (feasible) solution. The latter is important to constrained optimization problems, for which an optimization algorithm should produce, as a minimum requirement, a solution that satisfies all the constraints of the problem.

Normalized solutions (i.e. weighed relative to the value of the optimal solution) can give clearer view of the performance than that of the absolute values.

If the algorithm is tested on benchmark problems with known optimal solutions, then the distances of best found solution and first acceptable solution from the optimum can be used as additional measures of performance.

2.1.2. Algorithm efficiency

The running time, commonly reported as CPU time, is a primary indicator of an algorithm's efficiency. In stochastic optimization, it is often given as the total time spent to reach the best solution. If the first acceptable feasible solution is reported, then the total time spent to reach this solution should also be given. However, when used as a basis for algorithm comparison, CPU time becomes a source of unfairness, as it is influenced by several factors, including difference in coding skills of the programmers developing the programs and in machines running the programs [Hooker (1996); Khompatraporn *et al.* (2005)].

The number of evaluations of the objective function is often used as an alternative to CPU time as it is independent of the hardware used and less dependent on the programming skills. It is also more informative than CPU time, especially in real-world applications in which most of the running time is spent in solution evaluation.

The number of iterations of the algorithm is also used as a measure of time, though not for inter-algorithm comparison. The use of this measure makes sense in analyzing convergence of a particular algorithm since an iteration is the basic "building block" of the algorithm.

2.1.3. Quality-efficiency trade-off

Meta-heuristics sample the search space, alternating between searching the neighborhood of previously-visited promising regions and exploring new unvisited regions. In general, the longer the algorithm is run on a problem instance, the better the quality of the final solution; hence, the final conclusion on performance has to consider both solution quality and running time.

Graphs that plot solution quality against running time (in the form of number of algorithm iterations or objective function evaluations) help in assessing the algorithm's performances throughout the run. However, for inter-algorithm comparison they are not conclusive, since the plots are often intermingled prohibiting the determination of which of the competing algorithms is performing best overall.

Since the time allowed for an algorithm is limited in practice, a good policy is to limit it during experimentation too. In this way, the performance can be measured in terms of solution quality only. This is especially advantageous in inter-algorithm comparison since difference in algorithm performance tend to diminish as the results of each algorithm improve with time.

Many researchers use aggregate performance measures that assess both quality and time over the entire run. For example, off-line (*OFP*) performance measure [De Jong (1975)] is used extensively in evolutionary algorithms and can be used with other stochastic methods. *OFP* dynamically abstracts the evolving search into a single value that at any time assesses the algorithm performance up to that point.

$$OFP(T) = \frac{1}{T}\sum_{t=1}^{T} \varepsilon_t,\ \varepsilon_t = \max\{e_1, e_2, \ldots, e_t\}, \tag{1}$$

where e_t is the fitness of the solution evaluated at time step t. Another example is the efficiency index of the path planner used in J. Xiao and Trojanowski (1997). This index is the product of solution cost at the end of the run and the number of iterations spent by the algorithm to reach that solution. More discussion of performance measures can be found in Grefenstette (1999) and Morrison (2003).

2.2. Algorithm robustness

A robust algorithm can perform well over a wide range of problems without readjusting the user's settings of the algorithm parameters [Barr *et al.* (1995)]. The importance of algorithm robustness[1] is clear, since an algorithm that produces high quality solutions to only a few instances is of little use in practice. Yet, robustness is generally difficult to assess, since it is closely related to two problematic issues: test problems and parameter tuning.

The no free-lunch theorem [Wolpert and Macready (1997)] states that the performance of all search algorithms is the same when averaged over all possible functions. That is, no search method can be the best technique for all problems, and hence any claims of an algorithm robustness should be supported by specifying a set of problems that bound the scope of these claims.

Parameter tuning can counter claims of algorithm robustness. An algorithm whose parameters are tuned to every problem instance can hardly be called robust, regardless of the effectiveness of the result. (Several aspects of algorithm robustness are discussed under algorithm *generality* in Khompatraporn *et al.* (2005)).

2.3. Effects of stochastic sampling

Meta-heuristics incorporate some stochastic components and operators to sample the search space. One example is the mutation operator, which is utilized by many meta-heuristics to explore un-visited regions of the search space. The acceptance strategy with probability within simulated annealing is yet another stochastic component that promotes exploration.

These stochastic components produce alternative results with different seeds in the random function used. That is, running an algorithm several times on the same instance can give different results. Therefore, several trials (or runs) have to be performed and statistically analyzed to increase

[1] Care should be taken to distinguish algorithm robustness from solution robustness. The latter is not related to the optimization algorithm but to the objective function and the problem data; it is the insensitivity of a solution to changes in the environment or to noise in the decision variables.

confidence in the conclusions drawn from the results. Some of the commonly used methods of statistical analysis are given below.

- Statistics that show the distribution of results over the whole sample of runs are computed. For example, one can compute the best, average, worst, and standard deviation of the measures of performance over the conducted runs.
- Detailed statistical analysis might be necessary to support results that appear to be inconclusive. When comparing several algorithms, it is a good policy to perform statistical tests that build confidence intervals to help accept or reject hypothesis about superiority of one algorithm over others. Statistical t-tests that are used to compare the means of two samples can be used to compare the performance of the algorithms used. The typical t-test is performed to build a confidence interval that is used to either accept or reject a null hypothesis that both sample means are equal. In applying this test to compare the performance of two algorithms, the measures of performance are treated as sample means, the required replicates of each sample mean are obtained by performing several independent runs of each algorithm, and the null hypothesis is that there is no significant difference in the performance of both algorithms.
- When more than two samples are compared, the probability of multiple t-tests incorrectly finding a significant difference between a pair of samples increases with the number of comparisons [Hochberg and Tamhane (1987)]. Analysis of variance (ANOVA) overcomes this problem by testing the samples as a whole for significant differences. Therefore, ANOVA can be performed to test the hypothesis that measures the performance of all the models under considerations equally. Then, a multiple post ANOVA comparison test, such as Tukey's test, is carried out to produce 95% confidence intervals for the difference in the mean best of generation of each pair of models.

2.4. Parameter tuning

Deciding on the best set of parameter values for a specific implementation is a non-trivial task in meta-heuristic based approaches. Parameter tuning is a tedious task that can consume the user's time, promote unfair comparisons between algorithms, and counter claims of algorithm robustness.

For example, in a traditional evolutionary algorithm, the fundamental tunable parameters are population size, selection, mutation rate, and crossover rate, which seem too few to constitute any difficulty. However, they are interrelated in a complex way that does not enable tuning the parameters individually. This difficulty is aggravated in contemporary algorithms which — being hybrids of more than one technique — have more parameters tune their predecessors.

It is possible in general to tune the algorithm's parameters on a specific problem instance so that it produces good results for the particular problem instance. Such a practice raises several questions: What is the best way to perform tuning? Should it be done individually or collectively, randomly or deterministically? How much time and effort can be allocated to tuning? How far can one go on tuning without biasing fair comparison between algorithms and without prejudicing claims to algorithm robustness? Questions of this sort have been raised by many researchers Barr *et al.* (1995); Hooker (1996); Johnson (2002). Yet, there seems to be no satisfactory answers. The amount of parameter tuning that has to be performed is a subject of trade-off between solution quality from one side and experimentation time, algorithm robustness, and fairness of comparison on the other side. Ultimately, the amount of time that can be allocated to tuning is more or less decided by the user carrying out the experiments, who is still obliged, however, to report the value of these parameters and how they are determined.

To reduce the efforts and adverse effects of parameter tuning, one can use some statistic from the search process to control the parameters adaptively instead of fixing them *a priori*. Alternatively, one can fine-tune the parameters on a small set of problem instances that represent the benchmarks to be used.

3. Test Problems

Empirical experimentation is often used to demonstrate the effectiveness of non-exact algorithms since it cannot be proved theoretically. The algorithms are tested on specific (standardized) problem instances often referred to as benchmark problems. Results of such testing can then be used to attest to the superiority (or inferiority) of the algorithms under consideration. The suite of problem instances to be used has to be selected since any generalization of the results would be within the scope of the suite. In this section we overview sources of test problems and give guidelines for selecting test problems.

3.1. Classification of test problems

There has been considerable efforts to compile test problems for almost all applications. Today, we can find libraries of test problems used in global optimization, multiple-objective optimization, dynamic optimization, and combinatorial optimization to name just a few. One of the most comprehensive sources of global optimization test problem today are the volumes compiled by Floudas *et al.* (1999). Global optimization test problems are generally classified according to the type of the optimization problem; e.g. quadratic programming problems, general non-linear problems, mixed-integer nonlinear programming problems, combinatorial optimization problems. Even though most of these categories overlap, the classifications helps the user pick the test problems relevant to the scope of the algorithm under consideration. For a concise overview of these classification, interested readers may consult [Khompatraporn *et al.* (2005)]. For the purpose of this chapter, we classify test problems according to the source of problem data into two types.

The first type is made from synthetic (randomly generated) data. This type is easy to obtain and analyze, and enables the drawing of general conclusions about the algorithm performance. Synthetic data helps highlight strengths and expose weaknesses of the optimizing algorithm and aids in understanding the operation of the algorithmic ideas. It also helps the user to assess how far from optimality the solutions generated by the metaheuristic are, since the user can easily determine the optimal solution of the generated benchmark without resorting to an exact mathematical approach to solve the problem. However, they might produce artificial landscapes that typically bear little relation to real world problems. Accordingly, using a set of cleverly constructed benchmarks [Cong and Romesis (2003)] with known optima that match and resemble real world problem characteristics can help researchers close the gap and improve the performance of the developed meta-heuristics.

The second type is made from well known instances that are often based on real world problems.

An optimization algorithm will ultimately be applied to a real-world problem. Hence, any practical algorithm should be tested on some instances of the real-world problem for which it is designed. However, real-world problems are often large, complex, highly non-linear, and expensive to evaluate, and hence, their results are difficult to analyze and generalize. For this reason, such problems should be accompanied by artificial problems that help gain more insight. Of course, if the real-world problem is of exceptional significance, the success of the optimizing algorithm on such problem is a formidable goal in itself (without the need to generalize to other problems.).

The difference between the two types can be subtle and many researches end up using a single suite of problems for evaluation and comparison. Nevertheless, the choice of the final set of test problems is often a not trivial task.

3.2. Selecting the test suite

The developer of an algorithm is often spared the hardship of constructing the test problems; thanks to the vast number of benchmarks available in the literature. However, the algorithm developer is still faced with the difficulty of choosing the final suit of test problems for the developed algorithm. Combining the remarks of Crowder *et al.* (1979), Barr *et al.* (1995), Branke (2001), and Khompatraporn *et al.* (2005), we have compiled the following guidelines:

- Test problems should not be limited in terms of size and structure; they should challenge the optimization algorithm.
- Test problems should be generated from elementary components, so that problem difficulty can be controlled by the user.
- It is preferable to use test problems with known optimal values (and known optimal solutions if possible) so that the algorithm under consideration can be assessed even in the absence of results from competing algorithms.

- There should be adequate number of test problem instances to support general conclusions about the algorithm's performance. This number is generally limited by report size and the time needed for test the problems and compile their results.
- Whenever possible, test problems should be selected from benchmarks established in the literature, i.e. instances previously used to test similar algorithms.
- Test problems shold be diverse, within the scope of the algorithm. Diversity of test problems is the main proof of the algorithm's robustness. However, an algorithm tends to perform poorly on problems different from the problem it was designed for. Recall that stochastic algorithms guide heuristics that target specific problems.
- There should not be unrealistic gaps between the benchmarks and the ultimate real world problem. That is, the test problems should permit conjecture to real-world problems.

In summary, the choice of test problems and their diversity do not only affect the results of algorithm evaluation and comparison, but also affect the portability, conjecture, and credibility of the experiments.

4. Experiment Design and Parameter Settings

Keeping the previously mentioned considerations in mind, we recommend conducting the experiments in three phases: preliminary, basic and comparative.

4.1. Preliminary experimentation phase

The preliminary phase of experimentation includes testing of a wide range of algorithm parameters. Most of these experiments are rather trial and error involving a large number of parameter combinations and hence need not be reported. The main purpose of these experiments is to roughly establish *working* ranges of the algorithm parameters by excluding values that tend to produce unreasonable results (very low solution quality or excessive number of iterations). In this way, efforts of parameter tuning in experiments are reduced. Furthermore, biasing in tests is also reduced and any claims to algorithm robustness are better justified.

4.2. Basic experimentation phase

This phase constitutes the major experimental work that investigates the effect of the individual components within the developed algorithm by monitoring the evolution of several variables over time (number of iterations). Algorithm parameters are varied in a controlled predetermined manner, within ranges established in the literature or found during preliminary experiments. In this way, the effect of parameter setting on algorithm performance can be systematically investigated. To further reduce efforts of tuning when several parameters are investigated simultaneously, we propose testing three values only (high, medium, low) of each parameter when tested so that the number of combinations to test is reduced.

4.3. Comparative experimentation phase

This phase is concerned with comparing algorithms on benchmark problems. It uses settings that proved to work well in the previous phase. To ensure fair comparisons, no parameter tuning is carried out in this phase. Furthermore, when possible, comparisons should be based on aggregate performance measures that assess both quality and time over the entire run as was discussed in Sec. 2.1.3.

Needless to say that these phases do not constitute the only possible scheme of experiment design. There can be other alternatives. The main consideration, however, is to ensure that features motivating the use of the methods are adequately demonstrated.

5. Reporting

Authors need to report the full story without distracting the reader over minor issues. However, the typical experimental work in meta-heuristic approaches involves a number of test problems and large amount of input and output data. Thus, effective and efficient reporting of all results requires time, skill and common sense. A good report should have the methodology detailed, experimental design and settings described, and results analyzed so that the work can be appreciated and, if desired, reproduced. The literature contains several guidelines to what a report should contain (See Crowder *et al.* (1979) and Barr *et al.* (1995) as examples). In this section, we focus on the contents that are particularly relevant to work on stochastic algorithms.

5.1. Introductory material

At the outset, the report should clearly define the objectives of the work, and the targeted class of problems. It should review relevant research work with emphasis on recent developments.

5.2. Methodology

The general approach to the problem at hand should be described in detail and include any theory or rationale behind it. The structure of the algorithm can be given as pseudo code or flow chart. No details should be omitted, apart from those commonly known or easily found in the literature (in the latter case, proper references should be given).

5.3. Test problems

The report should describe the test problems used, including their importance and origin. It should explain how and why the particular set of test problems was chosen for the current experimentation. As well, if newly developed test problems are used, the way they are generated should be clearly described.

5.4. Parameter settings

Parameter settings, including termination criteria, should be given. The report should also explain how these settings were decided (e.g. by tuning on a particular set of test problems or by following some guidelines in the literature). If settings are varied adaptively (in response to some characteristics of the problem), then the underlying relationship should be explained.

5.5. Performance measures

The report should define the used performance measures (see Sec. 2.1.). It should display their final values as well as their progression over time. The effect of varying the algorithm's parameters on these measure should also be reported.

The three measures of time discussed in Sec. 2.1.2 can be reported together in the final report: CPU time should be routinely given since many readers are interested in it. The number of function evaluations is a fairer measure of time for algorithm comparison, as discussed earlier. The number of algorithm iterations, is more informative for convergence analysis of a particular algorithm. Note that if CPU time is reported, the report should indicate how the CPU time is computed, and should give the specification of the used machine (processor model and speed, and memory size), operating system, and the programming language.

5.6. Format

Authors need to draw the readers attention to the important results in their report and at the same time give them as much supporting details as possible. Results have to be clearly displayed and analyzed so that the final conclusions drawn from these results are justified.

Readers rarely appreciate large tables of raw data produced by executing several runs on several problems. Thus, rather than reporting the final solutions produced by fifty runs, for example, it is more informative and less distracting to report their best value, average value, worst value, and standard deviation. Moreover, important entries in the table should be highlighted so that a reader who skims the results can detect them quickly. Note that reporting the best run only can be misleading, since other (non-reported) runs might have lead to different interpretation of the results. If it is necessary to display large amounts of data, they can be attached in an appendix so that they do not disturb the flow of the report.

Graphs are good means to display large amounts of data and to highlight trends in data. They are often used to depict the evolution of results over time (e.g. to show how fast the solution value is changing with the number of iterations).

5.7. Discussion

The mere display of results (no matter how detailed) is seldom sufficient to give a clear picture to the readers. Authors should discuss what they have found out. They should highlight trends in results so that readers infer general facts about the algorithm behavior; such trends are usually difficult to detect by casual readers. Statistical analysis is particularly important for stochastic algorithms, as it helps to verify the significance of the findings.

The report should not be limited to the strong points of the presented algorithm, it should also point out weak points and give possible reasons for any limitations or failures the algorithm may have. Crowder *et al.* (1979) and Hooker (1996), among others who stress the importance of reporting negative results, note that publications tend to report positive results only whereas negative results can be as educational if not better.

5.8. Final summary and conclusions

The conclusion section is one of the most important sections in the report; in fact, many readers skip the main body of the report and go directly to the conclusion section, which summarizes the main points of the report. This section, thus, should be written in strong statements without unnecessary details. The conclusions section contains two essential parts: a summary of the report findings and directions for future work.

The findings' summary should be limited to statements that have a solid basis in the operation of the algorithm or in the results of experimentation. If authors elect to add any speculations, they should clearly mark them as opinions and avoid mixing them with appropriately justified statements.

The part on directions for future work, though important, is often hastily written or not at all. Many readers are interested in how to build on the work reported in the current report. This section should supply such reader with information on how to improve the current work; problems, other than the one covered by the report, that can be tackled by the current algorithm; and simply any recommendations for future research.

6. Conclusion

Meta-heuristics are a class of solution methods that guide and modify the operations of subordinate heuristics to efficiently produce high-quality solutions. They have been successfully applied to many optimization problems in business, economics, engineering and science. However, it is often impossible to obtain an analytical prediction of either the solution achievable within a given computation time or the time taken to find a solution of a given quality. The assessment of these methods is commonly based on experimental analysis. However, the lack of a methodology in these analyzes limits the scientific value of their results. In this chapter we presented different scenarios for the analysis and comparison of meta-heuristics by experimentation. We also described schemes for result reporting that take into consideration effects of stochastic sampling and fair algorithm comparison.

References

Barr, R., Golden, B., Kelly, J., Resende, M. and Stewart, W. (1995). Designing and reporting on computational experiments with heuristic methods. *J. of Heuristics*, **1**(1), pp. 9–32.

Branke, J. (2001). *Evolutionary Optimization in Dynamic Environments*, Environments. Kluwer Academic Publishers.

Cong, J. and Romesis, M. (2003). Optimality, scalability and stability study of partitioning and placement algorithms. *International Symposium on Physical Design* (IEEE, Monterey, CA, USA), pp. 88–94.

Crowder, H., Dembo, R.S. and Mulvey, J.M. (1979). On reporting computational experiments with mathematical software. *ACM Trans. Math. Softw.*, **5**(2), pp. 193–203.

De Jong, K.A. (1975). *An Analysis of the Behavior of a Class of Genetic Adaptive Systems*, PhD thesis, University of Michigan.

Floudas, C., Pardalos, P., Adjiman, C., Esposito, W., Gumus, Z., Harding, S., Klepeis, J., Meyer, C. and Schweiger, C. (1999). *Handbook of Test Problems for Local and Global Optimization*, Kluwer Academic Publishers.

Grefenstette, J.J. (1999). Evolvability in dynamic fitness landscapes: A genetic algorithm approach. *Congress on Evolutionary Computation*, (IEEE), **3**(1), pp. 2031–2038.

Hochberg, Y. and Tamhane, A.C. (1987). *Multiple Comparison Procedures*, Wiley.

Hooker, J. (1996). Testing heuristics: We have it all wrong. *Journal of Heuristics*, **1**(1), pp. 33–42.

J. Xiao, L.Z., Michalewicz, Z. and Trojanowski, K. (1997). Adaptive evolutionary planner/navigator for mobile robots. *IEEE Trans. Evolutionary Computation*, **1**(1), pp. 18–28.

Johnson, D.S. (2002). A theoretician's guide to the experimental analysis of algorithms, in Goldwasser, D.S.J.M. and McGeoch, C.C. (eds.), *5th and 6th DIMACS Implementation Challenges* (American Mathematical Society), pp. 215–250.

Khompatraporn, C., Pinter, J.D. and Zabinsky, Z.B. (2005). Comparative assessment of algorithms and software for global optimization. *Journal of Global Optimization*, **31**(4), pp. 613–633.

Morrison, R. (2003). Performance measurement in dynamic environments, in Branke, J. (ed.), *GECCO Workshop on Evolutionary Algorithms for Dynamic Optimization Problems*, pp. 5–8.

Osman, I. and Kelly, J. (1996). *Meta-Heuristics: Theory and Applications*. Kluwer Academic Publishers, Boston.

Wolpert, D.H. and Macready, W.G. (1997). No free lunch theorems for optimization. *IEEE Transactions on Evolutionary Computation*, **1**(1), pp. 67–82.

Chapter 11

A HYBRID APPROACH FOR CONSTRAINT HANDLING IN MINLP OPTIMIZATION USING STOCHASTIC ALGORITHMS

G. A. Durand*, A. M. Blanco†, M. C. Sanchez‡ and J. A. Bandoni§

Planta Piloto de Ingeniería Química-Plapiqui (UNS-Conicet)
Camino La Carrindanga Km. 7, 8000 Bahía Blanca, Argentina
**gdurand@plapiqui.edu.ar*
†ablanco@plapiqui.edu.ar
‡msanchez@plapiqui.edu.ar
§abandoni@plapiqui.edu.ar

1. Introduction

Global optimization deals with the calculation and characterization of global extrema of functions. Due to its outstanding importance in applied mathematics to science, an overwhelming amount of theoretical and computational contributions, categorized into deterministic and stochastic approaches, has been produced in the last decades.

Deterministic strategies can be classified into direct-search and gradient-based methods. Direct-search methods require only the evaluation of the involved functions. The objective function is locally improved from some starting point until a predefined precision is achieved. Since derivatives are not required, non-smooth functions can be naturally handled. Direct search algorithms include the Nelder–Meade simplex method, the Hooke–Jeeves procedure, the conjugate-directions technique due to Powell, etc. (Lewis *et al.*, 2000). On the other hand, gradient-based algorithms require derivatives information and perform satisfactorily when convex and smooth functions are involved in the optimization model. Many efficient algorithms have been developed to address linear and nonlinear programming problems and their mixed integer versions (Edgar *et al.*, 2001; Brooke *et al.*, 2003). Deterministic strategies are local in nature and are prone to get trapped in a local optimum depending on the starting point of the numerical method. Moreover, even remarkable improvements have been done in deterministic global optimization, the efficient solution of highly non-convex models remains a challenging issue.

Stochastic strategies such as evolutionary algorithms, simulated annealing and taboo search, showed to be successful in non-convex optimization due to their inherent parallel nature, which allows the simultaneous exploration of the whole search space, reducing the possibility of premature convergence to local extrema.

In essence, evolutionary algorithms are unconstrained optimization procedures. As real-world optimization problems involve constraints of certain kind, the issue of constraint handling in evolutionary optimization has received considerable attention in the last decade. A comprehensive discussion of the proposed approaches can be found in Michalewicz *et al.* (1996), Coello Coello (2002) and Tessema and Yen (2006). Accordingly to the categorization proposed in Coello Coello (2002), constraint handling methods can be divided into five groups: penalty functions, special representations and operators, repair algorithms, separation of objectives and constraints, and hybrid methods. In the following, these strategies are briefly described and the most relevant recent contributions highlighted to provide an overview on the topic.

A constrained optimization problem is frequently transformed into an unconstrained one by adding penalties to infeasible individuals. The degree of penalization is assigned as a function of the amount of constraint violation. Static and dynamic *penalty functions* have been proposed but they typically involve problem-specific tuning to perform well. To overcome this problem dependence, the challenge is to develop completely adaptive penalty functions that do not require user-provided parameters, and exploit helpful information acquired during the evolution to guide the search (Smith and Coit, 1997; Miettinen *et al.*, 2003).

Regarding adaptive penalty techniques, Runarsson and Yao (2000) presented a novel method to balance the dominance of the objective and penalty functions based on a stochastic ranking procedure. A single parameter that defines the probability of using only the objective function for comparisons in the infeasible regions of the search space, is selected. In this way, a convenient bias towards the objective function is specified in ranking individuals, which guides the search efficiently to the global optimum.

Tessema and Yen (2006) introduced a new fitness value, called distance value, and two penalty functions that allow an effective exploitation of the information hidden in infeasible individuals. Thus, proper individuals are selected at different stages of the search and under different conditions. The algorithm is very simple to implement and it is free of parameter tuning.

More recently, Wang *et al.* (2008) presented an adaptive scheme devoted to obtain an appropriate trade-off between the objective function and constraint violation at the different phases of the search. In phase I, the population contains no feasible individuals and a hierarchical non-dominated individual selection scheme is adopted with a search bias towards individuals with the less constraint violation. In the second phase, the population is made up of feasible and infeasible individuals. The trade-off scheme has the capability to adapt the fitness of each individual based on the proportion of feasible individuals of the last population maintaining therefore a proper balance between feasible and infeasible members. At last, the comparison of individuals is only based on their objective function values. No-user defined parameters are necessary for this strategy.

Another approach involves the formulation of *special representation schemes* to simplify the shape of the search space. Ad-hoc *genetic operators* are generally developed to preserve the feasibility of solutions. These procedures are very useful for the application for which they were devised, but their extension to similar problems is no straightforward. This approach also includes the techniques that work in the reduced space obtained after the elimination of the equality constraints and an equal number of variables from the optimization model (Michalewicz, 1996). The procedure performs well if it is possible to explicitly evaluate the dependent variables from the equality constraints, but this is hardly the case when nonlinear functions are present in the model.

Repair algorithms are useful to handle infeasible individuals when they can be easily transformed into feasible ones. There exist successful applications of these strategies to many combinatorial optimization problems (Michalewicz and Nazhiyath, 1995; Gobin *et al.*, 2007). The weaknesses of repair algorithms are the lack of standard heuristics to design repair operators and the bias in the evolutionary search that these operators may introduce (Smith and Coit, 1997).

Other approaches handle *constraints and objectives separately*. Among them, the procedure developed by Deb (2000) considers the superiority of feasible individuals that have a lower fitness than the infeasible ones, which are penalized by their constraint violation. The technique has the advantage that no parameters have to be set, but requires niching to maintain diversity in the population.

Recently, multi-objective optimization techniques were applied to solve single-objective constrained optimization problems. The methodology consists in reformulating the original model into another one, which takes into account constraints as additional objectives. Then a multiobjective optimization technique is utilized to solve the reformulated model. Mezura Montes and Coello Coello (2002) performed a numerical evaluation of four constraint-handling strategies based on multiobjective optimization. They concluded that the selection criterion of Pareto dominance gives better results than both Pareto ranking and population-based approaches. In Wang and Cai (2005), a bi-objective optimization problem for the minimization of the objective function and constraint violation was presented. The strategy is characterized by the replacement of a dominated individual of the population by a non-dominated one belonging to the offspring population, and the archiving of the best infeasible individual that returns to the population after some generations. The technique is computationally expensive and requires parameters tuning. Venkatraman and Yen (2005) presented a two-phase algorithm. At first it behaves as a constraint satisfaction problem and minimizes the constraint violations. Once a solution is obtained, a bi-objective optimization problem is solved. A non-dominated ranking is employed to rank individuals and a niching scheme preserves diversity. User-defined parameters are not required.

Hybrid approaches also constitute a promising alternative for constraint handling. In this group there exist numerous approaches that combine an evolutionary algorithm with another technique, for example: lagrangian multipliers (Pirkwieser *et al.*, 2008), fuzzy logic (Khorram and Hassanzadeh, 2008), cultural algorithms (Becerra and Coello Coello, 2005), immune systems (Bernardino *et al.*, 2007), ant colony optimization (Leguizamon and Coello Coello, 2007), etc.

A critical analysis of the reviewed methodologies indicates that each one presents advantages and disadvantages. To explicitly avoid the constraint handling issue in optimization, a novel methodology is proposed which combines the strengths of the deterministic optimality theory together with the ability of stochastic techniques as unconstrained model optimizers.

The following general nonlinear optimization problem is considered:

$$\begin{array}{ll} & \min_{\mathbf{x}} f(\mathbf{x}) \\ \text{s.t.} & g_j(\mathbf{x}) \geq 0 \quad j = 1, 2, \ldots, J \\ & h_k(\mathbf{x}) = 0 \quad k = 1, 2, \ldots, K^{\cdot} \\ & \mathbf{x} \in R^N \end{array} \tag{1}$$

For the purposes of this contribution, it is assumed that the functions involved in (1) are continuous and that first derivatives exist in all cases.

Although model (1) does not explicitly account for binary variables, a binary variable y, $y \in \{0, 1\}$, can be modeled within a continuous framework using the standard continuous reformulation (Ryoo and Sahinidis, 1995): $0 \leq y \leq 1$, $y(1 - y) = 0$, meaning that variable y is relaxed between 0 and 1 while the equality constraint forces y to be 0 or 1 at the solution.

The proposed methodology consists of an adequate reformulation of the constrained model, in order to pose an equivalent unconstraint version, which can be efficiently solved by stochastic optimization algorithms. The rationale is to formulate the Karush-Kuhn-Tucker system of problem (1), meaning its optimality conditions, whose stationary points are also the solutions of problem (1). An iterative scheme is adopted then to find the solutions of the resulting system one by one.

The chapter is structured as follows. In Sec. 2 the proposed methodology is introduced. The adopted strategy to solve the resulting unconstrained optimization problem using a sequential niche technique is described in

Sec. 3. A performance analysis of the proposed approach is presented in Sec. 4. A conclusions and future work section closes the article.

2. Proposed Methodology

A well known result from the theory of nonlinear constrained programming is that any local minimum, $\mathbf{x}^*$, of problem (1) must satisfy the following necessary conditions of optimality (Reklaitis *et al.*, 1983):

Kuhn-Tucker Conditions (KTC)

$$\nabla L(\mathbf{x}^*, \mathbf{u}^*, \mathbf{v}^*) = \nabla f(\mathbf{x}^*) - \sum_{j=1}^{J} u_j^* \nabla g_j(\mathbf{x}^*) - \sum_{k=1}^{K} v_k^* \nabla h_k(\mathbf{x}^*) = \mathbf{0}, \tag{2a}$$

$$g_j(\mathbf{x}^*) \geq 0, \quad j = 1, \ldots, J, \tag{2b}$$

$$h_k(\mathbf{x}^*) = 0, \quad k = 1, \ldots, K, \tag{2c}$$

$$u_j^* g_j(\mathbf{x}^*) = 0, \quad j = 1, \ldots, J, \tag{2d}$$

$$u_j^* \geq 0, \quad j = 1, \ldots, J, \tag{2e}$$

where $L(.)$ is the Lagrangian function defined as:

$$L(\mathbf{x}, \mathbf{u}, \mathbf{v}) = f(\mathbf{x}) - \sum_{j=1}^{J} u_j g_j(\mathbf{x}) - \sum_{k=1}^{K} v_k h_k(\mathbf{x}), \tag{3}$$

and u_j and v_k are the Lagrange multipliers associated with inequalities and equalities, respectively. Therefore it is equivalent to address the solution of problem (1) by solving the constrained system of Eq. (2).

Local gradient-based solvers are usually applied to solve constrained systems of equations, but the solutions highly depend on the starting point of the numerical method. Global deterministic methods were also proposed to enclose all the solutions of systems of nonlinear equations (Maranas and Floudas, 1995). To the best of our knowledge no stochastic-based methods have been systematically used to solve such systems. Advantage of stochastic techniques can be taken however to address the solution of systems of equations by proper reformulation, as proposed in Blanco *et al.* (2006).

Let us consider a general square system of nonlinear equations of dimension I:

$$p_i(\mathbf{x}) = 0 \quad i = 1, 2, \ldots, I \tag{4}$$

The solutions of system (4) can be obtained by solving the following optimization problem:

$$\begin{aligned} &\min_{x_1, x_2, \ldots, x_I} \phi = \textstyle\sum_i \|\varepsilon_i\| \\ \text{s.t.}\quad &\varepsilon_i = p_i(\mathbf{x}) \\ &i = 1, 2, \ldots, I \end{aligned}, \tag{5}$$

where ε_i represents the residual or error of each equation and $\|.\|$ stands for some convenient measure of distance, for example the absolute value or the square functions. A local minimizer of problem (5), for which the objective function value is zero, constitutes a solution of system (4).

In general, problem (5) is highly multi-modal and possibly non-smooth if non-derivable functions are adopted as objectives, e.g. the sum of absolute values of the errors. Moreover, due to the explicit nature of the residuals in terms of the optimization variables, problem (5) is an unconstrained optimization model that can be effectively addressed by stochastic techniques, such as GAs.

In summary, a nonlinearly constrained optimization model (1) can be transformed into an unconstrained optimization model (5) which be solved using stochastic techniques. The general procedure is detailed in Fig. 1.

The major advantage of this approach is that nonlinearly constrained optimization models of type (1), which are difficult to solve by both deterministic and stochastic techniques, can be addressed by means of a methodology that combines the strengths of nonlinear deterministic theory and those of stochastic algorithms.

Step 1.	Formulate the KTC system (2 a, b, c, d, e) of model (1)
Step 2.	Formulate problem (5) made up with equations (2 a, c and d)
Step 3.	Solve problem (5) using some stochastic strategy
Step 4.	Evaluate the objective function and inequality constraints at the found solutions and keep the best feasible as the solution of problem (1)

Figure 1. Proposed methodology.

The power of the deterministic component is the use of the theoretically sound Karush-Kuhn-Tucker optimality conditions of non-linear constrained models, which allows the transformation of the optimization problem into an algebraic system of nonlinear equations. Such a system is transformed back into an unconstrained optimization model whose solutions are those of the original constrained problem. In order to find those solutions, the ability of stochastic algorithms to solve nonlinear unconstrained multi-modal landscapes is exploited.

It should be noted that there exist some problems for which system (2) does not have a solution. In such cases the Lagrange multiplier method would fail, precluding the use of the proposed methodology. However, as pointed out in Reklaitis *et al.* (1983), such cases are rare in practice. In addition to the existence of a solution, the constraint qualification condition should also hold for optimality. Constraint qualification implies linear independence of equality constraints and active inequality constraints at $\mathbf{x}^*$. In general, it is difficult to verify the constraint qualification, since it requires that the optimum solution be known beforehand. However, for practical problems the constraint qualification usually holds (Reklaitis *et al.*, 1983).

In the following section the adopted strategy to address Step 3 of the proposed methodology is described. In particular a sequential niche algorithm, which shows attractive features for finding many solutions of unconstrained optimization models was implemented. However, any suitable technique could be used for this purpose.

3. Sequential Niche Technique for Multimodal Function Optimization

In order to illustrate the features of problem (2) and the proposed reformulation (5) consider the following system of equations, as described in Maranas and Floudas (1995), whose solutions are reported in Table 1:

$$\begin{cases} 4x_1^3 + 4x_1x_2 + 2x_2^2 - 42x_1 - 14 = 0 \\ 4x_2^3 + 2x_1^2 + 2x_1x_2 - 26x_2 - 22 = 0 \end{cases}. \tag{6}$$

The level surfaces of function $\phi = |\varepsilon_1| + |\varepsilon_2|$ are plotted in Fig. 2 where the nine solutions of Table 1 can be appreciated. This figure also shows how a random initial population of individuals (Fig. 2a) evolves by means

Table 1. Solutions of system (6).

Sol.	x_1	X_2
1	−0.2709	−0.9230
2	−0.1279	−1.9538
3	3.5844	−1.8481
4	3.3852	0.0739
5	3.000	2.000
6	0.0867	2.8843
7	−2.8051	3.1313
8	−3.0730	−0.0814
9	−3.7793	−3.2832

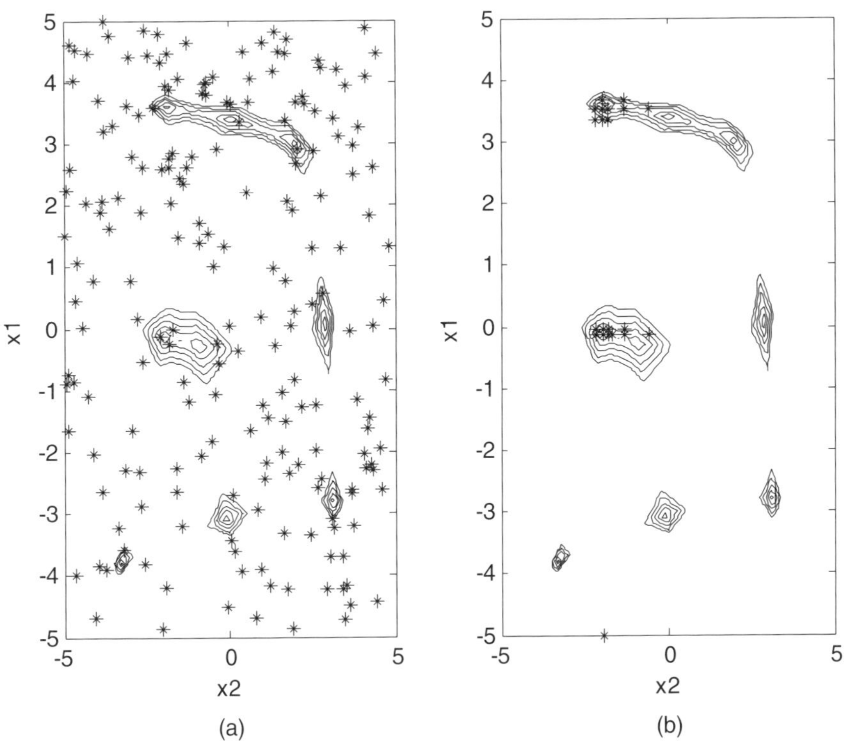

Figure 2. Initial and final populations for problem (6).

of a standard GA and generates clusters around some of the solutions (Fig. 2b).

Some general conclusions can be withdrawn from analyzing the example:

(i) A solution of system (5) is characterized by a zero value of function $\phi(\mathbf{x})$. Therefore, those individuals that present a function value close to zero, that is, below some small positive threshold, τ, constitute potential solutions of the system under study.

(ii) In order to converge to an *exact* root, a *hybrid* approach that combines the GA with a local search algorithm might be considered. The idea is to perform a local search from the most promising individuals of the GA.

(iii) There exist several possibilities for the objective function of problem (5). Typical choices would be the absolute value function and the square function. In Blanco *et al.* (2006) the MaxMin function was introduced, which also presents interesting features as objective function for problem (5). The adopted measure will determine the fitness landscape for the GA algorithm, influencing in that way the possibility of identifying the different solutions.

(iv) Despite the inherent *parallel* nature of the GA, there is no warranty to find all the solutions of the nonlinear system of equations under study by solving (5), although if enough diversity is ensured, there is a high probability of finding several of them in a single GA run as shown in the previous example.

Regarding remark (iv), the problem of finding all solutions of unconstrained optimization models has been quite studied and several methods were proposed (Gan and Warwick, 2000; 2001; Beasley *et al.*, 1993).

For the purposes of this research, a sequential niche technique was adopted (Beasley *et al.*, 1993). The rationale behind this approach is to identify one local optimum at a time until all expected optima are found. Once one optimum is identified, the objective function is modified in order to *exclude* that optimum from the fitness landscape and increase therefore the chance of convergence to other optima in posterior searches. This procedure is repeated until all expected optima are found.

The sequential niche technique is based on the use of derating functions. Let us consider that **s** is a particular point in the search space. A derating function G(**s**, **x**) performs the task of worsening the fitness of function $\phi(\mathbf{x})$ in some neighborhood of point **s** when both functions are multiplied. Based on this property of the derating functions, the algorithm of Fig. 3 is adopted to perform a sequential search of the solutions of multi-modal problem (5). This is a modified version of the algorithm proposed in Beasley *et al.* (1993).

In Step 2, the aim is to identify one of the many optima of the objective function under study by applying a GA. In Step 3, a local search algorithm is implemented in order to find the *exact* solution with a distance measure lower than τ. In Step 5 the function is derated in a neighborhood of the found solution to improve the chances of identifying new solutions in subsequent iterations. Step 6 involves the adoption of some termination criterion.

In this work, a constant derating function is adopted in Step 5:

$$G(\mathbf{s},\mathbf{x}) = \begin{cases} c \to \infty & \text{if } d_{\text{XS}} \le r \\ 1 & \text{otherwise} \end{cases}, \tag{7}$$

where r represents the niche radius, d_{XS} stands for the distance between individual x and solution s and c is a large positive constant. The effect of (7) on the optimization problem is to penalize the individuals that fall within radius r of optimum s. The selection of parameter r influences the performance of the technique. Small radiuses may not prune large enough areas of the search space, generating new local minima that are close to the previous solution. Large radiuses, on the other hand, may cover minima not

1. Set i=0, $\phi_0(\mathbf{x})=\phi(\mathbf{x})$ and $G_0(\mathbf{s}_0,\mathbf{x})=1$
2. Perform a genetic optimization on function $\phi_{i+1}(\mathbf{x})= G_i(\mathbf{s}_i,\mathbf{x}).\phi_i(\mathbf{x})$
3. Perform a local search from the best found individual to exactly identify the optimum $\mathbf{s}_i$
4. Set i=i+1
5. Construct function $G_i(\mathbf{s}_{i-1},\mathbf{x})$
6. If a solution was found in Step 3 (objective function value < τ), or the maximum number of unsuccessful iterations is not exceeded, return to Step 2
7. End

Figure 3. Sequential niche algorithm.

previously found, preventing their identification in the following searches. The optimal value of radius r would cover a large enough area above the minimum without overlapping with the valleys of other minima.

Since the optimum niche radius is not known *a priori*, some criteria should be adopted to establish the derating function domain. In our implementation the following niche radius policy was considered. Lets $\mathbf{s}_i^{\mathrm{GA}}$ be the best individual of the evolutionary optimization of Step 2 and $\mathbf{s}_i$ be the solution of the local optimization of Step 3. Then, the adopted niche radius is $r = \|\mathbf{s}_i^{\mathrm{GA}} - \mathbf{s}_i\|_2$, namely the Euclidean distance between the two points. This is a *self-adaptive* strategy, which avoids the requirement of providing a figure for parameter r beforehand.

Regarding Step 6, most niche-based techniques assume that the number of solutions of the optimization problem under study is known *a priori*. Therefore a straightforward termination criterion is to stop when this number is reached. In the general case the number of optima is hardly known and an alternative strategy is required. In our case, it is known that the solutions of problem (5) have a zero function value. Therefore, a practical termination criterion is to stop when no new solutions with a function value lower than τ are obtained from Step 3. A maximum number of unsuccessful trials is allowed however, since a potential solution can be skipped in one search but found in a later iteration.

It should be noted that a *dummy solution* may appear during the search. It is defined as an individual of the search space that shows a distance measure lower than the tolerance τ, but it is not included in the set of solutions considered as *expected*. A dummy solution may have two sources: (a) it can be a stationary point of system (2) which does not correspond to an expected minimum of problem (1), namely it can be a maximum or a saddle; (b) it can be a point of the search space that is not a stationary point of system (5) but has a small non-zero value of function ϕ ($\mathbf{x}$). A strategy for the identification and rejection of dummy solutions was not included in the current implementation of the sequential niche algorithm.

In order to illustrate the mechanics of the approach, the evolution of the algorithm for example (6) is illustrated in Figs. 4 to 6. In each figure, the asterisk and the gray circle represent the solution identified in Step 3 and the domain (niche) of the corresponding derating function (7), respectively. In the following iteration, the niche of a previously found solution is shown

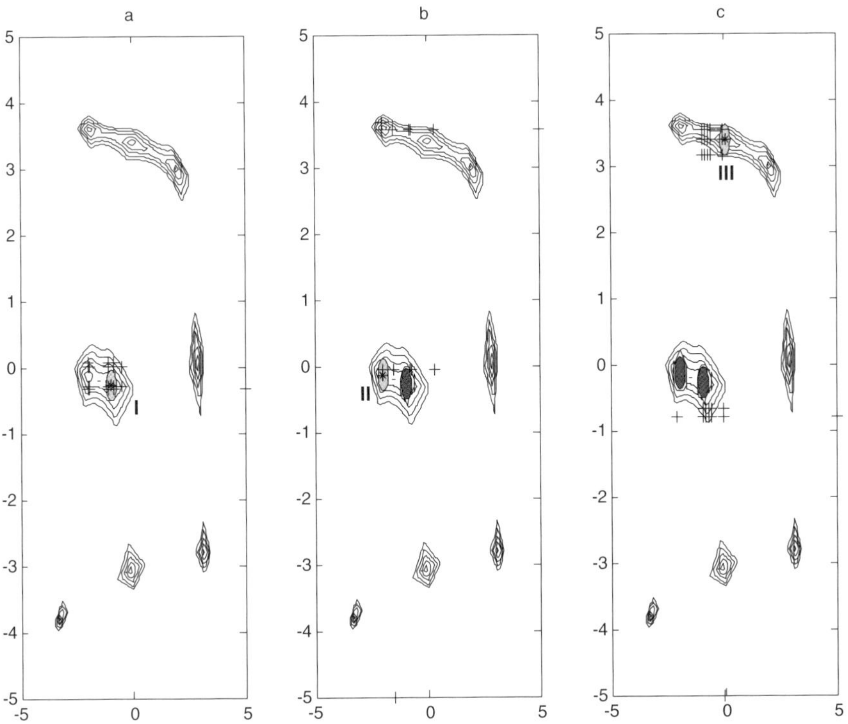

Figure 4. Sequential niche evolution 1.

in dark gray. Each solution is identified with a number that indicates the order in which it has been found by the algorithm.

It can be appreciated from Figs. 4a to Fig. 5e, that a different solution is found in each iteration of the algorithm.

In Fig. 5f it can be observed that two previously found solutions, I and II, present larger niches than the originally assigned. This is because before finding the new solution, VI, the GA got trapped in the neighborhood of previously found solutions I and II. The algorithm detects this situation and assigns larger radiuses to the niches of solutions I and II, covering in that way a larger portion of already explored space and increasing therefore the chances of identifying a new solution (solution VI in this case).

In Fig. 6g it can be seen that before finding solution VII, solutions I and V were further derated in some intermediate iterations of the algorithm.

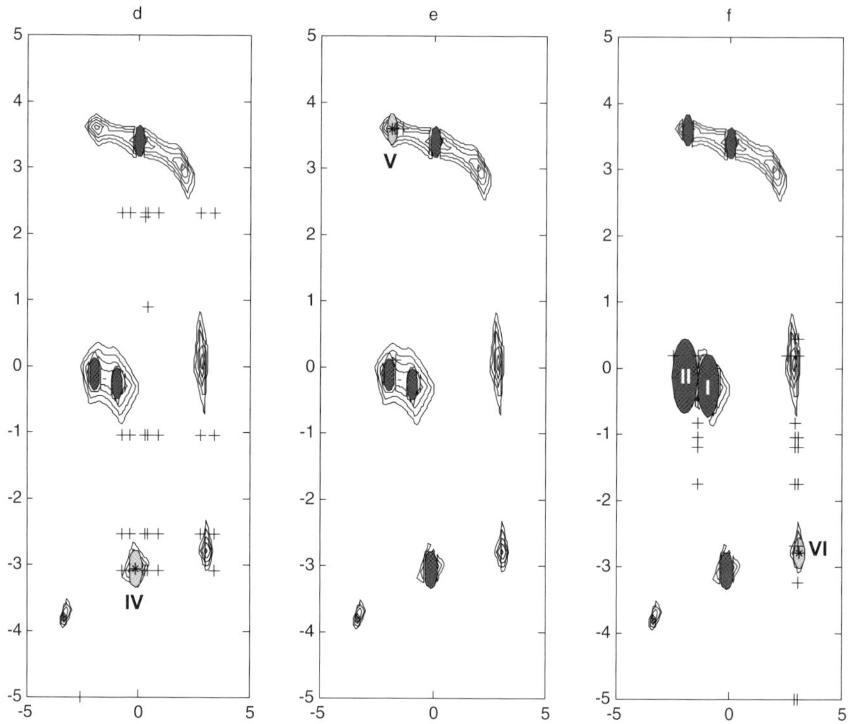

Figure 5. Sequential niche evolution 2.

Solution VIII (Fig. 6h) is immediately detected after solution VII. Finally, solution IX is found after increasing the niches of solutions IV and VI (Fig. 6i).

In the following section, the performance of the proposed approach for the resolution of test optimization problems is analyzed.

4. Results and Discussion

In Table 2 the main characteristics of the addressed test cases are summarized. The last column of the table reports the information provided in the original references for the total number of local and global optima.

All test cases are small-scale examples widely used in optimization studies that involve both stochastic and deterministic approaches. They were solved using the methodology summarized in Fig. 1. Step 2 of

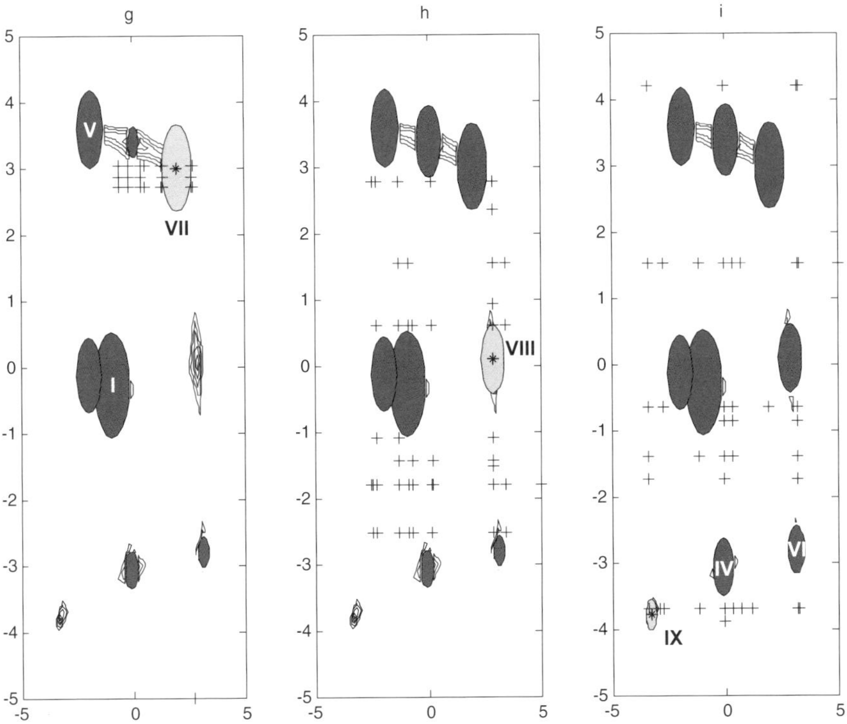

Figure 6. Sequential niche evolution 3.

the technique was tackled with the sequential niche algorithm described in Fig. 3.

The sequential niche procedure was implemented using the Matlab code, and the GA Toolbox of Matlab (gatool) was employed to solve Steps 2 and 3 (Fig. 3). For each example, the population size and maximum number of generations were set in $10^*(N + J + K)$ and $5^*(N + J + K)$, respectively (N: number of variables, J: number of equality constraints, K: number of inequality constraints). An initial range for the variables, including the Lagrange multipliers, was provided. In all cases, the genetic operators and parameters used are as follows: Population type: Double Vector; Fitness Scaling: Proportional; Selection: Roulette; Elitism: 0; Crossover: Single point; Crossover fraction: 0.75; Mutation: Uniform mutation; Mutation rate: 0.01; Hybrid function: fminsearch. Regarding Step 6 of the algorithm, the

Table 2. Test cases.

Example	Example number from Srinivas and Rangaiah (2007)	Type	Constraints		Variables (N)		Expected number of solutions
			Equality (J)	Inequality (K)	Binary	Real	
1	NLP-14	NLP	0	4	0	2	2
2	NLP-15	NLP	0	2	0	2	4
3	NLP-16	NLP	3	3	0	6	2
4	NLP-8	NLP	4	1	0	6	3
5	MINLP-1	MINLP	0	3	1	1	2
6	MINLP-2	MINLP	1	1	1	2	1
7	MINLP-3	MINLP	0	3	1	2	1

Table 3. Statistics.

	Proposed methodology					Differential evolution (RG) Srinivas and Rangaiah (2007)		
				Function evaluations			Function evaluations	
Example	Success rate 1	Success rate 2	Iterations	GA	Local search	Success rate	Method	Local search
1	100	96	2.18	1074	20	100	2399	10
2	89.5	99	11.07	2510	79	86	2100	14
3	90.5	97	2.75	2271	17	100	5419	8
4	100	94	5.32	5658	87	90	1682	28
5	98.5	93	4.01	1043	25	100	1427	8
6	100	100	1.17	981	9	65	1428	3
7	100	96	5.72	4485	50	94	2352	6

selected distance measure was the sum of the square errors, and the tolerance τ was set equal to 0.1.

In Table 3 the solution statistics are summarized. The results are average values obtained from 100 runs of the algorithm. The *Success Rate 1* column reports the success in finding the whole number of expected solutions of the problem under study. The *Success Rate 2* column indicates the success in finding the known global optimum. The average number of iterations of the sequential niche algorithm, and the average number of function evaluations for the GA and the local search optimizations are also reported in Table 3.

The *Success Rate 1* column shows that the expected number of solutions was found in a large extent for all the examples. The *Success Rate 2* results also indicate that the expected global optimum was found in most trials. However, it can be observed that the percentage of success in finding the global optimum is less than 100%, even in the cases where the success in finding the whole number of expected solutions is 100%. This is because dummy solutions were identified instead of the actual global optimum in some runs of the sequential niche algorithm.

It can be also observed that the average number of iterations of the sequential niche algorithm is larger than the expected number of solutions.

The difference is due to the fact that the optimization sometimes converges to previously found solutions or to none solution at all. In the first case, the current niche radius is increased before performing the following optimization. The second case arises when no individual with a distance measure lower than the tolerance was found. In this situation a new iteration is allowed if the maximum number of unsuccessful iterations was not reached.

For comparison purposes, the results from Srinivas and Rangaiah (2007) obtained using Differential Evolution (DE) are also reported for each example in Table 3. DE (Storn and Price, 1997) is also based on the evolution of a population of individuals with mutation, crossover and selection operations. In DE, a mutated individual is essentially a noisy replica of its parent individual. The noise added to the parent is a weighted contribution of the differences between randomly selected individuals. An offspring is created by crossover, and a one-to-one competence between parents and offspring is performed in the selection process. DE is reputed to be a simple and easy to use technique that converges faster than GAs.

Regarding the success rate in finding the global optimum, the average percentages taken into account all the examples are: 96.4% and 90.7% for the proposed technique and the DE based strategy, respectively. The other performance criterion considered is the number of function evaluations (stochastic technique + local search). The proposed method demanded in average (2574 + 41) function evaluations. The value of the same performance measure for the DE technique was (2401 + 11). Overall, the results obtained indicate that the methodology presented in this work is competitive compared with a current state of the art stochastic methodology, such as DE.

5. Conclusions and Future Work

In this work a general framework to address constrained nonlinear optimization problems is proposed. Since an NLP optimization model can be formulated as a system of nonlinear equations by posing the Karush-Kuhn-Tucker optimality conditions, the original problem is reformulated as an unconstrained optimization model, whose solutions are characterized by a zero objective function value. The methodology is also applicable when binary variables are present by performing a standard procedure that transforms MINLP models into NLP problems.

Since the resulting model is unconstrained and possibly non-smooth, full advantage can be taken from stochastic optimization techniques to find some (eventually all) the solutions of the system of nonlinear equations, and therefore some (eventually all) the local extrema of the original optimization problem. A specific implementation of a sequential niche based methodology was proposed to address the solution of the resulting nonlinear system by means of GAs. The algorithm uses a derating strategy to find the solutions one by one in an iterative fashion.

The methodology is applicable to any model whose Karush-Kuhn-Tucker system possesses solutions. This is certainly the case for a wide family of problems in engineering. Regarding its performance, the proposed reformulation plus the adopted sequential niche strategy for finding the solutions of the Karush-Kuhn-Tucker system shows to be competitive in comparison with current state of the art evolutionary techniques. This conclusion is drawn from the experimental evidence obtained after solving some typical examples, used to test optimization algorithms in Chemical Engineering, applying general-purpose implementations of GAs.

A comprehensive analysis of two implementation issues was beyond the scope of the present contribution and is consigned to future work. One of them is related with the implementation features of the sequential niche algorithm, such as the type of distance measures to be used for solving problem (5) and for calculating the niche radius, the tolerance to admit an individual as a solution of the system of equations, etc. The other issue has to do with the features of the particular GA implementation, such as selection, crossover and mutation methods along with their corresponding tuning parameters.

Future work will also consider the design of a strategy to deal with dummy solutions. A second order derivatives' analysis seems a practical option if these are also stationary points of the Karush-Kuhn-Tucker system. Other strategy should be adopted to handle the second type of dummy solutions, for example, some gradient-based local analysis in the neighborhood of the point.

References

Beasley, D., Bull, D.R. and Martin, R.R. (1993). A sequential niche technique for multimodal function optimization. *Evolutionary Computation*, **1**(2), pp. 101–125.

Becerra, R.L. and Coello Coello, C.A. (2005). Optimization with constraints using a cultured differential evolution approach. *GECCO 2005 — Genetic and Evolutionary Computation Conference*, pp. 27–34.

Bernardino, H.S., Barbosa, H.J.C. and Lemonge, A.C.C. (2007). A hybrid genetic algorithm for constrained optimization problems in mechanical engineering. *IEEE Congress on Evolutionary Computation, CEC 2007*, N 4424532, pp. 646–653.

Blanco, A.M., Sanchez, M.C. and Bandoni, J.A. (2006). On a novel genetic algorithmic based methodology for solving nonlinearly constrained systems of equations and MINLP problems. *Proceedings of XXII IACChE (CIIQ)/V CAIQ*.

Brooke, A., Kendrick, D., Meeraus, A. and Raman, R. (2003). *GAMS: A User's Guide*, GAMS Development Corporation.

Coello Coello, C.A. (2002). Theoretical and numerical constraint — handling techniques used with evolutionary algorithms: A survey of the state of the Art. *Comput. Methods Appl. Mech. Engrg.*, **191**, pp. 1245–1287.

Deb, K. (2000). An efficient constraint handling method for genetic algorithms. *Computer Methods in Applied Mechanics and Engineering*, **186**(2–4), pp. 311–338.

Edgar, T.F, Himmelblau, D.M. and Lasdon, L.S. (2001). *Optimization of Chemical Processes*, McGraw Hill, New York.

Gan, J. and Warwick, K. (2000). A variable radius niching technique for speciation in genetic algorithms. *Proceedings of the Genetic and Evolutionary Computation Conference*, pp. 96–103, Morgan Kaufmann.

Gan, J. and Warwick, K. (2001). Dynamic niche clustering: A fuzzy variable radius niching technique for multimodal optimization in GAs. *Proceedings of the 2001 Congress on Evolutionary Computation*, pp. 215–222.

Gobin, O.C., Martinez Joaristi, A. and Schuth, F. (2007). Multi-objective optimization in combinatorial chemistry applied to the selective catalytic reduction of NO with C_3H_6. *Journal of Catalysis*, **252**(2), pp. 205–214.

Khorram, E. and Hassanzadeh, R. (2008). Solving nonlinear optimization problems subjected to fuzzy relation equation constraints with max–average composition using a modified genetic algorithm. *Computers and Industrial Engineering*, **55**(1), pp. 1–14.

Leguizamón, G. and Coello Coello, C.A. (2007). A boundary search based ACO algorithm coupled with stochastic ranking. *2007 IEEE Congress on Evolutionary Computation (CEC'2007)*, pp. 165–172. IEEE Press, Singapore.

Lewis, R.M., Torczona, V. and Trosset, M.W. (2000). Direct search methods: then and now. *Journal of Computational and Applied Mathematics*, **124**, pp. 191–207.

Maranas, C.D. and Floudas, C.A. (1995). Finding all solutions of nonlinearly constrained systems of equations. *Journal of Global Optimization*, **7**(2), pp. 143–182.

Mezura-Montes, E. and Coello Coello, C.A. (2005). A simple multimembered evolution strategy to solve constrained optimization problems. *IEEE Transactions on Evolutionary Computation*, **9**(1), pp. 1–17.

Mezura-Montes, E. and Coello Coello, C.A. (2002). A numerical comparison of some multiobjective based techniques to handle constraints in genetic algorithms. *Tech. Rep. EVOCINV-03–2002. Evol. Comput. Group, CINVESTAV, Sección de Computación, CINVESTAV-IPN*, Dept. de Ingeniería Eléctrica. México D.F. México. Available on line: http://www.cs.cinvestav.mx/constraint/.

Michalewicz, Z. (1996). *Genetic Algorithms + Data Structures = Evolution Programs*, Third Edition, Springler, Berlin.

Michalewicz, Z., Dasgupta, D., Le Riche, R.G. and Schoenauer, M. (1996). Evolutionary algorithms for constrained engineering problems. *Computers and Industrial Engineering Journal*, **3**(2), pp. 851–870.

Michalewicz, Z. and Nazhiyath, G. (1995). Genocop III: A co-evolutionary algorithm for numerical optimization with nonlinear constraints, in Fogel, D.B. (ed.). *Proceedings of the Second IEEE International Conference on Evolutionary Computation*, IEEE Press, Piscataway, NJ, pp. 647–651.

Miettinen, K., Makela, M.M. and Toivanen, J. (2003). Numerical comparison of some penalty-based constraint handling techniques in genetic algorithms. *Journal of Global Optimization*, **27**(4), pp. 427–446.

Pirkwieser, S., Raidl, G.R. and Puchinger, J. (2008). A Lagrangian decomposition/evolutionary algorithm hybrid for the knapsack constrained maximum spanning tree problem. *Studies in Computational Intelligence*, **153**, pp. 69–85.

Reklaitis, G.V., Ravindran, A. and Ragsdell, K.M. (1983). *Engineering Optimization: Methods and Applications*, John Wiley and Sons.

Runarsson, T.P. and Yao, X. (2000). Stochastic ranking for constrained evolutionary optimization. *IEEE Transactions on Evolutionary Computation*, **4**(3), pp. 284–294.

Ryoo, H.S. and Sahinidis, N.V. (1995). Global optimization of nonconvex NLPs and MINLPs with applications in process design. *Comp. Chem. Eng.*, **19**(5), pp. 551–566.

Smith, A.E. and Coit, D.W. (1997). Constraint handling techniques — penalty functions. Thomas Bäck, David, B. Fogel, and Zbigniew Michalewicz, (eds.). *Handbook of Evolutionary Computation*, Chapter C 5.2. Oxford University Press and Institute of Physics Publishing.

Srinivas, M. and Rangaiah, G.P. (2007). Differential evolution with tabu list for solving nonlinear and mixed-integer nonlinear programming problems. *Ind. Eng. Chem. Res.*, **46**(22), pp. 7126–7135.

Storn, R. and Price, K. (1997). Differential evolution, a simple and efficient heurist for global optimization over continuous spaces. *J. Global Optimization*, **11** pp. 341–359.

Tessema, B. and Yen, G.G. (2006). A self adaptative penalty function based algorithm for constrained optimization. *2006 IEEE Congress on Evolutionary Computation* (*CEC'2006*), IEEE, Vancouver, BC, Canada, pp. 950–957.

Venkatraman, S. and Yen, G.G. (2005). A generic framework for constrained optimization using genetic algorithms. *IEEE Transactions on Evolutionary Computation*, **9**, pp. 424–435.

Wang, Y., Cai, Z., Zhou, Y. and Zeng, W. (2008). An adaptive tradeoff model for constrained evolutionary optimization. *IEEE Transactions on Evolutionary Computation*, **12**(1), pp. 80–92.

Wang, Y. and Cai, Z. (2005). A multiobjective optimization based evolutionary algorithm for constrained optimization. *Proceedings of Congress on Evolutionary Computation*, Edinburgh, UK, pp. 1081–1087.

Chapter 12

APPLICATION OF LUUS-JAAKOLA OPTIMIZATION PROCEDURE TO MODEL REDUCTION, PARAMETER ESTIMATION AND OPTIMAL CONTROL

Rein Luus

Department of Chemical Engineering
University of Toronto, 200 College Street
Toronto, ON M5S 3E5, Canada
rein.luus@utoronto.ca

1. Introduction

The Luus-Jaakola optimization procedure has been used successfully for optimization of a wide range of problems, including model reduction, parameter estimation, and optimal control (Luus, 1999, 2000a, 2001a, 2002). Recently Luus (2007) showed that by including a simple line search in the LJ optimization, the convergence is significantly improved. Therefore, here we also wish to show the advantages of including the line search, and examine the use of "greedy" approach in solving an optimal control problem.

The use of randomly chosen test points and making the search more intensive around the most likely point for the optimum as iterations proceed was suggested by Luus and Jaakola (1973). This procedure is outlined in some detail in Chapter 2 of this book and in references Luus (2000a,b; 2001a). In this chapter we show application of the Luus-Jaakola optimization procedure to problems requiring a significant amount of computational effort. In Chapter 2 we showed that the use of region size as determined by the change in variables after a pass as suggested by Luus (1998) is effective in reducing computational effort. Here we show that the convergence rate is improved and the optimum is accurately established by incorporating a line search into the optimization procedure, as suggested by Luus (2007).

Numerous attempts have been made to improve the convergence rate of the LJ optimization procedure. Instead of the best point, Nair (1976) proposed using a different center point for search after a number of iterations had been performed. Luus and Brenek (1989) followed up the work by showing that incorporation of the gradient into direct search optimization can improve the convergence rate by as much as a factor of two for some optimization problems. Recently, efforts to improve the convergence rate of the Luus-Jaakola optimization procedure have been directed towards the determination of region sizes (Luus, 1998), and with the latest developments, the LJ optimization procedure has been found to converge better than the genetic algorithm (Liao and Luus, 2005). Here, we illustrate the use of the LJ optimization procedure for model reduction, parameter estimation, and optimal control, and also examine the use of line search as suggested by Luus (2007).

Let us consider the optimization problem, where it is required to find the n variables $\mathbf{x} = (x_1, x_2, \ldots, x_n)^T$ to minimize a continuous function

$$I = f(x_1, x_2, \ldots, x_n), \tag{1}$$

subject to the inequality constraints

$$a_i \leq x_i \leq b_i, \quad i = 1, 2, \ldots, n, \tag{2}$$

and

$$g_j(x_1, x_2, \ldots, x_n) \leq 0, \quad j = 1, 2, \ldots, m. \tag{3}$$

The main idea here is to use a better point around which the search is performed. Therefore, after several passes, it is recommended to carry out a line search in the direction in which the best point has changed in the preceding two passes to provide the center point for the subsequent pass. The suggested algorithm for LJ optimization procedure and the incorporation of a line search (Luus, 2007) consists of the following steps:

(1) Choose a number of random points R in the n-dimensional space through the equation

$$\mathbf{x} = \mathbf{x}^* + \mathbf{D}\mathbf{r}^{(j)}, \tag{4}$$

where $\mathbf{x}^*$ is the best value (or initial value at start) of $\mathbf{x}$, $\mathbf{r}^{(j)}$ is the region size vector at the jth iteration ($j = 1$ at start) and $\mathbf{D}$ is a diagonal matrix with diagonal elements chosen at random between -1 and $+1$.

(2) The feasibility of each point is checked first with respect to Eq. (2) and clipping technique is used if the variable is outside the bounds: if $x_i < a_i$, then x_i is put equal to a_i; if $x_i > b_i$, then x_i is put equal to b_i. Next check the feasibility with respect to Eq. (3). If Eq. (3) is violated, then discard the point; if Eq. (3) is satisfied, then evaluate the performance index I from Eq. (1) and store the corresponding $\mathbf{x}$. After testing R points, replace $\mathbf{x}^*$ by the best feasible value of $\mathbf{x}$.

(3) An iteration is defined by Steps 1 and 2. At the end of iteration j reduce the size of the region $\mathbf{r}$ by a factor γ through

$$\mathbf{r}^{(j+1)} = \gamma\,\mathbf{r}^{(j)}, \tag{5}$$

where γ is the region reduction factor such as 0.95. This procedure is continued for N iterations which make up a pass.

(4) For the next few passes restore the region size to a fraction η (such as 0.85) of its value at the beginning of the previous pass.

(5) After a few passes, at the beginning of pass $q + 1$, use the region size as determined from the amount that the variable has changed as suggested by Luus (1998):

$$r_i = \left|x_i^{*(q-1)} - x_i^{*(q)}\right|, \quad i = 1, 2, \ldots, n. \tag{6}$$

If the value of r_i is less than ϵ then r_i is replaced by ϵ to prevent the region from collapsing. The convergence rate is dependent on the value of ϵ (Luus, 2003a), so a useful procedure is given in Step 6.

(6) Initially, the region collapse parameter ϵ is chosen to be of reasonable size, such as 10^{-3}. As iterations continue, if the performance index has not been improved by more than ϵ in 3 passes, then ϵ is reduced in size by a factor such as 0.9.

(7) Once ϵ reaches a very low value such as 10^{-11} or 10^{-12}, the iterations are stopped and the results are analyzed. This provides a very convenient way of terminating the iterative procedure.

To incorporate the line search, after q passes, we carry out the search in the direction

$$\mathbf{d} = \mathbf{x}^{*(q)} - \mathbf{x}^{*(q-1)} \tag{7}$$

to provide the best $\mathbf{x}^*$ for the subsequent pass. For all the problems here, we use 10 function evaluations to get the best value of $\mathbf{x}^*$ along the direction $\mathbf{d}$.

Note:

(1) For handling inequalities in Eq. (3), instead of simply discarding the infeasible points, we can identify the active inequalities in a preliminary run, and then use these active inequalities as equalities as has been suggested by Luus and Harapyn (2003) and Luus *et al.* (2006). This is especially useful approach for very complex systems.
(2) In Step 2 the standard procedure is to compare all the points before changing $\mathbf{x}^*$ to the best value at the end of the iteration. The "greedy" approach is to change $\mathbf{x}^*$ immediately to the better point once improvement in I is obtained. The usefulness of this approach is illustrated in the optimal control application.

2. Model Reduction of s-transfer Functions

The transfer function of a high-order system is given in the form

$$F(s) = \frac{b_0 + b_1 s + b_2 s^2 + \cdots + b_m s^m}{a_0 + a_1 s + a_2 s^2 + \cdots + a_n s^n}, \quad m < n, \tag{8}$$

where $a_0, a_1, \ldots, a_n, b_0, b_1, \ldots, b_m$ are constants. Let the transfer function of the reduced system be

$$G_q(s) = \frac{K(1 + d_1 s + d_2 s^2 + \cdots + d_p s^p)}{1 + c_1 s + c_2 s^2 + \cdots + c_q s^q}, \quad p < q < n, \tag{9}$$

where $K = b_0/a_0$ to preserve the steady state gain and q is the order of the reduced model. The model reduction problem is to choose the coefficients $c_1, c_2, \ldots, c_q, d_1, d_2, \ldots, d_p$ so that $G_q(s)$ represents $F(s)$ as closely as possible.

Establishment of the optimal reduced model may be carried out either in the frequency domain or in time domain. Comparison of these methods is given for three test problems by Luus (2006a).

(a) *Optimization in the frequency domain*

As suggested by Luus (1980), after choosing the order q and the number of zeroes p for the reduced model, we pick a number of frequencies N and choose the performance index to be minimized as

$$I = \sum_{i=1}^{N} |F(j\omega_i) - G_q(j\omega_i)|^2. \tag{10}$$

As was suggested in Luus (1980), here we choose $N = 100$ with $\omega_1 = 0.01$ and $\omega_{i+1} = 1.1\omega_i$ to cover the frequency range from 0.01 to 125. Although such a choice is arbitrary, for many systems it is quite reasonable.

(b) *Optimization in the time domain*

Let us denote the poles of F (roots of the denominator) by $\lambda_1, \lambda_2, \ldots, \lambda_n$, and the poles of G_q by $\beta_1, \beta_2, \ldots, \beta_q$. Then the output from the original system for step input is given by

$$f(t) = K + \sum_{i=1}^{n} f_i e^{\lambda_i t}, \tag{11}$$

where

$$f_i = \lim_{s \to \lambda_i} \frac{(s - \lambda_i)F(s)}{s}, \tag{12}$$

as given by Howitt and Luus (1990). Similarly, the output from the reduced system for step input is

$$g_q(t) = K + \sum_{i=1}^{r} g_{qi} e^{\beta_i t}, \tag{13}$$

where

$$g_{qi} = \lim_{s \to \beta_i} \frac{(s - \beta_i)G_q(s)}{s}. \tag{14}$$

If an impulse input is used, then the output from the original system is given by

$$f(t) = \sum_{i=1}^{n} h_i e^{\lambda_i t}, \tag{15}$$

where

$$h_i = \lim_{s \to \lambda_i} (s - \lambda_i)F(s), \tag{16}$$

and the output from the reduced system is

$$g_q(t) = \sum_{i=1}^{r} k_i e^{\beta_i t}, \tag{17}$$

where

$$k_i = \lim_{s \to \beta_i} (s - \beta_i)G_q(s). \tag{18}$$

The performance index to be minimized is

$$I = \int_0^T (f(t) - g_q(t))^2 dt, \tag{19}$$

where T is chosen to be sufficiently large.

A reasonable choice is $T = 10.0$, and in order to get a good approximation for the integral, to use Simpson's approximation with 100 intervals, each of length 0.10. This choice is quite arbitrary, but is found to be reasonable for many systems (Luus, 2006a).

The results in the time domain are dependent on the input signal. If we use as input a unit step function to establish the reduced model, then to test the reduced model, we should use a different input, such as the pulse input signal.

For all computations we used double precision in Fortran with the AMD3800/2.4 GHz personal computer which is approximately 1.5 times faster than Pentium4/2.4 GHz computer.

Example 1. Let us consider the eighth-order system of Krishnamurthy and Seshadri (1978) which has provided a very good test problem for model reduction investigation (Luus, 1980; Luus, 1999; Luus *et al.*, 1999; Wang *et al.*, 2005; Luus, 2006a; Luus, 2007)

$$F(s) = \frac{\begin{array}{c}194480 + 482964s + 511812s^2 + 278376s^3 + 82402s^4 \\ + 13285s^5 + 1086s^6 + 35s^7\end{array}}{\begin{array}{c}9600 + 28880s + 37492s^2 + 27470s^3 + 11870s^4 \\ + 3017s^5 + 437s^6 + 33s^7 + s^8\end{array}} \tag{20}$$

where we wish to get a reduced model of lower order. The zeroes (roots of the numerator) of the transfer function are $-1.0346 \pm 0.63102j, -2.6369, -3.6345, -4.9021, -7.8014$, and -9.7845. The poles (roots of the denominator) of the transfer function are $-1, -1 \pm j, -3, -4, -5, -8$, and -10.

Let us choose the third order reduced model in the form

$$G(s) = \frac{20.25833(1 + x_1 s + x_2 s^2)}{(1 + x_3 s)(1 + x_4 s + x_5 s^2)}, \tag{21}$$

where we have chosen the constant in the numerator to preserve the steady-state gain. We use the factored form for the reduced model to improve convergence in optimization as shown by Luus *et al.* (1999).

Model reduction in frequency domain

In frequency domain, the optimization was run with five different seeds for the random number generator with starting points for x_i chosen at random between 0 and 1, and the initial region size was taken as half of the value of the x_i. We chose $R = 25$ random points per iteration, and 20 iterations per pass. After every iteration the region size was reduced by $\gamma = 0.95$. For the first 10 passes, the region size was restored to $\eta = 0.85$ of the value used at the beginning of the previous pass. After that, the region was put equal to the extent of the change of the variables at the end of the pass, as stated in step 5 of the algorithm. The region collapse parameter ϵ was initially chosen to be 10^{-3} and reduced by the factor 0.9 whenever the performance index was not improved by ϵ in 3 passes. When ϵ was reduced to 10^{-11}, the optimization procedure was terminated and the results were analyzed.

Table 1. Convergence of the LJ optimization procedure in optimization in frequency domain, showing the number of passes required for convergence, performance index obtained, and the computation time on AMD Athlon3800/2.4 GHz.

	Without line search			With line search		
Case, number	Passes	I	CPU s	Passes	I	CPU s
1	1280	0.327871661	10.22	490	0.327871552	3.79
2	1363	0.327871609	10.93	516	0.327871557	4.17
3	1219	0.327871649	9.72	371	0.327871552	3.02
4	1351	0.327871616	10.77	346	0.327871552	2.80
5	1380	0.327871657	11.04	337	0.327871552	2.75

For line search we used a simple scheme of searching for the best value with the use of 10 points along the direction in which the performance index was improved in the present and previous pass.

The results in Table 1 show that convergence is readily obtained, and the use of line search speeds up convergence substantially and yields a lower value for the performance index I to be established.

Convergence to the minimum value $I = 0.3278716$ was easily reached without using the line search. However, to reach $I = 0.32787155$, line search had to be used. These results are very close to 0.32788 reported by Wang *et al.* (2005). The convergence was very regular as is seen in Fig. 1, and the advantage of the line search becomes obvious in Fig. 2. Line search provides better convergence and reduces the computational effort here by more than a factor of 3.

The reduced transfer function corresponding to $I = 0.327871552$ is

$$G_3(s) = \frac{20.25833(1 + 0.84080s + 0.55072s^2)}{(1 + 0.61770s)(1 + 0.73911s + 0.53532s^2)}, \quad (22)$$

with zeroes $-0.76336 \pm 1.1104j$, and poles -1.61891, $-0.69034 \pm 1.1796j$. It should be noted that to get 5-figure accuracy for all the x_i, we do need 9-figure accuracy for I.

Model reduction in time domain

In time domain, we used a step input and took $T = 10$, with 100 time intervals of 0.1. To calculate the integral of deviations, Simpson's rule was used.

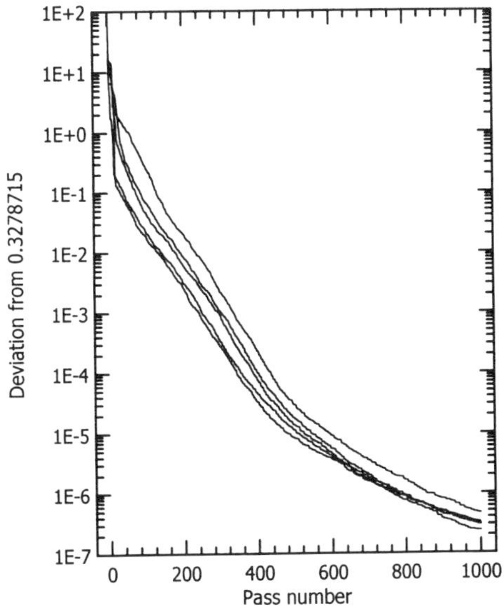

Figure 1. Convergence in frequency domain for the third order reduced model for Example 1 without line search, showing the deviation of the performance index from its minimum value.

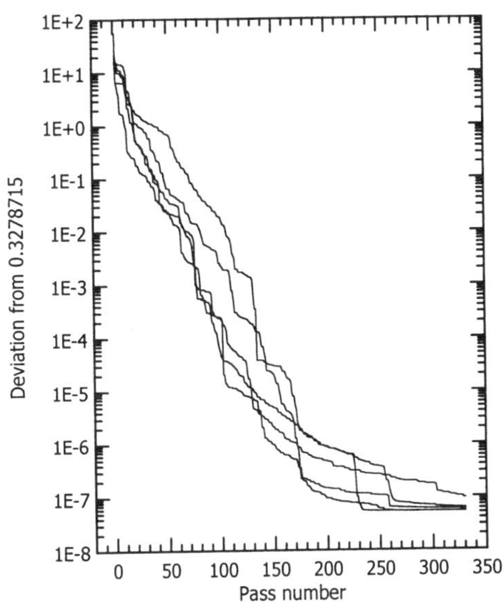

Figure 2. Convergence in frequency domain for the third order reduced model for Example 1 with line search, showing the deviation of the performance index from its minimum value.

Table 2. Convergence of the LJ optimization procedure in optimization in time domain, showing the number of passes required for convergence, performance index obtained, and the computation time on AMD Athlon3800/2.4 GHz.

	Without line search			With line search		
Case, number	Passes	I	CPU s	Passes	I	CPU s
1	1412	0.00317981231	25.70	646	0.00317981194	12.69
2	1187	0.00317981208	21.53	642	0.00317981197	12.63
3	1055	0.00317981233	19.97	619	0.00317981198	12.20
4	1092	0.00317981261	19.89	741	0.00317981194	14.55
5	1372	0.00317981272	24.88	673	0.00317981197	13.18

For optimization we used the same conditions as were used in frequency domain. The convergence to the minimum value $I = 0.003179812$, was obtained when 5 different seeds were used for the random number generator. Table 2 shows again the advantage of incorporating the line search. Figures 3 and 4 show the convergence profiles with and without the line search.

The reduced transfer function corresponding to $I = 0.0031798119$ is

$$G_3(s) = \frac{20.25833(1 + 1.16125s + 0.76826s^2)}{(1 + 0.69413s)(1 + 0.99059s + 0.67186s^2)}, \qquad (23)$$

with zeroes $-0.75658 \pm 0.85395j$, and poles -1.4406, $-0.73721 \pm 0.97208j$.

To test these two third order models, we used time response to an impulse input. We used 801 time steps, each of length 0.025. As is seen in Fig. 5, both models yielded good agreement, except at time zero, where the response from the reduced model in the frequency domain, shown as solid circles, gave 33.74 and the one in time domain, shown as solid triangles, gave 33.37 instead of 35.00 as obtained by the original system. Here the performance index showing the deviations for the frequency domain model is 0.07198 and for the time domain reduced model is a little larger with 0.12287. In each case, the approximation is excellent.

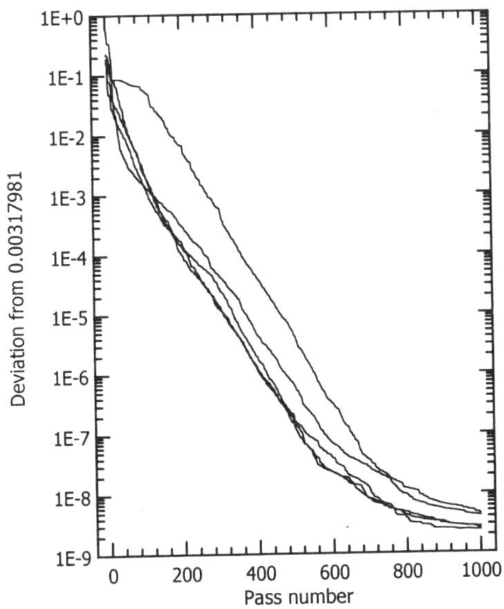

Figure 3. Convergence in time domain for the third order reduced model for Example 1 without line search, showing the deviation of the performance index from its minimum value.

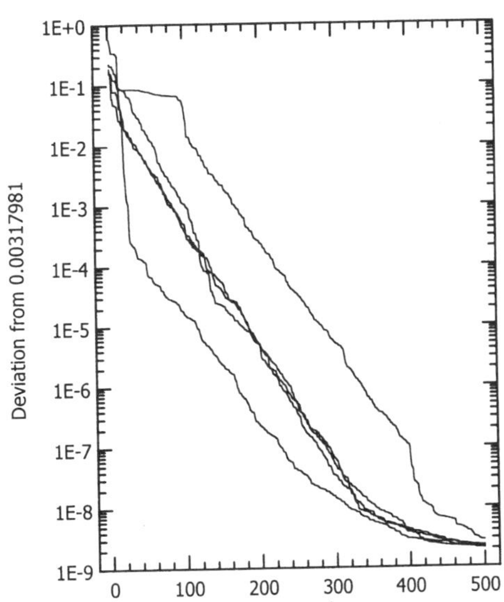

Figure 4. Convergence in time domain for the third order reduced model for Example 1 with line search, showing the deviation of the performance index from its minimum value.

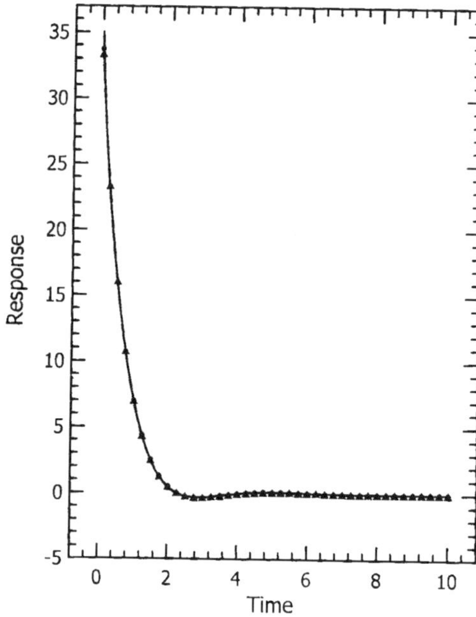

Figure 5. Comparison of models in frequency domain and the time domain for response to an impulse input for Example 1. Solid line gives the response to the original high order system.

3. Model Reduction of z-transfer Functions

The link between the z-transfer function and s-transfer function is

$$z = \exp(Ts), \tag{24}$$

where T is the sampling time. Thus, for frequency response locus we put

$$s = j\omega_i, \quad i = 1, 2, \ldots, N \tag{25}$$

to yield for frequency ω_i

$$z = \cos(\omega_i T) + j\sin(\omega_i T). \tag{26}$$

Therefore, the original transfer function $F(z)$ yields the complex number

$$F(z) = a(\omega_i T) + jb(\omega_i T), \tag{27}$$

and the reduced transfer function $G(z)$ gives the complex number

$$G(z) = c(\omega_i T) + jd(\omega_i T). \tag{28}$$

The coefficients in the reduced model $G(z)$ can then be chosen to minimize the sum of squares of deviations

$$I = \sum_{i=1}^{N} [(a(\omega_i T) - c(\omega_i T))^2 + (b(\omega_i T) - d(\omega_i T))^2]. \tag{29}$$

The problem here is actually simpler since the range for $\omega_i T$ should be taken from 0 to π, rather than from 0 to ∞. We illustrate this with an example.

Example 2. Yang and Luus (1983) used this approach to obtain an excellent reduced model for the model reduction problem of Shih and Wu (1973), but here we consider a more recent problem of Carrier and Stephanopoulos (1998), and commented on by Luus (2000c).

The original system consists of the fifth-order transfer function

$$F(z) = 10^{-5}\left[\frac{0.0052z^4 + 0.1316z^3 + 0.3307z^2 + 0.1302z + 0.005}{z^5 - 4.72z^4 + 9.14z^3 - 9.08z^2 + 4.63z - 0.969}\right]. \tag{30}$$

The poles of the transfer function are $0.87094 \pm 0.44323j$, $1.0374 \pm 0.21699j$, and 0.90323. The zeroes of the transfer function are -22.535, -2.2984, -0.4315, and -0.04302.

By using the LJ optimization procedure with 100 values for the frequency chosen between 0.01 and 3.20 through the increment factor of 1.06, we obtained the following reduced transfer function:

$$G(z) = 10^{-5}\left[\frac{-15.44026z^2 + 32.17736z - 13.77451}{z^3 - 2.97902z^2 + 2.99881z - 1.01488}\right]. \tag{31}$$

The poles of $G(z)$ are 0.89936, $1.0398 \pm 0.21726j$, and the zeroes are 0.60195, 0.14820.

This reduced model differs substantially from the one obtained by wavelets by Carrier and Stephanopoulos (1998):

$$G_w(z) = 10^{-5}\left[\frac{2.43z^2 - 3.25z + 3.39}{z^3 - 2.98z^2 + 2.97z - 0.986}\right], \tag{32}$$

with poles at 0.85617, $1.0619 \pm 0.15485j$, and zeroes at $0.66872 \pm 0.97359j$.

The differences between these two reduced models become apparent when the Nyquist plots in Figs. 6 and 7 are compared.

4. Parameter Estimation

To develop a quadratically convergent procedure for establishing parameters for nonlinear differential equations three procedures have been used. Kalogerakis and Luus (1983a) showed that the quasilinearization method where the sensitivity information is obtained from linearization of the

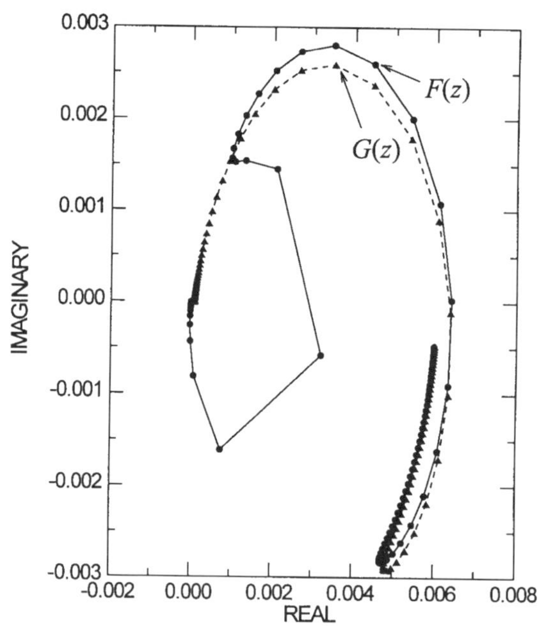

Figure 6. Nyquist plot of the original system $F(z)$ and the reduced model $G(z)$ obtained by LJ optimization procedure.

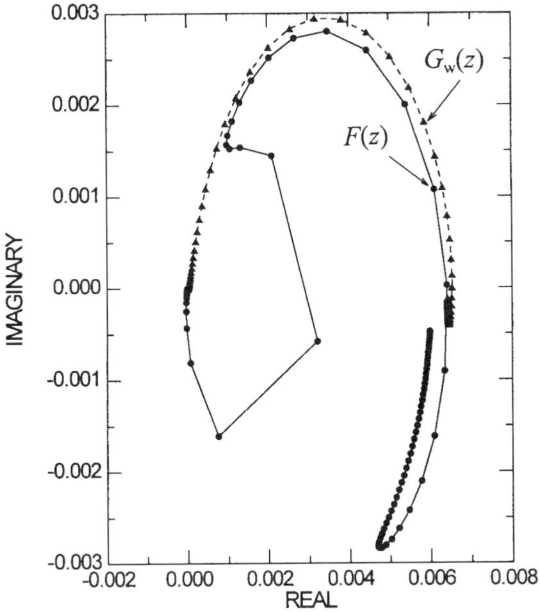

Figure 7. Nyquist plot of the original system $F(z)$ and the reduced model $G_w(z)$ obtained by the use of wavelets by Carrier and Stephanopoulos (1998).

equations and Gauss-Newton method where the output vector is expanded in terms of the parameter vector lead to the same algorithm. The third method, where the performance index is expanded in terms of changes in the parameter vector, as shown by Luus (1988), also yields the same equations for updating the parameter vector. However, with this approach, the approximations made in updating the parameter vector are clearer and it is easier to see why the region over which convergence can be obtained is quite small. To enlarge the region of convergence, one may use initially a smaller data length as suggested by Wang and Luus (1980), to use the information index (Kalogerakis and Luus, 1983b), or to use direct search optimization, such as the LJ optimization procedure, to get to the region of convergence and then switch to the quadratically convergent algorithm (Kalogerakis and Luus, 1982). With the recent changes in the LJ optimization procedure, especially in the use of the simple line search, it may very well be unnecessary to switch to a quadratically convergent procedure. Let us consider an example.

Example 3. For solving the ordinary differential equations we used the IMSL subroutine DVERK (Hull *et al.*, 1976) with local error tolerance of 10^{-8} to give reliable results. The Fortran listing of this subroutine is given in Luus (2000a).

A good parameter estimation problem is estimating parameters in toluene hydrogenation process presented by Belohlav *et al.* (1997), which was considered by Luus (2000a) and recently by Linga *et al.* (2006). The catalytic reaction

$$A \rightleftarrows B \rightarrow C$$

is described by three differential equations:

$$\frac{dx_1}{dt} = -r_1 + r_{-1}, \tag{33}$$

$$\frac{dx_2}{dt} = -r_{-1} + r_1 - r_2, \tag{34}$$

$$\frac{dx_3}{dt} = r_2, \tag{35}$$

where $r_1 = k_1\theta_A$, $r_{-1} = k_2\theta_B$, and $r_2 = k_3\theta_B$, and the surface coverages are given by

$$\theta_A = \frac{k_4x_1}{k_4x_1 + x_2 + k_5x_3}, \quad \theta_B = \frac{x_2}{k_4x_1 + x_2 + k_5x_3}. \tag{36}$$

The concentration of toluene (A) is given by x_1, of 1-methylcyclohexene (B) by x_2, and of methylcyclohexane (C) by x_3. Starting with pure toluene, the experimental measurements of the concentrations reported by Belohlav *et al.* (1997) are given as the 13 data points in Table 3.

The problem is to obtain the best estimate for the parameter vector $\mathbf{k} = (k_1, k_2, \ldots, k_5)^T$ by minimizing the sum of squares of deviations of the model values of the concentrations and the measured values of the concentrations at the 13 data points. Therefore, the performance index to be minimized is chosen to be

$$I = \sum_{i=1}^{13} [\mathbf{x}(t_i) - \hat{\mathbf{x}}(t_i)]^T [\mathbf{x}(t_i) - \hat{\mathbf{x}}(t_i)]. \tag{37}$$

Table 3. Experimental data for toluene hydrogenation as reported by Belohlav *et al.* (1997).

	Measured concentrations		
Time, min	$\hat{x}_1$	$\hat{x}_2$	$\hat{x}_3$
15	0.695	0.312	0.001
30	0.492	0.430	0.080
45	0.276	0.575	0.151
60	0.225	0.570	0.195
75	0.163	0.575	0.224
90	0.134	0.533	0.330
120	0.064	0.462	0.471
180	0.056	0.362	0.580
240	0.041	0.211	0.747
320	0.031	0.146	0.822
360	0.022	0.080	0.898
380	0.021	0.070	0.909
400	0.019	0.073	0.908

As in the previous example, we made 5 runs with and without the line search. Here we chose as initial values for the parameters $k_i = 0.5 \pm 0.2\text{rn}(0,1)$, where rn(0,1) denotes a random number between 0 and 1. The initial region sizes were chosen to be $r_i = 0.5k_i$. We chose $R = 25$, 20 iterations per pass, and allowed a maximum of 1,000 passes. Initial ϵ was chosen as 10^{-3}, with reduction factor of 0.9 if the performance index was not improved by ϵ in 3 passes, and iterations were stopped when ϵ reached 10^{-11}. The results are given in Table 4.

Without the line search, convergence accurately to the minimum performance index $I = 0.0159000308061$ could not be obtained and the convergence rate was considerably slower than with the line search, as is shown in Figs. 8 and 9. We need a large number of figures in I to get accurate parameter values to 5 figures. With the line search, convergence is better and there is also a small saving in computer time. The resulting optimum parameter vector is $\mathbf{k} = [0.0251136\ 0.00326078\ 0.00873002\ 1.272576\ 1.242295]^T$. These parameter values differ quite substantially from those given by Belohlav *et al.* (1997), but are in close agreement with the values reported by Luus (2000a) and Linga *et al.* (2006). For case 1, without line search, where $I = 0.015900030811$ is accurate only to 10 figures, the

Table 4. Convergence of the LJ optimization procedure with $R = 25$ and 20 iterations per pass in Example 3, showing the number of passes required for convergence, performance index obtained, and the computation time on AMD Athlon3800/2.4 GHz.

Case, number	Without line search			With line search		
	Passes	I	CPU s	Passes	I	CPU s
1	284	0.015900030811201	40.15	230	0.015900030806137	33.23
2	322	0.015900030810330	53.00	256	0.015900030806131	43.89
3	361	0.015900030806598	55.25	268	0.015900030806145	41.63
4	471	0.015900030806438	65.09	256	0.015900030806129	36.58
5	326	0.015900030806175	45.81	301	0.015900030806128	43.17

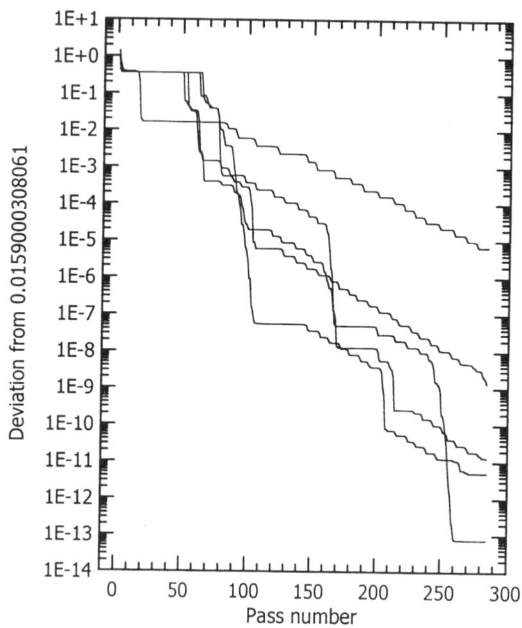

Figure 8. Convergence of the LJ optimization procedure without line search with $R = 25$ and 20 iterations per pass for Example 3 for the first 300 passes.

value of x_2 obtained is 0.00326067 which is accurate only up to 4 figures. So here we need 12 figure accuracy of I for reliable parameter estimates, and such accuracy was obtained with line search. Also, the computer time was reduced, as is seen in Table 4.

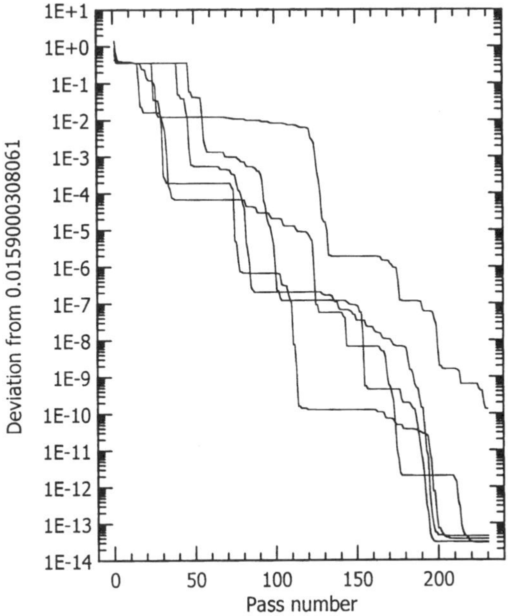

Figure 9. Convergence of the LJ optimization procedure with line search with $R = 25$ and 20 iterations per pass for Example 3 for the first 230 passes.

Table 5. Convergence of the LJ optimization procedure with $R = 50$ and 20 iterations per pass in Example 3, showing the number of passes required for convergence, performance index obtained, and the computation time on AMD Athlon3800/2.4 GHz.

Case, number	Without line search			With line search		
	Passes	I	CPU s	Passes	I	CPU s
1	349	0.015900030806151	103.20	231	0.015900030806132	74.15
2	311	0.015900030806129	90.24	225	0.015900030806128	65.58
3	280	0.015900030806129	78.27	225	0.015900030806148	63.73
4	242	0.015900030806156	75.36	235	0.015900030806144	74.81
5	317	0.015900030806128	88.10	242	0.015900030806128	68.55

The parameter estimation was repeated with the use of $R = 50$ instead of $R = 25$. As is shown in Table 5, we now get 12 figure accuracy for I without the line search. The results with the line search were not improved beyond the results obtained previously.

The LJ optimization procedure has been found especially useful in parameter estimation of errors-in-variables problems (Luus and Hernaez, 2000; Luus and Jammer, 2005). The reader is encouraged to read these papers and the paper by Esposito and Floudas (2000).

5. Optimal Control

Let us consider the optimal control problem in the standard form, where we have the system described by the differential equation

$$\frac{d\mathbf{x}}{dt} = \mathbf{f}(\mathbf{x}, \mathbf{u}), \tag{38}$$

where the initial state $\mathbf{x}(0)$ is given. The state vector $\mathbf{x}$ is an $(n \times 1)$ vector and $\mathbf{u}$ is an $(m \times 1)$ control vector bounded by

$$a_j \leq u_j \leq b_j, \quad j = 1, 2, \ldots, m. \tag{39}$$

It is required to find the control policy $\mathbf{u}(t)$ in the time interval $0 \leq t < t_f$ that maximizes (or minimizes) the performance index

$$I = \Phi(\mathbf{x}(t_f)), \tag{40}$$

where the final time t_f is specified.

The LJ optimization procedure was first used for optimal control by Luus (1974) in searching for the elements in the feedback gain matrix, and by Luus and Mutharasan (1974) in time suboptimal control by pole shifting. Nair (1976) used LJ optimization procedure through parametrization of the control policy by polynomial functions in t. As the digital computers became much faster, and when the personal computers became affordable and fast enough for meaningful computations, the LJ optimization procedure was found to be an easy and reliable method to use for challenging optimal control problems by Bojkov *et al.* (1993), and to provide a good means of tackling the simultaneous stabilization problem (Luus, 2003b). Although computationally more demanding than iterative dynamic programming, it was found to be well suited for singular control problems (Luus, 2001b; Luus, 2001c) and time optimal control problems (Luus, 2002). Dimensionality did not appear to be of concern, since it was applied successfully to high dimensional optimal control problems (Luus, 2000d; Luus and Rajagopalan, 2000; Hartig *et al.*, 1995; Luus *et al.*, 1995). It was also applied successfully to low sensitivity problems, such as fed-batch reactors by Luus and Hennessy (1999).

Although we are seeking a continuous control policy, for numerical solution the control policy is usually available only at discrete points in time. When the boundary value problem arising from the application of Pontryagin's maximum principle is solved by control vector iteration (Luus, 2001a), the state and the control policy are kept constant during each integration time step while the adjoint equations are solved backwards in time. The optimal control policy is thus dependent on the size of the integration time step. This happens even when we solve the Riccati equation for the linear quadratic optimal control problem. The results are dependent on the number of integration time steps used (Luus, 2000d). The numerical solution therefore involves parametrization where constant values for the state variables are used over an integration time interval.

When boundary condition iteration (BCI) is used to solve the boundary value problem resulting from the application of Pontryagin's maximum principle, and if the stationary condition for the Hamiltonian can be used to obtain a continuous expression for the control, then parametrization is not encountered. Both the state equation and the adjoint equation are solved simultaneously backward in time, and through iteration the required boundary conditions are satisfied (Luus, 2001d). Simple constraints on control can be readily handled and the computational effort is quite small for well-behaved optimal control problems, yielding accurate results (Luus, 2001e). However, the time horizon must be reasonably small in order to use the BCI method, and for optimal control of complex systems alternate procedures are preferred.

Another approach to solving the optimal control problem is to specify the number of time stages over which the control policy is parametrized. For example, one may break up the time interval into P time stages and seek a piecewise constant, or piecewise linear control at each of these time stages. These time stages may be of variable length to take full advantage of the nature of the expected optimal control policy. Here the approximation due to discretization is more obvious, and it is expected that the larger that P is, the better is the resulting control policy in approximating the continuous optimal control policy. With this type of parametrization, iterative dynamic programming (IDP) (Luus, 2000a) is well suited, since the problem is broken into stages and optimization proceeds from stage to stage. It was shown by Bojkov *et al.* (1993) that LJ optimization can also be used successfully even if the system of differential equations is highly nonlinear, and a very large number of stages with high-dimensional systems can be handled (Luus *et al.*, 1995; Luus, 2000d; Luus and Rajagopalan, 2000). It has been found, however, that IDP is more efficient than LJ optimization procedure in solving typical optimal control problems. Nevertheless, LJ optimization procedure is easier to program and can be used to check the results obtained with IDP or some other procedure. With the line search, the gap between IDP and LJ optimization procedureshould be narrowed. One can easily visualize incorporating a line search into IDP to remove any such advantage.

Example 4. The photochemical process introduced by Jensen (1964) and presented in detail by Lapidus and Luus (1967) has provided an excellent problem for testing optimal control procedures (Rao and Luus, 1972; Lopez Cruz *et al.*, 2000; Balsa-Canto *et al.*, 2001, Liao and Luus, 2005) and to examine the parametrization in solving optimal control problems (Luus, 2006b). This problem was also used by Luus (1991) to show the effects of choosing the final time t_f in optimal control.

The equations describing the system are:

$$\frac{dx_1}{dt} = q_1 - qx_1 - 17.6x_1x_2 - 23x_1x_6u_3, \tag{41}$$

$$\frac{dx_2}{dt} = u_1 - qx_2 - 17.6x_1x_2 - 146x_2x_3, \tag{42}$$

$$\frac{dx_3}{dt} = u_2 - qx_3 - 73x_2x_3, \tag{43}$$

$$\frac{dx_4}{dt} = -qx_4 + 35.2x_1x_2 - 51.3x_4x_5, \tag{44}$$

$$\frac{dx_5}{dt} = -qx_5 + 219x_2x_3 - 51.3x_4x_5, \tag{45}$$

$$\frac{dx_6}{dt} = -qx_6 + 102.6x_4x_5 - 23x_1x_6u_3, \tag{46}$$

$$\frac{dx_7}{dt} = -qx_7 + 46x_1x_6u_3, \tag{47}$$

$$\frac{dx_8}{dt} = 5.8(qx_1 - q_1) - 3.7u_1 - 4.1u_2 + q(23x_4 + 11x_5 + 28x_6 + 35x_7) - 5u_3^2 - 0.099, \quad (48)$$

where $q = q_1 + u_1 + u_2$ and $q_1 = 6$. The last differential equation gives the performance index. The problem is to maximize the profit function

$$I = x_8(t_f), \quad (49)$$

where the final time t_f is specified as 0.2 h.

The initial state is $\mathbf{x}(0) = [0.1883\ 0.2507\ 0.0467\ 0.0899\ 0.1804\ 0.1394\ 0.1046\ 0]^T$, and the control variables are constrained by

$$0 \leq u_1 \leq 20, \quad (50)$$

$$0 \leq u_2 \leq 6, \quad (51)$$

$$0 \leq u_3 \leq 4. \quad (52)$$

There are eight differential equations and three control variables. Recently it was used by Liao and Luus (2005) to compare LJ optimization procedure to genetic algorithm.

We initially chose $P = 50$ time stages of constant length to parametrize the optimal control problem and used piecewise constant control. This presents a 150-dimensional optimization problem for the LJ optimization procedure. We used 20 iterations per pass and allowed a maximum of 10,000 passes. The initial value for ϵ was taken as 10^{-3}. As before, we used the reduction factor 0.9 for reducing ϵ. When ϵ became less than 10^{-12}, the iterations were stopped. We chose $R = 37$ random points per iteration. The initial values for control at each time step were chosen at random in their allowable ranges. The initial region size was taken as half of the initial value. Five different runs were made by changing the seed number in the random number generator.

With line search, convergence to the maximum $I = 20.09564$ was obtained rapidly from five randomly chosen starting points as is seen in Fig. 10. The details of the runs are presented in Table 6, where the computation time is given on Athlon3800/2.4 GHz personal computer. As can be seen, each run converged in less than 400 s of computation time, and the total computation time for these 5 runs was 1850.5 s. The optimal control

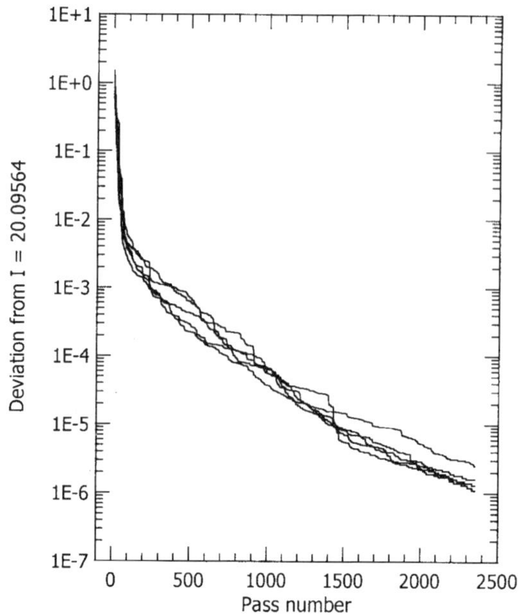

Figure 10. Convergence of LJ optimization procedure with line search for Example 4 with $P = 50$ time stages of constant length.

Table 6. Computational results with LJ optimization procedure with line search for Example 4 with $P = 50$ time stages of equal length and $R = 37$.

Run number	Number of passes	Performance index	CPU time, s
1	2949	20.095639	364.3
2	3024	20.095638	371.5
3	3261	20.095639	397.8
4	2999	20.095639	370.9
5	2780	20.095639	346.0

policy, given in Fig. 11, is the same as the one in Liao and Luus (2005). When the number of random points was reduced to $R = 25$, convergence to $I = 20.095639$ was obtained four times and $I = 20.095638$ once. It took a larger number of passes for convergence but the total computation time for the 5 runs was reduced to 1383.5 s. Thus, with line search the optimal control policy with $P = 50$ was easily established.

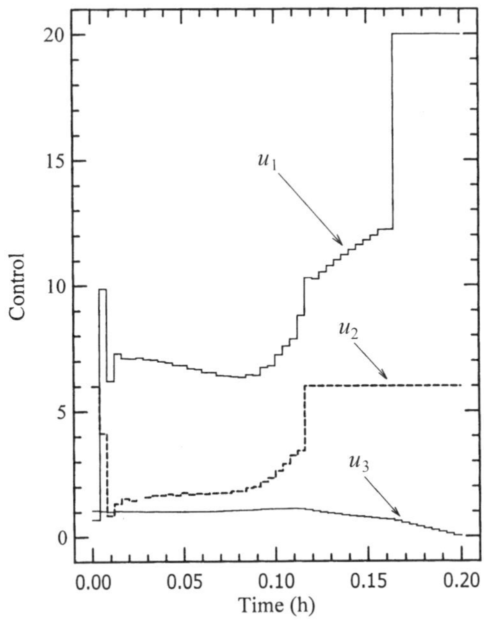

Figure 11. Optimal control policy for Example 4 with the use of $P = 50$ time stages of constant length.

Without the line search, however, the use of $R = 37$ resulted in premature collapse of the search region in each of the 5 cases with the best value $I = 20.094221$ in 707 passes and computation time of 79.5 s. Although this is within 0.007% of the optimum, the resulting control policy is unsatisfactory. Therefore, the so-called "greedy" approach was attempted, where the center point for search was changed immediately after a better point was found. As is seen in Table 7, although there was a premature collapse for one case, results close to the optimum could be readily obtained. When this greedy approach was used along with the line search results, as is shown in Table 8, the computation time is reduced to half, and convergence is very systematic for all the 5 cases. This shows the advantage of incorporating the line search into the LJ optimization procedure and that for this problem the "greedy" approach is advantageous.

We checked the results with iterative dynamic programming (IDP), using $R = 37$ random points at each stage in each iteration and allowed 100 passes, each consisting of 10 iterations. After every pass, the region

Table 7. Computational results with LJ optimization procedure without line search for Example 4 with $P = 50$ time stages of equal length and $R = 37$, and using a “greedy” approach.

Run number	Number of passes	Performance index	CPU time, s
1	1943	20.0956387	239.3
2	796	20.0955823	97.5
3	1853	20.0956362	230.2
4	1856	20.0956371	231.0
5	1779	20.0956322	221.5

Table 8. Computational results with LJ optimization procedure with line search for Example 4 with $P = 50$ time stages of equal length and $R = 37$, and using “greedy” approach.

Run number	Number of passes	Performance index	CPU time, s
1	1621	20.0956395	172.5
2	1484	20.0956393	158.4
3	1745	20.0956395	187.0
4	1685	20.0956393	178.3
5	1593	20.0956395	171.3

Table 9. Computational results with iterative dynamic programming for Example 4 with $P = 50$ time stages of equal length and $R = 37$.

Run number	Performance index	CPU time, s
1	20.0956395	146.1
2	20.0956393	146.1
3	20.0956394	145.9
4	20.0956394	146.1
5	20.0956393	146.1

size was restored to 0.90 of its value at the beginning of the previous pass. Same conditions as with LJ optimization procedure were used. The five runs required a total of 730 s of computation time. In each case $I = 20.095639$ was obtained, as is shown in Table 9. The same control policy was obtained

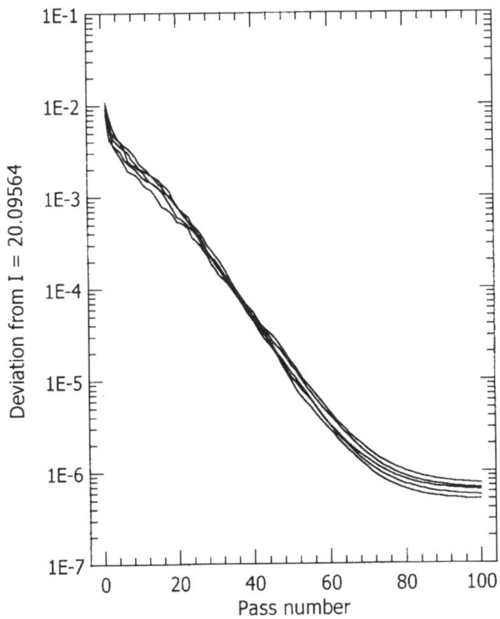

Figure 12. Convergence of IDP for Example 4 with $P = 50$ of constant length.

as with LJ optimization procedure with line search. The convergence profile is given in Fig. 12. Therefore, the LJ optimization procedure with "greedy" approach and line search rates very well with respect to IDP. In IDP, a three-dimensional optimization problem is carried out repeatedly in a stage-wise manner and one does not face the simultaneous 150-dimensional optimization problem. Thus one would expect IDP to be better suited for optimal control problems, but here the computation times for IDP and LJ optimization procedure with line search and "greedy" approach are very close.

Increasing the number of time stages to $P = 80$, presenting a 240-dimensional optimization problem for LJ optimization procedure, and using $R = 50$ gave similar results with the maximum $I = 20.0957401$. The optimal control policy is given in Fig. 13. With $P = 100$ the performance index is improved slightly to $I = 20.0957653$ and the oscillations in u_1 and u_2 become more pronounced, as is shown in Fig. 14. The oscillations become larger as P is increased further, as is shown in Figs. 15 and 16. As is seen in Fig. 17, the results with $P = 160$ are close to those reported by Luus (2006) where IDP was used.

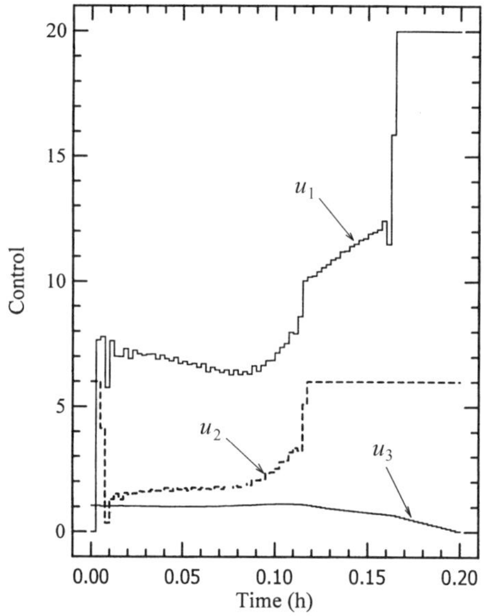

Figure 13. Optimal control policy for Example 4 with the use of $P = 80$ time stages of constant length, giving $I = 20.0957401$.

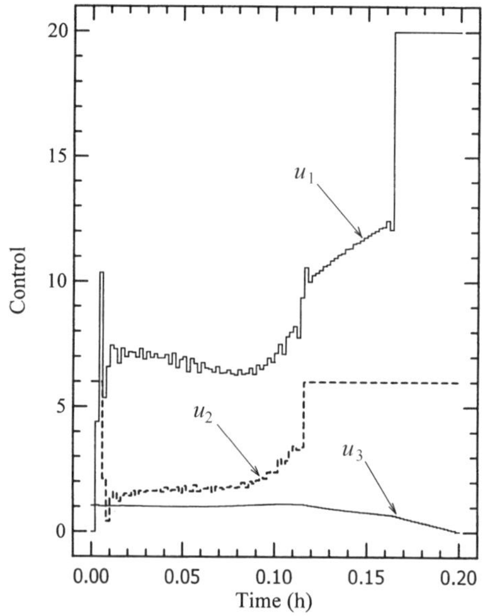

Figure 14. Optimal control policy for Example 4 with the use of $P = 100$ time stages of constant length, giving $I = 20.0957653$.

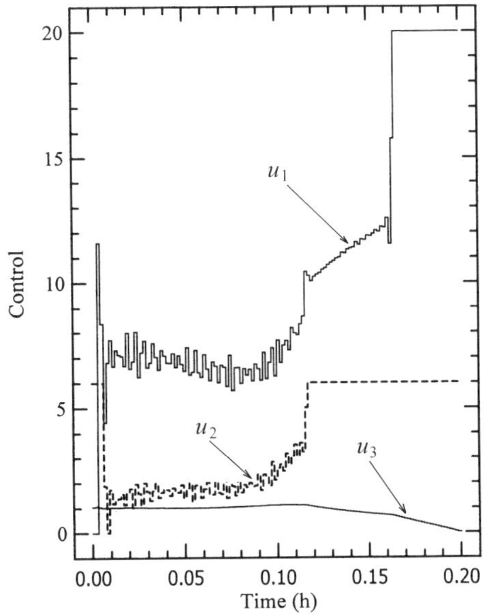

Figure 15. Optimal control policy for Example 4 with the use of $P = 130$ time stages of constant length, giving $I = 20.0957817$.

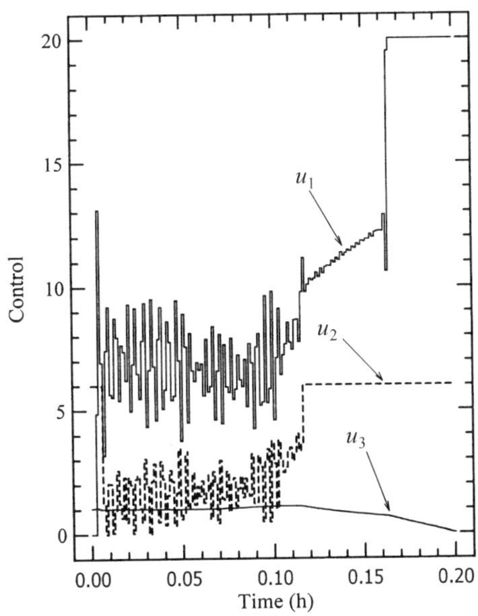

Figure 16. Optimal control policy for Example 4 with the use of $P = 160$ time stages of constant length, giving $I = 20.0957877$.

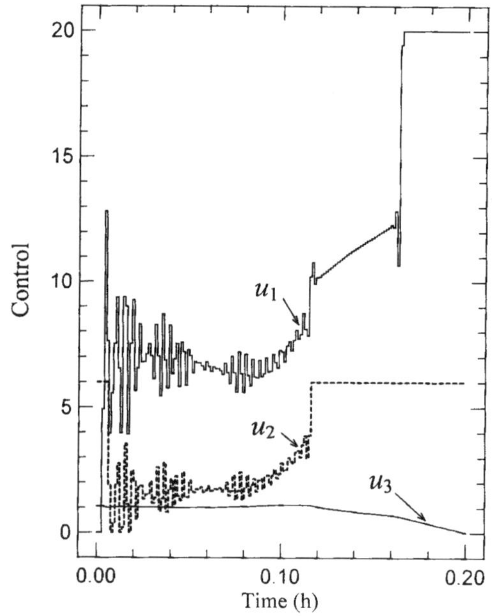

Figure 17. Optimal control policy for Example 4 with the use of $P = 160$ time stages of constant length, as obtained by IDP with $I = 20.09579$ (Luus, 2006).

Table 10. Convergence of the LJ optimization procedure with line search for Example 4 with $P = 200$, $R = 100$, showing the number of passes required for convergence, performance index obtained, and the computation time on AMD Athlon3800/2.4 GHz.

	Nongreedy approach			Greedy approach		
Case, number	Passes	I	CPU s	Passes	I	CPU s
1	6364	20.0957939	6557.48	7827	20.0958069	7776.23
2	7104	20.0957957	7316.56	1880	20.0957922	1871.81
3	7846	20.0957952	8078.48	1208	20.0957735	1191.45
4	7196	20.0957942	7412.57	3215	20.0958004	3170.65
5	6607	20.0957928	6806.67	8575	20.0958083	8451.81

The results with $P = 200$ are shown in Table 10, showing the LJ optimization procedure with line search with and without the greedy aspect. For successful runs, the greedy approach requires about the same amount of computation time as the nongreedy approach, but the end result is somewhat

Table 11. Computational results with iterative dynamic programming for Example 4 with $P = 200$ time stages of equal length and $R = 50$ after 100 passes.

Run number	Performance index	CPU time, s
1	20.0958036	5603.42
2	20.0958030	5600.83
3	20.0958007	5601.23
4	20.0958023	5576.60
5	20.0958029	5557.12

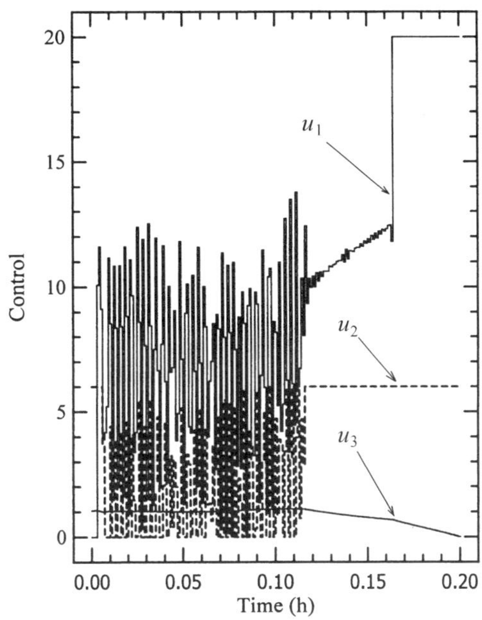

Figure 18. Optimal control policy for Example 4 with the use of $P = 200$ time stages of constant length, giving $I = 20.0958083$.

better. The use of IDP with $R = 50$ for 100 passes yielded consistent results for all the 5 cases as shown in Table 11, but the two successful runs with the greedy approach gave slightly better end result. The optimal control policy for $P = 200$ is given in Fig. 18.

The use of $P = 500$, gives a very slight increase in the performance index to $I = 20.09587$, and the oscillations are very large, as can be seen

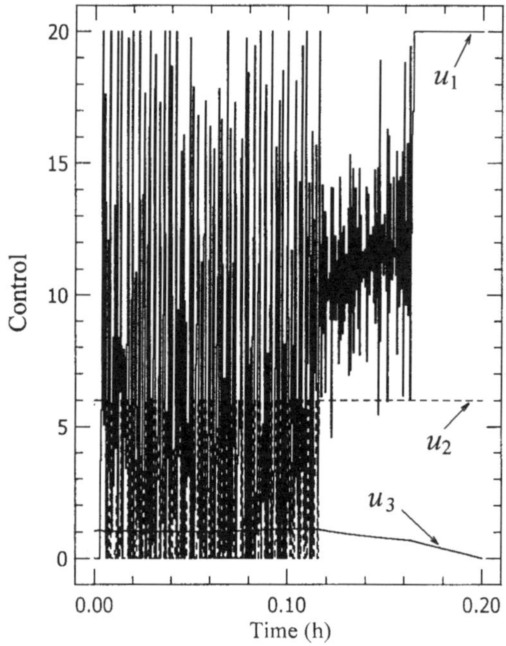

Figure 19. Optimal control policy for Example 4 with $P = 500$ time stages of constant length, giving $I = 20.09587$ (Luus, 2007).

in Fig. 19. The actual optimal control policy for this problem has not yet been established, but the use of stages of variable length reveals that the oscillations have a very regular pattern (Luus, 2006).

Here the oscillatory behavior of the inputs has very little effect in providing extra economic benefit. However, from the mathematical point of view, one should be curious to know what really is the optimal control policy for this problem. In other situations, however, oscillatory inputs can increase the benefits quite substantially as has been demonstrated with several systems by Luus (2008).

This example shows that line search improves the convergence rate very significantly and allows the LJ optimization procedure to yield very accurate results. The computational effort for the line search at the end of a pass to determine the best value for the optimization variables is quite insignificant with respect to the other computations. Here it was found that the "greedy" approach can reduce the computational effort quite significantly to make the LJ optimization procedure useful for solving complex optimal control problems. However, with the greedy approach, care must be taken to avoid premature collapse.

Thus the LJ optimization procedure with line search can be used for a wide variety of optimization problems, and it provides an attractive means for solving complex optimal control problems.

6. Conclusions

The application of the LJ optimization procedure has been illustrated for use in model reduction, parameter estimation and optimal control. The method is easy to apply and gives reliable results. The use of a simple line search to get a better centre point after a pass improves the convergence rate of the procedure, and allows the optimum values to be established more accurately. This is especially important in parameter estimation where a large number of figures are necessary in the performance index to get the parameter values to 5-figure accuracy.

The LJ optimization procedure is easy to program and to use. In the accompanying CD, sample programs with outputs are given to provide further detail to the interested reader.

References

Balsa-Canto, E., Banga, J.R., Alonso, A.A. and Vassiliadis, V.S. (2001). Dynamic optimization of chemical and biochemical processes using restricted second-order information. *Computers and Chem. Eng.*, **25**, pp. 539–546.

Belohlav, Z., Zamostny, P., Kluson, P. and Volf, J. (1997). Application of random-search algorithm for regression analysis of catalytic hydrogenation, *Canadian J. Chem. Eng.*, **75**, pp. 735–742.

Bojkov, B., Hansel, R. and Luus, R. (1993). Application of direct search optimization to optimal control problems. *Hungarian J. Ind. Chem.*, **21**, pp. 177–185.

Carrier, J.F. and Stephanopoulos, G. (1998). Wavelet-based modulation in control-relevant process identification. *AlChE Journal*, **44**, pp. 34–360.

Esposito, W.R. and Floudas, C.A. (2000). Global optimization for the parameter estimation of differential-algebraic systems. *Ind. Eng. Chem. Res.*, **39**, pp. 1291–1310.

Hartig, F., Keil, F.J. and Luus, R. (1995). Comparison of Optimization methods for a fed-batch reactor, *Hungarian J. Ind. Chem.*, **23**, pp. 141–148.

Howitt, G.D. and Luus, R. (1990). Model reduction by minimization of integral square error performance indices. *J. Franklin Inst.*, **327**, pp. 343–357.

Hull, T.E., Enright, W.D. and Jackson, K.R. (1976). *User guide to DVERK — A Subroutine for Solving Nonstiff ODE's*, Report 100, Department of Computer Science, University of Toronto, Canada.

Jensen, T. (1964). *Dynamic Control of large Dimension Nonlinear Chemical Processes*, PhD Dissertation, Princeton University, Princeton, NJ.

Kalogerakis, N. and Luus, R. (1982). Increasing the size of region of convergence for parameter estimation. *Proceedings of 1982 American Contr. Conf.*, Arlington, VA, pp. 358–362.

Kalogerakis, N. and Luus, R. (1983a). Simplification of quasilinearization method for parameter estimation. *AIChE Journal*, **29**, pp. 858–864.

Kalogerakis, N. and Luus, R. (1983b). Improvement of Gauss-Newton method for parameter estimation through the use of information index. *Industrial and Engineering Chemistry Fundam.*, **22**, pp. 436–445.

Krishnamurthy, V. and Seshadri, V. (1978). Model reduction using the Routh stability criterion. *IEEE Trans. Automat. Control*, **23**, pp. 729–731.

Lapidus, L. and Luus, R. (1967). *Optimal Control of Engineering Processes*, Blaisdell, Waltham, MA.

Liao, B. and Luus, R. (2005). Comparison of the Luus-Jaakola optimization procedure and the genetic algorithm. *Engineering Optimization*, **37**, pp. 381–398.

Linga, P., Al-Saifi, N. and Englezos, P. (2006). Comparison of the Luus-Jaakola optimization and Gauss-Newton methods for parameter estimation in ordinary differential equation models, *Ind. Eng. Chem. Res.*, **45**, pp. 4716–4725.

Lopez Cruz, I.L., Van Willigenburg, L.G. and Van Straten, G. (2000). Evolutionary algorithms for optimal control of chemical processes. *Proc. IASTED International Conference Control and Applications*, Acta Press, Anaheim, CA, pp. 155–161.

Luus, R. (1974). Optimal control by direct search on feedback gain matrix. *Chem. Eng. Sci.*, **29**, pp.1013–1017.

Luus, R. (1980). Optimization in model reduction. *Int. J. Control*, **32**, pp. 741–747.

Luus, R. (1988). Parameter estimation in systems described by nonlinear ordinary differential equations. *Hungarian J. Ind. Chem.*, **16**, pp. 23–28.

Luus, R. (1991). Effect of choice of final time in optimal control of nonlinear systems. *Canadian J. Chem. Eng.*, **69**, pp. 144–151.

Luus, R. (1998). Determination of the region sizes for LJ optimization procedure. *Hungarian J. Ind. Chem.*, **26**, pp. 281–286.

Luus, R. (1999). Optimal reduction of linear systems. *J. Franklin Inst.*, **336**, pp. 523–532.

Luus, R. (2000a). *Iterative Dynamic Programming*, Chapman and Hall/CRC, London.

Luus, R. (2000b). Luus-Jaakola optimization procedure. *Recent Res. Devel. Chemical Engng*, Transworld Research Network, India, **4**, pp. 45–64.
Luus, R. (2000c). To the editor. *AlChE J.*, **46**, pp. 427–428.
Luus, R. (2000d). Comparison of optimization procedures for optimal control of high-dimensional systems. *Hungarian J. Ind. Chem.*, **28**, pp. 217–223.
Luus, R. (2001a). Direct search Luus-Jaakola optimization procedure, LJ optimization procedure. *Encyclopedia of Optimization*, Floudas, C.A. and Pardalos, P.M. (eds.), Kluwer Academic Publishers, Dortrecht, The Netherlands, **1**, pp. 440–444.
Luus, R. (2001b). Use of Luus-Jaakola optimization procedure for singular optimal control problems. *Nonlinear Analysis*, **47**, pp. 5647–5658.
Luus, R. (2001c). Optimal control of singular control problems. *Proc. of IASTED International Conference Control and Applications*, June 27–29, 2001, Banff, Alberta, Canada, pp. 82–87.
Luus, R. (2001d). Boundary condition iteration, BCI. In *Encyclopedia of Optimization*, Floudas, C.A. and Pardalos, P.M. (eds.), Kluwer Academic Publishers, Dortrecht, The Netherlands, **1**, pp. 205–209.
Luus, R. (2001e). Further developments in the new approach to boundary condition iteration in optimal control. *Canadian J. Chem. Eng.*, **79**, pp. 968–976.
Luus, R. (2002). Comparison of LJ optimization procedure and IDP in solving time optimal control problems. *Recent Developments in Optimization and Optimal Control in Chemical Engineering*, Luus, R. (ed.), Research Signpost, Kerala, India, pp. 253–275.
Luus, R. (2003a). Effect of region collapse parameter on convergence rate of LJ optimization procedure. *Proceedings of IASTED International Conf. on Intelligent Systems and Control*, Salzburg, Austria, pp. 51–56.
Luus, R. (2003b). Time optimal control, pole shifting and simultaneous stabilization. *Proc. Indian Natn. Sci. Acad.*, **69, A**, pp. 301–315.
Luus, R. (2006a). Comparison of approaches for model reduction. *Proc. IASTED International Conference Intelligent Systems and Control*, August 14–16, 2006, Honolulu, pp. 62–67.
Luus, R. (2006b). Parametrization in nonlinear optimal control problems. *Optimization*, **55**, pp. 65–89.
Luus, R. (2008). Optimal control of oscillatory systems by iterative dynamic programming. *J. Industrial and Management Optimization*, **4**, pp. 1–15.
Luus, R. (2007). Use of line search in the Luus-Jaakola optimization procedure. *Proceedings of the 3rd IASTED International Conference on Computational Intelligence CI2007*, July 2–4, 2007, Banff, Alberta, Canada, pp. 128–135.
Luus, R. and Brenek, P. (1989). Incorporation of gradient into random search optimization. *Chem. Eng. technol.*, **12**, pp. 309–318.

Luus, R. and Harapyn, I. (2003). Two-step method for active inequality constraints for direct search optimization. *Proceedings of IASTED International Conf. on Intelligent Systems and Control*, Salzburg, Austria, pp. 443–448.

Luus, R., Hartig, F. and Keil, F.J. (1995). Optimal drug scheduling of cancer chemotherapy by direct search optimization. *Hungarian J. Ind. Chem.*, **23**, pp. 55–58.

Luus, R. and Hennessy, D. (1999). Optimization of fed-batch reactors by the Luus-Jaakola optimization procedure. *Ind. Eng. Chem. Res.*, **38**, pp. 1948–1955.

Luus, R. and Hernaez, H. (2000). Parameter estimation in errors-in-variables data. *Hungarian J. Ind. Chem.*, **28**, pp. 201–206.

Luus, R. and Jaakola, T.H.I. (1973). Optimization by direct search and systematic reduction of the size of search region. *AlChE Journal*, **19**, pp. 760–766.

Luus, R. and Jammer, M. (2005). Estimation of parameters in 3-parameter Weibull probability distribution functions. *Hungarian J. Ind. Chem.*, **33**, pp. 69–73.

Luus, R. and Mutharasan, R. (1974). Stabilizing of linear system behavior by pole shifting. *Int. J. Control*, **20**, pp. 395–405.

Luus, R. and Rajagopalan. (2000). Use of piecewise linear continuous control in Luus-Jaakola optimization procedure. *Proc. of IASTED International Conf. Control and Applications*, May 24–27, 2000, Cancun, Mexico, pp. 167–171.

Luus, R., Sabaliauskas, K. and Harapyn, I. (2006). Handling inequality constraints in direct search optimization. *Engineering Optimization*, **38**, pp. 391–405.

Luus, R., Wong, T.C.W. and Kostelnik, D. (1999). Importance of structure of the reduced model in model reduction. *Proc. IASTED International Conference Intelligent Systems and Control*, October 28–30, 1999, Santa Barbara, CA, USA, pp. 322–326.

Nair, G.P. (1976). Optimal control using LJ search method. *Automatic Control Theory and Applications*, **4**, pp. 1–6.

Rao, S.N. and Luus, R. (1972). Evaluation and improvement of control vector iteration procedures for optimal control. *Canadian J. Chem. Eng.*, **50**, pp. 777–784.

Shih, Y.P and Wu, W.T. (1973). Simplification of z-transfer functions by continuous fractions. *Int. J. Control*, **17**, pp. 1089–1094.

Wang, B.C. and Luus, R. (1980). Increasing the size of region of convergence for parameter estimation through the use of shorter data-length. *Int. J. Control*, **31**, pp. 947–972.

Wang, L., Li, L.L. and Tang, F. (2005). Optimal reduction of models using hybrid searching strategy. *Applied Mathematics and Computation*, **168**, pp. 1357–1369.

Yang, S.M. and Luus, R. (1983). A note on model reduction of digital control systems. *Int. J. Control*, **37**, pp. 437–439.

Exercises

(1) (a) Solve the model reduction problem where we take the 4th order reduced model

$$G_4(s) = \frac{20.25833(1 + x_1 s + x_2 s^2)(1 + x_3 s)}{(1 + x_4 s + x_5 s^2)(1 + x_6 s + x_7 s^2)} \tag{53}$$

by choosing the best values for $x_1, x_2, \ldots, x_7$ to minimize the deviation of G_4 from F as given by Eq. (10) in the Nyquist plot, where F is given in Eq. (20). *Hint:* At the optimum, $I = 0.0007325768$.

(b) To show the benefits of factoring, solve the model reduction problem in nonfactored form.

(2) In this chapter we solved the parameter estimation problem for toluene hydrogenation by choosing initial values of parameters $k_i = 0.5 \pm 0.2\,\mathrm{rn}(0, 1)$. Solve this problem by choosing initial values of parameters $k_i = 0.3 \pm 0.15\,\mathrm{rn}(0, 1)$.

(3) Solve the Jensen reactor optimal control problem with $P = 150$ stages of equal length.

(4) Solve the Jensen reactor optimal control problem by using $t_f = 0.4$ instead of 0.2.

Chapter 13

PHASE STABILITY AND EQUILIBRIUM CALCULATIONS IN REACTIVE SYSTEMS USING DIFFERENTIAL EVOLUTION AND TABU SEARCH

Adrián Bonilla-Petriciolet

Department of Chemical & Biochemical Engineering
Instituto Tecnológico de Aguascalientes
México, 20256
petriciolet@hotmail.com

Gade Pandu Rangaiah

Department of Chemical & Biomolecular Engineering
National University of Singapore, Singapore, 117576

Juan Gabriel Segovia-Hernández
and José Enrique Jaime-Leal

Department of Chemical Engineering
Universidad de Guanajuato, Mexico, 36050

1. Introduction

Reactive separations processes (RSPs), where separation and reaction units are combined, have received considerable interest from chemical engineers and have been used to develop new technologies for the chemical and petrochemical industries. They offer several technological, economic, and operational advantages over conventional systems (Taylor and Krishna, 2000). However, optimal performance of RSPs depends significantly on relevant process design issues, where the basis for most synthesis and analysis is phase equilibrium behavior. In the design of RSPs, it is often necessary to perform numerous phase stability and equilibrium calculations. The goal in phase equilibrium calculations is to determine the number and identity (composition, quantity and type) of phases existing at equilibrium for a mixture under specific conditions, while phase stability analysis helps to confirm if the global minimum of Gibbs free energy has been reached.

There have been many efforts to develop new techniques with the aim of reliably computing phase equilibrium in systems subject to chemical reactions (Seider and Widagdo, 1996). The existing techniques for these calculations can be divided into two main classes: (a) procedures involving minimization of a suitable objective function, and (b) strategies requiring solution of nonlinear equations obtained from the stationary conditions of that objective function. For the case of phase equilibrium calculations, the objective function is generally the Gibbs free energy whereas the stability analysis is performed using the tangent plane distance function. The available strategies can also be classified as either stoichiometric or non-stoichiometric, depending on the formulation of mass balance constraints (Stateva and Wakeham, 1997).

Reactive phase stability and equilibrium problems are non-linear, multivariable and may have multiple solutions. In particular, strong interactions among components, phases and reactions increase the complexity of these thermodynamic calculations (Xiao *et al.*, 1989; Stateva and Wakeham, 1997). Hence, reliable and efficient methods are necessary for solving phase equilibrium problems of reactive systems. Both deterministic and stochastic global solving methods have been proposed for computing chemical and phase equilibrium simultaneously. The former includes homotopy-continuation and interval methods for non-linear equations, and branch-and-bound optimization strategies. Unfortunately, they generally require large computational effort for multi-component mixtures and, in some cases, problem reformulation is needed (Wakeham and Stateva, 2004). On the other hand, stochastic optimization methods are attractive because they do not require continuity and other assumptions about the optimization problem, and are reliable and efficient. However, only a few studies have reported their use for reactive phase equilibrium modeling (e.g. simulated annealing by Reynolds *et al.*, 1997; Bonilla-Petriciolet *et al.*, 2006; and the random search method of Luus and Jaakola by Lee *et al.*, 1999). Results of these studies have shown the potential of stochastic optimization strategies.

In particular, both differential evolution (DE) and tabu search (TS) are capable of solving non-differentiable, non-linear and multi-modal objective functions to find the global minimum. DE is a population based method that mimics biological evolution by performing mutation, crossover and selection steps to find the global optimum. TS is a point-to-point method that uses an adaptive memory to avoid re-visits to the same place in the search region and to identify promising areas for optimization. Both these methods have been used for Chemical Engineering applications (see Chapters 4 and 5). Teh and Rangaiah (2003) have studied phase equilibrium calculations in non-reactive mixtures using TS, and its performance was compared with a genetic algorithm (GA). They concluded that TS has high reliability in locating the global minimum and converges faster than GA. Srinivas and Rangaiah (2007) have compared the performance of TS and DE in phase stability and Gibbs energy minimization problems for non-reactive mixtures; their results show that DE is more reliable compared than TS but the latter is computationally more efficient.

Even though both DE and TS are promising, they have not been applied to and tested for reactive phase equilibrium and stability problems. Therefore, in this chapter, we present the application of TS and DE for these important problems, and compare their performance for benchmark problems in this area. In particular, we have analyzed the reliability and efficiency of these methods using a variable transformation approach for modeling the phase equilibrium behavior of reactive mixtures. The remainder of this chapter is organized as follows. The basic concepts of reactive phase stability and equilibrium problems including available methods for their solution, and their formulation in terms of transformed composition variables are provided in Sec. 2. Section 3 describes the algorithms of DE and TS used in this study. Application and performance comparison of DE and TS for reactive phase equilibrium and stability problems, are presented in Secs. 4 and 5. Finally, conclusions of this study are provided in the last section.

2. Phase Equilibrium and Stability Problems in Reactive Systems

This section introduces the reader to the basic concepts and description of phase stability and equilibrium problems in reactive systems. Mathematically, both the problems can be stated as finding the global minimum, w^* and $f(w^*)$ of a non-linear function, $f(w)$ of n real decision variables, $w = (w_1, w_2, \ldots, w_n)$ subject to $w \in \Omega$ where Ω is the feasible region satisfying the governing constraints and bounds on decision variables.

2.1. Description of phase equilibrium problems

In phase equilibrium problems, given components present, temperature and pressure of a system or stream, the main objectives are to correctly establish the phase number and type, and the distribution of components among the phases at the equilibrium. Classical thermodynamics indicates that minimization of the Gibbs free energy is a natural approach for calculating the equilibrium state of a mixture. Most of the available methods for Gibbs free energy minimization in reactive mixtures have been proposed during the last two decades, and they include a variety of problem formulations and numerical techniques such as local search methods with and without decoupling strategies (Castillo and Grossmann, 1981; Castier *et al.*, 1989; Xiao *et al.*, 1989; Gupta *et al.*, 1991; Perez-Cisneros *et al.*, 1997; Stateva and Wakeham, 1997), branch-and-bound optimization methods (McDonald and Floudas, 1996; McKinnon and Mongeau, 1998), algorithms using homotopy continuation (Jalali and Seader, 1999; Jalali *et al.*, 2008), deterministic methods based on interval mathematics (Burgos-Solorzano *et al.*, 2004) and stochastic optimization methods (Reynolds *et al.*, 1997; Lee *et al.*, 1999; Bonilla-Petriciolet *et al.*, 2006; Bonilla-Petriciolet *et al.*, 2008).

In general, there are several challenges in computing the global minimum of the Gibbs free energy function. First, the number and types of phases, at which this thermodynamic function achieves the global minimum, are usually not known *a priori*. Hence, several equilibrium calculations must be performed using different phase configurations (adding or removing phases) to identify the stable equilibrium state. Moreover, for a fixed number of phases and components, Gibbs free energy function may have a local minimum value very comparable to the global minimum value, which makes it challenging to find the global minimum (Srinivas and Rangaiah, 2007). The poor conditioning of the Hessian matrix of the free energy for mixtures near phase boundaries (such as bubble, dew and critical points) may lead to failure of solving strategies (Seider and Widagdo, 1996). Trivial solutions, which satisfy the necessary equilibrium conditions and the mass constraints in the system, also exist but they are local optima of Gibbs free energy. Consequently, many strategies in the literature are local methods, depend on initialization and may converge to unstable solutions. These limitations have prompted the development of reliable algorithms for the global minimization of Gibbs free energy.

Many available methods for phase equilibrium problems in reactive systems, use conventional composition variables (mole numbers or fractions) as decision variables. A few studies have considered variable transformation approaches with the goal of developing a simpler thermodynamic framework for modeling reactive systems (e.g. Ung and Doherty, 1995b; Perez-Cisneros *et al.*, 1997). These approaches, in combination with appropriate methods such as stochastic optimization strategies, can be used to develop reliable techniques for computing phase equilibrium of reactive systems. Therefore, in this chapter we have used a suitable transformation scheme to formulate phase equilibrium and stability problems involving chemical reactions.

2.1.1. Formulation of the optimization problem

In a reactive mixture with c components and r independent chemical reactions (with $r < c$) that splits into π phases, the Gibbs free energy function can be written as

$$\Delta \hat{g} = \sum_{j=1}^{\pi} \sum_{i=1}^{c-r} \hat{n}_{i,j} \left(\frac{\Delta \mu_{i,j}}{RT} \right), \tag{1}$$

where $\Delta \hat{g}$ is the dimensionless transformed molar Gibbs free energy of mixing (Ung and Doherty, 1995b), $\hat{n}_{i,j}$ is the transformed mole number of component i in phase j and $\frac{\Delta\mu_{i,j}}{RT}$ is the chemical potential of component i in phase j. The transformed mole number and other transformed variables are defined below. Further, the following mass balances must be satisfied:

$$\hat{n}_{i,F} - \sum_{j=1}^{\pi} \hat{n}_{i,j} = 0 \quad \text{for } i = 1, \ldots, c-r, \tag{2}$$

where $\hat{n}_{i,F}$ is the transformed mole number of component i in the feed (or initial system).

The Gibbs free energy minimization problem is to minimize Eq. (1) with respect to $\hat{n}_{i,j}$ (for $i = 1, \ldots, c-r$ and $j = 1, \ldots, \pi$) subject to Eq. (2). Note that the chemical potentials in Eq. (1) are functions of transformed composition variables, which depend on the thermodynamic model used for predicting behavior of each phase.

Transformed mole numbers $\hat{n}_i$, for a system of c components (including both reacting and inert species) subject to r independent chemical reactions, are defined by selecting a set of r reference components (Ung and Doherty, 1995a; 1995b) as:

$$\hat{n}_i = n_i - v_i N^{-1} n_{\text{ref}}, \tag{3}$$

where n_i is the number of moles of component i, n_{ref} is a column vector of dimension r of the moles of each of the reference components, v_i is the row vector of stoichiometric coefficients of component i for each of the r reactions, and N is an invertible and square matrix formed from the stoichiometric coefficients of the reference components in the r reactions. Note that $v_{i,k} < 0$ for reactants, $v_{i,k} > 0$ for products, and $v_{i,k} = 0$ for inert components.

Consequently, the total amount of transformed moles $\hat{n}_T$ is

$$\hat{n}_T = \sum_{i=1}^{c-r} \hat{n}_i = n_T - v_{TOT} N^{-1} n_{\text{ref}}. \tag{4}$$

Here, n_T is the total number of moles present at any instant of time, and v_{TOT} is a row vector where each element corresponds to the sum of stoichiometric coefficients for all components that participate in each of the r reactions.

Dividing $\hat{n}_i$ by $\hat{n}_T$, the transformed mole fractions X_i are obtained:

$$X_i = \frac{\hat{n}_i}{\hat{n}_T} = \frac{x_i - v_i N^{-1} x_{\text{ref}}}{1 - v_{TOT} N^{-1} x_{\text{ref}}} \quad i = 1, \ldots, c - r, \tag{5}$$

where x_i is the mole fraction of component i and x_{ref} is the column vector of r reference component mole fractions. Due to mass balance restrictions, the sum of all transformed mole fractions must equal unity, or $\sum_{i=1}^{c-r} X_i = 1$ (Ung and Doherty, 1995a; 1995b).

The transformed composition variables ($\hat{n}_i$ and X_i) depend only on the initial composition of each independent chemical species and take the same value before, during and after reaction (Ung and Doherty, 1995a; 1995b). They also restrict the solution space to the compositions that satisfy stoichiometry requirements and reduce the dimension of the composition space by the number of independent reactions. The transformed variables allow all of the procedures used to compute phase equilibrium/stability of non-reactive mixtures to be extended to systems with chemical reactions (Ung and Doherty, 1995a; 1995b). They ($\hat{n}$ and X) in reactive systems play the same role as the usual composition variables (n and x) in non-reactive mixtures. However, transformed variables can be negative or positive depending on the reference components, number and type of reactions.

In Appendix A, we use a simple reactive mixture to illustrate the correspondence between the transformed composition variables ($\hat{n}$ and X) and the usual composition variables (n and x) at equilibrium, and how to calculate the previous ones by specifying the transformed variables and applying the chemical equilibrium constraints:

$$K_{eq,k} = \prod_{i=1}^{c} a_i^{v_{i,k}}, \quad k = 1, \ldots, r, \tag{6}$$

where $K_{eq,k}$ is the equilibrium constant for reaction k and a_i is the activity of component i. The formulation of phase equilibrium problem for this simple reactive mixture is also shown in this appendix.

For a reactive mixture, minimizing Gibbs free energy with respect to n is equivalent to minimizing the transformed Gibbs free energy with respect to $\hat{n}$ (Ung and Doherty, 1995b). The Gibbs free energy minimization problem has equality constraints (Eq. (2)). To perform an unconstrained minimization of free energy, we can use a set of new variables instead of transformed composition variables as decision variables. The introduction of these variables automatically satisfies the equality constraints (Eq. (2)) and reduces problem dimensionality further. Assuming that all transformed mole fractions have values in the range $X_{i,j} \in (0, 1)$, real variables $\lambda_{i,j} \in [0, 1]$ are defined and employed as decision variables using the following expressions:

$$\hat{n}_{i,1} = \lambda_{i,1}\hat{n}_{i,F} \quad \text{for } i = 1, \ldots, c-r, \tag{7}$$

$$\hat{n}_{i,j} = \lambda_{i,j}\left(\hat{n}_{i,F} - \sum_{m=1}^{j} \hat{n}_{i,m}\right) \quad \text{for } i = 1, \ldots, c-r \quad \text{and} \quad j = 2, \ldots, \pi - 1, \tag{8}$$

$$\hat{n}_{i,\pi} = \hat{n}_{i,F} - \sum_{m=1}^{\pi-1} \hat{n}_{i,m} \quad \text{for } i = 1, \ldots, c-r, \tag{9}$$

where $\hat{n}_{i,F} = n_{i,F} - v_i N^{-1} n_{\mathrm{ref},F}$, $\hat{n}_{T,F} = n_{T,F} - v_{TOT} N^{-1} n_{\mathrm{ref},F} = \sum_{i=1}^{c-r} \hat{n}_{i,F}$, $n_{T,F} = \sum_{i=1}^{c} n_{i,F}$ and $n_{i,F}$ is the number of moles of component i in the feed. Note that $Z_i = \hat{n}_{i,F}/\hat{n}_{T,F}$.

Using the formulation in Eqs. (7) to (9), all trial compositions will satisfy the mass balances (Eq. (2)), the optimization problem will have only bounds on decision variables and no other constraints, allowing the easy application of stochastic global optimization strategies. Recently, Bonilla-Petriciolet *et al.* (2008) have reported the application of simulated annealing (SA) for the minimization of the transformed Gibbs free energy minimization using the above formulation, and concluded that SA is reliable for this purpose. However, this method required significant computational effort. The study of Bonilla-Petriciolet *et al.* (2008) is the first attempt to apply a stochastic global method to the optimization of Gibbs free energy using transformed composition variables, and it is desirable to test other stochastic methods for this application. Hence, in this chapter, DE and TS are tested and compared for phase equilibrium calculations of reactive systems via Gibbs free energy minimization and transformed composition variables.

2.2. Description of phase stability problem

Phase stability analysis allows identification of the thermodynamic state that corresponds to the global minimum of the Gibbs free energy (i.e. globally stable equilibrium). A phase at a given temperature, pressure and composition, is stable if and only if the Gibbs free energy surface is at no point below the tangent plane to the surface at the given composition (Baker *et al.*, 1982; Michelsen, 1982). This statement is a necessary and sufficient condition for global stability. Generally, stability analysis in reactive systems is performed by the minimization of the distance between the Gibbs free energy surface and the tangent plane to the surface at the given composition (known as the tangent plane distance function, TPDF) with respect to all possible phase compositions, using mole numbers or fractions as decision variables. Non-negativity of the global minimum of this function implies that the given phase is stable. Note that stability calculations in reactive mixtures using TPDF must be performed only for phases that are chemically equilibrated (Michelsen, 1982; Castier *et al.*, 1989). This is because, in reactive systems, the total number of moles (n_T) present at any instant of time may not remain constant as the reactions proceed. Thus, it is not meaningful to directly minimize TPDF to study stability in reactive mixtures, and the proposed methods execute a chemical equilibration procedure before phase stability calculations.

Similar to the phase equilibrium calculations, phase stability analysis in reactive systems is a challenging global optimization problem because the objective function is multivariable, non-convex and highly non-linear. Many optimization methods have been tried for stability calculations in reactive systems using the classical tangent plane criterion (e.g. Gupta *et al.*, 1991; Stateva and Wakeham, 1997; Jalali and Seader, 1999; Jalali *et al.*, 2008). Traditional approaches, which generally utilize different initial estimates and local search methods, may converge to a local or trivial solution based on the initial guess. Several deterministic global optimization methods have also been tested for phase stability analysis in reactive systems; these include branch-and-bound methods (McDonald and Floudas, 1996; McKinnon and Mongeau, 1998), homotopy continuation algorithms (Jalali-Farahani and Seader, 2000) and the interval-Newton/general-bisection approach (Burgos-Solorzano *et al.*, 2004).

Since the tangent plane criterion is also applicable to chemically equilibrated phases, any method proposed for stability calculations in non-reactive systems can be extended to reactive mixtures. Hence, stochastic optimization methods such as SA, GA, TS and DE can be used for this purpose (Rangaiah, 2001; Teh and Rangaiah, 2003; Bonilla-Petriciolet *et al.*, 2006; Srinivas and Rangaiah, 2007). Further, it is possible to reformulate the tangent plane criterion in terms of transformed composition variables. This modified stability criterion, adopted for the present study, and its advantages are described below.

2.2.1. Formulation of the optimization problem

Based on the transformed variables of Ung and Doherty (1995a, 1995b), Wasylkiewicz and Ung (2000) introduced the reactive tangent plane distance function (RTPDF) for multi-component and multi-reaction systems, which is defined as

$$RTPDF = \sum_{i=1}^{c-r} X_i \left(\left.\frac{\Delta\mu_i}{RT}\right|_X - \left.\frac{\Delta\mu_i}{RT}\right|_Z \right), \tag{10}$$

where $\left.\frac{\Delta\mu_i}{RT}\right|_X$ and $\left.\frac{\Delta\mu_i}{RT}\right|_Z$ are the chemical potentials of component i calculated at the transformed mole fractions X and Z, respectively. The RTPDF represents the vertical displacement from the tangent plane at the given composition Z, to the transformed molar Gibbs free energy surface at the composition X.The necessary and sufficient condition for global phase stability is given by RTPDF ≥ 0.0 for any X in the whole transformed composition space. There is at least one solution for RTPDF ($= 0.0$) at the trivial stationary point $X = Z$. For a stable phase, this must be the global minimum of RTPDF.

RTPDF and its stationary points for arbitrary stable and unstable reactive mixtures are illustrated in Fig. 1. In the upper-left plot of this figure, the tangent to the transformed Gibbs free energy at Z lies below this free energy surface throughout the transformed composition space, and, as a consequence, a single phase with the transformed composition Z is stable (i.e. the global minimum of RTPDF is 0.0 at Z, as shown in the lower-left plot). In another case shown in the upper-right plot, the tangent at Z crosses the Gibbs free energy surface at several points; therefore, a phase with this transformed composition is unstable. This unstable mixture shows several stationary points for RTPDF (see the lower-right plot).

The minimization problem for phase stability via tangent plane criterion is to minimize Eq. (10) with respect to X_i (for $i = 1, \ldots, c - r$)

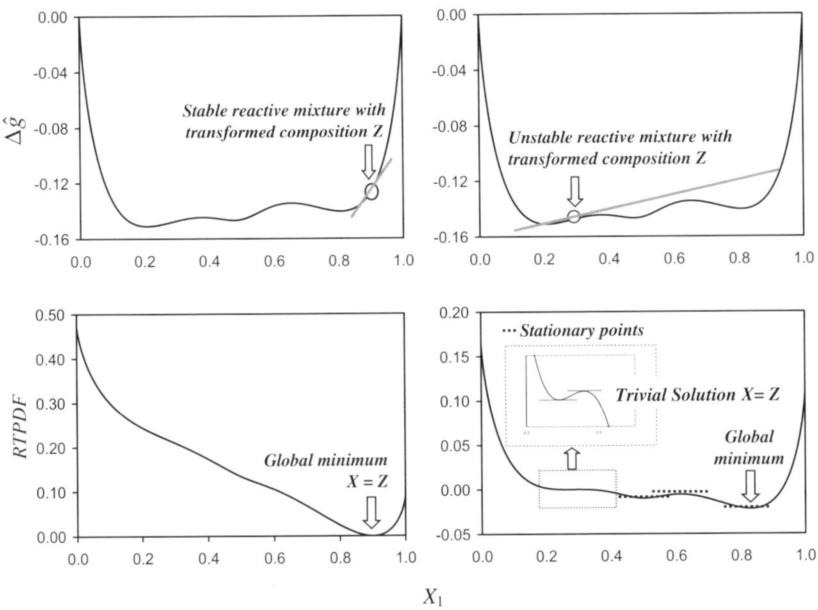

Figure 1. Gibbs free energy of mixing, $\Delta\hat{g}$ and RTPDF for a reactive mixture where $c - r = 2$, illustrating the transformed mole fraction Z of stable and unstable reactive mixtures, and their corresponding stationary points and global minimum of RTPDF. Transformed mole fraction of one component (X_1) is shown on the x-axis and that of the second component is $X_2 = 1.0 - X_1$.

subject to $\sum_{i=1}^{c-r} X_i = 1$. Instead of λ's used as decision variables in phase equilibrium problems to automatically satisfy the equality constraints, an alternate strategy is employed in phase stability problems. Here, $(c - r)$ decision variables, each in the range 0.0 to 1.0, are generated and then normalized so that their sum is equal to unity. The resulting variable values are used for calculating the objective, RTPDF. This strategy will not reduce the number of decision variables but this is not important since number of variables in phase stability problems is less than that in phase equilibrium problems. Note that the chemical potentials in Eq. (10) are functions of X_i, which depend on the thermodynamic model used for predicting the phase behavior. For illustration, phase stability problem for the simple case of an ideal ternary mixture is formulated in Appendix A.

The RTPDF offers two main advantages over the classical TPDF in the stability analysis of reactive systems (Ung and Doherty, 1995b;

Wasylkiewicz and Ung, 2000). First, given the initial composition of the reactive mixture, one can perform the stability test directly using the RTPDF, in contrast to the classical TPDF, where a preliminary chemical equilibration procedure is required. In fact, this equilibration procedure is implicit if the transformed variables are used. Another advantage of this approach is the significant reduction of problem dimensionality for multi-reaction systems.

A few studies have dealt with the global solution of RTPDF. Wasylkiewicz and Ung (2000) applied the homotopy continuation approach to locate all stationary points of RTPDF. Bonilla-Petriciolet *et al.* (2006) reported the application of stochastic optimization methods: SA, very fast SA, modified direct search SA and stochastic differential equations, for the global minimization of RTPDF. Among the methods tested, they found that SA is a reliable strategy for stability calculations in reactive mixtures; but its computational effort is very high for multi-component systems. These studies are the first attempts for the global solution of RTPDF. Hence, efficacy of TS and DE for the global minimization of RTPDF is studied in this chapter.

3. DE and TS, and Parameter Tuning

Introduction to and description of DE and TS are provided in earlier chapters. In the present study, FORTRAN codes developed by Teh and Rangaiah (2003), and Srinivas and Rangaiah (2007) for TS and DE algorithms, respectively, were used. Both methods have been implemented in combination with a local optimization technique for finding the global minimum accurately and efficiently. Figure 2 shows the flowchart of these hybrid algorithms; see Teh and Rangaiah (2003) and Srinivas and Rangaiah (2007) for the description of these algorithms. For local optimization, we have used a fast convergent quasi-Newton method implemented in the subroutine DBCONF from IMSL library. This subroutine calculates the gradient via finite differences and approximates the Hessian matrix (consisting of second derivatives of the objective function) according to BFGS formula. In brief, a quasi-Newton method is a modification of Newton method without the need for second derivatives of the objective function. For more details on this local strategy, see the optimization book by Dennis and Schnabel (1983). The default values of DBCONF parameters in the IMSL library

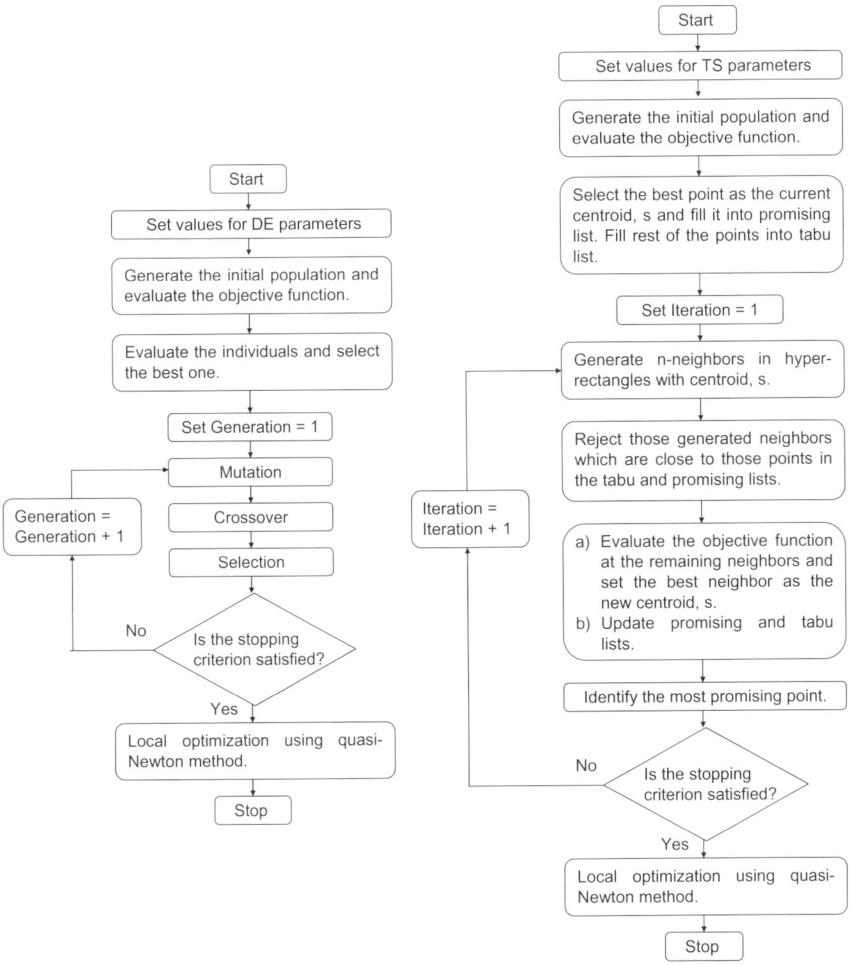

Figure 2. Flowcharts of DE (on the left) and TS (on the right).

were used in our calculations. The procedure used for tuning of DE and TS parameters is described below.

3.1. Tuning of TS and DE parameters using performance profiles

Like any stochastic method, TS and DE have a number of parameters that need to be heuristically tuned for any desired application. This tuning is a key step in the application and evaluation of a stochastic method, where a proper selection of its parameters is essential to achieve the best reliability and efficiency of the method. Parameter tuning is generally performed using benchmark problems and determining the average values of specific performance metrics. However, using this approach, a small number of the most difficult problems may tend to dominate the results for performance metrics and, as a consequence, lead to incorrect conclusions about the algorithm performance (Dolan and More, 2002). To address this shortcoming, we have employed the performance profile (PP) reported by Dolan and More (2002) to tune the parameters of both TS and DE methods.

Dolan and More (2002) introduced PP as an alternative tool for evaluating and comparing the performance of optimization software. In particular, PP has been proposed to compactly and comprehensively represent the data collected from a set of solvers for a specified performance metric. For instance, number of function evaluations (NFE) or computing time (CPUt) can be considered performance metrics for solver comparison. The PP plot allows visualization of the expected performance differences among several solvers and to compare the quality of their solutions by eliminating the bias of failures obtained in a small number of problems.

To introduce PP, consider n_s solvers (i.e. optimization methods) to be tested over a set $\Re$ of n_p problems. For each problem p and solver s, the performance metric t_{ps} must be defined. In our study, reliability of stochastic methods in accurately finding the global optimum of the objective function is considered as the principal goal, and hence the performance metric is defined as

$$t_{ps} = \hat{f}_{\text{calc}} - f^*, \tag{11}$$

where f^* is the known global optimum of the objective function and $\hat{f}_{\text{calc}}$ is the mean value of that objective function calculated by the stochastic method over several runs. In this study, $\hat{f}_{\text{calc}}$ is calculated from 100 runs to solve each test problem by each solver; note that each run is different because of random number seed used and the stochastic nature of the method. So, the focus is on the average performance of stochastic methods, which is desirable (Ali *et al.*, 2005).

For the performance metric of interest, the performance ratio, r_{ps} is used to compare the performance on problem p by solver s with the best performance by any solver on this problem. This performance ratio is given by

$$r_{ps} = \frac{t_{ps}}{\min\{t_{ps} : 1 \leq s \leq n_s\}}. \tag{12}$$

The value of r_{ps} is 1 for the solver s that performs the best on a specific problem p. For example, if $r_{ps} = 2$, the performance metric of solver s on problem p is twice the best value found by another solver on the same problem p.

To obtain an overall assessment of the performance of solver s on n_p problems, the following cumulative function for r_{ps} is used:

$$\rho_s(\varsigma) = \frac{1}{n_p}\, size\, \{p \in \Re : r_{ps} \leq \varsigma\}, \tag{13}$$

where $\rho_s(\varsigma)$ is the fraction of the total number of problems, for which solver s has a performance ratio r_{ps} within a factor $\varsigma \in (1, \infty)$ of the best possible ratio. The PP of a solver is a plot of $\rho_s(\varsigma)$ versus ς; it is a non-decreasing, piece-wise constant function, continuous from the right at each of the breakpoints (Dolan and More, 2002).

Note that $1.0 - \rho_s(\varsigma)$ is the fraction of problems that solver s failed to solve within a factor ς of the best method. This implies that, if the set of n_p problems is reasonably large and representative of circumstances that are likely to occur in the desired application, then solvers with larger $\rho_s(\varsigma)$ are better. If the objective is to identify the best solver, it is only necessary to compare the values of $\rho_s(1)$ for all solvers and to select the highest one, which is the probability that a specific solver will "win" over the rest of solvers used. For the performance metric in Eq. (11), the PP plot compares how accurately the stochastic methods can find the global optimum value relative to one another, and so the term "win" refers to the stochastic method that provides the most accurate value of the global optimum in reactive phase stability and equilibrium problems. We have calculated several PPs, and results are used for tuning TS and DE parameters for reactive phase stability and equilibrium calculations.

Parameters tuned in this study are: (a) DE: amplification factor (A), crossover constant (CR), population size (NP), maximum number of generations ($Gen_{\max}$), and maximum number of generations without improvement in the best function value ($Sc_{\max}$); and (b) TS: tabu list size (N_t), promising list size (N_p), tabu radius (ε_t), promising radius (ε_p), length of the hyper-rectangle (h_n), initial population size (NP_{init}), and maximum number of iterations ($Iter_{\max}$) and $Sc_{\max}$. Note that $Gen_{\max}$, $Iter_{\max}$ and $Sc_{\max}$ are the stopping criteria for DE and TS. The parameter tuning was performed using the stochastic methods without a local search algorithm. This procedure was carried out by varying one parameter at a time, while the remaining parameters were fixed at their nominal values. The tested and nominal values for each strategy are given in Table 1. Some algorithm parameters were associated to n_{var} (i.e. number of decision variables), and

Table 1. Parameters tested for tuning of TS and DE for the global minimization of $\Delta\hat{g}$ and RTPDF.

Method	Parameter	Tested values[a]	Nominal value
DE	Amplification factor A	0.3, 0.5, 0.7	0.5
	Crossover constant CR	0.1, 0.5, 0.9	0.1
	Population size NP	$5n_{var}$, $10n_{var}$, $25n_{var}$, $50n_{var}$	$10n_{var}$
	Maximum number of generations $Gen_{\max}$	50, 100, 200	50
	Maximum number of generations without improvement in the best function value $Sc_{\max}$	$6n_{var}$, $12n_{var}$	$12n_{var}$
TS	Tabu list size N_t	5, 10, 20	10
	Promising list size N_p	5, 10, 20	10
	Tabu radius ε_t	0.005, 0.01, 0.02	0.01
	Promising radius ε_p	0.005, 0.01, 0.02	0.01
	Length of the hyper-rectangle h_n	0.25, 0.5, 0.75	0.5
	Initial population size NP_{init}	$10n_{var}$, $20n_{var}$, $30n_{var}$	$20n_{var}$
	Maximum number of iterations $Iter_{\max}$	$25n_{var}$, $50n_{var}$, $100n_{var}$	$50n_{var}$
	Maximum number of iterations without improvement in the best function value $Sc_{\max}$	$2n_{var}$, $6n_{var}$, $12n_{var}$	$6n_{var}$
	Number of neighbors, N_{neigh} subject to a minimum of 10 and a maximum of 30.	$2n_{var}$	$2n_{var}$

[a] n_{var} is the number of decision variables in the optimization problem.

all values tested were chosen based on the results reported by Teh and Rangaiah (2003) and Srinivas and Rangaiah (2007).

For comparing the algorithm efficiency, NFE and CPUt were used as measures of computational effort. All calculations were performed on the Intel Pentium M 1.73 GHz processor with 504 MB of RAM. This computer performs 254 million floating point operations per second (MFlops) for the LINPACK benchmark program (available at http://www.netlib.org/; accessed in July, 2008) for a matrix of order 500.

4. Phase Equilibrium Calculations in Reactive Systems using DE and TS

4.1. Benchmark problems for parameter tuning

For parameter tuning, we have considered three reactive systems that are benchmarks in the studies on RSPs; they are: (a) the reaction for butyl acetate production at 298.15 K and 1 atm, (b) methyl *tert*-butyl ether (MTBE) reaction system with inert at 10 atm and 373.15 K, and (c) the reactive system for *tert*-amyl methyl ether (TAME) synthesis at 335 K and 1.5 atm. Thermodynamic models, parameters and transformed variables for these reactive systems are reported in Table 2, and Tables B1 - B3 of Appendix B. For variable transformation, the reference components are arbitrarily selected, and we have $c - r = 3$ transformed composition variables for all systems.

In the three problems, the number of phases existing at the equilibrium is assumed to be known *a priori*, but the phase type (vapor or liquid) is *unknown*. Bisection method is used to perform the composition transformation: $X \rightarrow x$ (see Appendix A), for objective function evaluation. Note that x obtained from this transformation satisfy the stoichiometry requirements and are chemically equilibrated. Further, the chemical potentials in $\Delta\hat{g}$ and RTPDF (Eqs. 1 and 10) are given by:

$$\frac{\Delta\mu_i}{RT} = \frac{\mu_i - \mu_i^0}{RT} = \ln\left(\frac{x_i\hat{\varphi}_i}{\varphi_i}\right) = \ln(x_i\gamma_i), \tag{14}$$

where μ_i^0 is the chemical potential of pure component i, $\hat{\varphi}_i$ is the fugacity coefficient of component i in the mixture, φ_i is the fugacity coefficient of pure component, γ_i is the activity coefficient of component i in the mixture, and x_i is the mole fraction of component i in the mixture.

Table 2. Details of the selected reactive mixtures for parameter tuning of TS and DE for reactive phase equilibrium problems.

Mixture	Phases and thermodynamic models (see Appendix B for parameter values)	Transformed mole fractions
$A_1 + A_2 \leftrightarrow A_3 + A_4$ (1) Acetic Acid (2) n-Butanol (3) Water (4) n-Butyl acetate	Liquid-liquid equilibrium, UNIQUAC model. $\ln K_{eq,1} = \frac{450}{T} + 0.8$ where T is in K. Model parameters are taken from Wasylkiewicz and Ung (2000).	$X_1 = x_1 + x_4$ $X_2 = x_2 + x_4$ $X_3 = x_3 - x_4 = 1 - X_1 - X_2$ Reference component: A_4
$A_1 + A_2 \leftrightarrow A_3$, and A_4 as an inert (1) Isobutene (2) Methanol (3) Methyl ter-butyl ether (4) n-Butane	Vapor-liquid equilibrium, Wilson model and ideal gas. $\Delta G^o_{rxs}/R = -4205.05 + 10.0982T - 0.2667T \ln T$ $\ln K_{eq,1} = \frac{-\Delta G^0_{rxs}}{RT}$ where T is in K. Model parameters are taken from Ung and Doherty (1995a).	$X_1 = \frac{x_1+x_3}{1+x_3}$ $X_2 = \frac{x_2+x_3}{1+x_3}$ $X_4 = \frac{x_4}{1+x_3} = 1 - X_1 - X_2$ Reference component: A_3
$A_1 + A_2 + 2A_3 \leftrightarrow 2A_4$ (1) 2-Methyl-1-butene (2) 2-Methyl-2-butene (3) Methanol (4) *Tert*-amyl methyl ether	Vapor-liquid equilibrium, Wilson model and ideal gas. $K_{eq,1} = 1.057 \cdot 10^{-04} e^{4273.5/T}$ where T is in K. Model parameters are taken from Bonilla-Petriciolet *et al.* (2008).	$X_1 = \frac{x_1+0.5x_4}{1+x_4}$ $X_2 = \frac{x_2+0.5x_4}{1+x_4}$ $X_3 = \frac{x_3+x_4}{1+x_4} = 1 - X_1 - X_2$ Reference component: A_4

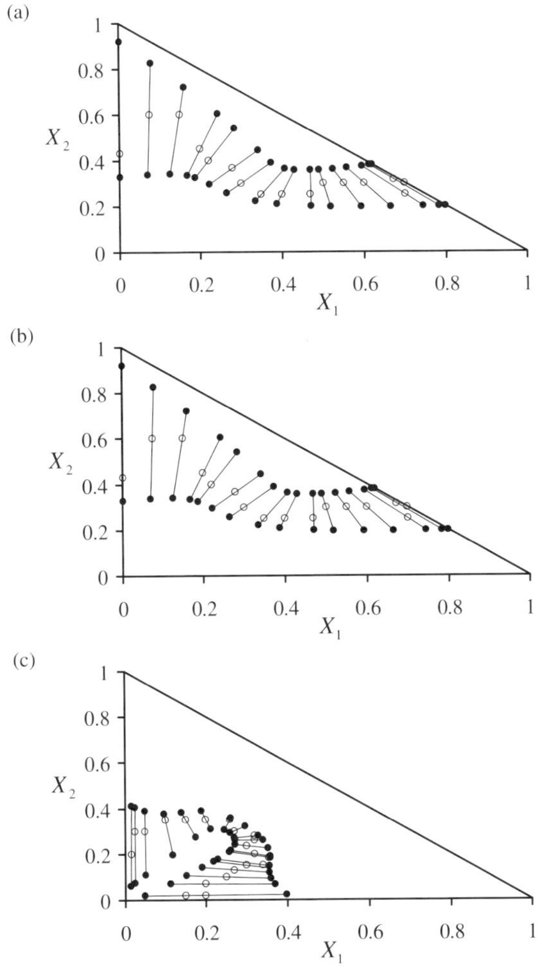

Figure 3. Transformed feed (○) and tie-lines (•) used for parameter tuning of TS and DE in reactive phase equilibrium problems. Reactive systems are: (a) acetic acid + *n*-butanol ↔ water + n-butyl acetate at 298.15 K and 1 atm, (b) isobutene + methanol ↔ methyl *tert*-butyl ether with *n*-butane as an inert at 10 atm and 373.15 K, and (c) 2-methyl-1-butene + 2-methyl-2-butene + 2 methanol ↔ 2 *tert*-amyl methyl ether at 335 K and 1.5 atm.

Feed compositions and tie-lines used to tune the parameters of TS and DE are shown in Fig. 3. We have utilized $n_p = 48$ different feeds, selected from the three reactive mixtures, to calculate PPs. The objective function has at least one local minimum, which corresponds to a trivial solution

($Z_i = X_i$ in phase stability and $Z_i = X_{i,1} = \cdots = X_{i,\pi}$ in equilibrium calculations), for all tested conditions. Note that different models are used in the calculation of thermodynamic properties for different phases (vapor and liquid) of reactive systems (b) and (c). The use of multiple thermodynamic models increases the complexity of phase stability and equilibrium calculations (Xu *et al.*, 2005); since there will be different $\Delta\hat{g}$ functions for different types of phases, evaluation of any trial composition must be done on the surface with the lowest value of transformed Gibbs free energy. Thus, the procedure of variable transformation is performed for each model at specified X and its corresponding $\Delta\hat{g}$ is evaluated. The model having the lowest value of transformed Gibbs free energy is then used for evaluating the objective function. Finally, the selected conditions involve feed compositions near phase boundaries, which are generally challenging for any algorithm. So, the tested conditions are appropriate for tuning the parameters of TS and DE.

In the three reactive systems considered for parameter tuning, $c = 4$ and $r = 1$ (see Table 2). Hence, there are three decision variables, $\lambda_{i,1} \in (0, 1)$ in the Gibbs energy minimization for phase equilibrium calculations. In the problems to test stability of the feed, depending upon the permissible values for X, we have considered different transformed variables and the equality constraint $\sum_{i=1}^{c-r} X_i = 1$. For the reactive system (a), RTPDF is minimized with respect to $X_1, X_2 \in (0, 1)$ only, and $X_3 = 1 - X_1 - X_2$. For phase stability calculations in reactive systems (b) and (c), three transformed fractions $X_i \in (0, 1)$ are used as decision variables, after their normalization to unity: $\bar{X}_i = X_i / \sum_{i=1}^{c-r} X_i$ where $\bar{X}_i$ is the normalized transformed mole fraction of component i.

4.2. Results of parameter tuning

Table 3 shows the performance summary for tuning the parameters of TS and DE; in this table, we report $\rho_s(1)$, NFE and CPUt obtained in the global minimization of RTPDF and $\Delta\hat{g}$. Here, NFE and CPUt are the mean of 100 runs for 48 feeds, and the range given for NFE and CPUt in each row of Table 3 is for the three problems solved. Depending on the parameter values, NFE varied from 390 to 7626 for stability calculations by DE, while it varied from 105 to 801 for stability calculations by TS; this computational

Table 3. Results for parameter tuning of TS and DE for the global minimization of $\Delta\hat{g}$ and RTPDF.

	Parameter tuning		Probability, $\rho_s(1)$ for		NFE for		CPUt (s) for	
Method	Parameter	Value[a]	RTPDF	$\Delta\hat{g}$	RTPDF	$\Delta\hat{g}$	RTPDF	$\Delta\hat{g}$
DE	A	0.3	0.542	0.625	769–1528	1471–1524	0.17–0.53	0.43–0.93
		0.5	0.208	0.292	779–1524	1457–1523	0.18–0.53	0.44–0.94
		0.7	0.250	0.083	772–1527	1462–1515	0.17–0.53	0.44–0.94
	CR	0.1	0.0	0.0	779–1524	1457–1523	0.18–0.53	0.44–0.94
		0.5	0.042	0.021	968–1530	1513–1530	0.16–0.53	0.44–0.94
		0.9	0.958	0.979	1020–1530	1511–1530	0.16–0.51	0.45–0.95
	NP	$5n_{var}$	0.0	0.0	396–762	730–758	0.09–0.28	0.21–0.48
		$10n_{var}$	0.0	0.0	779–1524	1457–1523	0.18–0.53	0.44–0.94
		$25n_{var}$	0.0	0.0	1830–3815	3666–3804	0.28–1.28	1.07–2.34
		$50n_{var}$	1.0	1.0	3672–7626	7239–7579	0.69–2.54	2.01–4.69
	Gen_{max}	50	0.021	0.0	779–1524	1457–1523	0.18–0.53	0.44–0.94
		100	0.292	0.396	805–2845	2058–2602	0.22–0.85	0.69–1.58
		200	0.688	0.604	825–5006	2160–3396	0.26–1.17	0.71–1.96
	Sc_{max}	$6n_{var}$	0.0	0.0	390–1291	1015–1220	0.09–0.48	0.42–0.72
		$12n_{var}$	1.0	1.0	779–1524	1457–1523	0.18–0.53	0.44–0.94

(*Continued*)

Table 3. (*Continued*)

Method	Parameter tuning		Probability, $\rho_s(1)$ for		NFE for		CPUt (s) for	
	Parameter	Value[a]	RTPDF	$\Delta\hat{g}$	RTPDF	$\Delta\hat{g}$	RTPDF	$\Delta\hat{g}$
TS	$N_t = N_p$	5	0.771	0.813	203–636	484–680	0.03–0.16	0.15–0.41
		10	0.083	0.083	194–455	385–521	0.04–0.13	0.11–0.30
		20	0.146	0.104	193–442	393–522	0.03–0.12	0.14–0.25
	$\varepsilon_t = \varepsilon_p$	0.005	0.875	0.938	226–498	408–681	0.03–0.14	0.14–0.35
		0.01	0.125	0.042	194–455	385–521	0.04–0.13	0.11–0.30
		0.02	0.0	0.021	172–418	347–408	0.03–0.12	0.11–0.26
	h_n	0.25	0.229	0.333	182–444	369–511	0.03–0.13	0.10–0.30
		0.5	0.396	0.375	194–455	385–521	0.04–0.13	0.11–0.30
		0.75	0.375	0.292	208–456	390–543	0.02–0.12	0.11–0.29
	NP_{init}	$10n_{var}$	0.250	0.250	183–430	361–519	0.03–0.12	0.09–0.27
		$20n_{var}$	0.438	0.458	194–455	385–521	0.04–0.13	0.11–0.30
		$30n_{var}$	0.313	0.292	214–462	417–541	0.03–0.12	0.11–0.30
	$Iter_{\max}$	$25n_{var}$,	0.333	0.229	192–443	375–487	0.03–0.12	0.08–0.30
		$50n_{var}$	0.396	0.417	194–455	385–521	0.04–0.13	0.11–0.30
		$100n_{var}$	0.271	0.375	196–459	393–532	0.03–0.11	0.12–0.42
	$Sc_{\max}$	$2n_{var}$	0.0	0.0	105–223	177–241	0.01–0.06	0.07–0.11
		$6n_{var}$	0.063	0.0	194–455	385–521	0.04–0.13	0.11–0.30
		$12n_{var}$	0.938	1.0	296–801	656–849	0.05–0.24	0.13–0.53

[a] Remaining parameters were fixed at nominal values, see Table 1.

effort corresponds to 0.09–2.54 s of CPUt for DE, and 0.01–0.24 s for TS (see Table 3). For Gibbs free energy minimization, NFE varied from 730 to 7579 (0.21–4.69 s) for DE, and from 177 to 849 (0.07–0.53 s) for TS, depending on the parameter values.

Since we are only interested in the optimal values of parameters of TS and DE that find the best solution of all tested conditions, only $\rho_s(1)$ is given in Table 3 and not the PPs. However, for illustrative purposes, several PPs obtained for TS and DE are shown in Figs. 4 and 5, which displays the fraction of phase stability and equilibrium problems solved by TS and DE with different parameter values, within a factor ς of the best found solution. These PPs show the effect of a parameter on the relative performance of TS and DE, and enable to find the optimal values of parameters. The results in Table 3 show that algorithm reliability increases with NP, $Gen_{\max}$, $Iter_{\max}$ and $Sc_{\max}$, but at the expense of computational effort. The PPs also show that optimal values of $A = 0.3$ and $CR = 0.9$ for DE agree with those

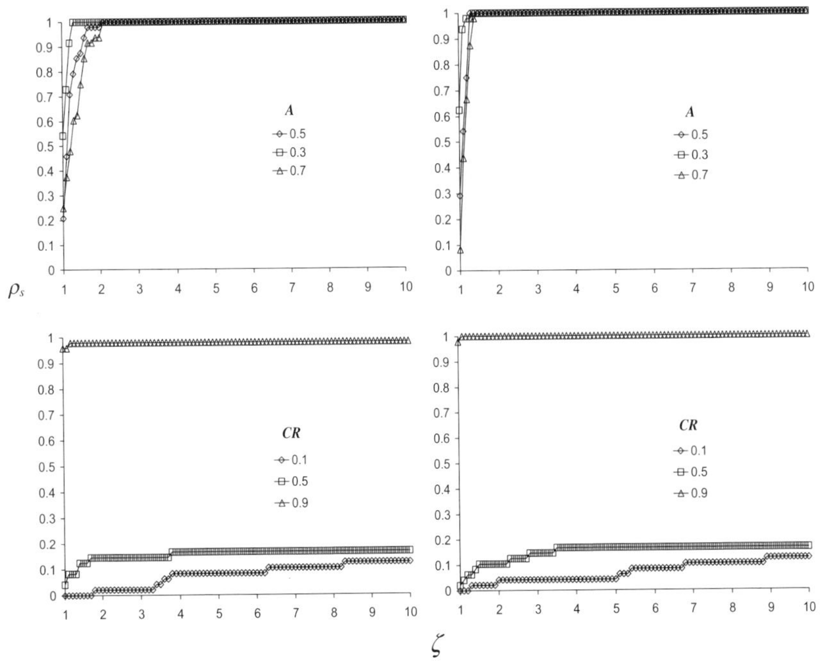

Figure 4. Performance profiles for algorithm parameters of DE in reactive phase stability (RTPDF: plots in the left column) and equilibrium calculations ($\Delta\hat{g}$: plots in the right column).

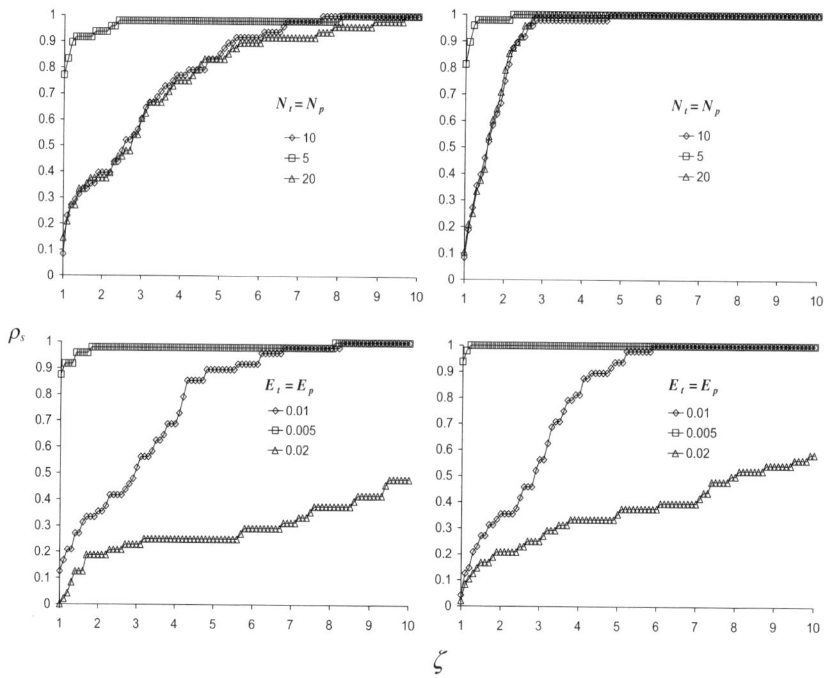

Figure 5. Performance profiles for algorithm parameters of TS in reactive phase stability (RTPDF: plots in the left column) and equilibrium calculations ($\Delta\hat{g}$: plots in the right column).

reported by Srinivas and Rangaiah (2007) for both the stability and equilibrium problems in non-reactive systems. Further, PPs indicate that *CR* is an important parameter to improve the performance of DE (Fig. 4). Finally, results on TS suggest that lower values for $N_t = N_p$ and $\varepsilon_t = \varepsilon_p$ favor the reliability of the method but increase the computational effort (see Table 3 and Fig. 5). The parameters: $h_n = 0.5$ and $NP_{\text{init}} = 20n_{var}$ appear to be optimal for reliability, and are in agreement with those reported by Srinivas and Rangaiah (2007) for the case of non-reactive systems.

5. Applications

Having selected the optimal values of parameters in TS and DE, two reactive systems are analyzed to test and compare their performance for reactive phase stability and equilibrium problems. First system has been selected to reflect different problem difficulties: a challenging objective function with a local minimum value very comparable to the global minimum value, and multi-reaction conditions with the presence of inert components. The second example is a reactive mixture widely used in the literature to test algorithms in the simultaneous calculation of chemical and phase equilibrium (e.g. Xiao *et al.*, 1989; Lee *et al.*, 1999; Jalali *et al.*, 2008).

Note that NP, NP_{init}, $Gen_{\max}$, $Iter_{\max}$, and $Sc_{\max}$ are problem-dependent parameters, significantly affect the performance of DE and TS, and determine the trade-off between efficiency and reliability. The selection of proper values for these is a key step, which depends on the goals of the user for a specific application. Hence, in example 1 we have studied the reliability of TS and DE at different levels of computational effort by changing the values of stopping criteria ($Gen_{\max}$, $Iter_{\max}$, and $Sc_{\max}$). Optimal values identified using PPs for the remaining parameters of both methods were used in these calculations: (a) DE: $A = 0.3$ and $CR = 0.9$; (b) TS: $N_t = N_p = 5$, $\varepsilon_t = \varepsilon_p = 0.005$ and $h_n = 0.5$, respectively. In example 2, all parameters are fixed based on the results on the previous example. Phase stability and equilibrium problems in each example were solved 100 times each, using random initial values via different random number seed. The local search method has been used to refine the solution obtained from the stochastic algorithm. For illustrative purposes, FORTRAN programs developed for example 1 are included in the folder of this chapter on the CD.

5.1. Example 1

Consider a hypothetical multi-component reactive mixture undergoing three reactions:

$$\begin{gathered} A_3 \Leftrightarrow A_4 \\ A_5 \Leftrightarrow A_4, \\ A_4 \Leftrightarrow A_6 \end{gathered} \tag{15}$$

with A_1 and A_2 as inert components. This mixture presents a vapor-liquid equilibrium at $P = 1$ atm and $T = 60°C$, where both phases are ideal (see thermodynamic data and models in Table B4 of Appendix B). For simplicity, equilibrium constants for the three reactions are: $K_{eq,1} = 1.5$, $K_{eq,2} = 0.15$ and $K_{eq,3} = 0.35$. Selecting A_3, A_4 and A_5 as the reference components, transformed mole fractions are defined as:

$$X_1 = x_1, \tag{16}$$

$$X_2 = x_2, \tag{17}$$

$$X_6 = x_3 + x_4 + x_5 + x_6, \tag{18}$$

where all $X_i \in (0, 1)$.

Due to the ideal phases assumed for modeling this system, the transformation procedure for $X \to x$ is straightforward. The chemical equilibrium constants for this mixture are given by

$$\begin{aligned} K_{eq,1} &= \frac{a_4}{a_3} \\ K_{eq,2} &= \frac{a_4}{a_5}, \\ K_{eq,3} &= \frac{a_6}{a_4} \end{aligned} \tag{19}$$

where the activity of component i is given by: $a_i = x_i \gamma_i$ for the liquid phase or $a_i = x_i P / P_i^{sat}$ for the ideal vapor phase, and P_i^{sat} is the saturation pressure of the pure component i. Expressing the mole fractions in terms of x_4 using Eqs. (16)–(18), and substituting into the chemical equilibrium constraints (Eq. (19)), mole fraction of this reference component for the set of specified transformed variables can be obtained from:

$$x_4 = \frac{X_6}{1 + \frac{\theta_{43}}{K_{eq,1}} + \frac{\theta_{45}}{K_{eq,2}} + \frac{K_{eq,3}}{\theta_{64}}}, \tag{20}$$

where $\theta_{ij} = \gamma_i / \gamma_j = 1$ for the ideal liquid phase or $\theta_{ij} = P_j^{sat} / P_i^{sat}$ for the ideal vapor phase. Remaining components come from Eqs. (16) and (17) for x_1 and x_2; while x_3, x_5 and x_6 are obtained using

$$x_3 = \frac{x_4 \theta_{43}}{K_{eq,1}}, \tag{21}$$

$$x_5 = \frac{x_4 \theta_{45}}{K_{eq,2}}, \tag{22}$$

$$x_6 = 1 - \sum_{i=1}^{5} x_i. \tag{23}$$

Based on preliminary trials, we have chosen a challenging condition to test and compare the stochastic methods in the global minimization of RTPDF and $\Delta\hat{g}$. Specifically, we have determined the equilibrium compositions for a feed that corresponds to a liquid phase with initial composition $n_{i,F} = (0.6305, 0.00355, 0.1, 0.1, 0.1, 0.06595)$ and $n_{T,F} = \sum_{i=1}^{c} n_{i,F} = 1$. In transformed variables, we have $Z_i = (0.6305, 0.00355, 0.36595)$ and $\hat{n}_{T,F} = 1$. This feed is near a phase boundary and, as consequence, the objective function of both phase stability and equilibrium problems has a local minimum that is comparable to the global minimum (i.e. function values at the local and global minima are close to each other).

The global minimum of RTPDF is -0.001591 at $X_i = (0.708954, 0.004042, 0.287004)$ versus 0.0 at Z; and the global minimum of $\Delta\hat{g} = -1.853908$ is at vapor-liquid equilibrium with transformed compositions of $X_{i,1} = (0.625858, 0.003521, 0.370621)$ and $X_{i,2} = (0.704855, 0.004015, 0.291130)$, versus the local minimum of $\Delta\hat{g} = -1.853861$ at $Z_i = X_{i,1} = X_{i,2}$. When the objective function value of a local minimum is very close to that at the global minimum, premature convergence of an optimization method is likely. As a consequence, low reliability may be obtained under these conditions.

For this reactive mixture, the reactive tangent plane for stability criterion is defined as:

$$RTPDF = X_1\left(\left.\frac{\Delta\mu_1}{RT}\right|_X - \ln(z_1\gamma_1)\right) + X_2\left(\left.\frac{\Delta\mu_2}{RT}\right|_X - \ln(z_2\gamma_2)\right) + X_6\left(\left.\frac{\Delta\mu_6}{RT}\right|_X - \ln(z_6\gamma_6)\right), \tag{24}$$

where z_i is obtained from $Z \rightarrow z$ using Eqs. (20)–(23). The expression for the chemical potential at X should be for the phase with lower Gibbs free energy of mixing, and is:

$$\left.\frac{\Delta\mu_i}{RT}\right|_X = \ln\left(\frac{x_i P}{P_i^{sat}}\right) \quad \text{if } \Delta\hat{g}_2 < \Delta\hat{g}_1 \quad \text{else } \left.\frac{\Delta\mu_i}{RT}\right|_X = \ln(x_i\gamma_i). \tag{25}$$

Here, the transformed Gibbs free energy expressions for liquid ($\Delta\hat{g}_1$) and vapor ($\Delta\hat{g}_2$) are:

$$\Delta\hat{g}_1 = \sum_{\substack{i=1 \\ i\neq 3,4,5}}^{6} X_i \ln(x_i\gamma_i), \tag{26}$$

$$\Delta\hat{g}_2 = \sum_{\substack{i=1 \\ i\neq 3,4,5}}^{6} X_i \ln\left(\frac{x_i P}{P_i^{sat}}\right), \tag{27}$$

where x_i is the mole fraction of component i that satisfy the chemical equilibrium, which is obtained from $X \rightarrow x$ using Eqs. (20)–(23). Three decision variables X_1, X_2 and X_6 are used, after their normalization to unity, $\bar{X}_i = X_i/(X_1 + X_2 + X_6)$, in the global optimization of RTPDF. Thus, the unconstrained minimization problem that must be solved for phase stability can be expressed as:

$$\min_X RTPDF(X)$$
$$X_1, X_2, X_6 \in (0, 1). \tag{28}$$

For phase equilibrium calculations, the transformed Gibbs free energy function is:

$$\Delta\hat{g} = \hat{n}_{1,1}\frac{\Delta\mu_{1,1}}{RT} + \hat{n}_{2,1}\frac{\Delta\mu_{2,1}}{RT} + \hat{n}_{6,1}\frac{\Delta\mu_{6,1}}{RT} + \hat{n}_{1,2}\frac{\Delta\mu_{1,2}}{RT} + \hat{n}_{2,2}\frac{\Delta\mu_{2,2}}{RT} + \hat{n}_{6,2}\frac{\Delta\mu_{6,2}}{RT}, \tag{29}$$

where the chemical potentials of phases 1 and 2 are also given by Eqs. (25)–(27). Using Eq. (7) gives the following expressions for the decision variables:

$$\begin{aligned} \hat{n}_{1,1} &= 0.6305\lambda_{1,1} \\ \hat{n}_{2,1} &= 0.00355\lambda_{2,1}, \\ \hat{n}_{3,1} &= 0.36595\lambda_{3,1} \end{aligned} \tag{30}$$

where $\hat{n}_{i,2}$ is obtained from Eq. (9). Thus, the unconstrained minimization problem that must be solved for phase equilibrium calculations is:

$$\begin{array}{l} \min\limits_{\lambda_{i,1}} \Delta\hat{g} \\ \lambda_{1,1}, \lambda_{2,1}, \lambda_{3,1} \in (0, 1) \end{array}. \tag{31}$$

Phase stability test and equilibrium calculations have been carried out using different values for NP, NP_{init}, iterations (*Iter*) and generations (*Gen*). To assess the quality of solutions obtained as the search progresses and to capture the improvement in the objective function values and their variation in each run performed, individual quartile sequential plots (Ali *et al.*, 2005) are prepared. A quartile is any of the three values which divide the sorted data set into four equal parts, so that each part represents 1/4th of the sampled population. In our case, this population is the set of objective values obtained by the stochastic method over 100 independent runs. Then, a quartile sequential plot displays the progress of the 25th, 50th and 75th percentiles of the objective values calculated by the stochastic method versus *Iter* or *Gen*. Here, quartile sequential plots were obtained using the best objective function values, without using the local search method, recorded at different values of *Iter* or *Gen*. To complement the quartile results, we also present the plots of success ratio (SR) versus *Iter* or *Gen*, where SR is defined as the number of runs out of 100 that satisfy the condition $|f_{\text{calc}} - f^*| \leq 10^{-06}$ but using both DE and TS in combination with the quasi-Newton method. For the case of SR plots, the switch to the local method takes place once stochastic method has performed the specified number of *Iter* or *Gen*. In these calculations, *Iter* or *Gen* are used alone as a stopping criterion (i.e. without $Sc_{\max}$).

Figures 6 and 7 show the quartile sequential and SR plots for TS and DE for phase stability and equilibrium calculations in this example. In the quartile plots, the vertical bar at the k-th *Iter* or *Gen* represents a range of objective function values between the 25th and the 75th percentiles, while the symbol ($\times$, $\diamond$ or Δ) corresponds to the 50th percentile (i.e. the median of the objective function values which are calculated by the stochastic method over 100 trials). As expected, these results indicate that the algorithm performance is highly dependent on the maximum number of *Gen* or *Iter*. For the case of stability calculations, TS and DE improve the solution quality as

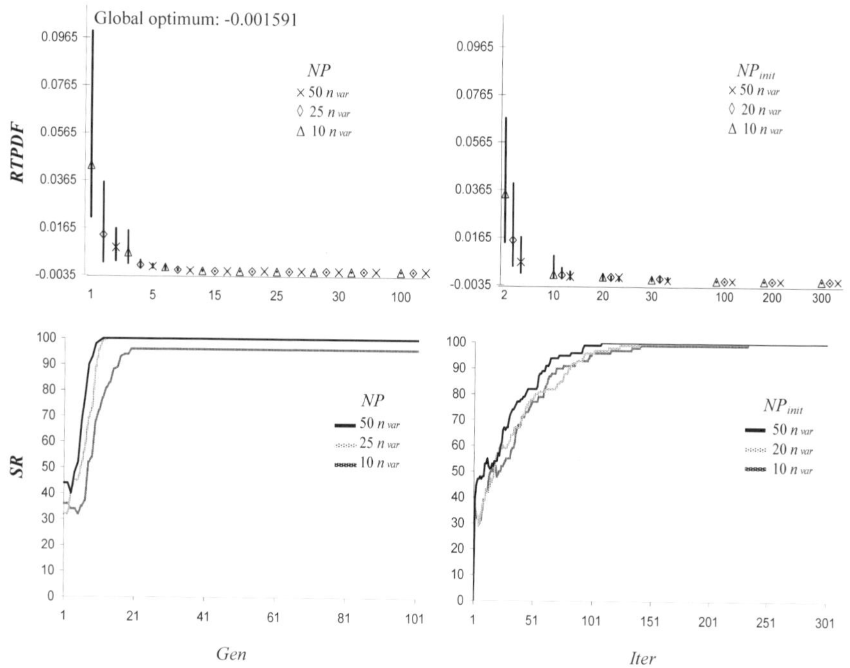

Figure 6. Quartile sequential and success rate plots of DE (in the left column) and TS (in the right column) in the global minimization of RTPDF for example 1.

Gen or *Iter* increases, irrespective of the values used for *NP* and NP_{init}. The variance of solutions obtained for the two stochastic methods is relatively large at the small values of *Iter* or *Gen*, and also for low values of *NP* and NP_{init} (see the broad percentile ranges in Fig. 6). As a consequence, both TS and DE, in combination with the local method, show low reliability (SR) in these conditions. However, with proper values for the maximum number of *Iter* or *Gen*, the global optimum of RTPDF can be found with a 100% SR.

Note that DE can reach a high SR (> 90%) at *Gen* > 15, unlike TS where more than 60 *Iter* are required to obtain an acceptable performance. Nevertheless, there are significant differences between the NFE of the two strategies because one generation in DE involves many more NFE compared to that in one iteration of TS (see Fig. 8). Comparing the reliability of both DE and TS followed by the quasi-Newton method in this stability problem at the same level of computational effort using NFE as reference

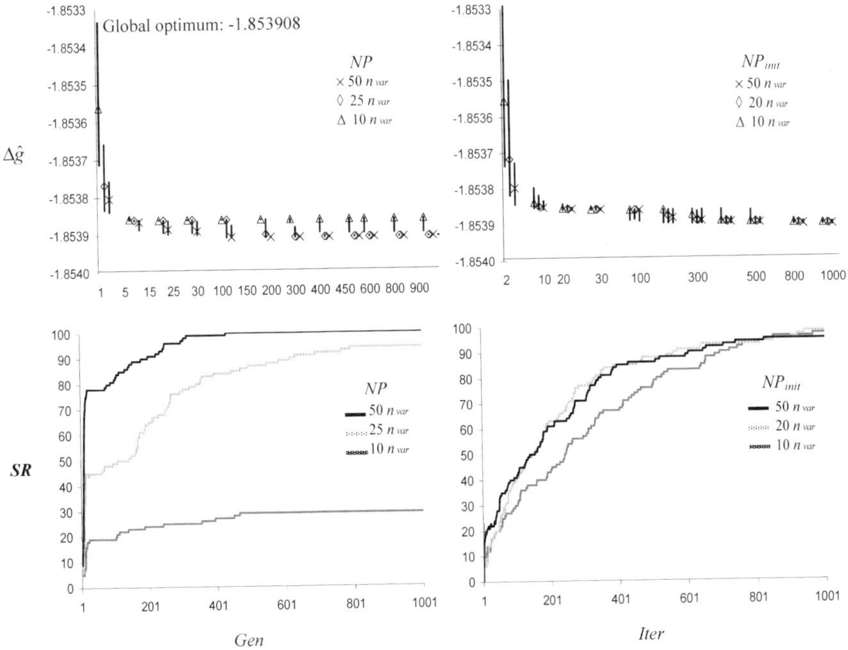

Figure 7. Quartile sequential and success rate plots of DE (in the left column) and TS (in the right column) in the global minimization of $\Delta \hat{g}$ for example 1.

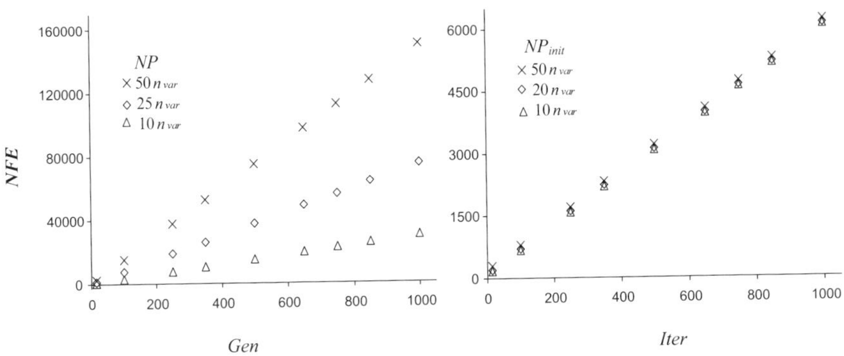

Figure 8. Mean computational effort of DE (right column) and TS (left column), each followed by the local method, in the global minimization of RTPDF and $\Delta \hat{g}$ for example 1.

(which includes the computational effort of both stochastic and local methods), reliability of TS is better than DE when NFE $< 1{,}000$, and vice versa for larger NFE. For the case of the stopping criterion $Sc_{\max}$, DE needs a mean value of 55, 119, and 156 generations to satisfy $Sc_{\max} = 2n_{var}$, $6n_{var}$ and $12n_{var}$, respectively. For the case of TS, these stopping conditions are reached in, on average, 18, 42 and 77 iterations. It is evident from Fig. 6 that the reliability of DE is better compared to TS if $Sc_{\max}$ is used alone as the convergence criterion.

Results for the minimization of $\Delta\hat{g}$ indicate that both DE and TS may be trapped at a local optimum depending on the algorithm parameters (see Fig. 7). In fact, this problem is very challenging due to the presence of trivial solution ($Z_i = X_{i,1} = X_{i,2}$) which has an objective value very near to the global minimum (-1.853861 and -1.853908, respectively). As before, the variance of solutions obtained from TS and DE is higher at low values of *NP*, NP_{init}, *Gen* and *Iter* and improves with increased values of these parameters. Quartile plots show that the objective function values are not improving throughout a significant amount of generations/iterations performed by TS and DE (Fig. 7). For the case of DE, combined with the local method, 100% SR is reached only for $NP = 50n_{var}$ and $Gen > 400$. In fact, this method can show improvement in the function values after getting stuck at a local optimum in early iterations. However, good performance is obtained at a significant computational cost: NFE $> 60{,}000$ (see Fig. 8). In this problem, *NP* plays an important role to determine the reliability of DE.

On the other hand, results show that TS followed by the quasi-Newton method gives poor performance at smaller values of *Iter* because it failed many times to locate the global minimum of the transformed Gibbs free energy (Fig. 7). The TS algorithm improves over further iterations and performs relatively well at $Iter > 600$ (NFE > 3700). However, it can only reach a maximum of 98% SR at 1,000 iterations and $NP_{\text{init}} = 20n_{var}$ (NFE $\approx 6{,}000$). There is no clear effect of NP_{init} on the reliability of TS, although it appears a large value for this parameter is better. Results show that DE requires an average of 38, 127, and 255 generations to fulfill $Sc_{\max} = 2n_{var}$, $6n_{var}$, and $12n_{var}$, while TS needs an average of 14, 37, and 70 iterations to satisfy these stopping conditions. As for phase stability problems, DE is more reliable than TS if $Sc_{\max}$ is used alone as the convergence criterion in phase equilibrium problems.

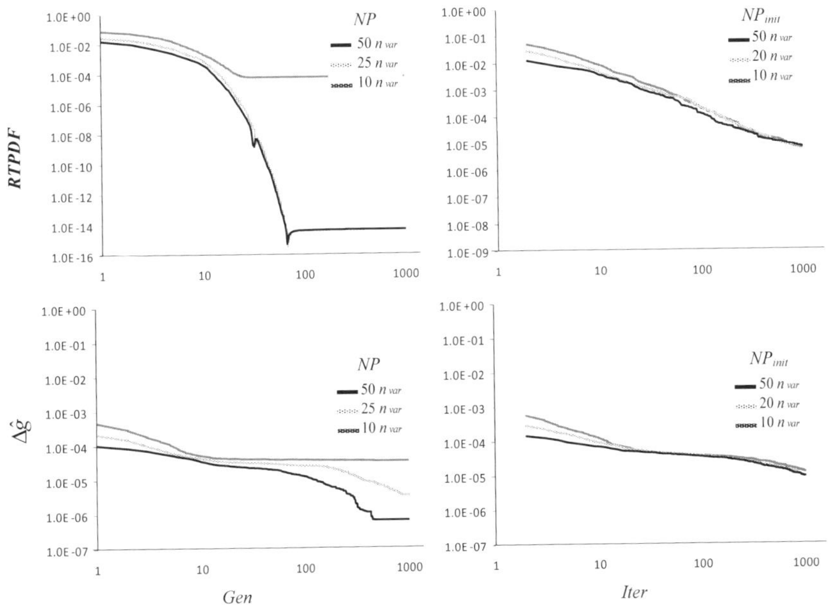

Figure 9. Convergence histories of the norm $|\hat{f}_{\text{calc}} - f^*|$ of DE (in the left column) and TS (in the right column) in the global minimization of RTPDF and $\Delta\hat{g}$ for example 1.

Figure 9 provides the convergence histories of the norm of $|\hat{f}_{\text{calc}} - f^*|$ for both TS and DE, *without* the local search method, in the global minimization of RTPDF and $\Delta\hat{g}$. This norm is based on the average (over 100 runs) of the best objective function $\hat{f}_{\text{calc}}$ recorded at each *Gen* and *Iter*. In general, the mean value of best solution obtained by TS and DE is nearer to the global minimum as the *Gen* and *Iter* increases. However, the norm reached by DE is lower than that achieved by TS in both stability and equilibrium calculations but at the expense of a significant computational effort (see Figs. 6–8).

For both TS and DE, the accuracy of final solutions is improved after the switch to the local optimization method. However, this accuracy is problem-specific because the objective function may be flat near the global minimum region, affecting the quality of solutions obtained by gradient-based methods combined with numerical derivatives (Srinivas and Rangaiah, 2007). In this reactive mixture, both TS and DE in combination with the local method can find very accurate solutions with an absolute error (i.e. $f_{\text{calc}} - f^*$) lower than 10^{-10} for phase stability calculations; in Gibbs free energy minimization, this absolute error is around 10^{-10} for DE and is slightly lower than 10^{-7} for TS. The use of direct search methods (e.g. Nelder–Mead simplex method) for local optimization at the end of TS and DE is a suitable option for handling objective functions that are flat near the global minimum. However, these local search strategies are known to be less efficient than the gradient-based methods. With respect to algorithm efficiency, CPUt ranged from 0.01 to 2.09 seconds for DE, and from 0.01 to 0.08 seconds for TS in all stability and equilibrium calculations performed.

Finally, we have solved the phase stability and equilibrium problems in this example using the quasi-Newton method alone with random initial values (100 runs for each problem). This local solver generally converged to trivial solutions, with only 42% SR for phase stability and 16% SR for equilibrium calculations. When the method converged to the global solution, it required an average of 351 NFE for phase stability and 156 NFE for equilibrium calculations (CPUt < 0.001 s). As expected, the quasi-Newton method is more efficient than DE and TS but less reliable for finding the global minimum and dependent on the initial values. Therefore, this and other local methods are not suitable for phase stability and equilibrium calculations in reactive systems.

5.2. Example 2

This example refers to the esterification reaction of ethanol and acetic acid to form ethyl acetate and water:

$$\text{ethanol (1)} + \text{acetic acid (2)} \leftrightarrow \text{ethyl acetate (3)} + \text{water (4)}. \tag{32}$$

This system is a benchmark problem for testing algorithms in phase equilibrium calculations for reactive systems (e.g. Castillo and Grossmann, 1981; Xiao *et al.*, 1989; McDonald and Floudas, 1996; Lee *et al.*, 1999; Jalali *et al.*, 2008). In this study, water is selected as the reference component for composition transformation, and transformed variables are given by:

$$X_1 = x_1 + x_4, \tag{33}$$

$$X_2 = x_2 + x_4, \tag{34}$$

$$X_3 = x_3 - x_4, \tag{35}$$

where $X_1, X_2 \in (0, 1)$ and $X_3 \in (-1, 1)$. We use the NRTL model for calculating the thermodynamic properties of the liquid phase, while the vapor phase is assumed to be ideal where the dimerization of acetic acid is not considered (see thermodynamic data in Table B5 of Appendix B).

Three feeds are analyzed at different temperatures and 1 atm. First, we consider an equi-molar feed of ethanol and acetic acid, $n_{i,F} = (0.5, 0.5, 0.0, 0.0)$ at 358 K. So, the transformed composition $Z_i = (0.5, 0.5, 0.0)$, where $\hat{n}_{i,F} = (0.5, 0.5, 0.0)$ and $\hat{n}_{T,F} = 1.0$, is tested for phase stability. In these conditions, the chemical equilibrium constant for this reaction is $K_{eq,1} = 18.154056$, which was calculated using the thermodynamic data reported by Lee *et al.* (1999), and this feed corresponds to a vapor phase (McDonald and Floudas, 1996; Lee *et al.*, 1999). Thus, RTPDF is defined as:

$$RTPDF = X_1\left(\left.\frac{\Delta\mu_1}{RT}\right|_X - \ln\left(\frac{z_1 P}{P_1^{sat}}\right)\right) + X_2\left(\left.\frac{\Delta\mu_2}{RT}\right|_X - \ln\left(\frac{z_2 P}{P_2^{sat}}\right)\right), + X_3\left(\left.\frac{\Delta\mu_3}{RT}\right|_X - \ln\left(\frac{z_3 P}{P_3^{sat}}\right)\right), \quad (36)$$

where the chemical potential at X is obtained from Eq. (25) with $\Delta\hat{g}_1$ and $\Delta\hat{g}_2$ given by

$$\Delta\hat{g}_1 = \sum_{i=1}^{3} X_i \ln(x_i\gamma_i), \quad (37)$$

$$\Delta\hat{g}_2 = \sum_{i=1}^{3} X_i \ln\left(\frac{x_i P}{P_i^{sat}}\right), \quad (38)$$

for liquid and vapor phases respectively. Note that z in Eq. 36 are the chemically equilibrated mole fractions, and are obtained from variable transformation $Z \rightarrow z$ (see Appendix A).

Phase stability analysis is performed using Eq. (36) and two decision variables: X_1, $X_2(0, 1)$; and X_3 is calculated using $X_3 = 1 - X_1 - X_2$. In these calculations, we use the following parameters for stochastic methods: (a) DE: $A = 0.3$, $CR = 0.9$, $NP = 50n_{var}$, $Gen_{\max} = 50$ and $Sc_{\max} = 6n_{var}$; and (b) TS: $N_t = N_p = 5$, $\varepsilon_t = \varepsilon_p = 0.005$, $NP_{\text{init}} = 20n_{var}$, $Iter_{\max} = 50n_{var}$, $h_n = 0.5$ and $Sc_{\max} = 12n_{var}$, where $n_{var} = 2$. Results of stability calculations are reported in Table 4. Since the global minimum of RTPDF is 0.0 at the trivial solution $Z_i = X_i$, this feed is stable. This result is consistent with that reported by Xiao *et al.* (1989), McDonald and

Floudas (1997) and Lee *et al.* (1999). Both TS and DE in combination with the local method, find the global solution of this stability problem with 100% SR, and TS is more efficient than DE (see Table 4).

Now, the same feed, $n_{i,F} = (0.5, 0.5, 0.0, 0.0)$ is analyzed at 355 K with $K_{eq,1} = 18.670951$ where both the liquid and vapor exists (Xiao *et al.*, 1989; McDonald and Floudas, 1996; Lee *et al.*, 1999). Again, $K_{eq,1}$ is determined from thermodynamic data reported by Lee *et al.* (1999). Stability calculations are performed using Eqs. (36) to (38) and the same decision variables; the results of stability analysis are reported in Table 4. Again, both TS and DE are reliable to determine that this mixture is unstable, and TS is more efficient than DE.

Considering the results of stability criterion, we have performed the minimization of transformed Gibbs free energy for calculating the phase equilibrium:

$$\Delta\hat{g} = \hat{n}_{1,1}\frac{\Delta\mu_{1,1}}{RT} + \hat{n}_{2,1}\frac{\Delta\mu_{2,1}}{RT} + \hat{n}_{3,1}\frac{\Delta\mu_{3,1}}{RT} + \hat{n}_{1,2}\frac{\Delta\mu_{1,2}}{RT} + \hat{n}_{2,2}\frac{\Delta\mu_{2,2}}{RT} + \hat{n}_{3,2}\frac{\Delta\mu_{3,2}}{RT}, \tag{39}$$

where the chemical potentials are defined by Eqs. (25), (37) and (38). For this feed, recall that $\hat{n}_{i,F} = (0.5, 0.5, 0.0)$ and $\hat{n}_{T,F} = 1.0$, and as a consequence $\hat{n}_{3,2} = -\hat{n}_{3,1}$. We can define the transformed moles of phase 1 as a fraction of total transformed moles in the feed (i.e. $\hat{n}_{T,1} = \hat{n}_{T,F}\lambda_F$ where $\lambda_F \in (0, 1)$). So, the global optimization of $\Delta\hat{g}$ is performed with respect to three decision variables: $\lambda_{1,1}, \lambda_{2,1} \in (0, 1)$ and $\lambda_F \in (0, 1)$. The transformed mole numbers are given by:

$$\begin{aligned} \hat{n}_{1,1} &= 0.5\lambda_{1,1} \\ \hat{n}_{2,1} &= 0.5\lambda_{2,1}, \\ \hat{n}_{3,1} &= \hat{n}_{T,F}\lambda_F - (\hat{n}_{1,1} + \hat{n}_{2,1}) \end{aligned} \tag{40}$$

and Eq. (9) is used to determine $\hat{n}_{i,2}$.

In these calculations, the parameters of DE and TS are the same as those for phase stability calculations except $Gen_{\max} = 75$, $Iter_{\max} = 100n_{var}$, and $Sc_{\max} = 12n_{var}$ where $n_{var} = 3$. These parameter values have been

Table 4. Results of phase stability and equilibrium calculations for example 2 using TS and DE.

Feed		Global optimum	SR (%) for		NFE(CPUt, s) for	
Z	T, K		DE	TS	DE	TS
(0.5, 0.5, 0)	358	RTPDF $= 0.0$ at $X_i = (0.5, 0.5, 0)$ or $x = (0.075325, 0.075325, 0.424675, 0.424675)$	100	100	5112 (1.4)	536 (0.14)
	355	RTPDF $= -0.032057$ at $X_i = (0.727391, 0.892842, -0.620233)$ or $x = (0.035139, 0.200590, 0.072019, 0.692252)$	100	99	5178 (1.3)	550 (0.14)
		$\Delta\hat{g} = -2.058125$ at $X_{i,1} = (0.716726, 0.878827, -0.595553)$ and $X_{i,2} = (0.488798, 0.480420, 0.030782)$; or $x_{i,1} = (0.078272, 0.069894, 0.441308, 0.410526)$ and $x_{i,2} = (0.039748, 0.201849, 0.081425, 0.676978)$	24	0	10623 (5.9)	—
(0.2, 0.4, 0.4)	355	RTPDF $= -0.223934$ at $X_i = (0.146452, 0.732001, 0.121547)$ or $x = (0.010354, 0.595904, 0.257644, 0.136098)$	100	100	5178 (1.3)	524 (0.13)
		$\Delta\hat{g} = -1.302752$ at $X_{i,1} = (0.163826, 0.575748, 0.260426)$ and $X_{i,2} = (0.217557, 0.314700, 0.467743)$; or $x_{i,1} = (0.023539, 0.435461, 0.400713, 0.140287)$ and $x_{i,2} = (0.029435, 0.126577, 0.655865, 0.188123)$	100	74	11592 (6.1)	780 (0.41)

modified because phase equilibrium problems are generally more difficult to solve compared to stability ones (Srinivas and Rangaiah, 2007). Results for these calculations are also reported in Table 4. In this case, both DE and TS face difficulties in finding the global minimum of $\Delta\hat{g}$; SR of DE is 24% whereas TS fails to find the global minimum. This problem is challenging due to the presence of a trivial solution ($Z_i = X_{i,1} = X_{i,2}$) which has an objective value (−2.057331) very comparable to the global minimum (−2.058125). Phase equilibrium compositions given in Table 4 are consistent with those reported by McDonald and Floudas (1996) and Lee *et al.* (1999). Note that there are differences between the objective function values in Table 4 and those given in these references because the previous works incorporated the Gibbs free energies of formation into the objective function (i.e. the total Gibbs free energy) while we have used the transformed Gibbs free energy of mixing as the objective. Both approaches are equivalent and can be used for phase equilibrium calculations in reactive systems.

Finally, consider a vapor mixture of ethanol, acid acetic, and ethyl acetate with initial composition of $n_{i,F} = (0.2, 0.4, 0.4, 0.0)$ at 355 K and 1 atm. At these conditions, $Z_i = (0.2, 0.4, 0.4)$, $\hat{n}_{i,F} = (0.2, 0.4, 0.4)$ and $\hat{n}_{T,F} = 1$, and this mixture also shows a vapor-liquid equilibrium with $K_{eq,1} = 18.670951$. Stability test is performed using Eqs. (36)–(38) and the same decision variables as in the previous feeds. Both TS and DE are capable of finding the global optimum of RTPDF without any difficulty (Table 4). Based on stability result, we have performed the minimization of transformed Gibbs free energy using Eqs. (25), (37)–(39). The global optimization of $\Delta\hat{g}$ is performed with respect to three decision variables $\lambda_{i,1} \in (0, 1)$. The transformed mole numbers are given by:

$$\begin{aligned} \hat{n}_{1,1} &= 0.2\lambda_{1,1} \\ \hat{n}_{2,1} &= 0.4\lambda_{2,1}, \\ \hat{n}_{3,1} &= 0.4\lambda_{3,1} \end{aligned} \tag{41}$$

and Eq. (9) is used to determine $\hat{n}_{i,2}$. The results of Gibbs energy minimization are summarized in Table 4. DE is 100% reliable, but it requires significant computational effort; although TS is very efficient, it may fail to find the equilibrium compositions as shown in Table 4.

5.3. Discussion

In this study, relative performance of TS and DE is in accordance with that reported for equilibrium calculations of non-reactive mixtures (Srinivas and Rangaiah, 2007). The differences between the performance of TS and DE may be associated with the efficacy of their escaping strategies. DE applies mutation and crossover strategies on a specific population to diversify the search and to escape from the local optimum. In this algorithm, the best solution that has been identified over all generations is reported after satisfying the convergence criterion. On the contrary, TS uses the best point in each iteration to create the new test points and to escape from the local solution, even though the current best solution may be worse than the best solutions found in the earlier iterations.

For comparison purposes, we have solved examples 1 and 2 using the SA method (Bonilla-Petriciolet *et al.*, 2006) with the following cooling schedule: initial annealing temperature = 10, number of iterations before reduction of annealing temperature = 5, and temperature reduction factor = 0.85. Each example was solved using 100 trials with random initial values via random number seed. Results of these calculations are summarized in Table 5. It appears that DE and TS are better than SA in terms of both NFE and CPUt but the order of reliability of these methods is: SA > DE > TS in both reactive phase stability and equilibrium calculations.

Table 5. Performance comparison of SA, DE and TS for reactive phase stability and equilibrium calculations for examples 1 and 2.

Objective function		SA			DE[a]			TS[a]		
		SR,%	NFE	CPUt	SR,%	NFE	CPUt	SR,%	NFE	CPUt
Example 1	*RTPDF*	100	23869	0.33	100	7677	0.07	86	595	0.004
	$\Delta\hat{g}$	92	22201	0.53	74	10841	0.15	26	578	0.008
Example 2	*RTPDF*[b]	100	18163	6.1	100	5112	1.4	100	536	0.14
	RTPDF[c]	100	17305	5.6	100	5178	1.3	99	550	0.14
	$\Delta\hat{g}^{c}$	44	26319	17.5	24	10623	5.9	0	—	—
	RTPDF[d]	100	17005	5.7	100	5178	1.3	100	524	0.13
	$\Delta\hat{g}^{d}$	100	27106	17.7	100	11592	6.1	74	780	0.41

[a] Parameters of TS and DE are those described in Example 2;
[b] Results for feed Z_i = (0.5, 0.5, 0.0) at 358 K and 1 atm;
[c] Results for feed Z_i = (0.5, 0.5, 0.0) at 355 K and 1 atm;
[d] Results for feed Z_i = (0.2, 0.4, 0.4) at 355 K and 1 atm.

6. Conclusions

This study shows, for the first time, that both TS and DE can be successful, if they are properly implemented, for solving phase stability and equilibrium calculations in multi-component and multi-reactive systems using transformed composition variables. Also, we demonstrate that performance profiles can be used to systematically tune the algorithm parameters of these methods. In general, the performance of DE is better than TS, but requires more NFE and CPUt. For both stability and equilibrium calculations, the maximum number of generations/iterations significantly affects the performance of these stochastic methods. Depending on problem characteristics, proper value for the termination criterion: $Sc_{\max}$ and the maximum number of generations/iterations should be selected to improve the performance of these stochastic methods.

References

Ali, M.M., Khompatraporn, C. and Zabinsky, Z.B. (2005). A numerical evaluation of several stochastic algorithms on selected continuous global optimization test problems. *Journal of Global Optimization*, **31**, pp. 635–672.

Baker, L.E., Pierce, A.C. and Luks, K.D. (1982). Gibbs energy analysis of phase equilibria. *Society of Petroleum Engineers Journal*, **22**, pp. 731–742.

Bonilla-Petriciolet, A., Bravo-Sanchez, U.I., Castillo-Borja, F., Frausto-Hernandez, S. and Segovia-Hernandez, J.G. (2008). Gibbs energy minimization using simulated annealing for two-phase equilibrium calculations in reactive systems. *Chemical and Biochemical Engineering Quarterly*, **22**, pp. 285–298.

Bonilla-Petriciolet, A., Vazquez-Roman, R., Iglesias-Silva, G.A. and Hall, K.R. (2006). Performance of stochastic optimization methods in the calculation of phase stability analyses for nonreactive and reactive mixtures. *Industrial and Engineering Chemistry Research*, **45**, pp. 4764–4772.

Burgos-Solorzano, G.I., Brennecke, J.F. and Stadtherr, M.A. (2004). Validated computing approach for high-pressure chemical and multiphase equilibrium. *Fluid Phase Equilibria*, **219**, pp. 245–255.

Castier, M., Rasmussen, P. and Fredenslund, A. (1989). Calculation of simultaneous chemical and phase equilibria in nonideal systems. *Chemical Engineering Science*, **44**, pp. 237–248.

Castillo, J. and Grossmann, I.E. (1981). Computation of phase and chemical equilibria. *Computers and Chemical Engineering*, **5**, pp. 99–108.

Dennis, J.E. Jr. and Schnabel, R.B. (1983). *Numerical Methods for Unconstrained Optimization and Nonlinear Equations*. Prentice-Hall, Englewood Cliffs, New Jersey.

Dolan, E.D. and More, J.J. (2002). Benchmarking optimization software with performance profiles. *Mathematical Programming Series A*, **91**, pp. 201–213.

Gupta, A.K., Bishnoi, P.R. and Kalogerakis, N. (1991). A method for the simultaneous phase equilibria and stability calculations for multiphase reacting and non-reacting systems. *Fluid Phase Equilibria*, **63**, pp. 65–89.

Iglesias-Silva, G.A., Bonilla-Petriciolet, A. and Hall, K.R. (2006). An algebraic formulation for an equal area rule to determine phase compositions in simple reactive systems. *Fluid Phase Equilibria*, **241**, pp. 25–30.

Jalali, F. and Seader, J.D. (1999). Homotopy continuation method in multi-phase multi-reaction equilibrium systems. *Computers and Chemical Engineering*, **23**, pp. 1319–1331.

Jalali-Farahani, J. and Seader, J.D. (2000). Use of homotopy-continuation method in stability analysis of multiphase, reacting systems. *Computers and Chemical Engineering*, **24**, pp. 1997–2008.

Jalali, F., Seader, J.D. and Khaleghi, S. (2008). Global solution approaches in equilibrium and stability analysis using homotopy continuation in the complex domain. *Computers and Chemical Engineering*, **32**, pp. 2333–2345.

Lee, Y.P., Rangaiah, G.P. and Luus, R. (1999). Phase and chemical equilibrium calculations by direct search optimization. *Computers and Chemical Engineering*, **23**, pp. 1183–1191.

McDonald, C.M. and Floudas, C.A. (1996). GLOPEQ: A new computational tool for the phase and chemical equilibrium problem. *Computers and Chemical Engineering*, **21**, pp. 1–23.

McKinnon, K. and Mongeau, M. (1998). A generic global optimization algorithm for the chemical and phase equilibrium problem. *Journal of Global Optimization*, **12**, pp. 325–351.

Michelsen, M.L. (1982). The isothermal flash problem. Part I. Stability. *Fluid Phase Equilibria*, **9**, pp. 1–19.

Perez-Cisneros, E.S., Gani, R. and Michelsen, M.L. (1997). Reactive separation systems-I. Computation of physical and chemical equilibrium. *Chemical Engineering Science*, **52**, pp. 527–543.

Rangaiah, G.P. (2001). Evaluation of genetic algorithms and simulated annealing for phase equilibrium and stability problems. *Fluid Phase Equilibria*, **187–188**, pp. 83–109.

Reynolds, D., Mulholland, A.J. and Gomatam, J. (1997). Calculation of chemical and phase equilibria via simulated annealing. *Journal of Mathematical Chemistry*, **22**, pp. 25–37.

Seider, W.D. and Widagdo, S. (1996). Multiphase equilibria of reactive systems. *Fluid Phase Equilibria,* **123**, pp. 283-303.

Srinivas, M. and Rangaiah, G.P. (2007). A study of differential evolution and tabu search for benchmark, phase equilibrium and phase stability problems. *Computers and Chemical Engineering*, **31**, pp. 760–772.

Stateva, R.P. and Wakeham, W.A. (1997). Phase equilibrium calculations for chemically reacting systems. *Industrial and Engineering Chemistry Research*, **36**, pp. 5474–5482.

Taylor, R. and Krishna, R. (2000). Modelling reactive distillation. *Chemical Engineering Science*, **55**, pp. 5183–5229.

Teh, Y.S. and Rangaiah, G.P. (2003). Tabu search for global optimization of continuous functions with application to phase equilibrium calculations. *Computers and Chemical Engineering*, **27**, pp. 1665–1679.

Ung, S. and Doherty, M.F. (1995a). Vapor-liquid phase equilibrium in systems with multiple chemical reactions. *Chemical Engineering Science*, **50**, pp. 23–48.

Ung, S. and Doherty, M.F. (1995b). Theory of phase equilibria in multireaction systems. *Chemical Engineering Science*, **50**, pp. 3201–3216.

Xiao, W., Zhu, K., Yuan, W. and Chien, H.H. (1989). An algorithm for simultaneous chemical and phase equilibrium calculations. *AIChE Journal*, **35**, pp. 1813–1820.

Xu, G., Haynes, W.D. and Stadtherr, M.A. (2005). Reliable phase stability analysis for asymmetric models. *Fluid Phase Equilibria*, **235**, pp. 152–165.

Wakeham, W.A. and Stateva, R.P. (2004). Numerical solution of the isothermal, isobaric phase equilibrium problem. *Reviews in Chemical Engineering*, **20**, pp. 1–56.

Wasylkiewicz, S.K. and Ung, S. (2000). Global phase stability analysis for heterogeneous reactive mixtures and calculation of reactive liquid-liquid and vapor-liquid-liquid equilibria. *Fluid Phase Equilibria*, **175**, pp. 253–272.

7. Exercises

(1) Consider a hypothetical reacting ternary mixture $A_1 + A_2 \Leftrightarrow A_3$ with liquid-liquid equilibrium (Iglesias-Silva *et al.*, 2006), where reaction equilibrium constant is independent of temperature, and thermodynamic properties are calculated using the Margules solution model $\frac{G^{ex}}{RT} = 3.6x_1x_2 + 2.4x_1x_3 + 2.3x_2x_3$. This system shows complex equilibrium behavior for some values of the reaction equilibrium constant. The third component is selected as the reference substance. For a transformed feed $Z_i = (0.6, 0.4)$, $\hat{n}_{T,F} = 1.0$ and $K_{eq,1} = 0.9825$, determine if this mixture is stable; if it is not stable, calculate the corresponding equilibrium compositions. Use both TS and DE, and compare their performance for different values of NP, NP_{init}, $Iter_{\max}$ and $Gen_{\max}$. [*Solution:* RTPDF $= -0.020055$ at $X_i = (0.83575, 0.16425)$; $\Delta\hat{g} = -0.144508$ at $X_{i,1} = (0.484538, 0.515462)$ and $X_{i,2} = (0.815044, 0.184956)$].

(2) Consider a six-component mixture with only one chemical reaction $2A_5 \Leftrightarrow A_6$ but with four inert components: A_1, A_2, A_3 and A_4. This mixture shows vapor-liquid equilibrium at 75°C and 1.5 atm where the reaction equilibrium constant is 5.0. Assume that the vapor and liquid phases behave as an ideal gas and ideal solution, thermodynamic data are given in Table B4, and choose component A_6 as the reference for calculating the transformed mole fractions. Establish the equilibrium compositions for a transformed feed $Z_i = (0.00497, 0.605, 0.0324, 0.31, 0.04763)$ and $\hat{n}_{T,F} = 1.0$ using both TS and DE. Construct the quartile sequential plots for different values of *NP* and NP_{init}. Perform at least 25 runs with random initial values for this purpose. Compare and discuss the results obtained from these trials. Which of the two methods is better, and why? [*Solution:* RTPDF = −0.004072 at $X_i = (0.005931, 0.736912, 0.033105, 0.180695, 0.043356)$; $\Delta\hat{g} = -0.144508$ at $X_{i,1} = (0.004923, 0.598568, 0.032360, 0.316323, 0.047827)$ and $X_{i,2} = (0.005899, 0.732032, 0.033198, 0.185129, 0.043742)$].

Appendix A

To illustrate the application of the transformation procedure of Ung and Doherty (1995a, 1995b), consider a ternary mixture that follows the reaction: $A_1 + A_2 \leftrightarrow A_3$. For this system, we have $c = 3$, $r = 1$ and $c - r = 2$ transformed composition variables. For variable transformation, it is necessary to select a set of r reference components such that N is an invertible matrix. This requires that all the reactions must involve at least one of the reference components, an inert component can not be a reference component and no two reference components can have identical stoichiometric coefficients in each reaction (Ung and Doherty, 1995a).

If A_3 is selected as the reference component, N is invertible and equal to unity. In this case, $v_1 = -1$, $v_2 = -1$, and $v_{TOT} = -1$, and so the transformed mole numbers and fractions are given by:

$$\hat{n}_1 = n_1 + n_3, \tag{A1}$$

$$\hat{n}_2 = n_2 + n_3, \tag{A2}$$

$$X_1 = \frac{\hat{n}_1}{\hat{n}_T} = \frac{n_1 + n_3}{n_T + n_3} = \frac{x_1 + x_3}{1 + x_3}, \tag{A3}$$

$$X_2 = \frac{\hat{n}_2}{\hat{n}_T} = \frac{n_2 + n_3}{n_T + n_3} = \frac{x_2 + x_3}{1 + x_3} = 1 - X_1, \tag{A4}$$

where $n_T = n_1 + n_2 + n_3$, $\hat{n}_T = \hat{n}_1 + \hat{n}_2$ and $X_1, X_2 \in [0, 1]$. Note that, for this mixture, any of the three components can be used as the reference component to define the transformed composition variables ($\hat{n}$ and X).

If the initial composition (mole numbers or fractions) of the reactive mixture is available, the corresponding transformed variables can be obtained by using the above equations. For example, consider a mixture with initial composition: $n_{i,F} = (0.3, 0.2, 0.5)$. Using Eqs. (A1)–(A4), transformed mole numbers are: $\hat{n}_{1,F} = 0.3 + 0.5 = 0.8, \hat{n}_{2,F} = 0.2 + 0.5 = 0.7$, and $\hat{n}_{T,F} = 0.8 + 0.7 = 1.5$; while transformed mole fractions are: $Z_1 = 0.8/1.5 \cong 0.5333$ and $Z_2 = 0.7/1.5 = 1 - Z_1 \cong 0.4667$.

The transformed composition variables take the same value before, during and after reaction (Ung and Doherty, 1995a) but depend on the initial composition. Therefore, the mole fractions that are chemically equilibrated, satisfy the stoichiometry requirements and corresponding to Z must be obtained. To do this, we need to find the composition of reference components that satisfy the chemical equilibrium constraints. In the general case, for a set of $c - r$ specified transformed variables X_i, the reference mole fractions x_{ref} are calculated using Eq. (5) and the equilibrium constants for each reaction $K_{eq,k}$; this requires solution of r nonlinear equations given by Eq. (6) because $K_{eq,k}$ are non-linear functions of x. After knowing x_{ref}, the corresponding mole fractions of $c - r$ non-reference components are calculated using Eq. (5) again. For this transformation procedure $X \rightarrow x$, any solver such as the bisection method (for systems with only one reaction) or the Newton method (for multi-reaction systems) can be used to find x_{ref}. Figure A1 presents the steps in the variable transformation between X and x. Note that if different models are used in the calculation of thermodynamic properties for the different phases, the procedure of variable transformation must be performed for each model at specified X. In these cases, the mole fraction x that corresponds to the lower value of transformed Gibbs free energy is selected.

For illustrative purposes, let us consider that this ternary mixture is an ideal liquid phase (i.e. $\gamma_i = 1.0$). The chemical equilibrium constant is

1. Start with the specified components, composition (mole numbers or fractions) of reactive mixture, T, P and the thermodynamic models.

2. Select r reference components, and determine $\boldsymbol{N}$ and $\boldsymbol{N}^{-1}$. Calculate the equilibrium constant, $K_{eq,k}$ of each of all reactions at the given T using relevant thermodynamic equations (e.g. see Table 2).

3. Calculate X using Eqs. (3)–(5), and choose initial values for $\boldsymbol{x}_{\text{ref}}$.

4. Using $\boldsymbol{x}_{\text{ref}}$, $c - r$ values of X_i obtained in step 3, and rearranged Eq. (5): $x_i = X_i\,(1 - \boldsymbol{\nu}_{TOT}\boldsymbol{N}^{-1}\boldsymbol{x}_{\text{ref}}) + \boldsymbol{\nu}_i\,\boldsymbol{N}^{-1}\boldsymbol{x}_{\text{ref}}$, determine mole fractions of non-reference components.

5. Calculate the activity, a_i of each of all components using the specified thermodynamic model for the phase and x_i found in Step 4.

6. Substitute values of equilibrium constants and activities in the r non-linear equations (Eq. 6). If these equations are not satisfied, use any solver such as the bisection method (for situations with only one reaction) or the Newton method (for multi-reaction systems) to find new values of $\boldsymbol{x}_{\text{ref}}$, and then go to step 4. Note that the gradient for the Newton method can be calculated via finite differences.

7. If the r non-linear equations are satisfied to the desired tolerance level, current values of $\boldsymbol{x}$ are chemically equilibrated and satisfy the stoichiometry and equilibrium requirements.

Figure A1. Steps in the composition transformation procedure $X \rightarrow x$.

defined as

$$K_{eq,1} = \frac{a_3}{a_1 a_2} = \frac{x_3\gamma_3}{x_1\gamma_1 x_2\gamma_2}. \tag{A5}$$

We can rearrange Eqs. (A3) and (A4) to obtain

$$x_i = X_i(1 + x_3) - x_3 \quad \text{for } i = 1, 2. \tag{A6}$$

Substituting Eq. (A6) into Eq. (A5), we find

$$f(x_3) = K_{eq}[X_1(1 + x_3) - x_3][X_2(1 + x_3) - x_3]\gamma_1\gamma_2 - x_3\gamma_3 = 0. \tag{A7}$$

Therefore, if we specify X, mole fractions x that simultaneously satisfy the chemical equilibrium equation and map onto the specified values of the transformed composition variables are determined by solving Eq. (A7) for the reference component x_3. For our example, if $K_{eq,1} = 1.0$, $\gamma_i = 1.0$ and $Z_i = (0.5333, 0.4667)$, solving $f(x_3)$ we find *chemically equilibrated* mole fractions in the given feed as $z_i = (0.4537, 0.3758, 0.1705)$. Note that these are not the given mole fractions in the feed but are chemically equilibrated mole fractions.

Now, consider this mixture but with the presence of liquid-liquid equilibrium while the reaction occurs at 323.15 K and $K_{eq,1} = 3.5$. In these conditions, we calculate the activity coefficient γ_i using the Margules solution equation:

$$T \ln \gamma_k = \frac{1}{2} \sum_{i=1}^{c} \sum_{j=1}^{c} (A_{ik} + A_{jk} - A_{ij}) x_i x_j, \tag{A8}$$

with $A_{ij} = A_{ji}$, $A_{ii} = 0.0$, $A_{12} = 478.6$, $A_{13} = 1074.484$ and $A_{23} = 626.9$. Note that Margules model is used to represent the liquid-phase non-idealities because there is no phase split (i.e. liquid-liquid phase equilibrium) for an ideal liquid (i.e. $\gamma_i = 1.0$). The chemical potential of each component and the transformed Gibbs free energy of mixing are respectively

$$\frac{\Delta \mu_i}{RT} = \ln(x_i \gamma_i), \tag{A9}$$

$$\Delta \hat{g}_1 = X_1 \ln(x_1 \gamma_1) + X_2 \ln(x_2 \gamma_2), \tag{A10}$$

where x_i results from $X \to x$ using Eqs. (A5)–(A7). Note that $f(x_3)$ is a non-linear function with x_3 as unknown and has to be solved for given X.

Figure A2 shows the plot of $\Delta \hat{g}_1$ versus X_1 for this mixture. This plot is obtained by varying X_1 from 0.0 to 1.0, calculating the associated mole fractions x for each X_1 (using Algorithm of Fig. A1), and evaluating $\Delta \hat{g}_1$ (Eq. A10). As shown in this figure, this reactive mixture exhibits a phase split for Z_1 between 0.497527 and 0.840050. Any trial composition inside this range will show phase equilibrium. For our analysis, we have selected the same feed $Z_i = (0.5333, 0.4667)$ and $\hat{n}_{T,F} = 1.0$. To evaluate phase stability, we need to minimize the reactive plane distance function, obtained from Eqs. (10) and (A9):

$$RTPDF = X_1(\ln(x_1 \gamma_1) - \ln(z_1 \gamma_1)) + X_2(\ln(x_2 \gamma_2) - \ln(z_2 \gamma_2)). \tag{A11}$$

Note that RTPDF can be optimized with respect to $X_1 \in (0, 1)$ because X_2 is obtained from the equality constraint: $X_2 = 1 - X_1$. The global optimum for this feed is -0.406606 at $X_i = (0.9527, 0.0473)$. As expected, the given feed is unstable.

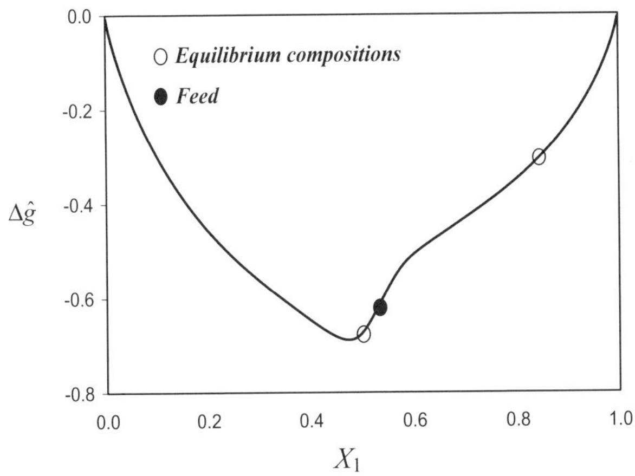

Figure A2. Transformed molar Gibbs free energy of mixing versus X_1 for reactive mixture: $A_1 + A_2 \leftrightarrow A_3$ at 323.15 K and $K_{eq,1} = 3.5$.

For performing phase equilibrium calculations, the Gibbs free energy function is obtained from Eqs. (1) and (A9) as:

$$\Delta\hat{g} = \hat{n}_{1,1}\ln(x_{1,1}\gamma_{1,1}) + \hat{n}_{2,1}\ln(x_{2,1}\gamma_{2,1}) + \hat{n}_{1,2}\ln(x_{1,2}\gamma_{1,2}) + \hat{n}_{2,2}\ln(x_{2,2}\gamma_{2,2}), \quad \text{(A12)}$$

where $\hat{n}_{i,j}$ is the transformed mole number of component i in liquid phase j. Note that the transformation procedure $X \to x$ is performed for each liquid phase, taking into account its corresponding transformed composition $\hat{n}_{i,j}$. We have two decision variables: $\lambda_{1,1}$ and $\lambda_{2,1}$ for this example. They are used to calculate $\hat{n}_{1,1} = 0.5333\lambda_{1,1}$ and $\hat{n}_{2,1} = 0.4667\lambda_{2,1}$ as well as the transformed composition of the second liquid phase by $\hat{n}_{1,2} = 0.5333 - 0.5333\lambda_{1,1}$ and $\hat{n}_{2,2} = 0.4667 - 0.4667\lambda_{1,1}$ (Eqs. 7 and 9). The global optimum of $\Delta\hat{g}$ is -0.639429 at $X_{i,1} = (0.497527, 0.502473)$ and $X_{i,2} = (0.840050, 0.15995)$, which are the transformed mole fractions at liquid-liquid equilibrium; the equilibrium compositions in conventional mole fractions are: $x_{i,1} = (0.056907, 0.066186, 0.876907)$ and $x_{i,2} = (0.829728, 0.105473, 0.064529)$.

Appendix B

Thermodynamic data and model parameters for all reactive systems used in this chapter are given in Tables B1–B5.

Table B1. Thermodynamic data for the methyl *tert*-butyl ether (MTBE) reaction system with inert at 10 atm and 373.15 K.

	Parameters of pure component				u_{ij} in the Wilson model (cal/mol)			
Component[1]	A_i	B_i	C_i	V_i	1	2	3	4
1	6.84132	923.201	239.99	93.33	—	169.9953	−60.1022	—
2	8.07372	1578.23	239.382	44.44	2576.8532	—	1483.2478	2283.8726
3	6.87201	1116.825	224.744	118.8	271.5669	−406.3902	—	—
4	6.80896	935.86	238.73	100.39	—	382.3429	—	—

$\log_{10} P_i^{sat} = A_i - \frac{B_i}{T+C_i}$ where P_i^{sat} in mmHg and T in °C.

$$\ln \gamma_i = 1 - \ln\left(\sum_{j=1}^{c} x_j \Lambda_{ij}\right) - \sum_{k=1}^{c}\left(x_k \Lambda_{ki} / \sum_{j=1}^{c} x_j \Lambda_{kj}\right)$$

$$\Lambda_{ij} = \frac{V_j}{V_i} \exp\left(\frac{-u_{ij}}{RT}\right)$$

$K_{eq,1} = \frac{a_3}{a_1 a_2}$ where $a_i = x_i \gamma_i$ for liquid phase and $a_i = x_i P / P_i^{sat}$ for vapor phase.

[1] 1: Isobutene, 2: Methanol, 3: MTBE, and 4: Butane.

Table B2. Thermodynamic data for the reactive system for *tert*-amyl methyl ether (TAME) synthesis at 335 K and 1.5 atm.

	Parameters of pure component					u_{ij} in the Wilson model (cal/mol)			
Component[1]	A_i	B_i	C_i	D_i	V_i	1	2	3	4
1	74.527	−5232.2	−8.1482	8.474E-06	0.10868	—	478.8	1376.5	−611.75
2	82.614	−5586.1	−9.4429	1.0858E-05	0.10671	−477.94	—	968.81	−386.04
3	23.5347	−3661.468	−32.77		0.04069	9772.3	10147	—	4826.3
4	20.9441	−2936.223	−47.70385		0.13345	951.33	712.33	−177	—

$\ln P_i^{sat} = A_i + \frac{B_i}{T} + C_i \ln T + D_i T^2$ for $i = 1, 2$

$\ln P_i^{sat} = A_i + \frac{B_i}{T+C_i}$ for $i = 3, 4$

where P_i^{sat} in Pa and T in Kelvin.

$\ln \gamma_i = 1 - \ln\left(\sum_{j=1}^{c} x_j \Lambda_{ij}\right) - \sum_{k=1}^{c}\left(x_k \Lambda_{ki} \Big/ \sum_{j=1}^{c} x_j \Lambda_{kj}\right)$

$\Lambda_{ij} = \frac{V_j}{V_i} \exp\left(\frac{-u_{ij}}{RT}\right)$

$K_{eq,1} = \frac{a_4^2}{a_1 a_2 a_3^2}$ where $a_i = x_i \gamma_i$ for liquid phase and $a_i = x_i P / P_i^{sat}$ for vapor phase.

[1] 1: 2-Methyl-1-butene, 2: 2-Methyl-2-butene, 3: Methanol, and 4: TAME.

Table B3. Thermodynamic data for the reaction for butyl acetate production at 298.15 K and 1 atm.

			u_{ij} in the UNIQUAC model (cal/mol)			
Component[1]	Q	R_u	1	2	3	4
1	2.072	2.2024	—	−131.7686	−343.593	−298.4344
2	3.052	3.4543	148.2833	—	68.0083	82.5336
3	1.4	0.92	527.9269	581.1471	—	394.2396
4	4.196	4.8724	712.2349	24.6386	756.4163	—

$$\ln \gamma_i = \ln \gamma_i^E + \ln \gamma_i^R$$

$$\ln \gamma_i^E = \ln \frac{\phi_i}{x_i} + 5Q_i \ln \frac{\phi_i}{x_i} + l_i - \frac{\phi_i}{x_i} \sum_{j=1}^{c} x_j l_j$$

$$\ln \gamma_i^R = Q_i \left(1 - \ln\left(\sum_{j=1}^{c} \theta_j \tau_{ji}\right) - \sum_{j=1}^{c} \left(\frac{\theta_j \tau_{ji}}{\sum_{l=1}^{c} \theta_l \tau_{li}}\right)\right)$$

$$\theta_i = \frac{Q_i x_i}{\sum_{j=1}^{c} Q_j x_j} \quad \phi_i = \frac{R_{u,i} x_i}{\sum_{j=1}^{c} R_{u,j} x_j}$$

$$l_i = 5(R_{u,i} - Q_i) - (R_{u,i} - 1) \quad K_{eq,1} = \frac{a_3 a_4}{a_1 a_2} \text{ where } a_i = x_i \gamma_i.$$

[1] 1: Acetic acid, 2: n-Butanol, 3: Water, and 4: n-Butyl acetate.

Table B4. Thermodynamic data for Example 1.

Component	A_i	B_i	C_i	γ_i
1	7.6313	1566.69	273.419	1.0
2	7.11714	1210.595	229.664	1.0
3	7.44777	1488.99	264.915	1.0
4	8.1122	1592.864	226.184	1.0
5	7.9701	1521.23	233.97	1.0
6	6.8664	1188.05	226.276	1.0

$$\log_{10} P_i^{sat} = A_i - \frac{B_i}{T + C_i}$$

where P_i^{sat} in mmHg and T in °C.

$$K_{eq,1} = \frac{a_4}{a_3} \quad K_{eq,2} = \frac{a_4}{a_5} \quad K_{eq,3} = \frac{a_6}{a_4}$$

where $a_i = x_i \gamma_i$ for liquid phase and $a_i = x_i P / P_i^{sat}$ for vapor phase.

Table B5. Thermodynamic data for Example 2.

	Parameters of pure component			τ_{ij} in the NRTL model			
Component[1]	A_i	B_i	C_i	1	2	3	4
1	9.95614	1440.52	−60.44	0.0	1.3941	0.6731	−0.2019
2	9.6845	1644.05	−39.63	−1.0182	0.0	0.007	−0.4735
3	9.22298	1238.71	−56.15	0.1652	0.5817	0.0	1.7002
4	10.09171	1668.21	−45.14	2.1715	1.6363	1.9257	0.0
	$\log_{10} P_i^{sat} = A_i - \frac{B_i}{T+C_i}$ where P_i^{sat} in N/m^2 and T in K.			$\alpha_{ij} = 0.3$ $\ln \gamma_i = \frac{\sum_{j=1}^{c} \tau_{ji} G_{ji} x_j}{\sum_{j=1}^{c} G_{ji} x_j} + \sum_{j=1}^{c} \frac{G_{ij} x_j}{\sum_{l=1}^{c} G_{lj} x_l} \left(\tau_{ij} - \frac{\sum_{l=1}^{c} \tau_{lj} G_{lj} x_l}{\sum_{l=1}^{c} G_{lj} x_l} \right)$ $G_{ij} = \exp(-\alpha_{ij} \tau_{ij})$			
	$K_{eq,1} = \frac{a_3 a_4}{a_1 a_2}$ where $a_i = x_i \gamma_i$ for liquid phase and $a_i = x_i P / P_i^{sat}$ for vapor phase.						

[1] 1: Ethanol, 2: Acetic acid, 3: Ethyl acetate, and 4: Water.

Chapter 14

DIFFERENTIAL EVOLUTION WITH TABU LIST FOR GLOBAL OPTIMIZATION: EVALUATION OF TWO VERSIONS ON BENCHMARK AND PHASE STABILITY PROBLEMS

Mekapati Srinivas* and Gade Pandu Rangaiah†

Department of Chemical & Biomolecular Engineering
National University of Singapore, Engineering Drive 4
Singapore 117576
†*chegpr@nus.edu.sg*

1. Introduction

Among the many stochastic global optimization techniques, differential evolution (DE) (Storn and Price, 1997; Price *et al.*, 2005) and tabu search (TS) (Glover, 1989; Chelouah and Siarry, 2000) are two of the promising methods. DE is a population based direct search method especially for nonlinear and non-differentiable continuous functions. It mimics biological evolution by performing mutation, crossover and selection steps as in genetic algorithm (GA). The main advantages of DE are its capability to escape from the local optima with a few parameters, and faster convergence to the global optimum compared to GA (Karaboga and Cetinkaya, 2004). TS was initially developed by Glover (1989) for combinatorial optimization and has been used for continuous optimization (Teh and Rangaiah, 2003)

*Current affiliation: ABB Indian Corporate Research Centre, Bangalore 560 048, India.

too. The main feature of TS is it avoids re-visits to the same place during the search by keeping track of the previous search using tabu and promising lists, thus providing good computational efficiency. Both DE and TS have been successfully applied to many applications in Chemical Engineering (see other chapters in this book).

Several modifications have been proposed to DE to improve its performance further. Lee *et al.* (1999) proposed two modifications to the original DE: implementing local search to calculate the optimal mutation parameter value and employing heuristic constraints which systematically reduce the size of the search space. The modified DE is then tested on a dynamic optimization problem, and the results show that its convergence rate is faster than that of DE. Hendtlass (2001) proposed a swarm DE which combines both particle swarm optimization and DE. It is tested on a few test functions, and the results show that its reliability in locating the global optimum is high compared to DE alone. Chiou *et al.* (2004) proposed ant direction hybrid differential evolution (ADHDE) for solving large capacitor placement problems. It utilizes the concept of ant colony search to explore a good mutation parameter value in hybrid differential evolution (HDE) proposed by Chiou and Wang (1999). The results show that ADHDE performs better than simulated annealing (SA) and HDE. Teo (2006) studied dynamic self-adaptive populations in DE. The results with a set of De-Jong's test functions show that the method is capable of locating the global minimum in addition to the reduction in number of parameters. Babu and Angira (2006) proposed modified differential evolution (MDE) for the optimization of nonlinear chemical processes. In MDE, the population is updated as and when a better member is found instead of waiting until all members of the new population are generated as in the original DE. In addition, MDE maintains only one set of population whereas DE holds two sets of population at any point of time during the search. Results in Babu and Angira (2006) show that the convergence rate of MDE is better than that of DE. Despite many successful studies and applications on DE, no proof exists for its convergence.

We proposed DE with tabu list (DETL), which combines the strong features of DE and TS to obtain good reliability as in DE along with good computational efficiency as in TS (Srinivas and Rangaiah, 2007b). In that study, tabu list was implemented in the *generation step* of the DE. The present study explores the possibility of implementing the tabu list in the *evaluation step* of the DE. The two implementations, referred as DETL-G and DETL-E, are first tested for several benchmark problems involving 2 to 20 variables and a few to thousands of local minima. The methods are then evaluated for challenging phase stability problems involving multiple components and comparable minima (i.e. objective function values at the local and global minima are close). A new benchmark problem with characteristics similar to those of phase stability problems is also proposed and used in the evaluation of the two versions of DETL.

2. Differential Evolution with Tabu List (DETL)

A continuous, unconstrained optimization problem can be written as:

$$\begin{aligned} &\text{Minimize} \quad f(\mathbf{x}) \\ &\text{Subject to} \quad x_i^l \le x_i \le x_i^u \text{ for } i = 1, 2, \ldots, N \end{aligned}$$

where N is the number of decision variables in the problem, $\mathbf{x}$ is the vector of continuous variables, and x_i^l and x_i^u are the lower and upper bounds of x_i respectively. This problem is considered for describing and testing the methods in this chapter.

DETL is developed by combining the reliable escaping mechanism (i.e. mutation and crossover) of DE with the concept of TS (i.e. avoiding revisits during the search) for reducing the number of objective function evaluations. DE was chosen instead of GA because the former has only a few parameters and computationally efficient compared to GA (Karaboga and Cetinkaya, 2004). Our extensive experience with DE and TS (Srinivas and Rangaiah, 2007a,b) shows that DE is more reliable in locating the global optimum compared to TS whereas the latter is computationally efficient than the former. These results motivated us to develop DETL with (i) good reliability as in DE, (ii) better computational efficiency as in TS, and (iii) less number of parameters (Srinivas and Rangaiah, 2007b). The concept of TS is implemented in DE by including the tabu list alone (with two additional parameters, namely, tabu radius and tabu list size, to be discussed in the next paragraph) instead of both tabu and promising lists as in TS. This is mainly to reduce the number of parameters in DETL. In this work, the concept of

TS is implemented in DE primarily to enhance the computational efficiency rather than to find the accurate solution at the end as in Liu *et al.* (2002).

In general, DE consists of three main steps: generation, evaluation and selection. The generation step produces the offspring (new individuals) by mutation and crossover operations whereas fitness (objective function value) of the new members of the population is calculated in the evaluation step. The selection step allows only those individuals of the current generation that have better fitness value, to the next generation. The cycle of generation, evaluation and selection is repeated until either the known global optimum is found or up to the specified maximum number of generations. In the present work, two implementations of tabu list in DE (i.e. DETL) to avoid re-visits are studied: (i) in the *evaluation step* (the corresponding method is known as DETL-E), and (ii) in the *generation step* (the corresponding method is known as DETL-G). Further, in DETL-G, population is immediately updated with the new member as in MDE, and hence it has only one population (Babu and Angira, 2006; Srinivas and Rangaiah, 2007b). On the other hand, algorithm of DETL-E is closer to the original DE compared to that of DETL-G.

2.1. Description of DETL-E and DETL-G

The flowcharts of DETL-E and DETL-G are given in Figs. 1 and 2 respectively. Both DETL-E and DETL-G start by selecting the algorithm parameters: population size (NP), amplification (also known as mutation) factor (A), crossover constant (CR), tabu list size (tls), tabu radius (tr), maximum number of generations (Gen_{max}) and maximum number of successive generations (Sc_{max}) without improvement in the best function value. The algorithms then generate the initial population of size NP over the entire feasible region using uniformly distributed random numbers. Each member or individual corresponds to a point in the search domain. The objective function is evaluated at each member of the initial population, and the best member is captured. The evaluated points are then sent to the tabu list in both the algorithms as shown in Figs. 1 and 2.

After the above initial steps, generations/iterations begin; in each generation, two important steps: mutation and crossover are performed on the members of the current generation (J) to produce offspring for the

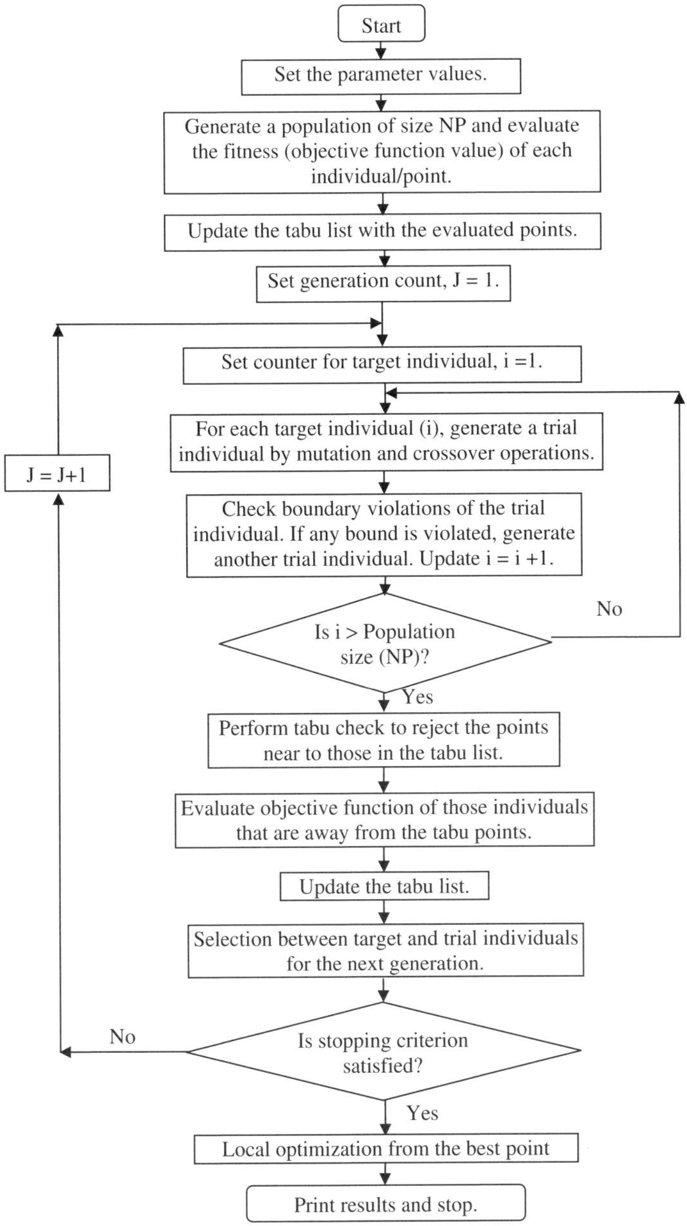

Figure 1. Flow chart of DETL-E.

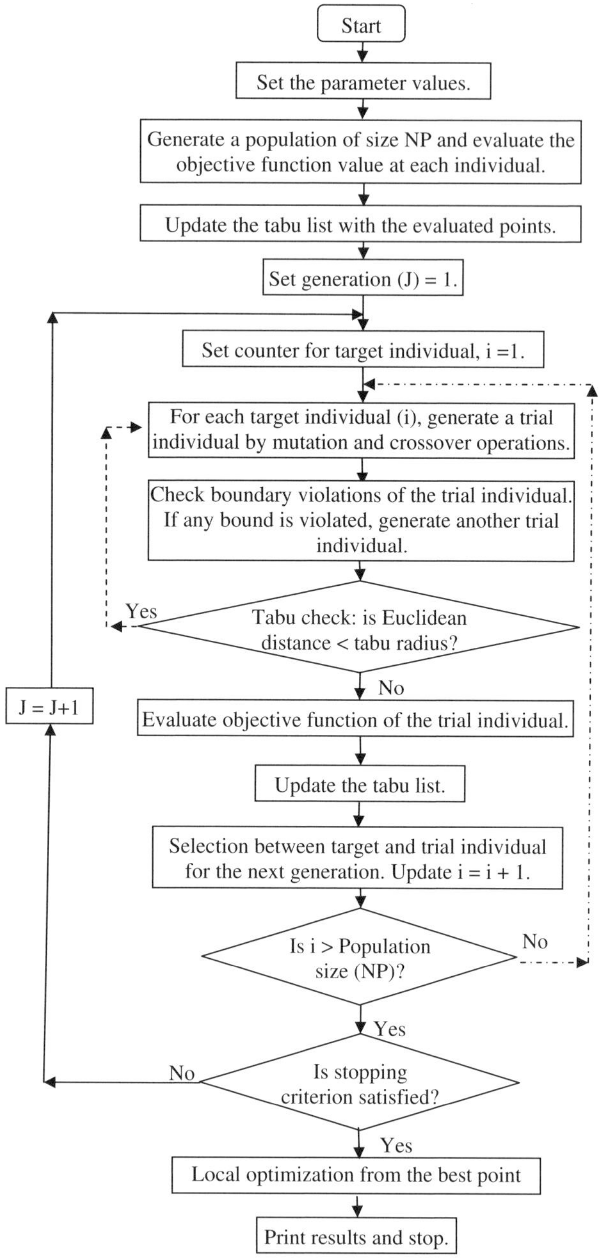

Figure 2. Flow chart of DETL-G.

next generation (J + 1) in both DETL-E and DETL-G. The mutation and crossover operations are performed for each member in the population as the target vector. Mutation is carried out for each *target* individual ($\mathbf{x}_{i,\,\mathrm{J}}$) in the population to generate *mutant* individual ($\mathbf{v}_{i,\,\mathrm{J}+1}$) using the equation:

$$\mathbf{v}_{i,\,\mathrm{J}+1} = \mathbf{x}_{r1,\,\mathrm{J}} + \mathrm{A}(\mathbf{x}_{r2,\,\mathrm{J}} - \mathbf{x}_{r3,\,\mathrm{J}}), \tag{1}$$

where $\mathbf{x}_{r1,\,\mathrm{J}}$, $\mathbf{x}_{r2,\,\mathrm{J}}$ and $\mathbf{x}_{r3,\,\mathrm{J}}$ are the three random individuals chosen from the population of the current generation (J). The random integers r_1, r_2 and r_3 should be distinct and also different from the running index, i, and hence NP should be 4 or more for mutation. The amplification factor, A has a value between 0 and 2, and controls the amplification of the differential variation between the two random individuals. In the crossover step, a *trial* individual is generated by copying some elements of the mutant individual to the target individual with probability of CR. A boundary violation check is performed to check the feasibility of the trial individual; if the trial individual is infeasible, it is replaced by generating a new individual by crossover operation.

The sequence of operations is different from the generation to the selection steps, in DETL-E and DETL-G as described below. In the case of DETL-E, all NP trial individuals for the next generation are produced using mutation, crossover and boundary violation check, as shown in Fig. 1. Next, tabu check is performed to see if any of the trial individuals is near to those in the tabu list by calculating the Euclidean distance between them. If the Euclidean distance is smaller than the specified value (tabu radius, tr), which indicates that the objective function value of trial individual and at one of the points in the tabu list (consisting of already evaluated points) are probably close, the corresponding trial individual is rejected to avoid objective function evaluation. The objective function values of the remaining trial individuals are computed. Then, selection between the respective trial individual (if not already rejected in the tabu check) and target individual is done (i.e. only after generating trial individuals for each and every target individual in the current population). The population is now updated with the selected individuals for the next generation.

In DETL-G (Fig. 2), a trial individual is produced by mutation, crossover and boundary check, and then compared with the points in the tabu list (i.e. tabu check is performed in the generation step itself in contrast to the evaluation step in DETL-E). If it is near to any of the points in the tabu list, the trial individual is rejected and another point is generated through mutation and crossover until it is away from all the points in the tabu list. The objective function is evaluated at the trial individual that is away from all the points in the tabu list, thus improving the diversity among the individuals in DETL-G. The tabu list is updated, and selection between the target and trial individuals is made based on objective value. If the trial individual is better than the target individual, it replaces the target individual in the current population immediately and may (depending on the random numbers) participate in the subsequent mutation and crossover operations. Thus, the population is updated as soon as a better individual is found and only one population is maintained in the iterations as in MDE (Babu and Angira, 2006).

The tabu list is updated dynamically to keep the latest point(s) in the list by replacing the earliest entered pint(s). For example, consider a tabu list of size 20 and the number of evaluated points is 30. The first 20 evaluated points will occupy the corresponding 20 locations in tabu list. The 21st evaluated point then replaces the 1st point in the tabu list and the subsequent points occupy the corresponding positions in the tabu list. In the selection step, a greedy criterion such as better objective value is used to select the better individual between the trial individual and the target individual.

The cycle of generation, evaluation and selection steps in both DETL-E and DETL-G, is repeated until the termination criterion such as maximum number of generations (Gen_{max}) or maximum number of successive generations (Sc_{max}) without improvement in the best function value, is satisfied (Figs. 1 and 2). The best point thus obtained over all the generations is further refined using a local optimization technique, and is considered as the best optimum found. The local optimization technique is used at the end of all the algorithms tested (namely, DE, TS, DETL-E and DETL-G) to find the final solution accurately and efficiently. In this study, a gradient-based algorithm, quasi-Newton method for unconstrained optimization is employed; alternatively, direct search methods such as Nelder–Mead Simplex method can also be used for local optimization. For applications where a near global solution is sufficient, local optimization is not required.

3. Implementation and Evaluation

The FORTRAN code of DE is taken from the website: www.icsi.berkeley.edu/~storn/code.html (accessed in January, 2006), and is modified for DETL-E and DETL-G by including tabu list, respectively at the evaluation and generation steps of DE. Other modifications are checking bound violations and correcting, and the normalization of decision variables to 0 to 1. DETL-E and DETL-G programs are given in the folder of this chapter on the CD. The FORTRAN code for TS is taken from Teh and Rangaiah (2003). For local optimization, subroutine DBCONF from IMSL software is used; it is based on the quasi-Newton method, and gradient of the objective function is calculated numerically.

All the four global optimization methods (DE, TS, DETL-E and DETL-G) are evaluated based on both reliability and computational efficiency in locating the global optimum. Reliability is measured in terms of success rate (SR) (i.e. number of times the algorithm located the global optimum out of 100 trials); and the computational efficiency is measured in terms of both number of function evaluations (NFE) and CPU time required to locate the global optimum. A trial is said to be successful only if the function value at the global optimum is obtained with an absolute error of 10^{-6} or less. This strict condition is used, partly because the function value at the local and global minima is similar up to 5 decimal points in one of the phase stability problems.

Both NFE and CPU time are the average values calculated based on the successful trials out of 100 trials. Unsuccessful trials are not included in NFE and CPU time in order to avoid under-estimating these performance measures. NFE includes both the function calls for objective value calculation and the function calls required for the numerical gradient in the local optimization. The computer system used in this study is Pentium 4 (CPU 2.8 GHz, 1 GB RAM), which can perform 282 million floating point operations per second (MFlops) for the LINPACK benchmark program (available at http://www.netlib.org/; accessed in February, 2007) for a matrix of order 500. Performance comparison in terms of NFE is useful for engineering applications where function evaluation is computationally expensive. On the other hand, performance comparison in terms of CPU time is appropriate for functions involving a few computations where contribution of the optimization algorithm computations to CPU time becomes significant.

The termination criterion used in each of the global optimization methods is reaching either the maximum number of generations (Gen_{max}) or maximum number of successive generations (Sc_{max}) without improvement in the best function value. This criterion is used instead of convergence to the global minimum as a stopping criterion since global minimum is unknown *a priori* in practical applications. Default settings in the DBCONF program are used for termination of the search.

4. Benchmark Problems

Both the versions of DETL are initially evaluated over several test functions involving 2 to 20 variables and a few to thousands of local minima. The test functions are divided into two groups: moderate and difficult based on the degree of difficulty observed in our trials. Functions having a few to several local minima are grouped under "moderate" whereas functions with hundreds to thousands of local minima are grouped under "difficult". This is because the probability of trapping into a local minimum for an algorithm is high for the function with numerous local minima compared to the one with a few local minima. The exceptions are Shubert (SH) and modified Himmelblau (MHB) functions. The former is grouped under moderate functions due to its many global minima (18) even though it has 760 local minima, whereas MHB is grouped under difficult functions even though it has only 4 local minima since it has small global region compared to the local minimum regions. It is important to recognize these differences among benchmark problems and use both moderate and difficult functions in evaluating new global optimization methods. The mathematical equations, global minima and the special features of both moderate and difficult functions are given in Table 1. Variables in all the test functions are normalized between 0 and 1 for consistency, in the optimization program.

4.1. Parameter tuning

The parameters of TS, DE, DETL-E and DETL-G are tuned using some test functions which are found to be relatively more difficult in the preliminary tests. GP, ES, ROS_5, ROS_{10} and ROS_{20} are chosen for moderate functions whereas RA_{10}, RA_{20}, GW_{10} and GW_{20} are selected for difficult functions. See Table 1 for details of these unctions. The parameters are tuned one at a

Table 1a. Details of the benchmark problems.

Function	Dimension and domain	Objective function to be minimized	Global minimum	Remarks
Goldstein and Price (GP)#	2; $-2 \le x_1, x_2 \le 2$	$[1 + (x_1 + x_2 + 1)^2(19 - 14x_1 + 3x_1^2 - 14x_2 + 6x_1x_2 + 3x_2^2)] \times [30 + (2x_1 - 3x_2)^2 (18 - 32x_1 + 12x_1^2 + 48x_2 - 36x_1x_2 + 27x_2^2)]$	3 at $x = \{0, -1\}$	Four local minima
Easom (ES)#	2; $-100 \le x_1, x_2 \le 100$	$-\cos(x_1)\cos(x_2)\exp[-((x_1 - \pi)^2 + (x_2 - \pi)^2)]$	-1 at $x = \{\pi, \pi\}$	Flat objective function
Shubert (SH)#	2; $-10 \le x_1, x_2 \le 10$	$\left\{\sum_{j=1}^{5} j\cos[(j+1)x_1 + j]\right\}\left\{\sum_{j=1}^{5} j\cos[(j+1)x_2 + j]\right\}$	-186.7309 at $x = \{0.0217, -0.9527\}$	760 local minima
Hartmann 3 variables (H_3)#	3; $0 \le x_1, x_2, x_3 \le 1$	$-\sum_{j=1}^{4} c_j \exp\left[-\sum_{i=1}^{3} a_{ji}(x_i - p_{ji})^2\right]$; The constants: c, a and p are given in Table 1b.	-3.86278215 at $x = \{0.114614, 0.555649, 0.852547\}$	4 local minima
Rosenbrock (ROS_N)#	5, 10, 15, 20 and 30; $-5 \le x_i \le 10$	$\sum_{i=1}^{N}[100(x_i^2 - x_{i+1})^2 + (x_i - 1)^2]$	0 at $x = \{1, \ldots, 1\}$	Several local minima

(Continued)

Table 1a. *(Continued)*

Function	Dimension and Domain	Objective Function to be Minimized	Global Minimum	Remarks
Zakharov (ZAK_N)#	5, 10, 15, 20 and 30; $-5 \le x_i \le 10$	$\left(\sum_{i=1}^{N} x_i^2\right) + \left(\sum_{i=1}^{N} 0.5 i x_i^2\right)^2 + \left(\sum_{i=1}^{N} 0.5 i x_i^2\right)^4$	0 at $x = \{0, \ldots, 0\}$	Several local minima
Modified Himmelblau (MHB)*	2; $-6 \le x_i \le 6$	$(x_1^2 + x_2 - 11)^2 + (x_1 + x_2^2 - 7)^2 + 0.1((x_1 - 3)^2 + (x_2 - 2)^2)$	0 at $x = \{3, 2\}$	Four local minima
Rastrigin (RA_N)*	2, 5, 10, 15, 20 and 30; $-600 \le x_i \le 600$	$10n + \sum_{i=1}^{N}(x_i^2 - 10\cos(2\pi x_i))$	0 at $x = \{0, \ldots, 0\}$	Thousands of local minima
Griewank (GW_N)*	10, 15, 20 and 30; $-600 \le x_i \le 600$	$\sum_{i=1}^{N} x_i^2/d - \prod_{i=1}^{N}\cos(x_i/\sqrt{i}) + 1$; where $d = 4000$	0 at $x = \{0, \ldots, 0\}$	Hundreds of local minima

#Moderate Functions; *Difficult Functions.

Table 1b. Constants in the Hartmann 3-variables function (Torn and Zilinskas, 1989).

j	a_{j1}	a_{j2}	a_{j3}	c_j	p_{j1}	p_{j2}	p_{j3}
1	3.0	10.0	30.0	1.0	0.3689	0.1170	0.2673
2	0.1	10.0	35.0	1.2	0.4699	0.4387	0.7470
3	3.0	10.0	30.0	3.0	0.1091	0.8732	0.5547
4	0.1	10.0	35.0	3.2	0.0381	0.5743	0.8828

time while keeping the remaining fixed at their nominal values. The nominal parameter values used are: $N_t = N_p = 10$, $\varepsilon_t = \varepsilon_p = 0.01$, $h_n = 0.5$, $\mathrm{NP_{init}} = 20\,\mathrm{N}$ (where N is the dimension of the problem), $N_{\mathrm{neigh}} = 2\,\mathrm{N}$ (subject to a minimum of 10 and maximum of 30), $\mathrm{Sc_{max}} = 5\,\mathrm{N}$ and $\mathrm{Iter_{max}} = 50$ for TS, and $A = 0.4$, $\mathrm{CR} = 0.1$, $\mathrm{NP} = 50$, $\mathrm{Sc_{max}} = 5\,\mathrm{N}$ and $\mathrm{Gen_{max}} = 50$ for DE. The nominal parameter values for DETL-E and DETL-G are same as those of DE except the additional parameters, tr and tls, which are chosen as 1×10^{-6} and 10 respectively. The nominal values for TS are chosen based on the optimal values available in Chelouah and Siarry (2000) whereas those of DE and DETL are chosen based on the preliminary numerical experience.

The optimal parameter values are obtained by fine tuning the nominal parameter values based on both the reliability and computational efficiency of the algorithms, and are given in Table 2. The optimal values obtained for many TS parameters are the same as the nominal parameter values, probably because they may have already been optimized by Chelouah and Siarry (2000). The optimal parameter values (A, CR, NP, $\mathrm{Gen_{max}}$ and $\mathrm{Sc_{max}}$) for DETL are kept the same as those obtained for DE except the additional parameters, tr and tls, which are tuned separately and the optimum values are given in Table 2. The optimal values of a few parameters in a method are different for each set of functions due to the different characteristics of each group: moderate functions have a few to several local minima whereas difficult functions have huge number of local minima.

4.2. Results and discussion

All the methods are evaluated for solving the moderate functions, and the average SR and NFE are given in Table 3. The SR of DE, DETL-E and

Table 2. Parameter values used in TS, DE, DETL-E and DETL-G.

Parameters	Benchmark problems: Moderate	Benchmark problems: Difficult	Phase Stability Problems
		TS	
N_t and N_p	10	10	10
ε_t and ε_p	0.01	0.01	0.02
NP_{init}	20 N	100 N	20 N
N_{neigh}	2 N	2 N	2 N
h_n	0.5	0.5	0.5
$Iter_{max}$	50 N	100 N	100 N
Sc_{max}	6 N	50 N	2 N
	DE, DETL-E and DETL-G		
A	0.5	0.2	0.3
CR	0.5	0.5	0.9
NP	20	20	Min {40 N, 120}
[a]Gen_{max}	30 N	60 N	50
[a]Sc_{max}	7 N	12 N	6 N
tr[b]	$N \times 10^{-3}$	$N \times 10^{-6}$	$N \times 10^{-3}$
tls[b]	50	50	50

Note: [a]For SH function, $Gen_{max} = 60$ N and $Sc_{max} = 15$ N, where N is the dimension of the problem. [b]These are the two additional parameters in DETL and not involved in DE.

Table 3. Results for benchmark problems — Moderate functions.

Function	DE SR	DE NFE*	TS SR	TS NFE*	DETL-E SR	DETL-E NFE*	DETL-G SR	DETL-G NFE*
GP	100	1082 + 25	99	275 + 32	100	662 + 30	100	636 + 30
ES	64	2420 + 10	85	419 + 14	63	1943 + 10	95	1409 + 171
SH	96	686 + 34	92	327 + 33	100	573 + 33	99	684 + 33
H_3	100	1820 + 26	100	341 + 60	100	800 + 48	100	719 + 47
ROS_5	97	2828 + 197	79	1830 + 283	97	2528 + 203	99	2671 + 196
ZAK_5	100	3020 + 34	100	1252 + 52	100	1147 + 93	100	1071 + 94
ROS_{10}	97	6020 + 657	78	7823 + 793	99	4459 + 684	97	4913 + 609
ZAK_{10}	100	6020 + 104	100	8392 + 100	100	2191 + 252	100	2194 + 252
ROS_{20}	95	12020 + 2306	75	19424 + 2845	100	6993 + 2492	98	9507 + 2363
ZAK_{20}	100	12020 + 607	100	19017 + 194	100	4577 + 652	100	4894 + 726

*The two numbers in each of these columns are NFE required by the method (DE, TS, DETL-E or DETL-G) + NFE required for the local optimization.

DETL-G is comparable, and is better than that of TS except for Easom (ES) function. This could be due to the different escaping mechanisms from the local minima associated with each method. DE, DETL-E and DETL-G employ mutation and crossover operations whereas TS uses the best point found in the current iteration to generate neighbors in the next iteration even though it is worse than those obtained in the earlier iterations. The SR of TS, DE and DETL-E is less for ES function because of the flat objective function. As the objective function is flat, most of the points generated will have the same function value leading to trapping of the search in that region. On the other hand, SR is 95% with DETL-G for ES function. This could be because of the implementation of TS concept in the generation step itself in DETL-G, which successfully explored the global minimum region for this function. SR of all the methods is high for SH function even though it has hundreds (760) of local minima since this function has 18 global minima and locating any one of them is sufficient to achieve the best (global) solution. SR of TS is low for Rosenbrock functions due to the narrow global minimum region associated with them.

The NFE of DE is more than that of TS for small variable problems (GP to ZAK_5 in Table 3), and is less than that of TS for 10 to 20 variable problems (ROS_{10} to ZAK_{20} in Table 3). On the other hand, DETL-E and DETL-G took less NFE though they have higher SR comparable to DE for moderate functions. This is because of avoiding re-visits during the search by the implementation of tabu list in these algorithms. NFE taken by local optimization is around 5% and 8% of the total NFE, respectively for DE and TS. NFE for local optimization in DETL-E and DETL-G is around 9% of the overall NFE, and is slightly higher compared to that of DE. This could be because of the approximate final solutions obtained after DETL alone (i.e. without local optimization) due to the implementation of tabu concept. The percentage reduction in NFE of DETL-E and DETL-G, on average, is about 37% and 35% compared to that of DE. These results clearly show the efficacy of implementing tabu list in DE.

The CPU time (in seconds) for solving the moderate functions by all the methods is given in Table 4. The CPU time taken by TS is less for the functions up to 5 variables and is more for 10 to 20 variables compared to all other methods, due to the associated smaller and higher NFE respectively. Though DETL-E and DETL-G took less NFE, their CPU times are more than that of DE due to the additional computational effort required

Table 4. CPU Time (in seconds) for benchmark problems — Moderate functions.

Function	DE	TS	DETL-E	DETL-G
GP	0.003	0.001	0.005	0.004
ES	0.006	0.003	0.016	0.009
SH	0.002	0.002	0.005	0.005
H_3	0.007	0.002	0.010	0.008
ROS_5	0.008	0.010	0.029	0.029
ZAK_5	0.010	0.008	0.018	0.015
ROS_{10}	0.030	0.115	0.095	0.108
ZAK_{10}	0.028	0.116	0.058	0.063
ROS_{20}	0.105	0.697	0.295	0.463
ZAK_{20}	0.099	0.683	0.232	0.342

(i) for comparing the newly generated members in the population to the points in the tabu list (tabu check step in DETL-E and DETL-G, Figs. 1 and 2), and (ii) for generating new members that are away from those in the tabu list (DETL-G only). Between DETL-E and DETL-G, the latter took slightly less CPU time for small problems (GP, ES, ZAK_2, H_3 and ZAK_5 in Table 4), and more CPU time for 10 to 20 variable problems (i.e. ROS_{10} to ZAK_{20} in Table 4) compared to the former. This could be because of several reasons: (i) associated number of function evaluations, (ii) DETL-G uses only one population set compared to the two sets of population in DETL-E, and (ii) DETL-G requires additional computational effort for generating new members that are away from those in the tabu list compared to DETL-E. The additional CPU time required in generating new members could be high in comparison with less CPU time required by using one set of population for 10 to 20 variable problems resulting more CPU time for DETL-G compared to DETL-E for these problems. Though both DETL-E and DETL-G took more CPU time compared to DE for the moderate benchmark functions, the additional computational effort for tabu checks is expected to be insignificant for application problems where the objective function evaluation takes considerable CPU time.

The performance results (average SR and NFE) of all the methods for difficult benchmark functions are given in Table 5. The SR of DE, DETL-E and DETL-G is high and is significantly better than that of TS. The SR of TS

Table 5. Results for benchmark problems — Difficult functions.

	DE		TS		DETL-E		DETL-G	
Function	SR	NFE*	SR	NFE*	SR	NFE*	SR	NFE*
MHB	89	1303 + 31	65	1507 + 24	89	637 + 31	93	1473 + 26
RA_2	81	2282 + 30	34	1927 + 33	81	1046 + 30	100	1421 + 26
RA_5	100	6020 + 28	—	—	100	3732 + 93	100	3592 + 92
RA_{10}	98	12020 + 56	—	—	98	7745 + 228	95	7544 + 221
RA_{15}	95	18020 + 86	—	—	95	12708 + 329	93	12532 + 323
RA_{20}	89	24020 + 124	—	—	93	19907 + 412	83	19156 + 423
GW_5	90	5972 + 125	12	4914 + 154	90	4764 + 127	98	5889 + 100
GW_{10}	97	12020 + 225	—	—	98	10816 + 282	100	11580 + 260
GW_{15}	100	18020 + 16	7	41217 + 488	99	10422 + 497	99	11105 + 492
GW_{20}	99	23978 + 21	31	55362 + 739	100	11053 + 775	98	11588 + 769

*The two numbers in each of these columns are NFE required by the method (DE, TS, DETL-E or DETL-G) + NFE required for the local optimization.

is zero for most of the functions (RA_5, RA_{10}, RA_{15}, RA_{20} and GW_{10}) and is low for the other functions (MHB, RA_2, GW_5, GW_{15} and GW_{20}) too. TS gets trapped in a local minimum due to the associated huge number of local minima in these functions. On the contrary, DE, DETL-E and DETL-G escaped from the local minima using crossover and mutation resulting in high SR (around 90%). SR of DE, DETL-E and DETL-G is almost comparable for these functions except for RA2, for which SR of DETL-G is 100% whereas it is 81% for both DE and DETL-E. This indicates that implementing the concept of TS in the generation step (i.e. DETL-G, Fig. 2) is marginally better for successfully exploring the global minimum region.

The NFE of DETL-E and DETL-G is less compared to both DE and TS. This is because of avoiding re-visits during the search by implementing the tabu list in them. TS took huge NFE for GW_{15} and GW_{20} because of the associated large number of local minima. NFE required for local optimization is negligible — around 1%, 2%, 3% and 3% in the overall NFE for DE, TS, DETL-E and DETL-G respectively. The percentage reduction in NFE of DETL-E and DETL-G compared to DE is around 34% and 24% respectively. DETL-G took more NFE, on average, compared to DETL-E. This is because in DETL-G, a new member is generated repeatedly until it is far away from the points in the tabu list (Fig. 2) and then evaluated,

Table 6. CPU Time (in seconds) for benchmark problems — Difficult functions.

Function	DE	TS	DETL-E	DETL-G
MHB	0.003	0.004	0.073	0.014
RA_2	0.006	0.006	0.120	0.015
RA_5	0.022	*	0.072	0.070
RA_{10}	0.069	*	0.183	0.288
RA_{15}	0.141	*	0.400	0.724
RA_{20}	0.239	*	0.762	1.429
GW_5	0.023	0.031	0.061	0.073
GW_{10}	0.075	*	0.234	0.267
GW_{15}	0.153	0.853	0.368	0.803
GW_{20}	0.261	1.506	0.578	1.470

*SR is zero and hence CPU time is not meaningful for these cases.

whereas DETL-E simply avoids the current member close to those in the tabu list from the evaluation step without generating again (Fig. 1).

The CPU time averaged over successful trials out of 100, required by all the methods for solving difficult benchmark problems is summarized in Table 6. As for the moderate benchmark problems, the CPU time taken by DETL-E and DETL-G is higher compared to DE due to the associated additional computational effort required for tabu check and generation steps (Figs. 1 and 2) in these methods. Between DETL-E and DETL-G, the latter took less CPU time for small variable problems (i.e. MHB, RA_2 and RA_5 in Table 6) and more CPU time for 10 to 20 variable problems (i.e. from RA_{10} to GW_{20} in Table 6) compared to the former. The possible reasons for these have already been mentioned earlier.

5. Phase Stability Problems

Phase stability problems play a significant role in the design of chemical processes. The objective is to determine whether a given phase is stable or not at a given feed composition, temperature and pressure. Phase stability is often tested using the tangent plane criterion, which states that a phase is thermodynamically stable provided the tangent plane generated at the given composition lies below the molar Gibbs free energy surface for all the compositions. The problem can be formulated as the minimization of tangent plane distance function (TPDF), which is a nonlinear and non-convex objective function requiring global minimization. A review of several works using tangent plane criterion for phase stability problems can be found in Rangaiah (2001).

Tessier *et al.* (2000) implemented the interval Newton technique for phase stability analysis. The examples are modeled by non-random two liquid (NRTL) and universal quasi-chemical (UNIQUAC) thermodynamic models. They also proposed two enhancements for the interval Newton method. The results indicate that the computational efficiency of the enhanced methods is better compared to that of the original one. Nichita *et al.* (2002) used a global optimization method, namely, tunnelling method for phase stability analysis. The problem has been formulated both in conventional approach (i.e. composition space) and in reduced variable approach. The results show that the method is reliable in solving the phase stability problems. Balogh *et al.* (2003) used a modified TPDF such that the zeros of the objective function become its minima, since it is advantageous to search for minima with known zero minimum value. They employed stochastic sampling and clustering method to locate the minima of the modified TPDF. The results show that the method is able to solve small to moderate size problems in an efficient and reliable way.

Xu *et al.* (2005) studied phase stability problems using interval-Newton approach for asymmetric modeling (i.e. different models are used for vapor and liquid phases). In their methodology, a binary variable is introduced in TPDF for accounting the presence of different liquid and vapor phase models. The methodology successfully identified the global minimum for several examples with NRTL as the liquid phase model and a cubic equation of state as the vapor phase model. Bonilla-Petriciolet *et al.* (2006) studied several stochastic methods: simulated annealing, very fast simulated annealing, a modified version direct search simulated annealing and stochastic differential equations, for phase stability analysis of both reactive and non-reactive mixtures. The results show that simulated annealing is more reliable compared to all other methods for the examples tested.

5.1. Problem formulation

Phase stability is often tested using the tangent plane criterion. This criterion simply states that a hypothetical phase is thermodynamically stable

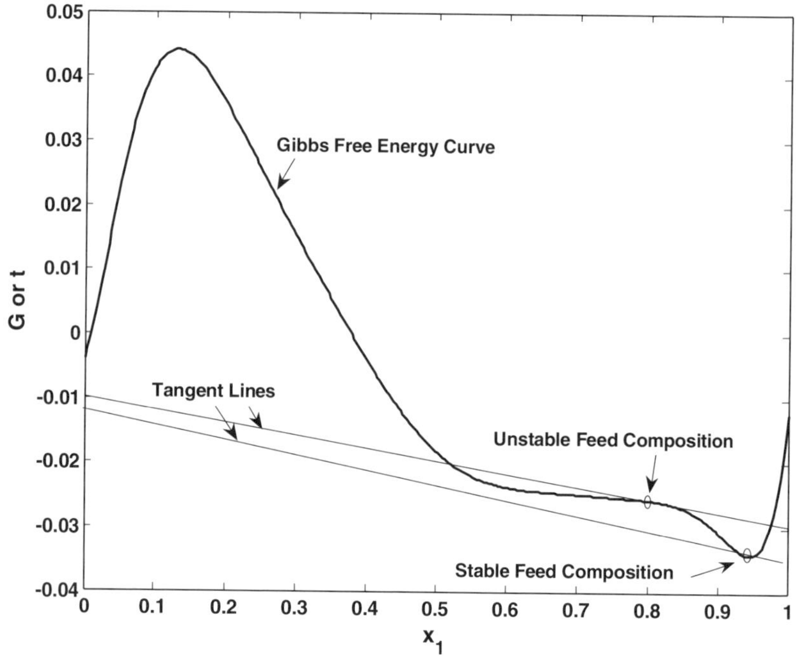

Figure 3. Gibbs free energy curve and tangent lines for unstable and stable feed compositions.

provided the tangent plane drawn at the given feed composition lies below the molar Gibbs energy surface for all compositions. Figure 3 shows geometric interpretation of this criterion for a binary system without reaction; here, molar Gibbs energy of a phase (G) is plotted against the mole fraction of component 1 (x_1), and $x_2 = 1 - x_1$. Consider the following 2 cases. For the given feed composition of $x_1^* = 0.8$, the tangent line is the upper straight line in Fig. 3. Since some portions of g lie below this tangent line, this feed is unstable as a single phase, according to the tangent plane criterion. In another case, the given feed composition is $x_1^* = 0.94$ where the tangent line (the lower straight line in Fig. 3) is entirely below the free energy curve for all values of x_1, and hence the given feed is stable.

One practical implementation of the tangent plane criterion for checking stability of a given phase composition $\mathbf{x}^*$, is to minimize TPDF, defined as the vertical distance between the molar Gibbs energy surface and the tangent plane, i.e. TPDF $= G(\mathbf{x}) - t(\mathbf{x}^*)$ where $\mathbf{x}$ is a trial composition. If the global minimum value of TPDF is found to be zero, then the specified phase is stable. For an unstable phase, global minimum of TPDF will be negative; in fact, a negative TPDF implies an unstable phase.

For a given temperature (T), pressure (P) and composition $\mathbf{x} = (x_1, x_2, x_3, \ldots, x_{nc})$, the molar Gibbs free energy, G of the system is given as the sum of the product of mole fraction (x_i) and partial molar Gibbs free energy ($\bar{G}_i$) for all components:

$$G = \sum_{i=1}^{nc} x_i \bar{G}_i. \tag{2}$$

The tangent plane, t at a specified composition $\mathbf{x}^* = \{x_1^*, x_2^*, x_3^*, \ldots, x_{nc}^*)$ is given as:

$$t(x^*) = \sum_{i=1}^{nc} x_i \bar{G}_i^*, \tag{3}$$

where superscript * represents evaluation at composition x^*. The TPDF can be expressed as:

$$H = G - t(x^*) = \sum_{i=1}^{nc} x_i (\bar{G}_i - \bar{G}_i^*). \tag{4}$$

Depending on the expressions of $\bar{G}_i$ and $\bar{G}_i^*$, different forms of H exist. If the non-ideality of the phase is described by fugacity approach, then the dimensionless H/RT can be expressed as:

$$\frac{H}{RT} = \sum_{i=1}^{nc} x_i [\ln(\phi_i x_i) - \ln(\phi_i^* x_i^*)], \tag{5}$$

where R is the universal gas constant, and ϕ_i represents the fugacity coefficient of component i in the given phase. If the excess Gibbs free energy approach is used to represent the non-ideality, then Eq. (4) becomes:

$$\frac{H}{RT} = \frac{g^E}{RT} + \sum_{i=1}^{nc} x_i [\ln x_i - \ln x_i^* \gamma_{iL}^*], \tag{6}$$

where g^E represents the excess Gibbs free energy and γ_{iL} represents the activity coefficient of component i in the liquid phase L.

Depending upon the approach, the objective function in the minimization of TPDF is either Eqs. (5) or (6), and the constraints are:

$$\sum_{i=1}^{nc} x_i = 1 \quad \text{and} \quad 0 \leq x_i \leq 1. \tag{7}$$

The decision variables in phase stability problems are x_i for $i = 1, 2, \ldots, nc$. The constrained problem can be transformed into an unconstrained one by introducing the variables, β_i (for $i = 1, 2, \ldots, nc - 1$) instead of x_i for $i = 1, 2, \ldots, nc$ (Rangaiah, 2001). The unconstrained problem consists of new decision variables: β_i for $i = 1, 2, \ldots, nc - 1$ instead of x_i for $i = 1, 2, \ldots, nc$ as the decision variables. Each β_i is bounded by 0 and 1, and is related to x_i by:

$$x_1 = \beta_1 x_1^*, \tag{8}$$

$$x_i = \beta_i \left(1 - \sum_{j=1}^{i-1} x_j^* \right) \quad i = 2, 3, \ldots, nc - 1, \tag{9}$$

$$x_{nc} = 1 - \sum_{j=1}^{nc-1} x_j. \tag{10}$$

Thus the equality constraints in Eq. (7), which are automatically satisfied due to Eqs. (8) to (10), need not be considered during minimization, and the number of decision variables is reduced to $nc - 1$. The bounds on decision variables are set as:

$$1 \times 10^{-15} \leq \beta_i \leq 1.0 \quad i = 1, 2, \ldots, nc - 1. \tag{11}$$

The lower bound of 1×10^{-15} instead of 0 is employed to avoid numerical difficulties in evaluating free energy when mole fraction of a component in a phase is zero.

The simplified problem can now be solved using an unconstrained optimization method. Five examples were considered for testing DE, TS, DETL-E and DETL-G for phase stability problems (Table 7a). These examples have multiple components and different thermodynamic models for describing the phase behavior. Number of components, temperature, pressure and the reference for each of the examples are summarized in Table 7a. Several compositions are considered for most of these examples; these and

Table 7a. Details of phase stability problems.

Example number	Components	Temperature and Pressure	Thermodynamic model*	Reference
1	n-Butyl acetate and water	298 K and 1 atm	NRTL	Rangaiah (2001)
2	Ethyl glycol, dodecanol and nitromethane	295 K and 1 atm	UNIQUAC	Rangaiah (2001)
3	Methane and hydrogen sulfide	190 K and 40.53 bar	SRK	Rangaiah (2001)
4	Nine hydrocarbons (C_1, C_2, C_3, iC_4, nC_4, iC_5, nC_5, nC_6 and iC_{15})	314 K and 19.84 atm	SRK	Rangaiah (2001)
5	Toluene, water and aniline	298 K and 1 atm	NRTL	Castillo and Grossmann (1981)

*Equations for thermodynamic models and the data for each example are given in Appendix A.

Table 7b. Details of Example 1 — n-Butyl acetate and water.

No.	Feed composition	Global solution: TPDF	Global solution: x
1	(0.5, 0.5)	−0.03246624	(0.004210, 0.995790)
2	(0.1, 0.9)	−0.21418620	(0.963452, 0.036548)
3	(0.2, 0.8)	−0.07427426	(0.003796, 0.996204)
4	(0.65, 0.35)	−0.00671171	(0.941306, 0.058694)
5	(0.93514, 0.06486)	−0.00070557	(0.594235, 0.405765)
6	(0.59199, 0.40801)	0.0	(0.591987, 0.408013)

the corresponding global minima are given in Tables 7b to 7f. Thermodynamic models and the corresponding data are presented in Appendix A.

5.2. Results and discussion

The parameters of all the methods are tuned based on example 2 (compositions 2, 4 and 5), which were found to be difficult in the preliminary trials.

Table 7c. Details of Example 2 — Ethyl glycol, dodecanol and nitromethane.

No.	Feed composition	Global solution TPDF	Global solution x
1	(0.4, 0.3, 0.3)	−0.11395074	(0.754252, 0.002219, 0.243529)
2	(0.27078, 0.47302, 0.25620)	−0.05876840	(0.023340, 0.001726, 0.974934)
3	(0.2, 0.3, 0.5)	−0.22827470	(0.012537, 0.001128, 0.986335)
4	(0.29672, 0.46950, 0.23378)	−0.02700214	(0.715399, 0.003359, 0.281242)
5	(0.27899, 0.49191, 0.22910)	0.0	(0.278990, 0.491910, 0.229100)

Table 7d. Details of Example 3 — Methane and hydrogen sulfide.

No.	Feed composition	Global solution TPDF	Global solution x
1	(0.9813, 0.0187)	−0.00395983	(0.923310, 0.076690)
2	(0.5, 0.5)	−0.08252179	(0.925382, 0.074618)
3	(0.112, 0.888)	−0.00246629	(0.920822, 0.079178)

Table 7e. Details of Example 4 — Mixture of 9 hydrocarbons.

Component no.	Feed composition	Global solution TPDF	Global solution x
1	0.61400	−1.48621570	0.946076
2	0.10259		0.043568
3	0.04985		0.007851
4	0.00898		0.000675
5	0.02116		0.001247
6	0.00722		0.000197
7	0.01187		0.000264
8	0.01435		0.000121
9	0.16998		0.000000

Table 7f. Details of Example 5 — Toulene, water and aniline.

No.	Feed composition	Global solution	
		TPDF	x
1	(0.29989, 0.20006, 0.50005)	−0.29454012	(0.000067, 0.996865, 0.003068)
2	(0.34674, 0.07584, 0.57742)	0.0	(0.346740, 0.075840, 0.577420)

The parameter values obtained for these problems are given in Table 2, and are slightly different from those of benchmark problems because these problems have a few but comparable minima (i.e. function values at the local and global minima are close to each other).

As before, all the examples are solved 100 times, each time with randomly generated initial estimates, and the computational efficiency of the methods is compared based on successful trials out of 100 trials (Table 8). The reliability of DE, DETL-E and DETL-G is close to 100%, whereas that of TS is less than 100% for several cases (example 2 with compositions 2, 4 and 5; example 3 with composition 3). TS trapped into the local minimum region for these examples due to the presence of comparable minima (function values at local and global minima are -4.11636×10^{-6} & -0.05876117, -3.09637×10^{-6} & -0.02700214, and 1.323587×10^{-6} & 0.0 respectively for 2nd, 4th and 5th compositions of example 2; 0.0 and -0.00246629 for 3rd composition of Example 3); on the other hand, DE, DETL-E and DETL-G successfully explored the global minimum regions giving almost 100% SR for all these cases.

Even though SR of DE is high, its NFE is higher compared to other methods tested for the phase stability problems. On the other hand, SR of both DETL-E and DETL-G is comparable to DE, and their NFE is less than that of DE, and comparable to that of TS. This is because of the implementation of the tabu list in DE. NFE required for the local optimization is small — around 1%, 5%, 2% and 2% of the total NFE for DE, TS, DETL-E and DETL-G respectively. The average percentage reduction in NFE of DETL-E and DETL-G compared to DE is 59% and 63% respectively for the phase stability problems. These results clearly show the potential of DETL-E and DETL-G, especially for problems involving computationally intensive objective functions.

Table 8. Results for phase stability problems.

	DE		TS		DETL-E		DETL-G	
Composition	SR	NFE*	SR	NFE*	SR	NFE*	SR	NFE*
				Example 1				
1	97	1266 + 13	99	519 + 48	100	353 + 11	97	487 + 6
2	100	1555 + 7	100	604 + 28	100	723 + 11	100	316 + 11
3	100	1574 + 9	100	417 + 48	99	439 + 13	100	374 + 15
4	100	1605 + 3	100	543 + 28	100	364 + 9	100	312 + 9
5	100	1436 + 2	100	564 + 19	100	354 + 9	100	310 + 8
6	100	1514 + 3	100	562 + 19	100	351 + 9	100	295 + 8
				Example 2				
1	100	4086 + 6	100	1157 + 39	100	1612 + 23	100	1480 + 23
2	99	3971 + 24	64	1205 + 58	100	2396 + 35	97	2275 + 34
3	100	4061 + 25	100	1196 + 76	100	1990 + 32	100	1895 + 32
4	100	3996 + 6	91	1197 + 38	100	1858 + 26	100	1686 + 25
5	99	4053 + 3	84	1259 + 30	99	1902 + 22	100	1675 + 22
				Example 3				
1	100	1397 + 3	97	555 + 16	100	402 + 9	100	348 + 8
2	100	1582 + 3	100	561 + 16	100	377 + 8	100	328 + 9
3	100	1484 + 3	80	558 + 16	100	473 + 9	100	433 + 8
				Example 4				
1	100	6120 + 556	100	3945 + 992	100	6116 + 542	100	6045 + 394
				Example 5				
1	100	4027 + 28	100	1223 + 83	100	2192 + 49	100	2006 + 52
2	100	4063 + 3	100	1250 + 30	100	1505 + 21	100	1351 + 21

Note: *The two numbers in each of these columns are NFE required by the method (DE, TS, DETL-E or DETL-G) + NFE required for the local optimization.

The CPU time taken by the methods, averaged over successful trials out of 100 trials, is given in Table 9; TS took less CPU time compared to other methods due to its smaller NFE. CPU time of DETL-E and DETL-G is comparable to that of DE even though the former methods require additional computational effort for the comparison and generation steps (Figs. 1 and 2). It is because of the about 60% less NFE required by both DETL-E and DETL-G compared to DE for these problems. DETL-G took less CPU time compared to DETL-E due to the associated smaller NFE except for Example 1, 2nd composition.

Table 9. CPU Time (in seconds) for phase stability problems.

Composition	DE	TS	DETL-E	DETL-G
		Example 1		
1	0.009	0.006	0.010	0.008
2	0.007	0.006	0.010	0.007
3	0.008	0.003	0.011	0.007
4	0.009	0.005	0.010	0.008
5	0.008	0.006	0.010	0.007
6	0.008	0.004	0.010	0.008
		Example 2		
1	0.040	0.014	0.029	0.019
2	0.040	0.019	0.040	0.028
3	0.042	0.013	0.034	0.025
4	0.042	0.014	0.032	0.021
5	0.041	0.013	0.032	0.021
		Example 3		
1	0.010	0.015	0.010	0.008
2	0.010	0.011	0.010	0.006
3	0.010	0.011	0.010	0.007
		Example 4		
1	0.176	0.175	0.253	0.214
		Example 5		
1	0.035	0.011	0.034	0.022
2	0.035	0.010	0.025	0.015

6. Benchmark Problems Similar to Phase Stability Problems

It is clear from the previous section that phase stability problems have special characteristics of a few but comparable minima, similar to phase equilibrium problems (Srinivas and Rangaiah, 2006); the local minimum in these problems is some times in a narrow valley. Although there are many benchmark problems in the literature with different characteristics (such as flat objective function and huge number of local minima), none of them represents the comparable minima as in phase stability problems. Motivated from the unique characteristic of phase stability problems, a new benchmark problem with similar characteristics is developed. This and the benchmark problem proposed in Srinivas and Rangaiah (2006) are used for evaluating the four methods. The proposed benchmark problem has only a few comparable minima whereas the test problem in Srinivas and Rangaiah (2006) has huge number of comparable minima. Owing to their simple and mathematical nature, these test functions can easily be used by all researchers in global optimization.

The benchmark problem is developed from the Rosenbrock function as it has a few minima as in phase stability and phase equilibrium problems. The minima in this function are made comparable by adding a multiplier (α_r/N) to the quadratic term:

$$f(x) = \sum_{i=1}^{N-1} \left[100(x_i^2 - x_{i+1})^2 + \frac{\alpha_r}{N}(x_i - 1)^2 \right], \qquad (12)$$

where α_r is a constant and N is the dimension of the problem. Each variable is bounded between −5 and 10 as in the Rosenbrock function (Table 1). Both Rosenbrock function and the modified one do not have local minima up to and including 3 variables but they have several local minima (many of them are constrained minima) beginning from 4 variables. The minima in the modified function are made comparable to the global minimum by decreasing the effect of the quadratic term via α_r, but the global minimum is unaffected (i.e. 0.0 at $x_i = 1.0$ for $i = 1, 2, \ldots, N$). These effects can be seen in Table 10 for the 5-variable case; as α_r decreases, the objective function becomes slightly flat and the minima become comparable but the global minimum and its location remain unaffected. The comparable minima for the modified Rosenbrock function with 4 to 20 variables are given in Table 11 for $\alpha_r = 1.5 \times 10^{-3}$. The difference in function value between the nearest local and global minimum is in the range of 1.39×10^{-3} to 3×10^{-4} for 4 to 20 variables, which is within the range (10^{-2} to 10^{-6}) of phase stability problems tested.

Table 10. Trend of comparable minima with α_r for the modified Rosenbrock function (5 variables case).

α_r	1	2.5×10^{-2}	1.5×10^{-2}	5×10^{-3}	1.5×10^{-3}
Local minimum	3.93084	1.972×10^{-2}	1.183×10^{-2}	3.944×10^{-3}	1.183×10^{-3}
Global minimum	0.0	0.0	0.0	0.0	0.0

Table 11. Function values at the comparable minimum for the modified Rosenbrock function with $\alpha_r = 1.50 \times 10^{-3}$. The global minimum is 0.0 at $x_i = 1.0$ for $i = 1, 2, \ldots, N$.

Number of variables (N)	Function value at the comparable minimum
4	1.394×10^{-3}
5	1.183×10^{-3}
6	9.968×10^{-4}
8	7.498×10^{-4}
10	5.999×10^{-4}
12	4.999×10^{-4}
14	4.2857×10^{-4}
16	3.75×10^{-4}
18	3.333×10^{-4}
20	3×10^{-4}

Table 12. Trend of comparable minima with α_n for the modified N-dimensional test function (4 variables case).

α_n	0.0	0.3	0.4304
Local minimum	−142.52794	−152.29792	−156.66095
Global minimum	−156.66466	−156.66466	−156.66466

Srinivas and Rangaiah (2006) developed a new benchmark problem from the N-dimensional test function reported in Cetin *et al.* (1993). Unlike Rosenbrock function, this function has huge number (2^N) of local minima. The minima in this function are made comparable by adding a quadratic term multiplied with a constant (α_n):

$$f(x) = \left(\frac{1}{2}\right)\sum_{i=1}^{N}(x_i^4 - 16x_i^2 + 5x_i) - \alpha_n \sum_{i=1}^{N}(x_i + 2.90353)^2, \quad (13)$$

where the search domain is $-5 \leq x_i \leq 5$. The global minimum of the modified unction for some values of α_n (e.g. 0.3, 0.42, 0.4304) is still located at $x_i = -2.90353$ for $i = 1, 2, \ldots, N$. As α_n value changes in Eq. (13), the minima become comparable and are given in Table 12 for 4

Table 13. Function values at the comparable minimum for the modified N-dimensional test function with $\alpha_n = 0.4304$.

Number of variables (N)	Function value at the comparable minimum	Function value at the global minimum ($x_i = -2.90353$ for $i = 1, 2, \ldots, N$)
2	−78.32862	−78.33231
4	−156.66095	−156.66466
5	−195.82712	−195.83082
6	−234.99328	−234.99699
8	−313.32562	−313.32932
10	−391.65795	−391.66165
12	−469.99028	−469.99398
14	−548.32261	−548.32631
16	−626.65494	−626.65865
18	−704.98727	−704.99098
20	−783.31960	−783.32331

variables case. At $\alpha_n = 0.4304$, the difference in function value between the nearest local and global minimum is around 3.70×10^{-3} for 2 to 20 variables (Table 13). This modified N-dimensional test function is more difficult than the modified Rosenbrock function since it has huge number of local minima compared to the latter.

DETL-E and DETL-G are further tested for the modified Rosenbrock and the modified N-dimensional problems with $\alpha_r = 1.5 \times 10^{-3}$ and $\alpha_n = 0.4304$, and the results are compared to those of DE and TS. The optimal parameter values used for DE, DETL-E and DETL-G are the same as those for moderate functions and difficult functions (Table 2) respectively for modified Rosenbrock function and modified N-dimensional function except for NP = 30, $\text{Gen}_{\text{max}} = 100\,\text{N}$ and $\text{Sc}_{\text{max}} = 12\,\text{N}$. For TS, the parameter values used are the same as those for difficult functions (Table 2).

The performance results (SR and NFE) averaged over successful trials out of 100, are given in Tables 14 and 15 for the modified Rosenbrock and the modified N-dimensional test functions respectively. SR of DE, DETL-E and DETL-G is comparable, and is better than that of TS for both the functions. SR of the methods tried either improves or is unaffected as the number of variables increases for the modified Rosenbrock function. This is because the number of minima in this function is only a few although the minima are comparable as in phase stability problems. On the other hand, SR of

Table 14. Results for the modified Rosenbrock function with $\alpha_r = 1.5 \times 10^{-3}$.

	DE		TS		DETL-E		DETL-G	
Dimension	SR	NFE	SR	NFE	SR	NFE	SR	NFE
4	82	10,832	76	4749	86	9988	85	10,631
5	90	17,235	76	7111	94	14,459	81	15,438
10	96	47,530	81	25,838	99	33,528	98	41,484
15	97	86,995	81	64,360	98	60,425	98	76,104
20	100	138,865	78	98,849	96	99,522	100	125,085

Table 15. Results for the modified N-dimensional test function with $\alpha_n = 0.4304$.

	DE		TS		DETL-E		DETL-G	
Dimension	SR	NFE	SR	NFE	SR	NFE	SR	NFE
2	70	4503	21	1483	70	2000	93	2301
4	85	11463	5	2779	78	4868	81	4167
5	73	13198	3	5524	79	5998	73	5640
10	44	21843	0	—	42	11905	35	11714
15	19	33908	0	—	20	20316	5	20283
20	2	47541	0	—	3	33854	4	28941

all the methods decreases with the number of variables for the modified N-dimensional function (Table 15) due to the increase in the number of minima with the number of variables according to 2^N. SR of DETL-G is high (93%) for the 2 variable N-dimensional test function compared to DE and DETL-E. This could be due to the implementation of TS concept in the generation step itself, which in turn is able to explore the global minimum region successfully. Compared to DE, NFE of DETL-E and DETL-G is respectively 22% and 9% less for the modified Rosenbrock function, and 47% and 49% less for the modified N-dimensional test function.

In order to see the effect of comparable minima when they are in large number, DETL-E and DETL-G are evaluated for $\alpha_n = 0$, 0.3 and 0.42 for the modified N-dimensional test function, and the results are presented in Table 16. At $\alpha_n = 0$, SR of DETL-E and DETL-G is 100%, and decreases as α_n increases. NFE of both DETL-E and DETL-G increases as the minima

Table 16. Effect of α_n value on the performance of DETL-E and DETL-G for modified N-dimensional test function.

	SR and NFE (in brackets) of DETL-E for the function with α_n =				SR and NFE (in brackets) of DETL-G for the function with α_n =			
Dimension	0.0	0.3	0.42	0.4304	0.0	0.3	0.42	0.4304
2	100 (1689)	100 (1665)	98 (1736)	70 (2000)	100 (1804)	100 (1872)	100 (2027)	93 (2301)
4	100 (3402)	100 (3559)	98 (4022)	78 (4868)	100 (3287)	100 (3421)	97 (3912)	81 (4167)
5	100 (4053)	100 (4324)	95 (4943)	79 (5998)	100 (3934)	100 (4226)	98 (4762)	73 (5640)
10	100 (7106)	100 (7764)	89 (9304)	42 (11905)	100 (7180)	100 (7817)	83 (9330)	35 (11714)
15	100 (9772)	100 (11301)	68 (14694)	20 (20316)	100 (10245)	98 (11570)	67 (15023)	5 (20283)
20	100 (12431)	97 (15211)	46 (21737)	3 (33854)	99 (13249)	97 (15755)	41 (22154)	4 (28941)

become comparable (i.e. α_n changes from 0 to 0.4304). These results clearly show the challenging nature of the comparable minima to a global optimization algorithm, particularly when the dimension of the problem is high.

Overall, the performance of DETL-E and DETL-G is better than that of DE and TS in terms of NFE and SR for the examples tested. In terms of CPU time, TS performs better for small variable problems (up to 5 variables) whereas DE performs better for 10 to 20 variable problems compared to all other methods. The relative performance of DETL-E and DETL-G is comparable; this is consistent with the comparable performance of DE and MDE (Srinivas and Rangaiah, 2007b) since the former is similar to DETL-E whereas the latter is similar to DETL-G except for the inclusion of the tabu list and checks. In the present study, DETL-E and DETL-G are evaluated and compared with DE and TS for continuous problems only. They need to be tested for non-differentiable and constrained problems.

7. Conclusions

This study describes and evaluates two methods, namely, DETL-E and DETL-G with tabu list and check in the evaluation and generation step of

DE respectively. Initially, the methods are applied to two sets of benchmark problems, namely, moderate and difficult functions, which involve 2 to 20 variables and a few to thousands of local minima. The reliability of DETL-E and DETL-G is found to be comparable to that of DE, and is better than that of TS for both moderate and difficult functions. NFE of DETL-E and DETL-G is around 37% and 35% less, and 34% and 24% less compared to DE, respectively for moderate and difficult functions. The methods are then tested for challenging phase stability problems which include several components. Both DETL-E and DETL-G located the global minimum successfully for these problems with almost 100% reliability similar to DE, and about 60% less NFE compared to DE. Overall, the performance of DETL-E and DETL-G is found to be better than that of DE and TS. A new benchmark problem (modified Rosenbrock function) with characteristics similar to phase stability problems is proposed. This facilitates the development and testing of global optimization algorithms for phase stability type of problems.

Nomenclature

A	Amplification factor
CR	Crossover constant
g^E	Excess Gibbs free energy
G	Gibbs free energy
$\bar{G}$	Partial molar Gibbs free energy
Gen_{max}	Maximum number of generations
h_n	Length of the hyper rectangle
H	Tangent plane distance function
Iter_{max}	Maximum number of iterations
nc	Number of components
N	Number of decision variables in the problem
NP	Population size
NP_{init}	Initial population size
N_t and N_p	Number of tabu and promising points
N_{neigh}	Number of neighbors in each iteration
P	System pressure

R	Universal gas constant
Sc_{max}	Maximum number of successive generations without improvement in the best function value
t	Tangent plane distance function
tls	Tabu list size
tr	Tabu radius
T	System temperature
$\mathbf{v}_{i,\text{J}+1}$	Mutant individual i for generation J + 1
$\mathbf{x}_{i,\text{J}}$	Target individual i in generation J
$\mathbf{x}$	Vector of continuous variables
x_i^l and x_i^u	Lower and upper bounds on decision variable, x_i

Greek letters

α_r	Constant in the modified Rosenbrock function
α_n	Constant in the modified N-dimensional test function
β_i	A new decision variable introduced in place of x_i
γ_i	Activity coefficient of component i
ϕ_i	Fugacity coefficient of component i
ε	Radius

References

Anderson, T.F. and Prausnitz, J.M. (1978). Application of the UNIQUAC equation to calculation of multicomponent phase equilibria: 1. vapor-liquid equilibria. *I and EC Proc. Des. Dev.*, **17**, pp. 552–567.

Babu, B.V. and Angira, R. (2006). Modified differential evolution (MDE) for optimization of non-linear chemical processes. *Computers and Chemical Engineering*, **30**, pp. 989–1002.

Balogh, J., Csendes, T. and Stateva, R.P. (2003). Application of a stochastic method to the solution of the phase stability problem: cubic equations of state. *Fluid Phase Equilibria*, **212**, pp. 257–267.

Bonilla-Petriciolet, A., Vazques-Roman, R., Iglesias-Silva, G.A. and Hall, K.R. (2006). Performance of stochastic global optimization methods in the calculation of phase stability analyses for non-reactive and reactive mixtures. *Industrial and Engineering Chemistry Research*, **45**, pp. 4764–4772.

Castillo, J. and Grossmann, I.E. (1981). Computation of phase and chemical equilibria. *Computers and Chemical Engineering*, **5**, pp. 99–108.

Cetin, B.C., Barhen, J. and Burdick, W.J. (1993). Terminal repeller unconstrained subenergy tunneling for fast global optimization. *Journal of Optimization Theory and Applications*, **77**, pp. 97–126.

Chelouah, R. and Siarry, P. (2000). Tabu search applied to global optimization. *European Journal of Operational Research*, **123**, pp. 256–270.

Chiou, J.P., Chang, C.F. and Su, C.T. (2004). Ant direction hybrid differential evolution for solving large capacitor placement problems. *IEEE Transactions on Power Systems*, **19**, pp. 1794–1800.

Chiou, J.P. and Wang, F.S. (1999). Hybrid method of evolutionary algorithms for static and dynamic optimization problems with application to fed-batch fermentation process. *Computers and Chemical Engineering*, **23**, pp. 1277–1291.

Glover, F. (1989). Tabu search: part 1. *ORSA Journal on Computing*, **1**, pp. 190–206.

Hendtlass, T. (2001). A combined swarm differential evolution algorithm for optimization problems. *Lecture Notes in Artificial Intelligence*, **2070**, pp. 11–18.

Karaboga, N. and Cetinkaya, B. (2004). Performance comparison of genetic and differential evolution algorithms for digital FIR filter design. *3rd International Conference on Advances In Information Systems, Lecture Notes in Computer Science*, **3261**, pp. 482–488.

Lee, M.H., Han, C. and Chang, K.S. (1999). Dynamic optimization of continuous polymer reactor using modified differential evolution algorithm. *Industrial and Engineering Chemistry Research*, **38**, pp. 4825–4831.

Liu, Y., Ma, L. and Zhang, J. (2002). Reactor power optimization by GA/SA/TS combined algorithms. *Electrical Power and Energy Systems*, **24**, pp. 765–769.

Nichita, D.V., Gomez, S. and Luna, E. (2002). Phase stability analysis with cubic equations of state by using a global optimization method. *Fluid Phase Equilibria*, **194–197**, pp. 411–437.

Price, K.V., Storn, R.M. and Lampinen, J.A. (2005). *Differential evolution.* Germany: Springer-Verlag.

Rangaiah, G.P. (2001). Evaluation of genetic algorithms and simulated annealing for phase equilibrium and stability problems. *Fluid Phase Equilibria*, **187–188**, pp. 83–109.

Renon, H. and Prausnitz, J.M. (1968). Local compositions in thermodynamic excess functions for liquid mixtures. *AlChE*, **14**, pp. 135–144.

Soave, G. (1972). Equilibrium constants from a modified Redlich-Kwong equation of state. *Chemical Engineering Science*, **27**, pp. 1197–1203.

Srinivas, M. and Rangaiah, G.P. (2006). Implementation and evaluation of random tunneling algorithm for chemical engineering applications. *Computers and Chemical Engineering*, **30**, pp. 1400–1415.

Srinivas, M. and Rangaiah, G.P. (2007a). A study of differential evolution and tabu search for benchmark, phase equilibrium and phase stability problems. *Computers and Chemical Engineering*, **31**, pp. 760–772.

Srinivas, M. and Rangaiah, G.P. (2007b). Differential evolution with tabu list for global optimization and its application to phase equilibrium and parameter estimation problems. *Industrial and Engineering Chemistry Research*, **46**, pp. 3410–3421.

Storn, R. and Price, K. (1997). Differential evolution — A simple and efficient heuristic for global optimization over continuous spaces. *Journal of Global Optimization*, **11**, pp. 341–359.

Teh, Y.S. and Rangaiah, G.P. (2003). Tabu search for global optimization of continuous functions with application to phase equilibrium calculations. *Computers and Chemical Engineering*, **27**, pp. 1665–1679.

Teo, J. (2006). Exploring dynamic self-adaptive populations in differential evolution. *Soft Computing*, **10**, pp. 673–686.

Tessier, S.R., Brennecke, J.F. and Stadtherr, M.A. (2000). Reliable phase stability analysis for excess Gibbs energy models. *Chemical Engineering Science*, **55**, pp. 1785–1796.

Torn, A. and Zilinskas, A. (1989). *Global Optimization*. Berlin, Springer-Verlag.

Xu, G., Haynes, W.D. and Stadtherr, M.A. (2005). Reliable phase stability analysis for asymmetric models. *Fluid Phase Equilibria*, **235**, pp. 152–165.

Appendix A

1. Nonrandom Two–Liquid (NRTL) Model (Renon and Prausnitz, 1968)

For multi-component systems, the excess Gibbs free energy (g^E) is as follows:

$$\frac{g^E}{RT} = \sum_{i=1}^{nc} x_i \frac{\sum_{j=1}^{nc} \tau_{ji} G_{ji} x_j}{\sum_{j=1}^{nc} G_{ji} x_j}, \tag{A.1}$$

where

$$G_{ji} = \exp(-\alpha_{ji}\tau_{ji}) \quad \text{for} \quad i, j = 1, \ldots, nc, \tag{A.2}$$

$$\tau_{ji} = \frac{g_{ji} - g_{ii}}{RT} \quad \text{for} \quad i, j = 1, \ldots, nc. \tag{A.3}$$

In the above equations, τ_{ji} is a non–symmetric binary interaction parameter (with $\tau_{ii} = 0$) and α_{ji} is a non–randomness parameter, which is another adjustable binary parameter (with $\alpha_{ii} = 0$ and $\alpha_{ij} = \alpha_{ji}$). The parameter, g_{ji} is the interaction energy between an $i - j$ pair of molecules.

The activity coefficient of component i for NRTL model is:

$$\ln \gamma_i = \frac{\sum_{j=1}^{nc} G_{ji}\tau_{ji}x_j}{\sum_{k=1}^{nc} G_{ki}x_k} + \sum_{j=1}^{nc} \frac{G_{ij}x_j}{\sum_{k=1}^{N} G_{kj}x_k}\left[\tau_{ij} - \frac{\sum_{\ell=1}^{nc} G_{\ell j}\tau_{\ell j}x_\ell}{\sum_{k=1}^{nc} G_{kj}x_k}\right] \quad \text{for } i = 1, \ldots, nc. \tag{A.4}$$

Data for Example 1 (n-Butyl Acetate and Water): $\tau_{1,1} = 0.0$, $\tau_{1,2} = 3.00498$, $\tau_{2,1} = 4.69071$, $\tau_{2,2} = 0.0$, $\alpha_{1,1} = 0.0$, $\alpha_{1,2} = 0.391965$, $\alpha_{2,1} = 0.391965$ and $\alpha_{2,2} = 0.0$.

Data for Example 5 (Toluene, Water and Aniline): $\tau_{1,1} = 0.0$, $\tau_{1,2} = 4.93035$, $\tau_{1,3} = 1.59806$, $\tau_{2,1} = 7.77063$, $\tau_{2,2} = 0.0$, $\tau_{2,3} = 4.18462$, $\tau_{3,1} = 0.03509$, $\tau_{3,2} = 1.27932$, $\tau_{3,3} = 0.0$, $\alpha_{1,1} = 0.0$, $\alpha_{1,2} = 4.93035$, $\alpha_{1,3} = 1.59806$, $\alpha_{2,1} = 7.77063$, $\alpha_{2,2} = 0.0$, $\alpha_{2,3} = 4.18462$, $\alpha_{3,1} = 0.03509$, $\alpha_{3,2} = 1.27932$ and $\alpha_{3,3} = 0.0$.

2. Universal Quasi–Chemical (UNIQUAC) Model (Anderson and Prausnitz, 1978)

The excess Gibbs free energy (g^E) of a system having *nc* components is given by:

$$\frac{g^E}{RT} = \frac{g^E_{\text{Combinatorial}}}{RT} + \frac{g^E_{\text{Residual}}}{RT}, \tag{A.5}$$

where
$$\frac{g^E_{\text{Combinatorial}}}{RT} = \sum_{i=1}^{nc} x_i \left(\ln \frac{\phi_i}{x_i}\right) + \frac{z}{2}\sum_{i=1}^{nc} x_i q_i \ln \frac{\theta_i}{\phi_i} \quad \text{for } i = 1, \ldots, nc, \tag{A.6}$$

$$\frac{g^E_{\text{Residual}}}{RT} = -\sum_{i=1}^{nc} x_i q'_i \ln\left(\sum_{j=1}^{nc} \theta'_j \tau_{ji}\right) \quad \text{for } i = 1, \ldots, nc. \tag{A.7}$$

Here, τ_{ji} is the binary interaction parameter (with $\tau_{ii} = 0$), ϕ_i is the average segment fraction, θ and θ' are the area fraction and the volume fraction respectively, which are given by:

$$\tau_{ji} = \exp\left[-\frac{u_{ji} - u_{ii}}{RT}\right] \quad \text{for } i, j = 1, \ldots, nc, \tag{A.8}$$

$$\phi_i = \frac{r_i x_i}{\sum_{j=1}^{nc} r_j x_j} \quad \text{for } i = 1, \ldots, nc, \tag{A.9}$$

$$\theta_i = \frac{q_i x_i}{\sum_{j=1}^{nc} q_j x_j} \quad \text{for } i = 1, \ldots, nc, \tag{A.10}$$

$$\theta_i' = \frac{q_i' x_i}{\sum_{j=1}^{nc} q_j' x_j} \quad \text{for } i = 1, \ldots, nc. \tag{A.11}$$

In the above equations, u_{ji} is the interaction energy, r_i is the pure component volume parameter and q_i is the pure component area parameter. The parameter, q_i' is the adjusted value of q_i for water and alcohols; for components other than water and alcohols, $q_i = q_i'$.

The activity coefficient for component i is defined as:

$$\ln \gamma_i = \ln \gamma_i^{\text{Combinatorial}} + \ln \gamma_i^{\text{Residual}} \quad \text{for } i = 1, \ldots, nc, \tag{A.12}$$

where $$\ln \gamma_i^{\text{Combinatorial}} = \ln \frac{\phi_i}{x_i} + \frac{z}{2} q_i \ln \frac{\theta_i}{\phi_i} + \ell_i - \frac{\phi_i}{x_i} \sum_{j=1}^{nc} x_j \ell_j \quad \text{for } i = 1, \ldots, nc, \tag{A.13}$$

$$\ln \gamma_i^{\text{Residual}} = q_i' \left[1 - \ln\left(\sum_{j=1}^{nc} \theta_j' \tau_{ji}\right) - \sum_{j=1}^{nc} \left(\frac{\theta_j' \tau_{ij}}{\sum_{k=1}^{nc} \theta_k' \tau_{kj}}\right)\right] \quad \text{for } i = 1, \ldots, nc, \tag{A.14}$$

$$\ell_i = \frac{z}{2}(r_i - q_i) - (r_i - 1) \quad \text{for } i = 1, \ldots, nc. \tag{A.15}$$

The parameter, z is the lattice coordination number (usually taken as 10). Numerical results for $\ln \gamma_i$ are insensitive to the value of z, as long as a reasonable value ($6 \leq z \leq 12$) is chosen. For consistency, $z = 10$ is chosen in this work.

Data for Example 2 (Ethyl Glycol, Dodecanol and Nitro-methane): $r_1 = 2.4088$, $r_2 = 8.8495$, $r_3 = 2.0086$, $q_1 = 2.2480$, $q_2 = 7.3720$, $q_3 = 1.8680$, $q_1' = 2.2480$, $q_2' = 7.3720$, $q_3' = 1.8680$, $u_{1,1} = 0.0$, $u_{1,2} = 247.2$, $u_{1,3} = 54.701$, $u_{2,1} = 69.69$, $u_{2,2} = 0.0$, $u_{2,3} = 305.52$, $u_{3,1} = 467.88$, $u_{3,2} = 133.19$ and $u_{3,3} = 0.0$.

3. Soave–Redlich–Kwong (SRK) Equation of State (Soave, 1972)

For a mixture, SRK equation of state is given by:

$$P = \frac{RT}{v-b} - \frac{a}{v(v+b)}, \tag{A.16}$$

where

$$a = \sum_{i=1}^{nc}\sum_{j=1}^{nc} x_i x_j a_{ij}, \tag{A.17}$$

$$b = \sum_{i=1}^{nc} x_i b_i, \tag{A.18}$$

$$b_i = 0.08664\frac{RTc_i}{Pc_i} \quad \text{for} \quad i = 1, \ldots, nc, \tag{A.19}$$

$$a_i = 0.42747\alpha_i\frac{(RTc_i)^2}{Pc_i} \quad \text{for} \quad i = 1, \ldots, nc, \tag{A.20}$$

$$\alpha_i = [1 + (1 - \mathrm{Tr}_i^{0.5})(0.480 + 1.574\omega - 0.176\omega^2)]^2, \quad \text{for } i = 1, \ldots, nc. \tag{A.21}$$

where P_{c_i}, T_{c_i} and T_{r_i} are respectively the critical pressure, critical temperature and reduced temperature of component i. A mixing rule, which is known as van der Waals one–fluid–theory classical mixing rule, is used. The mixing rule is expressed as:

$$a_{ij} = (1 - k_{ij})\sqrt{a_i a_j} \quad \text{for } i, j = 1, \ldots, nc, \tag{A.22}$$

where k_{ij} is the binary interaction coefficient.

The partial fugacity coefficient of component i in a mixture can be calculated from the following equation:

$$\ln\hat{\phi}_i = \frac{B_i}{B}(Z-1) - \ln(Z-B) - \frac{A}{B}\left(\frac{2\sum_{j=1}^{nc} x_j a_{ij}}{a} - \frac{B_i}{B}\right)\ln\left(1 + \frac{B}{Z}\right), \tag{A.23}$$

where $A = \frac{aP}{(RT)^2}$, $A_i = \frac{a_i P}{(RT)^2}$ for $i = 1, \ldots, nc$, $B = \frac{bP}{RT}$ and $B_i = \frac{b_i P}{RT}$ for $i = 1, \ldots, nc$. The compressibility factor, Z, is given by $Z^3 - Z^2 + (A - B - B^2)Z - AB = 0$, which must be solved first to get the value of Z before proceeding to calculate the partial fugacity coefficient using Equation (A.23).

Data for Example 3 (Methane and Hydrogen Sulfide): $Tc_1 = 190.6$, $Tc_2 = 373.2$, $w_1 = 0.008$, $w_2 = 0.1$, $Pc_1 = 46.0/1.01325$, $Pc_2 = 89.4/1.01325$, $k_{11} = 0$, $k_{12} = 0.08$, $k_{21} = 0.08$, $k_{22} = 0.0$.

Data for Example 4 (Mixture of 9 Hydrocarbons): $Tc_1 = 190.6$, $Tc_2 = 305.4$, $Tc_3 = 369.8$, $Tc_4 = 408.1$, $Tc_5 = 425.2$, $Tc_6 = 460.4$, $Tc_7 = 469.6$, $Tc_8 = 507.4$, $Tc_9 = 707.0$, $Pc_1 = 45.4$, $Pc_2 = 48.2$, $Pc_3 = 41.9$, $Pc_4 = 36.0$, $Pc_5 = 37.5$, $Pc_6 = 33.4$, $Pc_7 = 33.3$, $Pc_8 = 29.3$, $Pc_9 = 15.0$, $w_1 = 0.008$, $w_2 = 0.098$, $w_3 = 0.152$, $w_4 = 0.176$, $w_5 = 0.193$, $w_6 = 0.229$, $w_7 = 0.251$, $w_8 = 0.296$, $w_9 = 0.706$ and $k_{ij} = 81$ for $i = 1, 2, \ldots, nc$ and $j = 1, 2, \ldots, nc$.

Chapter 15

APPLICATION OF ADAPTIVE RANDOM SEARCH OPTIMIZATION FOR SOLVING INDUSTRIAL WATER ALLOCATION PROBLEM

Grzegorz Poplewski and Jacek M. Jeżowski*

Department of Chemical and Process Engineering
Rzeszów University of Technology, Rzeszów, Poland
**ichjj@prz.edu.pl*

1. Introduction

Scarcity of raw water resources and stringent ecological regulations for wastewater discharge are demanding the reduction of both water usage and wastewater generation in the process industry. Water is required in many processes in various branches of the industry. Certain compounds, called contaminants, present in water streams transferred in water-using processes have to be removed from wastewater in treatment processes before they are discharged into the environment. In chemical and related industries, various impurities, harmful for both humans and the environment, are transferred to water streams and, hence, water treatment plays vital role for these industries. Moreover, wastewater treatment processes are expensive. Reduction of total freshwater usage results in an equivalent drop of wastewater generation in most cases. Hence, minimization of freshwater usage, and in consequence, minimization of wastewater generation in the process industry is of great environmental and economic importance. Significant benefits can be achieved by designing optimal total *water network*, that involves both water-using processes as well as regeneration and/or treatment units — this is also called "(industrial) *water allocation*". Zero water discharge, i.e. closed water circuit, is the ultimate aim to be reached.

In chemical and related industries, water is used in typical separation processes such as extraction, absorption and distillation with steam. Additionally, water is necessary for the washing of equipment in cyclones and filtration. Great amount of make-up water is also consumed in steam boilers and in cooling water circuits. Many of the operations are mass transfer type processes or can be treated as such operations. However, not all water-using processes may be treated as mass transfer operations, e.g. chemical reactions. Hence, this results in different mathematical models. However, it can be shown that by employing a model of mass transfer type process for all water-using operations we will be able to develop a generic approach since non-mass transfer operation forms a special and easier case. This chapter focuses on water networks with mass transfer type operations. Note that the solution approach for WN problem with only non-mass transfer processes is addressed in many other papers such as Jacob *et al.* (2002), El-Halwagi *et al.* (2003), Manan *et al.* (2004), Prakash and Shenoy (2005a,b), Agrawal and Shenoy (2006), Pillai and Bandyopadhyay (2007), and Wałczyk *et al.* (2007). The solution approaches made use of linear character of the problem by applying insight-based and, sometimes also optimization-based methods. This linearity of the problem supports our claim that the design of WN with non-mass transfer processes are easier than that with mass transfer ones since the latter is inherently nonlinear.

The means to reach substantial reduction of both freshwater consumption and resulting wastewater generation without changing technology of underlying processes are:

- water reuse in water-using processes,
- application of regeneration processes.

A total water network consists of water-using processes, regeneration and treatment operations. It can be treated as a single system with environmental limits on contaminant concentrations at its final effluent and identical models of regeneration and treatment operations — see e.g. Alva-Argaez *et al.* (2007) and Ng *et al.* (2007a,b). However, most often, total water network is considered as the system consisting of two parts: water usage network (WUN) and wastewater treatment network (WWTN). In this chapter we will address the former. It is worth noting that WWTN problem can be solved using very similar concepts and tools as those addressed in this chapter. The reader is referred to Poplewski and Jeżowski (2007) for a brief description of a stochastic optimization-based method for WWTN. To find more detailed information and on other approaches, the reader is referred to Kuo and Smith (1997), Galan and Grossmann (1998), Hernandez-Suarez *et al.* (2004), Meyer and Floudas (2006), and Statyukha *et al.* (2008) to mention a few.

To summarize, we will limit the presentation to water network with water-using processes and regenerators. This is usually referred to as water network with reuse and regeneration (WNRR). Alternatively, we can name it: water allocation with application of regenerators. Solution approach addressed in this chapter is able to deal with multiple contaminants. Also, it can deal with cost goal function. These possibilities are hardly accessible with insight-based methods.

In order to explain how reuse and regeneration is applied, let us compare the example of WNRR in Fig. 2 with a standard facility without reuse and regeneration in Fig. 1. In the latter freshwater is supplied in parallel to each water-using process. The water usage network in Fig. 2 applies water reuse, for instance water stream from process p_1 is reused in p_2. Additionally, regeneration r cleans the water before sending it to processes p_2 and p_3.

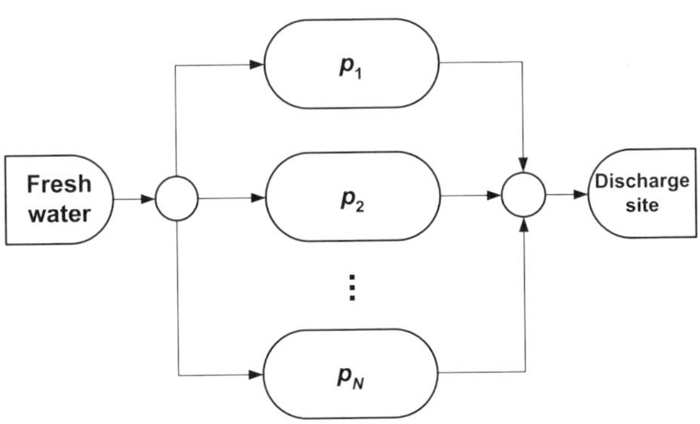

Figure 1. Traditional industrial water usage network.

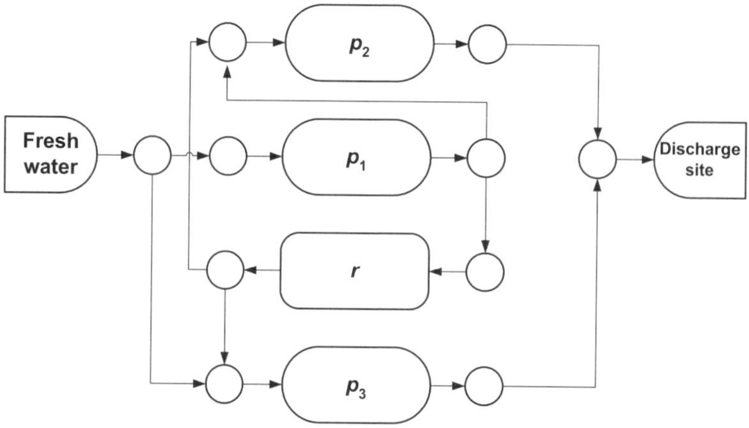

Figure 2. Example of water (usage) network with reuse and regeneration (WNRR).

It is evident that the traditional network consumes more freshwater and, also, generates more wastewater than WNRR though the latter is most often more complex in structure. In order to design optimal or even "good" WNRR, it is necessary to calculate both the structure and parameters of WNRR. Hence, it is a complex process design problem. Note its close resemblance to heat exchanger network synthesis and to mass exchanger network design. These similarities were explored in many insight-based methods for water allocation problem. For instance, water pinch concept (e.g. Wang and Smith, 1994) can be evolved directly from heat pinch (see Linnhoff and Hindmarsh, 1983 for the seminal paper). Also, other techniques for WN, particularly those for non-mass transfer processes, have their roots in heat and mass integration tools described in numerous papers and several books such as Smith (2005), Mann and Liu (1999) and El-Halwagi (1997, 2006).

The rest of this chapter is organized as follows. First we will present problem formulation and mathematical models of the WNRR elements. Then, we will proceed to network superstructure and its optimization model. The subsequent section will present a solution technique to the model. Some numerical hints will be given together with examples. The summary section concludes the chapter. The software used for solving WNRR problems reported in this chapter is available on the accompanying compact disc (CD).

2. Design of Water Network with Reuse and Regeneration — Problem Statement

The general problem is stated as follows. Given water-using processes, contaminants, available (or potential) regeneration processes, and the number of freshwater sources and their basic parameters, design a water network with reuse and regeneration that optimizes specified performance index.

In the following, we will explain models of processes and goal functions commonly applied. First, the variables and the parameters of the problem are described together with basic components of the water network.

- Contaminants

Contaminants are chemicals and compounds (e.g. solid phase suspension) that are transferred to water streams in water-using processes and/or are removed in regeneration operations. The number and types of contaminants are normally given.

- Freshwater sources

It is assumed that the number of available freshwater sources is known. Contaminant concentrations as well as the unit cost of freshwater are given for each source. Upper limits on freshwater flow rates are also known.

- Water-using processes

The number of water-using processes is fixed. As mentioned above the discussion is limited to the typical mass transfer processes as shown in Fig. 3. The water-using process is modeled as a counter-current mass exchanger (p) in which given loads of contaminants i (L^i_p) are transferred to the water stream from a real (or a fictitious) process stream. A good example for water-using process with real process stream is extraction process with water as solvent. In equipment washing a fictitious process stream is used to model water-using process.

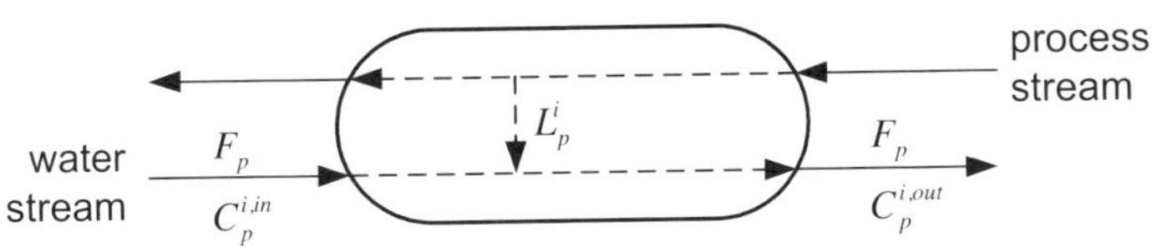

Figure 3. Scheme of water-using process p modeled as mass exchanger.

Since contaminant concentrations are normally very small, it is acceptable to assume a constant flow rate of water stream. Thus, mass balance equation of contaminant i in the water stream of process p is given as:

$$L_p^i = F_p(C_p^{i,\text{out}} - C_p^{i,\text{in}}). \tag{1}$$

The concentrations of contaminants at the inlet and outlet of process stream are known. Hence, maximum allowable contaminant concentrations in the water streams can be estimated from equilibrium conditions, with some additional tolerances to account for proper operation and design. In certain cases such as equipment washing, these concentrations depend on solubility, fouling or corrosion limits. Identical model has been applied for generic mass exchanger networks, see books by El-Halwagi (1997, 2006).

Thus, for mass transfer type water-using processes, the following data are required for water-using process p and for each contaminant i: mass load of contaminant L_p^i, maximum permissible inlet concentration of contaminant $C_p^{i,\text{in,max}}$ and maximum permissible outlet concentration of contaminant $C_p^{i,\text{out,max}}$.

Finally, the model of mass transfer process consists of the balance equation (1) and inequality constraints (2) and (3) for each contaminant.

$$C_p^{i,\text{in}} \leq C_p^{i,\text{in,max}}, \tag{2}$$

$$C_p^{i,\text{out}} \leq C_p^{i,\text{out,max}}. \tag{3}$$

The basic model has been extended in some works by the inclusion of water gains and losses. The losses and gains are assumed as fixed and, hence, the extension of the model does not influence the solution algorithm in the method presented. To keep the model compact, we will not account for water gains and losses.

- Regeneration processes

Mass balance of species i for regenerator r is given by Eq. (4) — see Fig. 4 for symbols.

$$L_r^i = F_r(C_r^{i,\text{in}} - C_r^{i,\text{out}}). \tag{4}$$

The constant flow rate is assumed similar to the water-using process with mass balance by Eq. (1). Two simple design equations are commonly

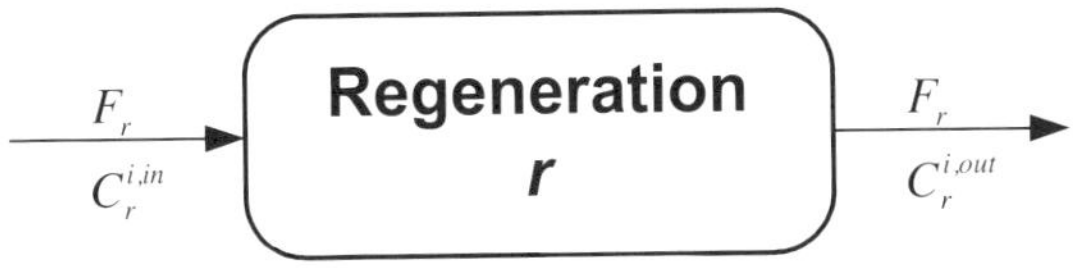

Figure 4. Scheme of regeneration process.

used for regeneration processes; Eqs. (5) and (6). Equation (5) fixes concentrations of some or all contaminants at the outlet of the regeneration. These outlet concentrations C_r^{i,out^*}, given in the data, are independent of flow rate and the inlet concentrations. Alternatively, Eq. (6) defines recovery ratio of some or all contaminants in a process. These ratios are assumed to be known.

$$C_r^{i,\mathrm{out}} = C_r^{i,\mathrm{out}^*}, \tag{5}$$

$$\psi_r^i = \frac{F_r C_r^{i,\mathrm{in}} - F_r C_r^{i,\mathrm{out}}}{F_r C_r^{i,\mathrm{in}}} = \frac{L_r^i}{F_r C_r^{i,\mathrm{in}}}. \tag{6}$$

Some other conditions can be included for both cases. For instance, inequality (7) limits inlet concentration of selected or all substances. Also, total flow rate of wastewater stream passing through the regeneration unit can be limited by Eq. (8).

$$C_r^{i,\mathrm{in}} \le C_r^{i,\mathrm{in,max}}, \tag{7}$$

$$F_r \le F_r^{\mathrm{max}}. \tag{8}$$

It is most often assumed that regeneration technologies are known. Number of units for the technology and their arrangement are not treated as decision variables. One unit is usually used for a single regeneration process. It is possible to apply heuristic rules developed in Chang and Li (2005) for estimating a number of units and their arrangement. Both regeneration technology and regeneration train arrangement can also be treated as decision variables; but this is rarely the case in existing approaches.

- Goal functions

The most often applied goal function is total freshwater flow rate. The more general criterion is total cost of the WNRR. In order to include the cost of pipelines, it is necessary to use binary variables for connections between elements of a network. This inevitably results in the mixed-integer programming (MIP) formulation.

• Mixers and splitters

In order to handle water reuse and to connect water-using processes with regeneration units, it is necessary to employ splitters and mixers. It is reasonable to assume isothermal operation for both mixers and splitters since thermal issue is not included in the typical formulation of the problem. However, mass balances have to be used for splitters and mixers. These balances will be presented with WNRR optimization model in Sec. 3.

3. Foundations of the Method for WNRR Problem

Approaches to WNRR problem proposed to date can be roughly divided into two categories:

(1) Insight-based/thermodynamics/heuristic methods,
(2) Optimization-based/mathematical methods.

Though the methods from group 1 have many advantages such as user driven philosophy, they are not able to deal efficiently with simultaneous structure and parameter optimization, particularly for multiple contaminants. Majority of systematic methods use superstructure concept — the generic approach in process systems engineering. The approach requires the creation of a superstructure specific for the system that should embed all feasible solutions. Then, optimization techniques are applied to find the optimal solution to the design problem. The approach in this chapter belongs to the category of mathematical/optimization methods. First, the WNRR superstructure and the optimization model will be described. Next, the solution technique by stochastic optimization will be presented in Sec. 4. We have employed adaptive random search (ARS) technique since it has performed well in many NLP benchmark and processing engineering problems. For instance, the paper by Liao and Luus (2005) reports its superiority over genetic algorithms.

3.1. WNRR superstructure

The superstructure and its model embed all water-using processes modeled as mass transfer unit, regeneration operations and accounts for multiple contaminant case. In order to simplify the presentation, a single freshwater source and a single wastewater discharge site are assumed. Extension for multiple freshwater sources and multiple discharge sites is straightforward. Also, it is assumed that the number of regeneration units for a technology is fixed and the technology is known, and that each regeneration process is performed in a single unit (module).

To build the superstructure, numerous splitters and mixers are necessary. A mixer is added at the inlet and a splitter at the outlet of all water-using processes as well as regeneration processes. Freshwater is redistributed by a splitter (freshwater splitter). Connections between freshwater splitter and regenerators are explicitly excluded since they have no economical and technical sense. Notice that the wastewater stream from each process p and the outlet stream from each regeneration unit r can be sent to final treatment, to environment or to other sites such as irrigation fields. We call this the "discharge site". Splitters of mass transfer water-using processes and of regenerations are connected to the mixers of mass transfer processes, the regeneration units and to the discharge site. Self recycles are also included for both mass transfer water-using processes and regeneration operations. These recycles are commonly inserted in regeneration processes only. We embedded them for the water-using processes also to account for cases of fixed flow rate (retrofit scenario) following the remarks of Wenzel *et al.* (2002) and Dunn and Wenzel (2001). Also, Wang and Smith (1995) applied such self-recycles streams to maintain fixed flow rates for water-using processes. The items of the WNRR superstructure are shown in Fig. 5.

3.2. Optimization problem

To formulate optimization problem, the following indices, parameters and variables are used.

Index sets:

$$I = \{i \mid i \text{ is contaminant in WNRR}\}$$

$$P = \{p \mid p \text{ is mass transfer water-using process}\}$$

$$R = \{r \mid r \text{ is regeneration process}\}$$

It is convenient to divide set R into two subsets. Subset $R1$ contains indices of treatment units modeled by (5) while $R2$ contains those modeled by (6).

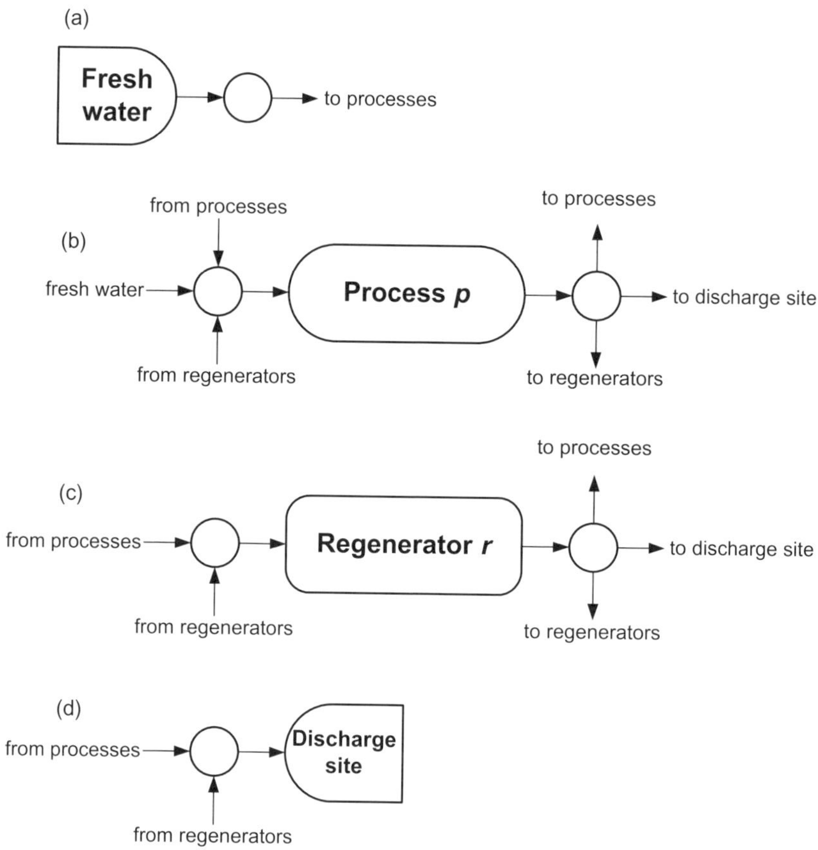

Figure 5. Elements of WNRR superstructure: (a) freshwater splitter; (b) mass transfer process; (c) regeneration process; and (d) discharge site.

The subsets satisfy conditions (9) and (10).

$$R1 \bigcup R2 = R, \tag{9}$$

$$R1 \bigcap R2 = \emptyset. \tag{10}$$

Parameters:

AR — annualization factor for investment on regeneration operations
C^{i}_{fw} — contaminant i concentration in freshwater source
$C_p^{i,\text{in},\max}/C_p^{i,\text{out},\max}$ — maximum value of inlet/outlet contaminant i concentration in processes p

$C_p^{i,\text{in,max}} / C_p^{i,\text{out,max}}$ — maximum value of inlet/outlet contaminant i concentration in processes p

$C_r^{i,\text{in,max}}$ — maximum permissible inlet concentration of contaminant i at the inlet to regeneration processes r

C_r^{i,out^*} — outlet concentration of contaminant i for regeneration process r modeled by (5)

$F_r^{\max}$ — maximum permissible flow rate via regeneration process r

F^* — sufficiently large number, higher than any possible flow rate in the network

H — time of operation per year

L_p^i — mass load of contaminant i transferred to water in process p

α_{fw} — unit cost of freshwater

$\beta_{fw,p}$ — fixed cost of piping section from freshwater splitter fw to process p

$\beta_{r,r'} / \beta_{r,p} / \beta_{r,m}$ — fixed cost of piping section from regeneration process r to regeneration process r'/ process p/ discharge site m

$\beta_{p,p'} / \beta_{p,r} / \beta_{p,m}$ — fixed cost of piping section from process p to process p'/ regeneration r/ discharge site m

γ_r / η_r — cost parameters in operation/investment expenses on regeneration in (13)

κ_r — parameter in (13) for investment cost of regeneration

ψ_r^i — removal ratio of contaminant i for regeneration processes defined by (6)

For each piping section for the parameters β listed above, minimum flow rate values are given too. They are denoted by $F_{i,j}^{\min}$ where i, j are indices of mixers and splitters at the operations r, p, source fw and discharge m.

Variables:

$C_p^{i,\text{in}} / C_p^{i,\text{out}}$ — inlet/outlet contaminant i concentration in processes p

$C_r^{i,\text{in}}$ — inlet concentration of contaminant i to regeneration processes r

C_r^{i,out^*} — outlet concentration of contaminant i for regeneration process r modeled by (5)

$F_{fw,p}$ — flow rate from freshwater splitter fw to process p

$F_{r,r'} / F_{r,p} / F_{r,m}$ — flow rate from regeneration process r to regeneration process r'/ process p/ discharge site

$F_{p,p'} / F_{p,r} / F_{p,m}$ — flow rate from process p to: process p'/ regeneration r/discharge site

Additionally, binary variables are defined to control fixed cost of piping sections and, in consequence, to simplify structure and to reduce investment cost of WNRR. For each connection, a binary variable has to be used. In order to make the description concise, we define them in a general way:

$$y_{i,j} = \begin{cases} 1 & \text{if the connection between } i \text{ and } j \text{ exist} \\ 0 & \text{if not} \end{cases}, \tag{11}$$

where i, j are indices p, r, fw, m similarly as in parameters β and $F^{\min}$.

Total cost of WNRR consists of cost of freshwater, cost of regeneration operations and cost of piping. Additionally, one can take into account cost of stream transportation and cost of water disposal but they are not considered here for simplicity sake. Cost of freshwater is given by:

$$\Phi_{fw} = \alpha_{fw} \sum_{p \in P} F_{fw,p}. \tag{12}$$

Cost of regeneration processes consists of operation and investment expenses according to:

$$\Phi_{\text{treat}} = H \sum_{r \in R} \gamma_r F_r + AR \sum_{r \in R} \eta_r (F_r)^{\kappa_r}. \tag{13}$$

In order to reduce the number of variables in the optimization problem, variable flow rates F_r in (13) were calculated via basic variables from:

$$F_r = \sum_{p \in P} F_{p,r} + \sum_{r' \in R} F_{r',r}; \quad r \in R. \tag{14}$$

Finally, the fixed cost of piping sections are calculated from:

$$\begin{aligned} \Phi_{\text{pipe}} = AR \Bigg[& \sum_{p \in P} \beta_{fw,p} y_{fw,p} + \sum_{p \in P} \sum_{p' \in P} \beta_{p,p'} y_{p,p'} \\ & + \sum_{p \in P} \beta_{p,m} y_{p,m} + \sum_{p \in P} \sum_{r \in R} \beta_{p,r} y_{p,r} + \sum_{r \in R} \sum_{r' \in R} \beta_{r,r'} y_{r,r'} \\ & + \sum_{r \in R} \beta_{r,m} y_{r,m} + \sum_{r \in R} \sum_{p \in P} \beta_{r,p} y_{r,p} \Bigg]. \end{aligned} \tag{15}$$

In addition to fixed charges, it is possible to account for operation cost, i.e. cost of transportation through pipelines.

The optimization problem of the superstructure for the WNRR problem is given in the following. To model processes, mixers and splitters, we have adapted formulations of Bagajewicz (2000). The equations are written for arrangements: mixer-process-splitters, mixer-regeneration-splitter rather than for individual equipment. Also, flow rates were employed in modeling splitters instead of split ratios. This was also advised by Karuppiah and Grossmann (2006). Additionally, application of flow rates is advantageous in stochastic optimization approach addressed in this chapter as we will show in the following section.

WNRR optimization model:

$$\text{minimize} : \text{TAC} = \Phi_{fw} + \Phi_{\text{treat}} + \Phi_{\text{pipe}}, \tag{16}$$

s.t.

$$F_{fw,p} + \sum_{p' \in P} F_{p',p} + \sum_{r \in R} F_{r,p} - F_{p,m} - \sum_{p' \in P} F_{p,p'} - \sum_{r \in R} F_{p,r} = 0; \quad p \in P, \tag{17}$$

$$F_{fw,p} C^i_{fw} + \sum_{p' \in P} (F_{p',p} C^{i,\text{out}}_{p'}) + \sum_{r \in R2} (F_{r,p} C^{i,\text{out}}_r) + \sum_{r' \in R1} F_{r',r} C^{i,\text{out}^*}_{r'} + L^i_p - C^{i,\text{out}}_p \left(F_{p,m} + \sum_{p' \in P} F_{p,p'} + \sum_{r \in R} F_{p,r} \right) = 0; \quad i \in I,\ p \in P, \tag{18}$$

$$\sum_{p' \in P} (F_{p',p} C^{i,\text{out}}_{p'}) + \sum_{r \in R2} (F_{r,p} C^{i,\text{out}}_r) + \sum_{r \in R1} (F_{r,p} C^{i,\text{out}^*}_r) - \left(\sum_{p' \in P} F_{p',p} + \sum_{r \in R} F_{r,p} + F_{fw,p} \right) C^{i,\text{in}}_p = 0 \quad i \in I,\ p \in P, \tag{19}$$

$$\sum_{p\in P} F_{p,r} + \sum_{r'\in R} F_{r',r} - F_{r,m} - \sum_{p\in P} F_{r,p} - \sum_{r'\in R} F_{r,r'} = 0; \quad r \in R, \tag{20}$$

$$\sum_{r'\in R2} (F_{r',r} C_{r'}^{i,\text{out}}) + \sum_{r'\in R1} (F_{r',r} C_{r'}^{i,\text{out}^*}) + (1 - \psi_r^i) \sum_{p\in P} (F_{p,r} C_p^{i,\text{out}})$$
$$- C_r^{i,\text{out}} \left(F_{r,m} + \sum_{p\in P} F_{r,p} \right) = 0; \quad i \in I, r \in R2, \tag{21}$$

$$\sum_{p\in P} (F_{p,r} C_p^{i,\text{out}}) + \sum_{r'\in R2} (F_{r',r} C_{r'}^{i,\text{out}}) + \sum_{r'\in R1} (F_{r',r} C_{r'}^{i,\text{out}^*})$$
$$- \left(\sum_{p\in P} F_{p,r} + \sum_{r'\in R} F_{r',r} \right) C_r^{i,\text{in}} = 0; \quad i \in I, r \in R, \tag{22}$$

$$F_{fw,p} \sum_{p'\in P} F_{p',p} + \sum_{r\in R} F_{r,p} \le F_p^{\max}; \quad p \in P, \tag{23}$$

$$\sum_{p\in P} F_{p,r} + \sum_{r'\in R} F_{r',r} \le F_r^{in,\max}; \quad r \in R, \tag{24}$$

$$F_{p,p'} \le F^* y_{p,p'}; \; F_{p,m} \le F^* y_{p,m}; \; F_{p,r} \le F^* y_{p,r}; \; p \in P, \tag{25}$$

$$F_{r,r'} \le F^* y_{r,r'}; \; F_{r,m} \le F^* y_{r,m};$$
$$F_{r,p} \le F^* y_{r,p}; \; r, r' \in R, \; p \in P, \tag{26}$$

$$F_{fw,p} \le F^* y_{fw,p}; \quad p \in P, \tag{27}$$

$$F_{r,r'} \ge F_{r,r'}^{\min} y_{r,r'}; \; F_{r,m} \ge F_{r,m}^{\min} y_{r,m};$$
$$F_{r,p} \ge F_{r,p}^{\min} y_{r,p}; \; r, r' \in R, \; p \in P, \tag{28}$$

$$F_{fw,p} \ge F_{fw,p}^{\min} y_{fw,p}; \quad p \in P, \tag{29}$$

$$C_p^{i,\text{in}} \le C_p^{i,\text{in},\max}; \quad i \in I, p \in P, \tag{30}$$

$$C_p^{i,\text{out}} \le C_p^{i,\text{out},\max}; \quad i \in I, p \in P, \tag{31}$$

$$C_r^{i,\text{in}} \le C_r^{i,\text{in},\max}; \quad r \in R. \tag{32}$$

Equations (17), (18) and (19) are mass balances of mass transfer water-using processes, associated mixers and splitters. Equation (17) is the general mass balance of the arrangement: mixer-process-splitter while Eq. (18) is mass balance of contaminants. Equation (19) defines mass balances of contaminants in the mixer. Balances for regenerators are organized in the same way. General mass balance of the sequences: mixer-regeneration-splitter is defined by (20) while Eq. (21) is the mass balance of contaminants. Note that the latter is necessary only for those regeneration processes which belong to subset $R2$. Contaminants mass balances of mixers attached to regeneration processes are given by (22).

Inequalities (23) and (24) impose the upper limits on flow rates via mass transfer water-using processes and regeneration processes. Inequalities (25) to (27) are logical conditions that force flow rates via piping sections to zero if associated binary variable is zero. They are inactive if binary variables are equal to zero. Inequalities (28) and (29) ensure that flow rates via piping sections will be not smaller than lower limits or will be equal to zero. Conditions on maximum permissible contaminant concentrations in water-using processes and regeneration operations are defined by inequalities (30) to (32).

Additional conditions can also be included. For instance, conditions on forbidden and compulsory connections have simple form: $y_{ij} = 0$ and $y_{ij} = 1$, respectively, which are very useful in retrofit scenario. The binary variables are also helpful in simplifying network topology. For instance, the number of branches from a splitter or/and the number of inlet connections for an operation can be restricted with these variables (see Gunaratnam *et al.*, 2005).

4. Solution Approach by Stochastic Optimization for Superstructure Model

The crucial difficulty with solving the optimization problems is its nonlinearity caused by bilinear terms (i.e. products of concentration and flow rate, such as (18), (19), (21)). Cost goal function is additional source of nonlinearity, though this can be avoided under some circumstances by employing approximate performance index (e.g. cost of freshwater). Obviously, binary variables add next source of difficulty resulting in MINLP formulation. In general, existing deterministic methods are not able to guarantee global optimum. For the MINLP case, even reaching a feasible solution may be impossible in reasonable CPU time.

The proposed methods for solving these problems can be grouped into four classes:

(1) Direct linearization of the optimization problem
(2) Sequential solution procedure with successive linearization
(3) Use of convex under-estimators (with spatial branching for NLP in some cases) — this also requires a sequence of optimization runs
(4) Application of stochastic/meta-heuristic optimization methods

Techniques (1) and (2) rely on eliminating nonlinearities caused by bilinear terms; i.e. the products of flow rate and concentration. Concentrations are commonly fixed at the maximum allowable limits since such settings ensure minimization of freshwater usage as was proven by Savelski and Bagajewicz (2000, 2003). For single contaminant, direct linearization produces LP or MILP models — see Savelski and Bagajewicz (2001). For multiple contaminants, the optimization problem is also linear but only for fixed network topology. Hence, Bagajewicz *et al.* (2000) developed a tree search procedure to allow solution of linear problems. However, the generation of water network structures is time consuming, particularly for cases of reuse and regeneration. More recently, Wałczyk and Jeżowski (2008) overcame the difficulty by proposing a linearization-based method which can be performed in one stage. The issue is that direct linearization by fixing concentrations does not guarantee cost optimum network because it is aimed at freshwater minimization. However, a good local optimum can be generated. Successive linearization procedure in sequence of optimization steps — approaches from 2nd group — was the basis of the method presented in Alva-Argaez *et al.* (1998), Gunaratnam *et al.* (2005) and Alva-Argaez *et al.* (2007). This procedure generated a good initialization for solving finally original NLP or MINLP formulation.

Sophisticated and complex optimization procedures (belonging to group 3) were developed in propositions. They are claimed to be capable of locating the global optimum. The key concept is the use of convex underestimators with a sequence of optimization runs. The convergence of the methods depends heavily on underestimating functions. Additionally, branching can be applied even for NLP problems (spatial branching). The optimization algorithms are aimed at general process system synthesis problems and were used, among others, for water network design in Lee and Grossmann (2003) and Bergamini *et al.* (2005, 2008).

In regards to application of stochastic optimization to solve the WNRR superstructure model, genetic algorithms (GA) were employed most often, for instance by Xue *et al.* (2000), Tsai and Chang (2001), Shafiei *et al.* (2004), Prakotpol and Srinophakun (2004), and Lavric *et al.* (2005, 2007). The use of adaptive random search (ARS) technique was reported for wastewater treatment network by Poplewski and Jeżowski (2007). Particle swarm optimization approach was used for water networks in the works of Hul *et al.* (2007) and Tan *et al.* (2008). The design was limited to non-mass transfer water-using processes and water network with water re-use. GA approach was able to deal with the WNRR problem but required a long CPU time or/and some simplifications of the problem.

Analysis of the works published to date allows concluding the following in case of deterministic optimization approaches:

- direct linearization does not rigorously account for cost function
- sequential linearization only locates feasible starting point for NLP/MINLP solver
- approaches with under-estimators need reliable NLP solver and expert use
- GA method calculates "good" results but there are problems with computation times for larger problems.

In the following section, we will present application of an ARS technique, which is a modification of Luus-Jaakola method (Luus and Jaakola, 1973; Luus, 1973). This approach is able to find a good local optimum in a reasonable time for the WNRR problems in the literature.

5. Application of Adaptive Random Search to WNRR Problem

The detailed algorithm of the ARS optimizer that was applied to solve the WNRR problem is presented in Chapter 3 and a brief explanation is given in the Appendix of this chapter. The solver is able to efficiently solve NLP problems with a moderate number of variables and inequality constraints. To cope with large-scale constrained MINLP cases using the ARS technique, it was necessary to develop solution methods for the following sub-problems on how to:

(1) deal with equality constraints
(2) find feasible starting point and "good" initial search sizes of variables (search sizes are changeable ranges from which the ARS method looks for solution — see Appendix).
(3) handle numerous binaries

The methods are explained in seriatim fashion below.

In regards to the first problem Poplewski (2004) developed a way of dealing with equality constraints of the WNRR model by sequentially solving them as single linear equations and sets of simultaneous linear equations. This is the most robust and efficient way to solve constrained optimization tasks with the help of stochastic optimizers. This conclusion was drawn from our numerical experiments — see Jeżowski and Bochenek (2002), and also works of Luus, for instance Luus *et al.* (2002), who applied this for the case of "simple" equations. As noted in the preceding section, all nonlinearities in the constraints of water allocation problem are caused by the bilinear terms — products of flow rates and concentrations. In the case of ARS approach it was convenient to fix certain flow rates to achieve linearization. Hence, flow rates of selected water streams were chosen as the decision variables and the others, dependent variables were calculated from the equality constraints. The proper selection of the decision variables and appropriate sequencing of equality constraints, allow developing a solution procedure in the ARS approach such that all equality constraints became linear ones in regards to the dependent variables. Luus (2002) used similar strategy and called such cases "simple constraints" since there are cases where such procedure cannot be applied. This is applicable for models with mass and heat balances that are common in synthesis of process systems.

It is also important that the number of decision variables was reduced to the number of degrees of freedom of the superstructure model. This is shown below.

(a) Number of variables in the optimization problem is:
   - no. of streams $= P^2 + fwP + Pm + PR + Rm + R^2$,
   - number of concentration variables $= 2(PI + RI)$.

(b) Number of equalities equals to: $P + 2PI + R + 2RI$.

For $fw = 1$ and $m = 1$ the number of degrees of freedom (NDF) is equal to the difference of (a) and (b) and is given by:

$$\text{NDF} = P^2 + P + P + PR + R + R^2 + 2(PI + RI) \\ -[P + 2PI + R + 2RI] = P^2 + P + PR + R^2$$

Note that the total number of streams in the superstructure is equal to:

$$P^2 + P + PR + R^2 + [P + R].$$

Thus, the number of degrees of freedom equals the number of all streams within the superstructure minus the number of those streams that leave processes and regenerators to the discharge site. Let us choose flow rates of the streams, except those that are sent to discharge sites, as the decision variables, whose values are generated by the ARS. The rest of the variables: flow rates of streams to the discharge sites and all concentrations are dependent variables and are calculated by solving the equality constraints of the model. It is only necessary to organize calculations in such a way so as to ensure that all equalities are linear with respect to the dependent variables.

The appropriate sequence of solving the model equations is:

(i) calculate flow rates of streams to the discharge site by solving general mass balances of processes and regenerators from Eqs. (17) and (20), respectively
(ii) calculate outlet concentrations by solving set of simultaneous linear Eqs. (18) and (21)
(iii) calculate inlet concentrations to processes and regenerators by solving mass balances of mixers (19) and (22)

The following remarks should be accounted for when organizing the calculations:

(a) it is necessary, prior to step (iii), to check that there is no forbidden operation of division by zero when calculating concentrations from Eqs. (19) and (22)
(b) for regenerators modeled by Eq. (5) the outlet concentrations are not variables

(c) inlet concentrations to regenerators should be calculated if and only if upper limits on them are imposed in the model

(d) feasibility of calculated variables after each step (i), (ii) and (iii) should be checked; for instance outlet concentrations determined in step (i) are checked for upper and lower limits (lower limit is zero). If the values are infeasible, the death penalty is employed, i.e. a new set of decision variables is generated.

With regards to the second problem of the determination of initial point and choice of major control parameters, the ARS algorithm is sometimes able to locate global optimum even if it starts from an infeasible point but it is seldom sufficient in larger problems. Additionally, the WWRN model is strongly constrained, i.e. the space of feasible solutions is relatively small in comparison to a "safe" initial search space which the user has to use for stochastic solver in order not to eliminate the optimum. Therefore, a good feasible initial point increases the robustness of optimization. Also, it reduces CPU time since a relatively small number of goal function evaluations (NFE) is required. We have developed a simple algorithm for locating an initial point that proved its validity in all tests for WNRR problem we have performed to date.

The decision variables for the initial point are calculated as follows:

- Flow rates of streams from freshwater source to processes are determined for the network of parallel structure such as that in Fig. 1.
- Flow rates via regenerators are assumed to be zero, i.e. the regenerators are not applied in the initial point.

The procedure for calculating the starting point is as follows.

(1) For each water-using process, flow rates of freshwater streams are calculated for each contaminant from:

$$F^{i}_{fw,p} = \frac{L^{i}_{p}}{C^{i,\text{out,max}}_{p}}; \quad p \in P, i \in I. \tag{33}$$

(2) The value of the flow rate for process p (for the initial point) is taken as the maximum value of the flow rates calculated in point 1:

$$F_{fw,p} = \max_{i \in I}(F^{i}_{fw,p}). \tag{34}$$

The algorithm ensures that the maximum concentration is reached by (usually) a single contaminant. This is the condition of reaching the minimum freshwater consumption in the network of water-using processes — see Savelski and Bagajewicz (2003) for the proof.

Similar to the starting point, the initial search regions for decision variables influence largely the optimization efficiency. To increase efficiency, the regions should be preferably small but they also should contain the optimum. As the lower bounds for flow rates we have applied zero or the given minimum value denoted by $F^{\min}$. The upper bound for each flow rate is determined as the sum of freshwater flow rates $F_{fw,p}$ determined from (34) for all water-using processes $p \in P$. Most likely the limits are too high in most cases. However, the calculation procedure is straightforward and the limits ensure that the optimum values of variables will not be lost.

With regards to dealing with binary variables (problem 3), they were treated as the dependent ones. This was necessary since ARS techniques are not efficient tools for discrete valued models of large scale. For every flow rate, which is the decision variable, and was given by the algorithm a value less than the fixed minimum number $F^{\min}$, the value of zero is assigned to the associated binary variable. In the opposite case, the binary variable is set at 1. This was coded in the optimization method by replacing constraints (28) and (29) by a logical statement as follows:

$$\textbf{IF } F_{i,j} < F_{i,j}^{\min} \textbf{ THEN } y_{i,j} \wedge F_{i,j} \Rightarrow 0 \textbf{ ELSE } y_{i,j} \Rightarrow 1. \tag{35}$$

If flow rates $F_{i,j}$ are not the decision variables but dependent ones such relations can not be applied since invalid mass balances would result. Hence, for the dependent variables the logical statement has the form:

$$\textbf{IF } F_{i,j} = 0 \textbf{ THEN } y_{i,j} \Rightarrow 0 \textbf{ ELSE } y_{i,j} \Rightarrow 1. \tag{36}$$

Note that original logical conditions (25) to (27) are sources of difficulties in the solution of MINLP or MILP problems by deterministic techniques (see for instance Biegler *et al.*, 1995), since it is difficult to find tight estimation for F^* parameter in logical conditions (25) to (27) and usually a very large number is employed causing large "gaps" in the optimization problem.

6. Examples, Results and Discussion

The method has been tested with numerous case studies from the literature. They addressed networks with water-using processes only as well as with the processes and regeneration units. There are relatively few examples in the literature for the networks with regeneration units since their design is very difficult. Note that a goal function for WNRR has to include cost of regeneration since otherwise all wastewater would be sent to regeneration to minimize cost (or consumption) of freshwater. Because cost of regeneration depends nonlinearly on flow rate, the overall goal function will be nonlinear too, increasing difficulty of optimization. In this section we will show three examples of solving the WNRR design problem.

6.1. Example 1

This case study is taken from Wang and Smith (1994), Alva-Argaez *et al.* (1999) and Doyle and Smith (1997). It is the WNRR problem for simplified industrial atmospheric crude oil distillation train. The network consists of three water-using processes and single regeneration operation. The data for the water-using processes are gathered in Table 1. The regeneration is defined by removal ratio — Eq. (6). The ratio is equal to 0.99 for H_2S and 0.0 for two other contaminants. The cost goal function consists of freshwater cost, regeneration cost and expenses on final treatment as defined by Eq. (37). The costs of regenerator and final treatment consist of both operational and investment expenses as shown by Eq. (37). The superstructure

Table 1. Data for example 1.

Process	Contaminants	L [kg/h]	C_{in}^{max} [ppm]	C_{out}^{max} [ppm]
1. Distillation	Hydrocarbon	0.675	0	15
	H_2S	18.000	0	400
	Salt	1.575	0	35
2. Hydrodesulphurization	Hydrocarbon	3.400	20	120
	H_2S	414.800	300	12,500
	Salt	4.590	45	180
3. Desalter	Hydrocarbon	5.600	120	220
	H_2S	1.400	20	45
	Salt	520.800	200	9500

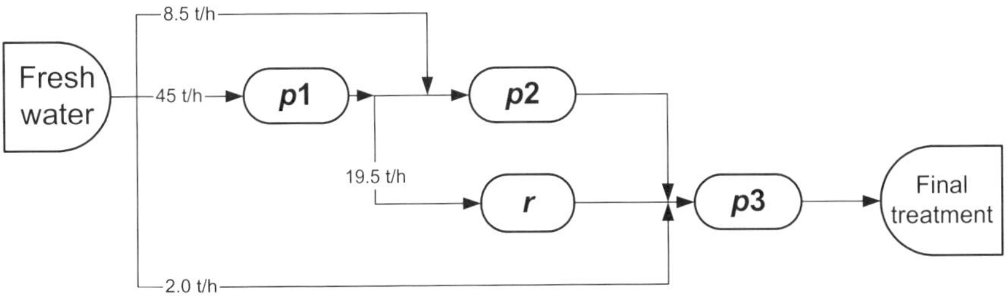

Figure 6. The optimal solution for example 1.

model applied for this case contains conditions (35) on lower limits of flow rates with $F^{\min} = 2\,\text{t/h}$. The best solution obtained with ARS costs 1.1676 million \$/year. It is identical to the literature optimum. The network is shown in Fig. 6. The basic control parameters of the ARS for solving example 1 were: NEL = 100 and NIL = 10,000. CPU time for these parameters is around 15 seconds per optimizer run on a personal computer AMD 1700 MHz processor.

Minimize:

$$GF = \left(\begin{array}{l} 8{,}600 \sum_{p\in P} F_{fw,p} + 34{,}200 \left(\sum_{p\in P} F_{p,m} \right)^{0.7} \\ +860 \left(1.0067 \sum_{p\in P} F_{p,m} + \sum_{p\in P, r\in R} F_{p,r} \right)^{0.7} \\ +16{,}800 \left(\sum_{p\in P, r\in R} F_{p,r} \right) \end{array} \right). \tag{37}$$

For this example, we show the robustness of the ARS algorithm in regards to locating the optimum. The results in the following are for 100 runs of the subroutine. The worst result was 1.2966 and the average was 1.1951 million \$/year. The distribution of the results is illustrated in Fig. 7 with the plot showing the dependence between success ratio (in %) vs. relative error (in %). The error was calculated for the best result, i.e. 1.1676 million \$/year. It is worth noting that the values of the success ratio for small relative errors can be improved by increasing NFE.

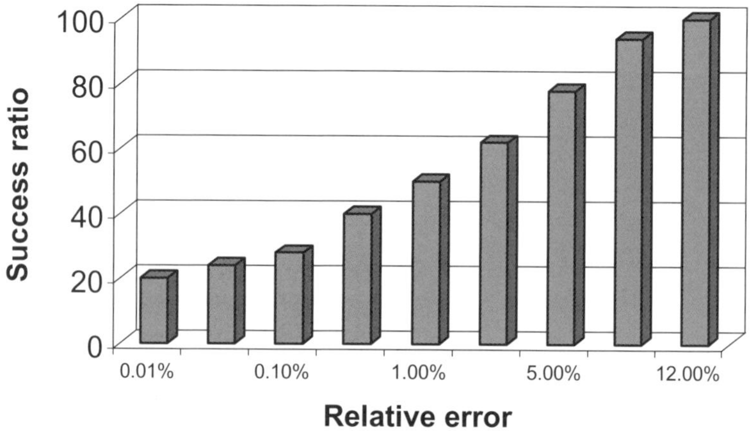

Figure 7. Success ratio versus relative error for example 1 from 100 runs of optimization.

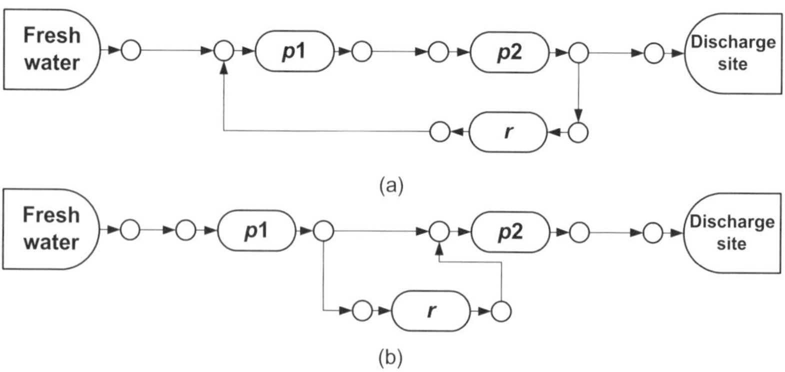

Figure 8. Illustration of regeneration recycle: (a) example of WNRR with regeneration recycle; (b) example of WNRR without regeneration recycle.

The important feature of this example is that the final solution should not contain so called "regeneration recycles". By forbidding regeneration recycling, the user eliminates the recycle of treated water from regenerator to a water-using process from where the water stream comes from. Figure 8(a) shows the example of regeneration recycle, and Fig. 8(b) shows regeneration without recycle. The aim of forbidding regeneration recycle is to eliminate accumulation of harmful substances in the processes involved in the regeneration loop. Generally, elimination of regeneration recycles is a difficult problem if it is to be solved by a deterministic approach as one may see by analysing the solution technique developed in Alva-Argaez *et al.* (2007).

The use of ARS or other stochastic approaches requires a different technique. The main idea of the developed method relies on loop breaking. The loop is equivalent to the regeneration recycle. First, all loops within a network have to be identified. Binary variables have to be employed to identify the loops. Let binary variable "Z_p" denotes whether process p is in the loop ($Z_p = 1$ means that process p is involved in the loop). The identification procedure for a single regeneration process in a network is as follows:

- Set $Z_p \to 0$ for $p \in P$
- Find all water-using processes that send wastewater stream to regeneration r

$$\textbf{IF } F_{p,r} > 0 \textbf{ THEN } Z_p \to 1; \quad p \in P \tag{38}$$

- Find all processes that are connected directly or indirectly with the processes selected in the previous point. To do this it is necessary to check $P - 1$ times the following condition for all processes:

$$\textbf{IF } (F_{p',p} > 0 \text{ AND } Z_p = 1) \textbf{ THEN } Z_{p'} \to 1;\ p', p \in P,\ p' \neq p. \tag{39}$$

- Break the loop by elimination from the solution such sequences: regeneration — process p for which Z_p equals 1 using the relation:

$$\textbf{IF } (F_{r,p} > 0 \text{ AND } Z_p = 1) \textbf{ THEN } F_{r,p} \to 0;\ p \in P. \tag{40}$$

Note that binary variables Z_p applied in the loop breaking procedure are not decision variables. Hence, they do not influence optimization. The only effect on calculations is an increase of CPU time, which is not substantial in general. In order to illustrate the influence of loop breaking procedure, we present here Table 2 with results for example 1. The results were calculated for 50 runs of optimizer. Additionally, numbers in column 2 show robustness of the ARS algorithm in this example.

6.2. Example 2

This case study involves 10 water-using processes and one regeneration operation, and is taken from Savelski and Bagajewicz (2001). Data for

Table 2. Comparison of the results for example 1 with and without use of the loop breaking procedure (for 50 runs).

Loop breaking procedure applied?	No. of times the optimal solution was found	No. of solutions without regenerator recycle
NO	6	22
YES	10	50

Table 3. Data for example 2.

Process no.	L [kg/h]	$C_{in}^{\max}$ [ppm]	$C_{out}^{\max}$ [ppm]
1	2.0	25	80
2	2.88	25	90
3	4.0	25	200
4	3.0	50	100
5	30.0	50	800
6	5.0	400	800
7	2.0	400	600
8	1.0	0	100
9	20.0	50	300
10	6.5	150	300

water-using processes are in Table 3. Regeneration is modeled by Eq. (5) with outlet concentration set at 5 ppm. We have applied freshwater consumption as the goal function and reached the solution, which requires 10 t/h of freshwater. The same goal function was calculated by Savelski and Bagajewicz (2001). However, our network features high flow rate of wastewater to the regeneration. Hence, operation cost of this solution is high due to expenses on regeneration. It is sufficient to add penalty for regeneration use such as that in Eq. (41) to substantially decrease regeneration flow rate. This case proves our claim that to design "good" WNRR cost of regeneration has to be included in the goal function.

$$GF = \sum_{p \in P} F_{fw,p} + 0.1 \sum_{p \in P} F_{p,r} \tag{41}$$

Table 4 presents the optimal solution found for the goal function defined by Eq. (41). The goal function at the optimum is 27.509 and the network

Table 4. Optimal solution found by the present approach for example 2.

Outlets	Inlets											
	p_1	p_2	p_3	p_4	p_5	p_6	p_7	p_8	p_9	p_{10}	r	Discharge site
Fresh water	—	—	—	—	—	—	—	10	—	—	—	—
p_1	—	5.225	—	6.786	7.279	—	5.520	—	8.172	—	2.003	—
p_2	—	—	—	17.330	10.754	—	—	—	13.999	9.918	—	—
p_3	—	—	—	—	2.375	—	4.478	—	2.000	5.543	6.420	2.101
p_4	—	7.560	5.104	—	—	8.167	—	—	13.674	18.760	—	—
p_5	—	—	—	—	—	6.431	5.559	—	—	—	26.589	1.381
p_6	—	—	—	—	—	—	21.869	—	—	2.400	16.496	0.698
p_7	—	—	—	—	—	2.806	—	—	—	—	43.654	0.770
p_8	3.716	—	—	—	—	—	—	—	4.095	2.071	—	0.118
p_9	—	—	—	—	—	14.033	9.803	—	—	—	52.679	3.485
p_{10}	—	—	—	—	—	10.026	—	—	—	—	27.249	1.416
r	31.271	39.241	17.813	29.150	19.552	—	—	—	38.061	—	—	—

has small flow rates to regenerators. The number of function evaluations necessary to reach the solution was 12,000,000 (NEL = 1,000 and NIL = 12,000). For PC with AMD 1,700 MHz processor, CPU time is approximately 380 seconds per run. It is relatively high because this example is large scale in regards to number of water-using processes.

6.3. Example 3

Water network in this example is for a paperboard mill. The problem is taken from Koppol *et al.* (2003). The mill consists of chests for pulping and diluting paper fibers, paper machine, showers and added treatments. Paper fibers are the single contaminant of water, which is measured by total suspended solids (TSS). The data for processes in the mill are gathered in Table 5. In the paper machine, the paper fibers are removed from the water stream and whitewater of two different concentrations exits the process. This machine is considered as consisting of two processes: one rich and one lean. They are modeled in our approach as water-using processes but with opposite mass transfer direction, i.e. the contaminant loads have negative sign in the balances (18). The rich white-water has a high concentration of fibers and exits the paper machine into a silo. The lean white-water of lower concentration of fibers comes from suction boxes. To reach the required throughput of 600 ton/day, flow-rates through rich and lean processes of paper machine have to be kept fixed. The other processes with numbers 1, 4, 5 and 6 in Table 5 are typical water-using processes.

The typical treatment processes applied in pulp and paper industry are: dissolved air flotation (called physical treatment), and combination of physical treatment and membrane cleaning up (ultrafiltration). Both regeneration processes are modeled by Eq. (5) and the limiting outlet concentrations are: 30 ppm for physical regeneration and 2 ppm for combined physical and membrane treatment.

The objective is to design cost optimal WNRR. Operation cost consists of cost of freshwater and cost of regeneration and final treatment. The cost of treatment by physical regenerator is 0.15 $/ton and by combined physical and membrane regenerator is 0.9 $/ton. The cost of final treatment of wastewater which passed through a regenerator is 0.35 $/ton. The cost of freshwater and final treatment (without regeneration) is 1.65 $/ton. Note that

Table 5. Data for processes in the paper mill according to Koppol *et al.* (2003).

Process	Load [kg/h]	C_{in}^{max} [ppm]	C_{out}^{max} [ppm]
1. Pulping/Dilution	24800	500	5000
2. Paper Machine (Rich)	−18225	5000	500
3. Paper Machine (Lean)	−6480	5000	200
4. Deckle Showers	21.67	100	500
5. Cylinder Showers	37.50	300	600
6. Felt Showers	3.33	20	100
Treatment type		$C_{t,out}^{max}$ [ppm]	**Cost [$/ton]**
1. Physical		30	0.15
2. Physical & Membrane		2	0.9

only final treatment cost is important for the WNRR problem since treatment operations are not included into the superstructure. Operating time was assumed 350 days per year. All these parameters were taken from Koppol *et al.* (2003). For later comparison, we calculated the network with parallel splitting of freshwater stream supplied to water-using processes. This consumes 1492 t/h of freshwater and has operating cost of 20.68 million $/year.

First, we solved the case with water reuse only, i.e. we assumed that regeneration processes are not used and all the wastewater is discharged to the final treatment. Hence, a reduction of freshwater usage can be reached only via wastewater reuse. The best network obtained by our method uses 363.9 t/h of freshwater and has operating cost of 5.04 million $/year (versus 20.68 million $/year for parallel structure). This shows the large economic advantage of water reuse. Next, we added physical treatment (dissolved air flotation) as regeneration process. The optimal network for such case features largely reduced freshwater consumption, i.e. the freshwater needed drops down to 13.875 t/h. Wastewater of flow rate 311.8 t/h is processed by physical regenerator and the discharge to final treatment amounts to 13.875 t/h. The network is shown in Fig. 9.

Finally, we solved the case with water reuse and combined physical and ultrafiltration treatment as the single regeneration process. The consumption of freshwater falls to zero, i.e. closed circuit of water is reached. This network is shown in Fig. 10. The stream of wastewater to physical and

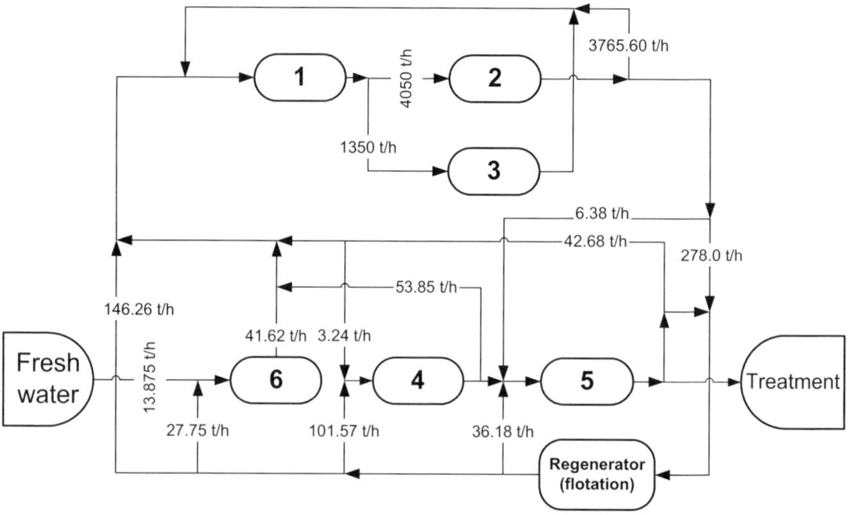

Figure 9. WNRR for the paper mill example; reuse and physical regeneration.

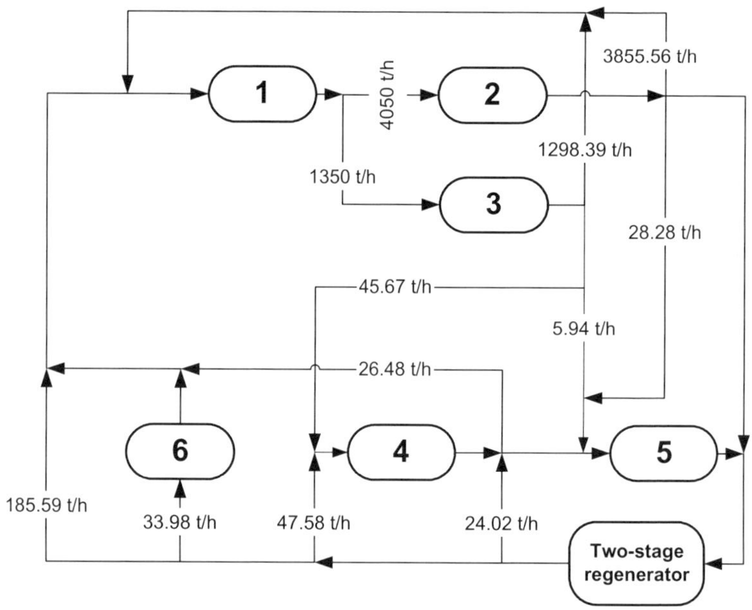

Figure 10. WNRR for the paper mill example; zero water discharge with combined physical treatment and ultrafiltration.

membrane treatment is 293.9 t/h. Due to high cost of the regeneration, the operation cost of this network is 2.2 million $/yr. Koppol *et al.* (2003) found a network of identical cost but of a slightly different structure with the use of sequential method by tree search and optimization at each tree node.

This case study has interesting features that are faced often in the industry. They are:

(1) forbidden and compulsory connections
(2) fixed (i.e. given) flow rate via some or all water-using processes

To account for forbidden connection, we suggest eliminating it directly from the initial superstructure. In case of compulsory connection, our suggestion is to apply appropriate lower limit for the range the variable is generated from in the ARS method. For compulsory connection, the lower limit for the flow rate through it should be higher than zero. With regards to fixed flow rate, we have applied self-recycles in water-using processes within the superstructure.

7. Summary

This chapter addresses the design of water network with reuse and regeneration, which is the most important case of general water allocation problem. The proposed solution method can efficiently deal with mass transfer type water-using processes and multiple contaminants. The approach uses the generic superstructure concept. MINLP optimization problem of the superstructure with economic goal function has been developed. ARS algorithm, developed by the authors, has been applied to solve this complex optimization problem. Certain enhancements have been applied in order to reach satisfactory robustness and efficiency of the optimization. We have proposed the scheme of solving equality constraints of the optimization problem, which generates linear equations in regards to dependent variables. Also, we have suggested and applied the technique of treating binary variables as dependent ones, and have calculated them via continuous variables using simple logical statements. The enhancements allowed successful solution of large problems, and can be used in solving general MINLP problems, particularly problems of designing process systems such as mass exchanger networks.

The proposed solution method allows accounting for various specific requirements that can be met in industry. The examples solved illustrate how one can deal with forbidden and compulsory connections in the optimization by stochastic optimizers. Also, a simple way of breaking loops to elimnate regeneration recycles has been developed. It does not require applying binary variables as decision variables. Hence, it is simpler than techniques that were used with deterministic optimizers in the literature. The experience gathered from solving WNRR design problems allows us to state that simple ARS algorithm, when applied together with certain problem specific enhancements, is easy to use and capable of designing optimal or near-optimal process systems.

Symbols/Abbreviations

(Here, only symbols used often in the text are listed since others are explained in Sec. 3.2)

ARS — adaptive random search
C — concentration of contaminant
F — flow rate
L — contaminant mass load
MILP — mixed-integer linear programming
MINLP — mixed-integer nonlinear programming
NDF — number of degrees of freedom
NEL — number of external loops
NFE — number of (goal) function evaluations
NIL — number of internal loops
NLP — nonlinear programming

Superscripts

min — minimum
max — maximum

Subscripts

i — contaminant
in — inlet
out — outlet

p — water-using process
r — regeneration

Acknowledgment

The authors would like to convey their thanks to Dr D.C.Y. Foo (University of Nottingham, Malaysia) for his great help in the preparation of the final revision of this chapter.

References

Agrawal, V. and Shenoy, U.V. (2006). Unified conceptual approach to targeting and design of water and hydrogen networks. *AIChE J.*, **52**(3), pp. 1071–1082.

Alva-Argaez, A., Kokossis, A.C. and Smith, R. (1998). Wastewater minimisation of industrial systems using an integrated approach. *Comput. Chem. Eng.*, **22**(Suppl.), pp. S741–S744.

Alva-Argaez, A., Kokossis, A.C. and Smith, R. (2007). A conceptual decomposition of MINLP models for the design of water-using systems. *Int. J. Environment and Pollution*, **29**(1/2/3), pp. 177–205.

Alva-Argaez, A., Vallianatos, A. and Kokossis, A.C. (1999). A multi-contaminant transhipment model for mass exchange networks and wastewater minimisation problems. *Comput. Chem. Engng*, **23**, pp. 1439–1453.

Bagajewicz, M. (2000). A review of recent design procedures for water networks in refineries and process plants. *Comput. Chem. Eng.*, **24**, pp. 2093–2113.

Bagajewicz, M.J., Rivas, M. and Savelski, M.J. (2000). A robust method to obtain optimal and sub-optimal design and retrofit solutions of water utilization systems with multiple contaminants in process plants. *Comput. Chem. Eng.*, **24**, pp. 1461–1466.

Bergamini, M.L., Aguirre, P. and Grossmann, I. (2005). Logic-based outer approximation for globally optimal synthesis of process networks. *Comput. Chem. Engng*, **29**, pp. 1914–1993.

Bergamini, M.L., Grossmann, I., Scenna, N. and Aguirre, P. (2008). An improved piecewise outer-approximation algorithm for the global optimization of MINLP models involving concave and bilinear terms. *Comput. Chem. Engng*, **32**, pp. 477–493.

Biegler, L.T., Grossmann, I.E. and Westerberg, A.W. (1997). *Systematic Methods of Chemical Process Design*, New Jersey, Prentice Hall PTR.

Chang, C.-T. and Li, B.-H. (2005). Improved optimization strategies for generating practical water-usage and treatment network structures. *Ind. Eng.Chem. Res.*, **44**, pp. 3607–3618.

Doyle, S.J. and Smith, R. (1997). Targeting water reuse with multiple contaminants, *Trans IChemE*, Part B, **75**(3), 181–189.

Dunn, R.F. and Wenzel, H. (2001). Process integration design methods for water conservation and wastewater reduction in industry. Part1: Design for single contaminants. *Clean Products Process.*, **3**, pp. 307–318.

El-Halwagi, M.M., Gabriel, F. and Harell, D. (2003). Process design and control: Rigorous graphical targeting for resource conservation via material recycle/ reuse networks. *Ind. Eng. Chem. Res.*, **42**, pp. 4319–4328.

El-Halwagi, M.M. (1997). *Pollution Prevention Through Process Integration. Systematic Design Tools.*, Academic Press, San Diego, USA.

El-Halwagi, M.M. (2006). *Process Integration.* Elsevier, Academic Press, Amsterdam.

Galan, B. and Grossmann, I.E. (1998). Optimal design of distributed wastewater treatment networks. *Ind. Eng. Chem. Res.*, **37**, pp. 4036–4048.

Gunaratnam, M., Alva-Argáez, A., Kokossis, C.A., Kim, J.-K. and Smith, R. (2005). Automated design of total water systems. *Ind. Eng. Chem. Res.*, **44**, pp. 588–599.

Hernandez-Suarez, R., Castellanos-Fernandez, J. and Zamorra, J.M. (2004). Superstructure decomposition and parametric optimization approach for the synthesis of distributed wastewater treatment networks. *Ind. Eng. Chem. Res.*, **42**, pp. 2175–2191.

Hul, S., Tan, R.R., Auresenia, J., Fuchino, T. and Foo, D.C.Y. (2007). Synthesis of near-optimal topologically constrained property-based water network using swarm intelligence. *Clean Techn Environ Policy*, **9**, pp. 27–36.

Jacob, J., Kaipe, H., Coudrec, F. and Paris, J. (2002). Water network analysis in pulp and paper processes by pinch and linear programming techniques. *Chem. Eng. Commun.*, **189**(2), pp. 184–206.

Jeżowski, J.M. and Bochenek, R.J. (2002). Experiences with the use of the Luus-Jaakola Algorithm and its modifications in optimization of process engineering problems. *Recent Developments in Optimization and Optimal Control in Chemical Engineering.* Rein Luus (ed.), Research Signpost, Trivandrum, India., pp. 89–114.

Karuppiah, R. and Grossmann, I.E. (2006). Global optimization for the synthesis of integrated water systems in chemical processes. *Comput. Chem. Eng.*, **30**, pp. 650–673.

Koppol, A.P.R., Bagajewicz, M.J., Dericks, B.J. and Savelski, M.J. (2003). On zero water discharge solutions in the process industry. *Adv. Environ. Research*, **8**(2), pp. 151–171.

Kuo, W.C.J. and Smith, R. (1997). Effluent treatment system design. *Chem. Eng. Sci.*, **52**, pp. 4273–4290.

Lavric, V., Iancu, P. and Pleşu, V. (2005). Genetic algorithm optimisation of water consumption and wastewater network topology. *J. Cleaner Prod.*, **13**, pp. 1405–1415.

Lavric, V., Iancu, P. and Pleşu, V. (2007). Cost-based design of wastewater network optimal topology. Resources, *Conservation and Recycling*, **50**, pp. 186–201.

Lee, S. and Grossmann, I.E. (2003). Global optimization of nonlinear generalized disjunctive programming with bilinear equality constraints: applications to process networks. *Comput. Chem. Eng.*, **27**, pp. 1557–1575.

Liao, B. and Luus, R. (2005). Comparison of the Luus-Jaakola optimization procedure and the genetic algorithm. *Eng. Optim.*, **37**(4), pp. 381–398.

Linnhoff, B. and Hindmarsh, E. (1983). The pinch design method for heat exchanger networks. *Chem. Eng. Sci.*, **38**(5), pp. 745–763.

Luus, R. (1973). A direct approach to optimization of a complex system. *AIChE J.*, **19**(3), pp. 645–646.

Luus, R., Iyer, R.S. and Woo, S.S. (2002). Handling equality constraints in direct search optimization. *Recent Developments in Optimization and Optimal Control in Chemical Engineering*. Rein Luus (ed.), Research Signpost, Trivandrum, India. pp. 129–153.

Luus, R. and Jaakola, T.H.I. (1973). Optimization by direct search and systematic reduction of the size of search region. *AIChE J.*, **19**(4), pp. 760–766.

Manan, Z.A., Tan, Y.L. and Foo, D.C.Y. (2004). Targeting the minimum water flow rate using water cascade analysis technique. *AIChE J.*, **50**(12), pp. 3169–3183.

Mann, Y.G. and Liu, Y.A. (1999). *Industrial Water Reuse and Wastewater Minimization*, McGraw-Hill, New York.

Meyer, C.A. and Floudas, C.A. (2006). Global optimization of a combinatorially complex generalized pooling problem. *AIChE J.*, **52**(3), pp. 1027–1037.

Pillai, H.K. and Bandyopadhyay, S. (2007). A rigorous targeting algorithm for resource allocation networks. *Chem. Eng. Sci.*, **62**, pp. 6212–6221.

Poplewski, G. and Jeżowski, J. (2007). A simultaneous approach for designing optimal wastewater treatment network. *Chemical Engineering Transactions*, **12**, pp. 321–326.

Poplewski, G. (2004). *Optymalizacja Sieci Wody Procesowej* (Water network optimisation). PhD Thesis. Rzeszow University of Technology, Rzeszow Poland (in Polish language, under the supervision of Jeżowski, J.).

Prakash, R. and Shenoy, U.V. (2005a). Targeting and design of water networks for fixed flowrate and fixed contaminant load operations. *Chem. Eng. Sci.*, **60**, pp. 255–268.

Prakash, R. and Shenoy, U.V. (2005b). Design and evolution of water networks by source shifts. *Chem. Eng. Sci.*, **60**, pp. 2089–2093.

Prakotpol, D. and Srinophakun, T. (2004). GAPinch: genetic algorithm toolbox for water pinch technology. *Chem. Eng. Process.*, **43**, pp. 203–217.

Savelski, M.J. and Bagajewicz, M.J. (2000). On the optimality conditions of water utilization systems in process plants with single contaminants. *Chem. Eng. Sci.*, **55**(21), pp. 5035–5048.

Savelski, M.J. and Bagajewicz, M.J. (2003). On the necessary conditions of optimality of water utilization systems in process plants with multiple contaminants. *Chem. Eng. Sci.*, **58**(23–24), pp. 5349–5362.

Savelski, M.J. and Bagajewicz, M.J. (2001). Algorithmic procedure to design water utilization systems featuring a single contaminant in process plants. *Chem. Eng. Sci.*, **56**(5), pp. 1897–1911.

Shafiei, S., Domenech, R., Koteles, R. and Paris, J. (2004). System closure in pulp and paper mills: network analysis by genetic algorithm. *J. Cleaner Production*, **12**, pp. 131–135.

Smith, R. (2005). *Chemical Process Design and Integration*. Chichester, England: John Wiley & Sons.

Statyukha, G., Kvitka, O., Dzhygyrey, I. and Jeżowski, J. (2008). A simple sequential approach for designing industrial wastewater treatment networks. *Journal of Cleaner Production*, **16**, pp. 215–224.

Tan, R.R., Col-Iong, K.J., Foo, D.C.Y., Hul, S. and Ng, D.K.S. (2008). A methodology for the design of efficient resource conservation networks using adaptive swarm intelligence. *J. Cleaner Prod.*, **16**, pp. 822–832.

Tsai, M.-J. and Chang, C.-T. (2001). Water usage and treatment network design using genetic algorithms. *Ind. Eng. Chem. Res.*, **40**, pp. 4874–4888.

Wałczyk, K. and Jeżowski, J.M. (2008). A single stage approach for designing water networks with multiple contaminants. *Computer-Aided Chemical Engineering*, **25**, ESCAPE 18, (edited by Braunschweig, B., Joulia, X.), Elsevier, Amsterdam, pp. 719–724.

Wałczyk, K., Poplewski, G., Jeżowski, J., Jeżowska, A. and Shakhnovsky, A. (2007). Optimization of water network with models of non-mass transfer processes. *Chemical and Process Engineering*, **28**, pp. 515–525.

Wang, Y.P. and Smith, R. (1994). Wastewater minimization. *Chem. Eng. Sci.*, **49**(7), pp. 981–1006.

Wang, Y.P. and Smith, R. (1995). Wastewater minimization with flowrate constraints. *Trans IChemE.*, **73**, Part A, pp. 889–904.

Wenzel, H., Dunn, R.F., Gottrump, L. and Kringelum, J. (2002). Process integration design methods for water conservation and wastewater reduction in industry. Part 3: Experience of industrial application. *J. Clean Technologies Environ. Policy*, **4**(1), pp. 16–25.

Xue, D., Li, S., Yuan, Y. and Yao, P. (2000). Synthesis of waste interception and allocation networks using genetic-alopex algorithm. *Comput. Chem. Eng.*, **24**, pp. 1455–1460.

Exercises

(1) Most often a decrease of freshwater usage causes a drop of wastewater generation in a system. What are the reasons from exceptions from the rule? What processes typical for chemical industry can cause such effect? Hint: consider water gains and losses.

(2) There are two types of non-mass transfer water-using processes: water sources (donors) and water sinks (acceptors). The data needed for them are as follows (assuming a single contaminant):

- for sources (s): water flow rate F_w^s and contaminant concentration C^s,
- for sinks (d): water flow rate F_w^d and max. allowable concentration C^d.

Explain how one can transform the data for the typical model of mass transfer water-using process into data for pair: source — sink created from this process. Use the model of water process given by (1), (2) and (3). Hint: for mass transfer model, maximum inlet and outlet concentrations ensures minimum freshwater in the network.

(3) Assume water network consisting of m sources and n sinks with the data described in Sec. 2. Develop a superstructure for sinks and sources following the superstructure in Fig. 5 with limitation to a single freshwater source.

(4) Based on superstructure in Fig. 5, develop general superstructure for both mass transfer and non-mass transfer water-using operations. In order to keep the problem relatively simple, assume a single contaminant. What has to be added and changed in the optimization problem? First, specify additional indices, parameters and variables that should be included. Then, formulate balance equations for sinks and sources. Hint: apply the equations developed for solving Exercise 3.

(5) Formulate simple and general equations for cost of transporting streams in a network. Remark: account for both fixed and operation expenses.

(6) Calculate the number of degrees of freedom of the model defined by Eqs. (17) to (32) for the case of 2 sources of freshwater, 3 water-using processes and 2 regeneration processes.

(7) Reformulate the WNRR superstructure model by applying split ratios for splitters instead of flow rates in branches from splitters.

Appendix

This appendix gives the brief description of the algorithm of adaptive random search optimization applied to the WNRR problem. The algorithm is a modified version of the procedure developed first by Luus and Jaakola (1973). The solver we have used and some of its variants are addressed in full in Chapter 3. Here, we focus on NLP with J inequality constraints.

Min $F(\mathbf{x})$ subject to $x_i^l \leq x_i \leq x_i^u$ for $i = 1, \ldots, P$ and $g_j(\mathbf{x}) \geq 0$ for $j = 1, \ldots, J$

Given: initial point $\mathbf{x}^0$, final search regions δ_i^f, initial search region $\delta_i^0 = x_i^u - x_i^l$, number of external loops–*NEL*, and number of internal loops–*NIL*

The main steps of ARS algorithm are:

(1) Calculate from data search region contraction parameter β_i:

$$\beta_i = \left(\frac{\delta_i^f}{\delta_i^0}\right)^{1/NEL} \tag{A.1}$$

(2) Set external loop counter k at 1

(3) Set internal loop counter l at 1

(4) Calculate $\mathbf{x}^l$ from (A.2) (with $\mathbf{x}^* = \mathbf{x}^0$, $\delta_i^k = \delta_i^0$ for $k = 1$)

$$x_i^k = x_i^* + r_i \delta_i^{k-1}; \quad i = 1, \ldots, p \tag{A.2}$$

where r_i is random number of uniform distribution from the range $(-0.5, 0.5)$

(5) Increase counter l by 1. If $l = NIL + 1$ go to step 9.

(6) Check the feasibility of inequality constraints $g_j(\mathbf{x})$. If at least 1 constraint is not met go to step 4.

(7) If $GF(\mathbf{x}^l)$ is better than $GF(\mathbf{x}^*)$, update $\mathbf{x}^* = \mathbf{x}^l$.

(8) Go to step 4.

(9) Update δ_i^k according to:

$$\delta_i^k = \beta_i \delta_i^{k-1}. \tag{A.3}$$

(10) Increase counter k by 1, and go back to step 3 until *NEL*.

Note that the above ARS algorithm differs from the original one by Luus and Jaakola (1973) mainly in the definition of parameter β_i that contracts search region gradually. We proposed to calculate it from (A.1) instead of fixed value identical for all variables. Hence, β_i depends on search region sizes: δ_i^0, δ_i^f of each variable. Final region sizes, δ_i^f can be estimated by the user based on physical interpretation of corresponding variables; good results are obtained with the use of δ_i^f of the order 10^{-3} to 10^{-4} for all variables.

Chapter 16

GENETIC ALGORITHMS FORMULATION FOR RETROFITTING HEAT EXCHANGER NETWORK

Roman Bochenek, Jacek M. Jeżowski*

Department of Chemical and Process Engineering
Rzeszów University of Technology, Rzeszów, Poland
**ichjj@prz.edu.pl*

1. Introduction

Heat integration in a plant is performed within a centralized heat exchanger network (HEN), which serves as heat recovery subsystem. Hence, HEN design is crucial for heat integration. Investigations on heat recovery and HEN design have started at the end of 1960s. Huge number of papers has been published and numerous solution methods have been proposed. However, even a brief overview of approaches is beyond the scope of this work. The reader is referred to textbooks and monographs such as those by: Shenoy (1995), Floudas (1995), Biegler *et al.* (1997), Seider *et al.* (2004), Smith (2005). The most recent review paper by Furman and Sahinidis (2000) cites more than 600 papers published before 2000 year. The examples of more recent works that focused on HEN synthesis and retrofit are as follows: Björk and Westerlund (2002), Sorsak and Kravanja (2002, 2004), Frausto-Hernandez *et al.* (2003), Lin and Miller (2004), Ravagnani *et al.* (2005), Pettersson (2005), Barbaro and Bagajewicz (2005), Pettersson and Söderman (2007), Bergamini *et al.* (2007), Ravagnani and Caballero (2007), Ponce-Ortega *et al.* (2007), Yoon *et al.* (2007), Ma *et al.* (2008) to mention a few. It is important to notice that majority of the recent works applies mainly systematic optimization based approaches. They have potential to cope with large scale industrial cases. Moreover they offer a possibility to include more detailed models of apparatus (heat exchanger in particular) than those employed in previous works.

The majority of papers and methods addressed HEN synthesis. HEN retrofit was considered in relatively few works though it is of industrial importance and also is a difficult scientific problem. Some of the synthesis approaches can be applied after some modifications for HEN retrofit. The example is revamp design method by Sorsak and Kravanja (2004). It is also worth to notice that such extension complicates substantially the optimization model which was originally developed for HEN synthesis in Sorsak and Kravanja (2002). We can claim, as the rule of thumb, that the retrofit is more combinatorial than synthesis. This important feature was first shown in seminal work by Gundersen (1989). Also, Jeżowski *et al.* (2007) pointed out a highly combinatorial character of the problem when analyzing the choice of stochastic optimization techniques to solve the HEN retrofit problem. The combinatorial complexity is clear even from a simple qualitative analysis of HEN retrofit problem formulation, which is given below.

The topology of existing HEN is known in regards to location of heat exchangers, splitters and mixers. The parameters of apparatus, initial and final parameters of process streams and utilities are known. The objective is to re-design the HEN at minimum total annual cost or other cost functions.

For comparison, the data for synthesis problem includes only initial and final parameters of process streams. In synthesis, there are, most often, no rigorous limits on topology such as locations of apparatus. Also, the cost goal function does not force re-use of existing apparatus.

In contrast HEN retrofit is performed by structural and parameter changes on existing networks, which are as follows:

(1) Structural changes: heat exchanger relocations i.e. changing both heat exchanging streams, one of the two streams, shifting an apparatus to another place without changing streams, splitter adding or deleting, inserting a new heat exchanger.

(2) Parameter changes: change of split ratio and change of heat exchanger surface area (that obviously is accompanied by changes of other parameters)

Note that the number of structural changes, which are discrete decisions, is high. This results in a highly combinatorial nature of the problem. To account for possible structural changes numerous binary variables have to be applied in optimization approaches — e.g. Yee and Grossmann (1991), Ciric and Floudas (1989, 1990), Briones and Kokossis (1999), Sorsak and Kravanja (2004). Moreover, the changes are interrelated, i.e. a specific change often excludes or requires another one. In result, complex logical conditions have to be applied in the optimization models. To code them, the additional binary variables are needed. In addition to numerous binary variables the model is nonlinear. Altogether this results in a complex and difficult mixed-integer nonlinear programming (MINLP) formulation. The nonlinearities are caused mainly by bi-linear terms in mass balances: flow rate (*CP* in particular) multiplied by temperature. Investment cost term in a goal function may be also the source of nonlinearity. Next, if one intends to account for the use of standard heat exchangers (e.g. TEMA standards), having discrete values of heat transfer areas as well as other parameters, has to apply much more additional binary variables. Finally, more rigorous models should be accounted for not only simple rating equation with fixed film heat transfer coefficients. The highly nonlinear character of the optimization model with numerous binaries contributes to the situation that the application of widely available deterministic optimization approaches will not ensure successful solution — see for instance analysis and examples in Björk and Westerlund (2002). Stochastic or modern meta-heuristic techniques seem to suit better for combinatorial problems, see Jeżowski *et al.* (2007) for deeper discussion. Notice that such optimizers have been used to HEN synthesis though with certain simplifications of the general problem. The examples are simulating annealing (SA) application in Dolan *et al.* (1989, 1990) and genetic algorithms (GA) in Androulakis and Venkatasubramanian (1991), Lewin *et al.* (1998), Lewin (1998) and Ravagnani *et al.* (2005). Ma *et al.* (2008) employed a hybrid strategy — GA and simulated annealing. It is also characteristic that GA or similar evolutionary strategies are quite often employed to design and rate single heat exchangers and HENs of fixed structure. The reader is referred to Chaudhuri and Diwekar (1997), Tayal and Fu (1999), Selabs *et al.* (2006), Bochenek and Jeżowski (2007), Őzcelik (2007), Babu and Munawar (2007), Xie *et al.* (2008) to list a few. Summarizing, one can conclude that GA or similar stochastic optimization technique is well suited to solve HEN design task and HEN retrofit in particular.

In this work we will address the use of GA for retrofitting heat recovery subsystem consisting of standard heat exchangers. It is worth noting that the approach is able to cope with HEN synthesis, too. This chapter describes the use of short cut model. However, due to advantageous features of GA discussed above the extension to more detailed design equations should be straightforward.

The rest of the work is structured as follows. The next section will address the general solution strategy that is performed at two stages with structural and, then, parameter optimization. Then, we will explain briefly structural optimization issues. Structural genetic operators will be also presented. The next section will contain detailed algorithm for parameter optimization of HEN with given topology. A brief description of GA procedure applied for parameter optimization will follow. Case studies will be shown in the next section and, traditionally, summary section will end the main body of the chapter.

2. The Foundation of HEN Retrofit Approach

The approach consists in two-level iterative procedure with the use of GA at both levels. Structural optimization is performed at a primary level while in a slave level parameters of heat exchangers and splitters are optimized. The second level of the algorithm is executed for each HEN structure generated at the primary level to find values of parameters that have an influence on HEN operation and performance. The algorithm starts from an initial parent population. For HEN retrofit a user defined superstructure based on the existing topology is recommended to generate randomly other members of the initial population. In case of synthesis more general superstructure can also be used. All members of parent populations are defined only at the topology level. Then, all the members are processed at the second level. Selected, best-fitted members of the parent population are, then, modified

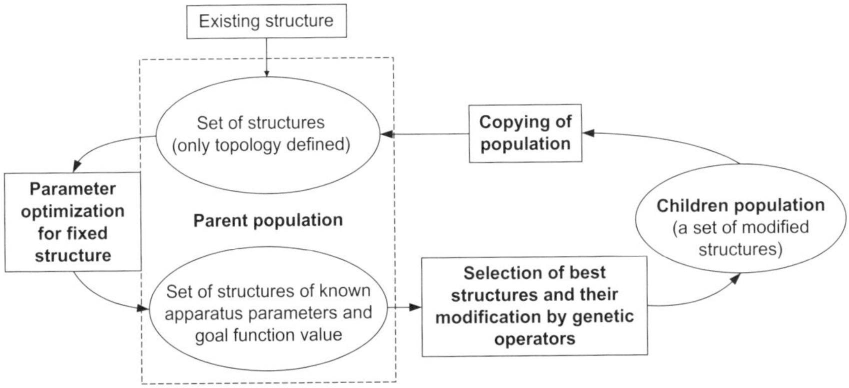

Figure 1. Flow-sheet of a single cycle of GA for retrofitting HEN.

by crossover and mutation operators and become new parent population for next cycle of GA. The general flow sheet of the design algorithm is illustrated by Fig. 1.

Though GA can be applied as a general purpose optimizer the best results are achieved if GA is applied to solution space (representation and codes) that is "natural" and specific for a problem. Hence, we applied a chromosome that is a direct representation of network topology.

Instead of using classical superstructure model in form of complex equalities and inequalities with large number of binary variables, the problem of structure optimization is transformed to processing a single multi-variable representation, which is, in fact the variable in structural GA optimization. The variable encapsulates all topological features of a HEN. It will be referred to as structural matrix **SM**. Figure 2 illustrates general structure of the **SM** matrix. The nodes were employed in this matrix. A pair: hot node-cold node defines a match (heat exchanger).

Each row is related to a single heat exchanger. The entries of a row, which are addresses of nodes, define uniquely heat exchanger location within a HEN in regards to streams exchanging heat. Feasible space for all possible structural changes in **SM** matrix is bounded by superstructure defined by vector of hot nodes ($\mathbf{NOD^H}$) and cold nodes ($\mathbf{NOD^C}$) and, also, by split matrices ($\mathbf{SPL^H}$, $\mathbf{SPL^C}$). This space ensures that structural changes performed directly on HEN structure, by modifying its codes with specialized genetic operators, generating only structurally feasible solutions.

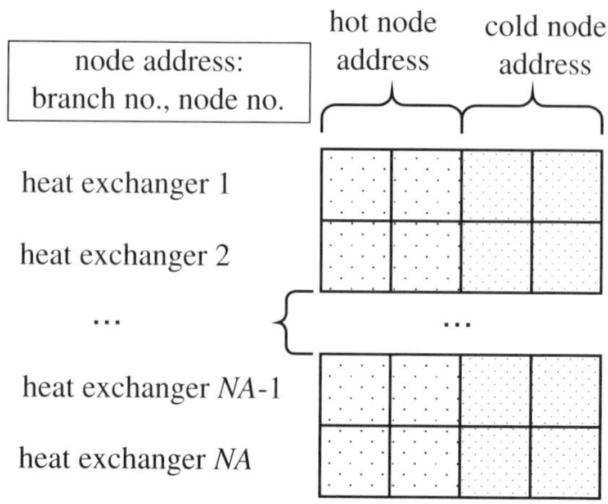

Figure 2. A general structure of matrix **SM**.

The deeper explanation of structure optimizations level is given in the next section.

3. Structure Optimization Level

3.1. Representation and codes for superstructure and structures

HEN superstructure concept in this work is primarily based on stream superstructure. Stream superstructure consists of main- and side branches. The latter define potential splitters and mixers. The nodes for the matches are, then, imposed on stream superstructure to create a space of heat exchanger superstructure. In contrast to classical superstructure, potential matches are not included into it but only space is defined for the location of heat exchangers' nodes. In consequence equation oriented optimization model to find structural changes in existing HEN is not needed.

All process and utility streams in existing HEN are inserted into the superstructure. For each stream its configuration has to be developed in the form of superstream. Superstream encapsulates all possible flow paths for the stream from its inlet to outlet from HEN. It is only required that superstream always consists of a single main branch spanning inlet and outlet of the stream. The side streams can be added to the main branch in frame of superstream. Inclusion of side streams creates splitters and mixers at the main branch. It is worth mentioning that without side branches superstream is equivalent to stream and main branch as well.

Thus, to construct the superstructure it is sufficient to:

- Add side branches to the existing streams in existing HEN — superstreams consisting of main and side branches are created
- Distribute hot and cold nodes for the potential heat exchangers

All branches and nodes in the superstructure are numbered, separately for hot and cold streams. The only rule for numbering of branches is that main branches are given numbers first and, then, side branches. Nodes are given numbers separately for each branch, starting from the inlet of a stream to its outlet. Figure 3 illustrates the scheme of numbering for branches and nodes in superstructure. Each node is given unique address consisting of two items: *branch no. — node no.* A structure is generated from the superstructure by assigning to heat exchangers the addresses of two nodes: a hot stream node and a cold stream node.

The code for defining both the superstructure and structure is based on matrix representation. The information on the superstructure is stored in

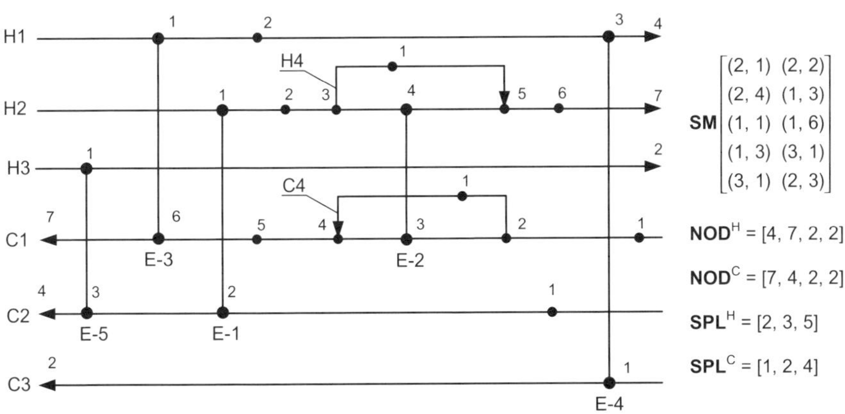

Figure 3. Structure of the existing HEN within the applied superstructure. H1-H3, C1-C3 are main branches, H4, C4 are side branches. These symbols correspond to standard names in the literature and are not used in the method. Black dot not connected to another one denotes hot or cold node of potential heat exchanger. For instance, dot number 6 on H2 is hot node because it lies at hot stream. Construction of matrices **SM**, **NOD** and **SPL** is explained in Secs. 3.1 and 3.2.

node vectors $\mathbf{NOD^H}$, $\mathbf{NOD^C}$ and split matrices $\mathbf{SPL^H}$, $\mathbf{SPL^H}$ for hot and cold streams. Vectors **NOD** and matrices **SPL** have the following structure:

- $\mathbf{NOD} = [n_1, \ldots, n_i, \ldots, n_{NB}]$; the vector has *NB* elements, where *NB* is equal to the total number of branches; element n_i equals the number of nodes at *i*th branch.
- $\mathbf{SPL} = [S_{ij};\ i = 1, \ldots, NS;\ j = 1, 2, 3]$; the matrix has *NS* rows (where *NS* is equal to total number of side branches) and three columns ($j = 1, 2, 3$). The number of main branch from which a side branch is created is in the 1st column ($j = 1$), split node number at main branch in the 2nd column ($j = 2$), mixing node number at main branch is in the 3rd column ($j = 3$).

The example of vectors **NOD** and matrices **SPL** is shown in Fig. 3. This is the example of retrofit superstructure. The existing structure has no splitters and mixers and five heat exchangers with symbols E-1 to E-5.

To summarize, HEN retrofit superstructure is built on the basis of the existing HEN topology by adding new potentially possible splits (side branches) and new potentially possible locations of heat exchangers (nodes).

3.2. Generation of structures and genetic operators

A selection of structures from parent population is performed proportionally to a value of evaluation criterion with repetitions — see Michalewicz and Fogel (2002) for details of selection mechanism. The children population created from the parent one becomes parent population in the next evolution cycle after parameter optimization. All the genetic operations, i.e. mutation and crossover, are performed on codes of structures, i.e. structural matrix **SM**. Each structure embedded in the retrofit superstructure is created by assigning to each heat exchanger within a structure unique addresses of hot and cold nodes from the superstructure. Structural matrix **SM**, shown also in Fig. 3, is as follows:

$\mathbf{SM}[i = 1, \ldots, NA;\ j = 1, \ldots, 4]$; the matrix has *NA* rows (where *NA* equals maximal number of heat exchangers which can exist in a HEN) and four columns ($j = 1, \ldots, 4$). The first two columns ($j = 1, 2$) involve hot node address while columns 3 and 4 — cold node address.

It is important to note that information on splits (i.e. side branches) is extracted from superstructure code (matrices $\mathbf{SPL^H}$, $\mathbf{SPL^C}$). Hence, matrix **SM** provides the necessary and sufficient information on HEN structure. The algorithm extracts all other information that is needed for calculations from the superstructure code, i.e. vectors $\mathbf{NOD^H}$, $\mathbf{NOD^C}$ and matrices $\mathbf{SPL^H}$, $\mathbf{SPL^C}$. These vectors and matrices are not subjected to any changes. In order to perform structural retrofit changes a set of genetic operators was developed that operate on structural matrix **SM**. The applied genetic operators allow performing all structural changes that are possible in HEN revamp design.

Two crossover operators and seven mutation operators were applied in the GA optimization at first level. Single-point crossover and multi-point crossover operators were developed. The first crossover operation is illustrated in Fig. 4, where the dashed horizontal line shows the crossover point. The locations of heat exchangers that are below crossover point are changed by this operator. It is important to note that the single-point crossover operator can produce solutions that are structurally infeasible. Such a case is shown in Fig. 4 where two heat exchangers are assigned to the single node

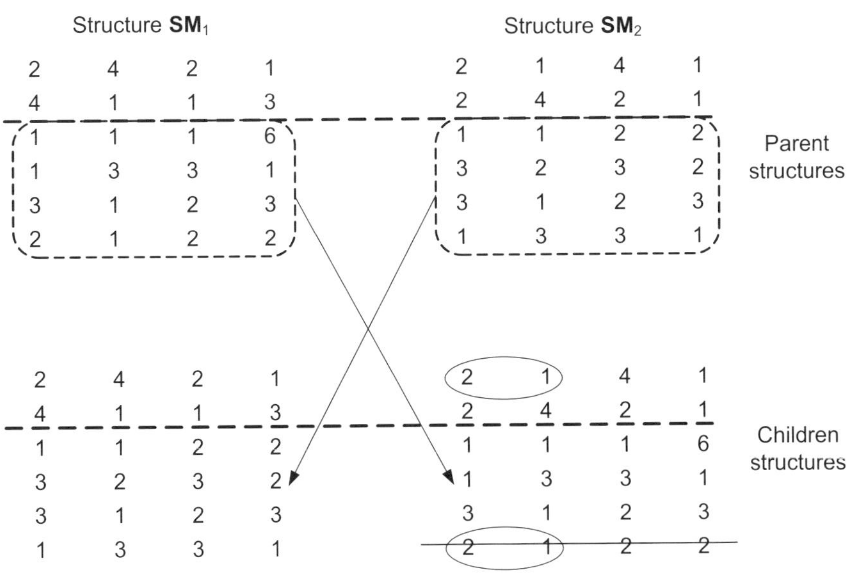

Figure 4. Illustration of single-point crossover.

(hot node 2.1). Hence, we developed repairing mechanism to convert such solutions into structurally feasible ones. The repairing mechanism removes in the second (right-side) children solution one exchanger — see Fig. 4. Multi-point crossover requires more complex procedure because its application yields feasible structures — no reparation is needed. Conceptually the operation consists in a change of a certain (given) number n of rows of matrix **SM** with total number of rows equals m. It is required that $m > n$ and n is the parameter of the operator.

The procedure of multi-point crossover is as follows:

(1) choose randomly two parent matrices: $(\mathbf{SM_1})_{\text{parent}}$ and $(\mathbf{SM}_2)_{\text{parent}}$
(2) create children structures $(\mathbf{SM}_1)_{\text{children}}$ and $(\mathbf{SM}_2)_{\text{children}}$ by copying directly $(\mathbf{SM}_1)_{\text{parent}}$ and $(\mathbf{SM}_2)_{\text{parent}}$
(3) set counter k of copied rows at 0: $k = 0$
(4) for $i = 1$ to m do steps $5 \div 7$
(5) copy ith row of $(\mathbf{SM}_2)_{\text{parent}}$ into ith row of $(\mathbf{SM}_1)_{\text{children}}$ providing that rows from $(i + 1)$ till m do not contain the nodes identical to those in ith row of $(\mathbf{SM}_2)_{\text{parent}}$
(6) set $k = k + 1$ if row was copied (i.e. if operation of (5) was successfully completed)
(7) if $k = n$ then go to (8)
(8) set counter k of copied rows at 0: $k = 0$
(9) for $i = 1$ to m do steps $10 \div 12$
(10) copy ith row of $(\mathbf{SM}_1)_{\text{parent}}$ into ith row of $(\mathbf{SM}_2)_{\text{children}}$ providing that rows from $(i + 1)$ till m do not contain the nodes identical to those in ith row of $(\mathbf{SM}_1)_{\text{parent}}$
(11) set $k = k + 1$ if row was copied (i.e. if operation of (10) was successfully completed)
(12) STOP if $k = n$

The final results of the algorithm are two children structures $(\mathbf{SM}_1)_{\text{children}}$ and $(\mathbf{SM}_2)_{\text{children}}$ in which, at most n rows have been exchanged. Notice, that number of exchanged rows can be less than n. Also, they do not necessarily form tight block as it was in single point crossover. To give additional explanation we show in Fig. 5 the illustration of multi-point crossover for $n = 3$. One can notice, that in case of structure $(\mathbf{SM}_1)_{\text{children}}$ the exchange

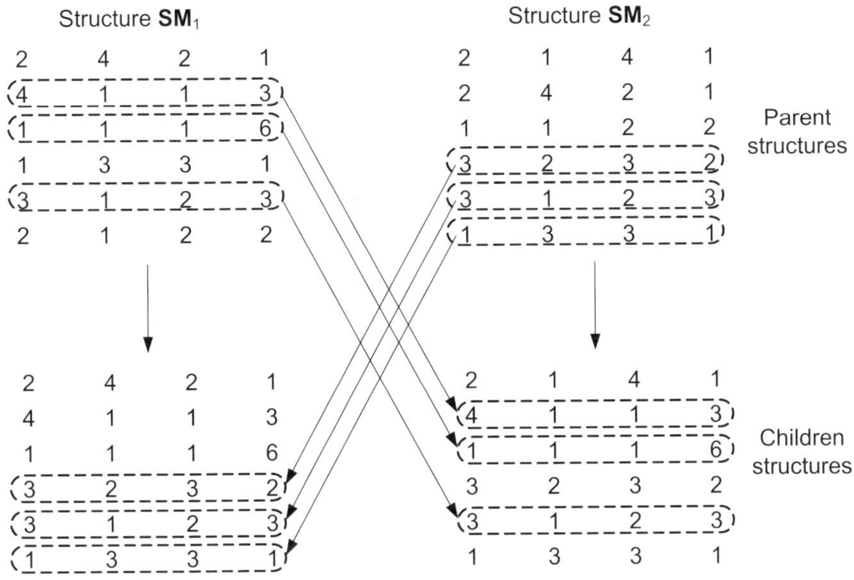

Figure 5. Illustration of multi-point crossover.

was possible for last three rows while for $(\mathbf{SM}_2)_{\text{children}}$ rows 2, 3 and 5 were changed (they are not blocked together).

The following mutation operators were applied in the GA method: relocation of heat exchanger — hot node, relocation of heat exchanger — cold node, relocation of a heat exchanger — both nodes, exchange of hot nodes between two heat exchangers, exchange of cold nodes between two heat exchangers, adding of a new heat exchanger, removing of a heat exchanger. They are the simple operations and we do not present them in detail. Algorithm of GA optimizer together with standard operators for solving general parameter optimization task will be explained in point 4.

Applied genetic operators allow performing all structural changes that are possible in HEN revamp design. They can generate all the structures within the retrofit superstructure. It is worth noting that due to the superstructure representation applied in this method operators that act on superstream (i.e. changing stream splits) are not needed. Seemingly, genetic operators change only heat exchanger locations. In fact, each operator can also change superstream structure by adding or deleting splits (side branches). It results from the fact that stream topology is involved within the superstructure defining locations of all possible side branches. A structure of each HEN, coded by structural matrix **SM**, involves the information on location of all heat exchangers. If there is no heat exchanger at a side branch the branch is inactive in this structure, i.e. it is eliminated. However, if a heat exchanger is located by any genetic operator on the branch this will result in re-activation of a side branch in the structure.

4. Parameter Optimization Level

4.1. General concept of the parameter optimization level

Structures generated at the first level are defined only in regards to topology. To select the children population the knowledge of optimization criterion value is necessary. In consequence, basic parameters of apparatus and streams are needed to calculate HEN goal function. Costs of structural changes are calculated during the first optimization stage. The costs caused by the changes of heat exchanger surface area and operation cost can be determined after executing the second stage of the method. General concept of the algorithm for parameter optimization at 2nd level of HEN retrofit method is shown in Fig. 6.

The problem which is to be solved at the parameter optimization level can be formulated as follows: *For a HEN of fixed structure calculate heat exchanger surface areas and split ratios in splitters in order to minimize or maximize applied goal function.*

It is an NLP problem providing that non-standard heat exchangers are to be used in a HEN or MINLP one for standard apparatus. It is also worth noting that this problem or simpler one of rating a single heat exchanger is common in industrial practice. Due to non-continuous goal function it is very difficult to solve it using the deterministic solvers even for non-standard heat exchangers. Therefore quite often stochastic techniques are applied. We have mentioned some works from this field in Sec. 1. Finally, let us notice that majority of works on HEN parameter optimization does not account for split ratios or, alternatively, flow rates in branches. In our method the flow rates are decision variables. It can finally result in elimination of a splitter and associated mixer. In this respect the topology is not fixed because some changes can occur. However, the method does not allow for

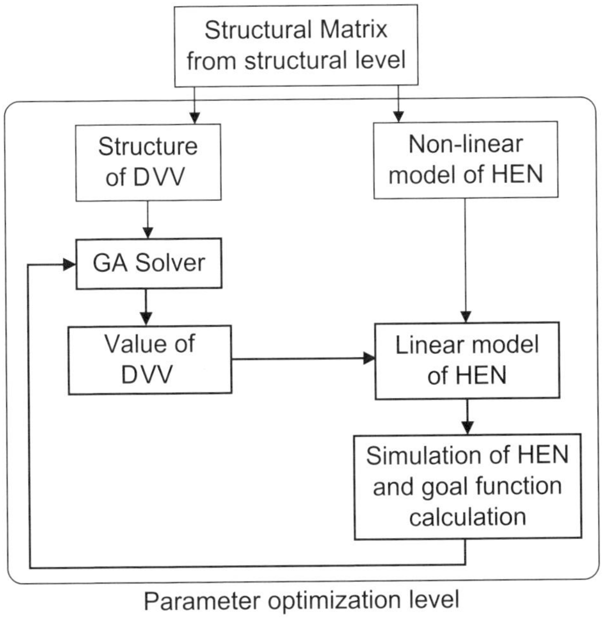

Figure 6. General concept of the algorithm for parameter optimization at 2nd level of HEN retrofit method. DVV means decision variable vector — see also symbols.

creating new splitters or new branches of existing splitters. Similarly, heat exchanger area can be reduced to zero but new heat exchanger cannot be created. To summarize, it is of importance that parameter optimization can affect in certain structural changes.

The constraints of optimization model of a HEN are as follows:

(1) Equality constraints of HEN model (heat and mass balances, design equations).
(2) Equality constraints on required values of outlet temperatures of streams leaving a HEN.
(3) Inequality constraints on permissible temperature approaches in heat exchangers.
(4) Various other technological and economic constraints that are often case dependent.

Some of the equality constraints from group (1) are nonlinear. The cost goal function is usually also nonlinear and, additionally, it may contain the discontinuous terms, such as the economic correlations for calculating investment cost of heat exchangers.

They have the general form:

if *parameter* $\leq$ *parameter*$_{\min}$ **then** (formula 1)
else (formula 2)

Equality constraints cause serious difficulties for all classes of stochastic optimization approaches, including GA. In order to deal efficiently with constraints from group (1) we developed the scheme which allows solving them as a set of linear equations in regards to dependent variables. Equality constraints from group (2) were relaxed as follows:

$$g(x) = 0 \rightarrow \begin{array}{l} g(x) + \varepsilon \geq 0 \\ g(x) - \varepsilon \leq 0 \end{array}, \tag{1}$$

where ε is a small number given by the designer.

The relaxation has also physical meaning. In our opinion imposing "hard" equality constraints on outlet temperatures is often unnecessary since in many cases they are allowed to vary within some ranges. Relaxation (1) is the replacement of equality conditions on outlet temperatures by inequalities defining allowable final temperature variation, which is problem dependent.

Relaxed constraints from group (2) as well as inequalities from groups (3) and (4) were accounted for in augmented penalty function.

The penalty terms require the use of penalty coefficients that are most often determined basing on the heuristics and/or error-and-trial procedure. For equalities of type (2) we applied penalty terms that have natural physical interpretation. For hot streams that do not reach required final temperature (i.e. are too "hot") the penalty term is equal to operation and investment cost on cooler, i.e. cost of cooling utility plus cost of cooler such that cools down the stream to given temperature. If hot stream temperature is too low (the stream has been precooled) we applied very high penalty coefficients typical for the barrier penalties. Analogous procedure was used for penalty terms of the cold streams. We noticed from tests that constraints from group (2) were met with high preciseness. Notice that, additionally, tedious choice of penalty coefficients was eliminated. Additional information on penalty terms mechanism in GA solver were addressed in Sec. 4.3.

4.2. Mathematical model of the HEN with fixed structure

Usually HEN models are based on heat balances of matches with heat loads of heat exchangers as variables. The mean temperature differences are applied in design equations. The model applied in this work is based on work of Kotjabasakis and Linnhoff (1986) where heat transfer surface area instead of heat load is the parameter defining heat exchanger. The model does not use mean temperature differences. Notice that this is consistent with standard methods of rating heat exchangers. The assumptions imposed on the heat exchanger model, are similar to those used in other works:

- heat exchangers are of shell-and-tube type,
- strictly counter-current flow scheme is applied in all heat exchangers, i.e. mean temperature difference is approximated by the logarithmic mean temperature difference,
- enthalpy changes of streams are linear function of temperature, i.e. *CP* values are constant in each heat exchanger

Under these assumptions the typical basic model of a heat exchanger applied in HEN design is given by:

$$Q = CP(T_{\text{in}}^{H} - T_{\text{out}}^{H}), \tag{2}$$

$$Q = CP^{C}(T_{\text{out}}^{C} - T_{\text{in}}^{C}), \tag{3}$$

$$A = \frac{Q}{U \cdot \Delta T_m}, \tag{4}$$

$$\Delta T_m = \frac{(T_{\text{in}}^{\text{H}} - T_{\text{out}}^{\text{C}}) - (T_{\text{out}}^{\text{H}} - T_{\text{in}}^{\text{C}})}{\ln \frac{T_{\text{in}}^{\text{H}} - T_{\text{out}}^{\text{C}}}{T_{\text{out}}^{\text{H}} - T_{\text{in}}^{\text{C}}}}. \tag{5}$$

By combining Eqs. (2), (3) and (4) one easily yields, after elementary algebraic operations, Eqs. (6), (7) for the outlet temperatures of streams from heat exchangers in a HEN.

$$\alpha^{\text{H}} T_{\text{in}}^{\text{H}} + (1 - \alpha^{\text{H}}) T_{\text{in}}^{\text{C}} = T_{\text{out}}^{\text{H}}, \tag{6}$$

$$(1 - \alpha^{C}) T_{\text{in}}^{\text{H}} + \alpha^{C} T_{\text{in}}^{\text{C}} = T_{\text{out}}^{\text{C}}. \tag{7}$$

Parameters α^{H} and α^{C} are defined as follows:

$$\alpha^{\mathrm{H}} = \frac{(\gamma^{\mathrm{C}} - \gamma^{\mathrm{H}})e^{\gamma^{\mathrm{C}}}}{\gamma^{\mathrm{C}}e^{\gamma^{\mathrm{C}}} - \gamma^{\mathrm{H}}e^{\gamma^{\mathrm{H}}}}, \tag{8}$$

$$\alpha^{\mathrm{C}} = \frac{(\gamma^{H} - \gamma^{\mathrm{C}})e^{\gamma^{\mathrm{H}}}}{\gamma^{H}e^{\gamma^{\mathrm{H}}} - \gamma^{\mathrm{C}}e^{\gamma^{\mathrm{C}}}}, \tag{9}$$

$$\gamma^{H} = \frac{UA}{CP^{H}}, \tag{10}$$

$$\gamma^{C} = \frac{UA}{CP^{C}}. \tag{11}$$

Formulas (6) to (11) form alternative model to that defined by (2) to (5). The former is seldom applied to rate single apparatus and to design HEN. Thus, we will analyze it in the following to show clearly its specific features and advantages over popular model given by (2) to (5). Each heat exchanger in HEN requires two equations only, Eq. (6) for hot node and Eq. (7) for cold one. For the typical data in simulation mode: A, CP^{H}, CP^{C} and U, parametrs α^{H} i α^{C} are fixed (i.e. they can be calculated from the data). Thus, Eqs. (6) and (7) are linear. It is convenient to use for them matrix notation:

$$\begin{bmatrix} T_{\mathrm{in}}^{H} \\ T_{\mathrm{in}}^{C} \end{bmatrix} \begin{bmatrix} \alpha^{H} & 1-\alpha^{H} \\ 1-\alpha^{C} & \alpha^{C} \end{bmatrix} = \begin{bmatrix} T_{\mathrm{out}}^{H} \\ T_{\mathrm{out}}^{C} \end{bmatrix}. \tag{12}$$

The analysis of the above equation — see Bochenek and Jeżowski (1999) — proved that it is unnecessary to differentiate hot and cold streams.

The model of heat exchangers is given by set of $2n$ identical equations for the HEN without splitters containing n heat exchangers

$$T_{\mathrm{out}}^{i} = \alpha^{i}T_{\mathrm{in}}^{i} + (1-\alpha^{i})T_{\mathrm{in}}^{i^*};\ i = 1, \ldots, 2n, \tag{13}$$

$$\alpha^{i} = \frac{(\gamma^{\mathrm{i}^*} - \gamma^{i})\, e^{\gamma^{\mathrm{i}^*}}}{\gamma^{\mathrm{i}^*}e^{\gamma^{\mathrm{i}^*}} - \gamma^{i}e^{\gamma^{\mathrm{i}}}}, \tag{14}$$

$$\gamma^{i} = \frac{UA}{CP^{i}}, \quad \gamma^{i^*} = \frac{UA}{CP^{i*}}. \tag{15}$$

Notice that superscript (*) denotes one from the two streams exchanging heat (hot or cold) while parameters of the second stream do not have any superscript. The distinct advantages of the model are:

(a) it is based on heat exchanger surface areas instead of heat loads,
(b) it does not require thermodynamic constraints on temperature approaches.

The first feature is clearly visible. However, to our knowledge the second feature has not been proved in the literature. In order to prove it here we will show first the other relations and conclusions that can be deduced from the analysis of (13) to (15). They are also essential and should be considered when constructing algorithm for solving constraints to calculate temperatures.

(1) For identical CP values in heat exchanger "i" parameter α^i is indefinite: since if $CP^i = CP^{i^*}$ then $\gamma^i = \gamma^{i^*}$, and, thus, $\alpha^i = (\gamma - \gamma)e^{\gamma}/[\gamma\, e^{\gamma} - \gamma\, e^{\gamma}] = 0/0$
(2) If $A \to 0$ then γ^i and $\gamma^{i^*} \to 0$, and, thus α^i is indefinite: $\alpha^i = 0/0$
(3) If $A \to \infty$ then γ^i and $\gamma^{i^*} \to \infty$, and, thus α^i is indefinite: $\alpha^i = 0/0$
(4) If $k \to 0$ then γ^i and $\gamma^{i^*} \to 0$, and, thus α^i is indefinite: $\alpha^i = 0/0$
(5) If $CP^i = 0$ then γ^i is indefinite: $\gamma^i = UA/0$
(6) From analysis of Eq. (13) the conclusions could be drawn out:

- for $\alpha_i = 1$ Eq. (13) becomes the equality: $T^i_{\text{out}} = T^i_{\text{in}}$ what is valid for $A \to 0$
- for $\alpha_i = 0$ Eq. (13) becomes the equality: $T^i_{\text{out}} = T^{i^*}_{\text{in}}$ what is valid for $A \to \infty$

However, from points (2) and (3) results that α^i is indefinite for $A \to 0$ and $A \to \infty$. Hence, Eq. (13) is valid for $0 < \alpha^i < 1$. Let us note that Kotjabasakis and Linnhoff (1986) stated that the equation is valid for $0 \leq \alpha^i \leq 1$ what is misleading and may result in invalid conditions used in calculation procedure.

It is however important that the features shown above do not spoil generality and applicability of the model. The case from point (1) — the same CP of streams - is quite rare in industrial practice. To prevent indefinites in operations of points (2) to (5) obvious conditions (16) should be imposed into model of each heat exchanger:

$$A,\ CP,\ U > 0. \tag{16}$$

Simple algebraic manipulations of (13) to (15) with restriction (16) gives us the following useful conditions:

$$(0 < \alpha^i < 1), \tag{17}$$

$$\gamma^{i^*}(1-\alpha^i) = \gamma^i(1-\alpha^{i^*}), \tag{18}$$

$$\alpha^i = (T^i_{\text{out}} - T^{i^*}_{\text{in}})/(T^i_{\text{in}} - T^{i^*}_{\text{in}}), \tag{19}$$

$$\gamma^i/\gamma^{i^*} = CP^{i^*}/CP^i. \tag{20}$$

We will prove that for all physically and technically meaningful values of A, CP, U the model does not require constraints on temperature differences at the ends of each heat exchanger — Δt_1, Δt_2. In HEN optimization model, which is based on Eqs. (2) to (5), condition (21) has to be applied for each heat exchanger.

$$\Delta T_1 > 0 \quad \text{and} \quad \Delta T_2 > 0, \tag{21}$$

where ΔT_1, ΔT_2 — temperature differences at heat exchanger inlet and outlet.

For the purpose of the proof they are substituted by the following:

$$\Delta T_1/\Delta T_2 > 0. \tag{22}$$

Inequality (22) is clearly equivalent to (21) if both ΔT_1 and ΔT_2 are positive. However, it is unnecessary in the model to differ between hot and cold streams. Hence, it is possible that both Δt_1 and Δt_2 are negative if the differences are calculated for each heat exchanger always in the same manner. However, if (22) is met both temperature approaches have the same sign and this means that a hot stream in a heat exchanger has always higher temperature than a cold one. Hence, condition (22) is always equivalent to (21) in the model applied.

It is necessary to add that constraints (21) are in practice formulated as:

$$\Delta T_1 > \Delta T_{\min} \quad \text{and} \quad \Delta T_2 > \Delta T_{\min}, \tag{23}$$

where $\Delta T_{\min}$ is given beforehand.

We will prove that (22) is always met for the applied HEN model. Let's reformulate Eq. (13) for both nodes of heat exchangers to:

$$\Delta T_1 = T^i_{\text{out}} - T^{i^*}_{\text{in}} = \alpha^i (T^i_{\text{in}} - T^{i^*}_{\text{in}}), \tag{24}$$

$$\Delta T_2 = T^i_{\text{in}} - T^{i^*}_{\text{out}} = \alpha^{i^*} (T^i_{\text{in}} - T^{i^*}_{\text{in}}), \tag{25}$$

From (24) and (25), it will result in:

$$\Delta T_1/\Delta T_2 = \alpha^i/\alpha^{i^*} \tag{26}$$

Because $0 < \alpha^i,\ \alpha^{i^*} < 1$, for obvious condition (16), see (17), thus:

$$\Delta T_1/\Delta T_2 > 0 \tag{27}$$

for all heat exchangers in the HEN. Hence, condition (21) or (22) is a superfluous one in the HEN model applied.

The features shown above make heat exchanger model defined by (13) to (15) simpler and easier to use in HEN optimization model than typical model defined by (2) to (5).

Equations (13) to (15) are nonlinear with respect to heat transfer surface areas and, hence, the optimization model of HEN with fixed topology is also nonlinear, independently from a goal function applied. The model presented in this point was limited to known fixed values of split ratio because *CP* values were fixed. In our retrofit approach split ratios are variables, so we had to add balances of splitters and mixers. Notice that in fact we did not use directly split ratios but flow rates in the branches. The use of the flow rates was advantageous in organizing calculations to determine the temperature. Hence, in addition to Eqs. (13) to (15) the mathematical model of the HEN contains also the following equations (symbols applied in the equations are explained in Fig. 7):

(a) mass balance of splitters

$$G_{\text{in}} = G^1_{\text{out}} + G^2_{\text{out}}, \tag{28}$$

(b) mass balance of mixers

$$G_{\text{out}} = G^1_{\text{in}} + G^2_{\text{in}}, \tag{29}$$

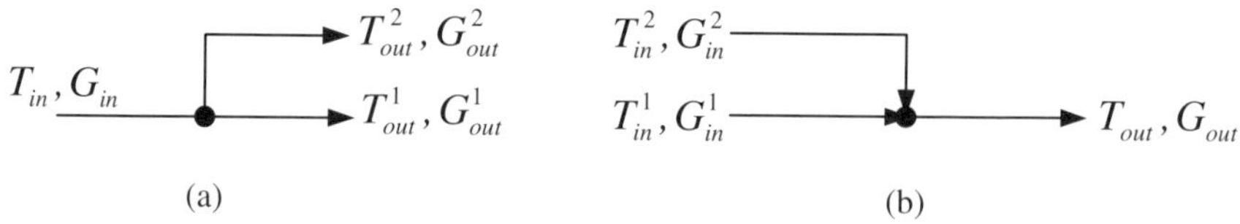

Figure 7. Explanations for symbols used in equations for splitters and mixers: (a) splitter, (b) mixer.

(c) heat balance for splitters

$$T^1_{\mathrm{out}} = T_{\mathrm{in}}, \quad T^2_{\mathrm{out}} = T_{\mathrm{in}}, \tag{30}$$

(d) heat balance for mixers

$$\frac{T^1_{\mathrm{in}} CP^1_{\mathrm{in}}}{CP^1_{\mathrm{in}} + CP^2_{\mathrm{in}}} + \frac{T^2_{\mathrm{in}} CP^2_{\mathrm{in}}}{CP^1_{\mathrm{in}} + CP^2_{\mathrm{in}}} = T_{\mathrm{out}}, \tag{31}$$

(e) continuity conditions for "dummy" nodes (dummy node is such hot or cold node which is inactive in the structure under optimization)

$$T_{\mathrm{out}} = T_{\mathrm{in}}, \quad G_{\mathrm{out}} = G_{\mathrm{in}}, \tag{32}$$

(f) conditions for inlet nodes, i.e. assignments of inlet temperature to inlet node of the branches

$$T_{\mathrm{in}} = \text{inlet temperature}, \tag{33}$$

Variable flow rates in branches of splitted streams are sources of new nonlinearities. Hence, there are two sources of nonlinearities: Eqs. (13) and (31). It is also important that the model may contain discrete variables if standard heat exchangers are to be used, because heat transfer areas are discrete. It is of importance to note that Eq. (13) becomes linear if heat exchanger surface area, overall heat transfer coefficient and heat capacity flow-rates are fixed. Equation (31) becomes linear one if heat capacity flow-rates are fixed. In the frames of stochastic optimization approaches such as GA, which solves the optimization model sequentially, it is possible to make use of the fact that some equations can be transformed into linear ones by fixing the values of the selected variables. This can be achieved by dividing the variables in the optimization model into two subsets: a subset of the decision (independent) variables and a subset of the dependent (state) ones.

The values of decision variables are generated by stochastic optimization algorithm and, thus, are known when solving constraints of optimization model. Hence, the equations can be solved as sets of linear equations in regards to dependent variables if decision variables were property chosen and constraints property sequenced.

To solve the HEN parameter optimization problem we have chosen heat transfer surface areas and mass flow rate values of side branches as decision variables. The temperatures are, then, dependent variables calculated from model equations, which are linear in regards to these temperatures.

The implementation of the HEN model given by Eqs. (13) to (15) and (28) to (33) requires an appropriate method that automatically generates model equations from codes of the superstructure and of the structure, i.e. from vectors **NOD** and matrices **SPL** and **SM**. Also, the method of model generation has to account for the data: stream inlet and outlet temperatures, overall heat transfer coefficients. The developed method transforms information on each node from codes of structural optimization level into relevant equation of the HEN model. Thus, the equation based model for parameter optimization of HENs with given structure is fully integrated with the developed representation of superstructure and structures.

4.3. Genetic Algorithms solver for parameter optimization

To solve both NLP and MINLP problems that arise in parameter optimization level we applied the GA solver, which is a version of general solver called GenCom — see Bochenek and Jeżowski (2008). The solver was first tested for benchmark optimization problems — general as well as chemical and process engineering ones. The tests proved its efficiency and robustness — see Bochenek *et al.* (2005), Bochenek and Jeżowski (2008). The characteristic feature of the method consists in the fact that in each evolution cycle three populations are processed: parent, intermediate and children population in contrast to classical GA where only parent and children populations are used. In order to increase a chance of survival of the best fitted individuals and to decrease an effect of premature degeneration (i.e. local optimum trap) we applied the so-called "genetically modified sub-population". An application of genetic operators to selected members of parent population creates members of this sub-population. Then, an intermediate population is created that is a superset of parent population and genetically modified sub-population. A selection mechanism chooses the individuals of parent population of next generation. The sub-population has a smaller number of members than the parent population and the number of its members is determined by control parameter called modification ratio *u_mod*. The algorithm of GenCom illustrated in Fig. 8, is explained below. Notice that "simulation module" in the figure is a procedure to solve equality constraints and to check inequalities in optimization model. For HEN parameter optimization it solves HEN model and calculates dependent variables according to the algorithm described in the preceding paragraphs. The chromosome of a solution is coded as real number vector (decision variables vector **DVV**) containing heat exchanger surface area of all heat exchangers and flow rate in branches of all splitters in HEN.

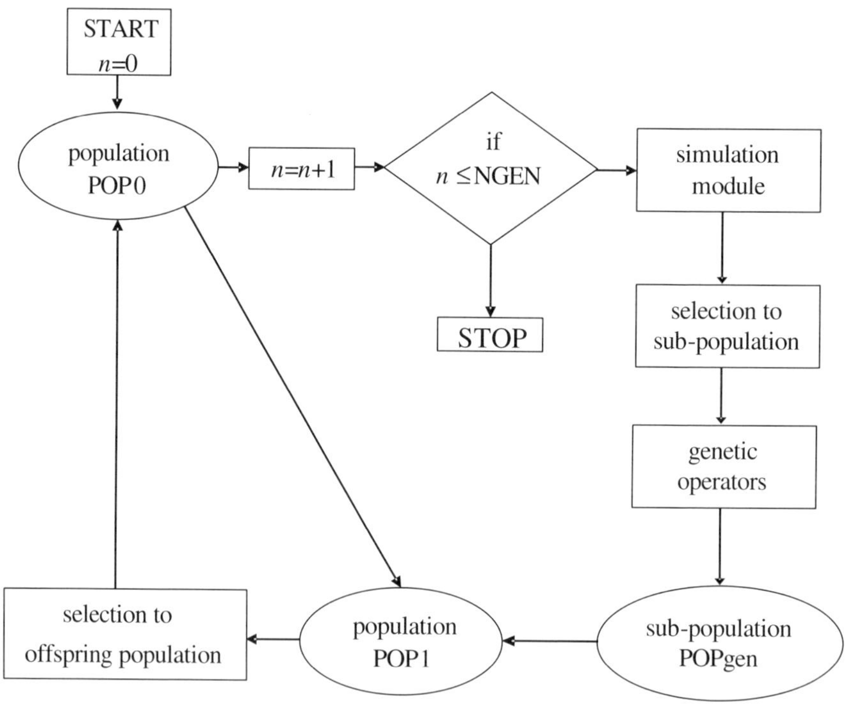

Figure 8. Flowsheet of GenCom algorithm; n denotes number of generation, other symbols explained in text.

The algorithm of parameter optimization by GenCom is as follows:

(1) Creation of decision variables vector **DVV** from structural matrix **SM** of HEN
(2) Transformation of the matrix **SM** into equation oriented mathematical model of the HEN
(3) Input values of control parameters: size of parent and offspring population NPOP, size of sub-population NPOPgen, total number of generations NGEN, number of genetic operators NOPER to be applied, modification ratio u_mod, probabilities of using the operators p_gen_m for $m = 1, \ldots,$ NOPER
(4) Creation of initial population POP0
(5) Simulation of HEN and calculation of fitness function $FP_i; i = 1, \ldots,$ NPOP
(6) Calculation of selection probabilities $p0_i$; $i = 1, \ldots,$ NPOP
(7) Creation of genetically modified sub-population
   (a) selection of genetic operator m with probability p_gen_m
   (b) selection of individual/individuals i for genetic modifications with probability $p0_i$
   (c) creation of modified individual / individuals j by using selected operator
   (d) points (a) and (b) are performed until given size of sub-population NPOPgen is reached
(8) Calculation of fitness function FP_j; $j = 1, \ldots,$ NPOPgen for members of sub-population POPgen.
(9) Calculation of selection probabilities $p1_k$; $k = 1, \ldots,$ (NPOP + NPOPgen) for individuals from the superset POP1 consisting of parent population and sub-population.
(10) Creation of offspring population having NPOP members by choosing members from the superset POP1 consisting of parent population and sub-population with probability $p1_k$ (the superset is intermediate population)
(11) The population from previous step is copied into parent population of next generation
(12) Points 5–11 are performed until generation number is greater than NGEN

To deal with both inequality and equality constraints of optimization problem two types of penalty mechanisms are available: death penalty and penalty terms in the augmented goal functions with user specified penalty coefficients. Penalty term is given by Eq. (34).

$$\text{Penalty term} = \text{penalty coefficient} \times [\text{ABS(sum of unsatisfied constraints)}]^p, \quad (34)$$

where p is a parameter given by the user.

Generally, a major difficulty when adjusting parameters and calculation options of Evolutionary Algorithms (in general) is a problem how to find a compromise between two opposite tendencies but both giving negative effects such as:

- premature degeneration
- purely random moves of algorithm

Therefore, of great importance is proper choice of selection pressure that influences on both effects. The use of high selection pressure by, for instance, reproducing only the best members will finally result in premature degeneration, i.e. locating poor local optimum. Too weak selection pressure will considerably increase computation time and, also, will cause pure random selections. In consequence, both extreme cases decrease a chance of calculating global or good local optimum within reasonable CPU time.

Algorithm GenCom uses selection mechanism twice:

Selection I — the choice for sub-population, i.e. selection from basic population to create sub-population of genetically modified members.

Selection II — the choice for children population, i.e. selection from both the basic population and the sub-population to create children population, which becomes basic population in next generation.

To define overall selection mechanism one should to choose a type of probability distribution and method of member selection.

Three various probability distributions can be specified in GenCom:

- proportional, defined by Eq. (35).

$$p_i = \frac{FP_i}{\sum_{i=1}^{N} FP_i} \quad (35)$$

- deterministic, defined by Eq. (36), with condition (37).

$$p_i = \frac{Q(1-Q)^{i-1}}{\sum_{i=1}^{N} Q(1-Q)^{i-1}}, \quad (36)$$

$$\sum_{i=1}^{N} p_i = 1. \quad (37)$$

- uniform, i.e. such that each member is given identical probability p_i independent of its fitness function value. Notice that this distribution is not applied for selection for the new population since effects in a purely stochastic mechanism.

Parameter Q in (35) is an additional control parameter given in data. Index i in (35) is a position in the list of the members, where $i = 1$ is kept for the best member. Parameter N in (34)–(36) states for number of members in population/sub-population.

Proportional distribution is most often used in EA. However, it can result in premature degeneration effect such that members would feature almost identical fitness and the algorithm would fail to locate an optimum. For very small values of parameter Q the deterministic distribution (33) becomes similar to uniform one. In opposite case, with a high value of Q, the distribution provides a high preference for the best fitted members. Thus, it becomes similar to pure elitist mechanism. The setting of Q has to be done by the user from the range: 0.0 to 1.0. Additionally, the specific conditions can be imposed on selection mechanism. They are used to control how many times a member can be chosen for sub-population or offspring population.

Two mechanisms are available in program GenCom for selecting a member:

- Multiple choice of a member (with repetitions) — no constraints are imposed on the number of choices for a member
- Expected value method (without repetitions) — a member can be selected *eve* times, where parameter *eve* is called expected value. After each selection the value of *eve* is reduced by 1 if the member has been chosen for crossover or by 0.5 if for mutation operation. The member with parameter *eve* equals zero cannot be selected any further.

Genetic operators — crossovers and mutations — are often aimed at processing continuous or discrete variables. However, those operators that process discrete valued chromosome can be directly changed into objects, which can be applied for continuous variables. It is enough to change only the way of generating random numbers, which are used in the operators. The user of GenCom has access to various popular operators. To describe operators we will apply the following notation for chromosome of ith members: $\mathbf{M}_i = \{m_{i1}, m_{i2}, \ldots, m_{ik}, \ldots, m_{\text{in}}\}$.

The following operators are available in GenCom: four crossover operators (simple, arithmetic, heuristic, mixed) and four mutation operators (uniform, non-uniform, local, range limit). We have found in tests performed to date that only two of them are sufficient for HEN parameter optimization: simple crossover and uniform mutation. However, the tests are still continued. Though the operators are well known we give brief explanation in the following.

Simple crossover — acts for continuous and discrete variables. Two parents $\mathbf{M}_1$ and $\mathbf{M}_2$ exchange information by interchanging segments of their chromosomes. For the selected parents and randomly chosen cutting position "k" in both chromosomes the operation is defined as:

$$\begin{aligned}&\{m_{11}, m_{12}, \ldots, m_{1k}, \ldots, m_{1n}\} + \{m_{21}, m_{22}, \ldots, m_{2k}, \ldots, m_{2n}\}\\&\quad\to \{m_{11}, m_{12}, \ldots, m_{1k}, m_{2k+1}, \ldots, m_{2n}\}\\&\quad+ \{m_{21}, m_{22}, \ldots, m_{2k}, m_{1k+1}, \ldots, m_{1n}\}. \qquad (38)\end{aligned}$$

Uniform mutation — processes continuous and discrete variables. One parent M is chosen and, then, randomly chosen position "k" of chromosome is mutated in such the way that its value is changed by the value chosen from the entire range. Selection of the position in chromosome and choice of the value is performed using uniform distribution.

5. Examples

Example 1

This case study was addressed in some other papers such as Asante and Zhu (1996,1997), Ahmad and Poley (1990), Nielsen *et al.* (1996). The example concerns typical debottlenecking scenario. That means that HEN

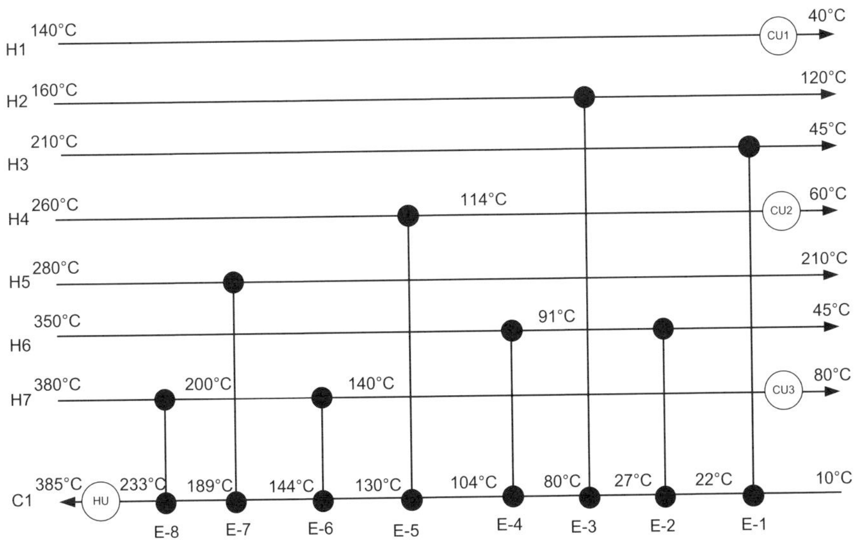

Figure 9. Structure of existing network for example 1. E-1 ÷ E-8 — process heat exchangers, CU1 to CU3 — coolers, HU — heater.

retrofit is forced by an increase of throughput — in this case in petroleum refinery installation for crude oil atmospheric distillation.

The existing HEN is shown in Fig. 9. Crude oil for distillation column is cold stream denoted by C1. Hot streams, H1 to H7, are outputs from the column. It is assumed that the increase of C1 flow rate by 10% results in identical increase of flow rates of all hot streams. The basic parameters of heat exchangers in the HEN are gathered in Table 2, while parameters of streams in Table 1.

The restriction is imposed on existing heater that its allowable heat load should not be higher than 100 kW. Steam (HU) of temperature 500°C is the single heating utility and water is cooling one (CU). The temperature range of cooling water is 20°C–40°C. Following the literature case study we will assume minimum temperature approach $\Delta T_{\min}$ of 10 deg for all process and utility heat exchangers. Finally, the prices of utilities are: 57.6 \$/kW year for heating utility and 9.6 \$/kW year for cooling utility.

Heat recovery targets (heat loads on utilities) for increased flow rates are: $Q^{HU}_{\min,glob} = 60.46$ MW, $Q^{CU}_{\min,glob} = 39.18$ MW, pinch point is 160/150°C. Notice that $Q^{HU}_{\min,glob}$ and $Q^{CU}_{\min,glob}$ are targets for synthesis case, usually

Table 1. Streams data for Example 1.

Stream	T_{in} [°C]	T_{out} [°C]	CP [kW/K]	CP [kW/K] 10% increased	h [kW/m^2 K]
H1	140	40	0.470	0.517	0.20
H2	160	120	0.750	0.825	1.50
H3	210	45	0.042	0.047	0.80
H4	260	60	0.100	0.110	0.70
H5	280	210	0.357	0.393	1.00
H6	350	45	0.052	0.058	0.45
H7	380	80	0.140	0.154	0.35
HU	500	499			0.80
C1	10	385	0.565	0.622	0.70
CU	10	25			0.80

Table 2. Heat exchangers areas of existing network in Example 1.

Heat exchanger	Area [m^2]
1	255
2	595
3	1744
4	133
5	605
6	1288
7	1733
8	1092

difficult to reach for retrofit due to high investment cost (therefore they have subscript "glob"). For comparison, existing HEN requires for the same data: $Q^{HU} = 94.73$ MW, $Q^{CU} = 66.53$ MW. It shows a large scope for structural improvements. The goal function is annual cost of utilities. Additional economic condition is imposed that total investment cost for retrofit should not exceed \$557000. All retrofit changes listed above will be considered in this example. Cost for the changes are calculated from the equations below taken from Asante and Zhu (1996,1997):

(a) cost of area added to the existing heat exchanger [\$]: $670\Delta A^{0.83}$
(b) area depended cost of a new heat exchanger [\$]: $670 A^{0.83}$

(c) fixed cost of a new heat exchanger [$]:	8 600
(d) expenses on moving heat exchanger [$]:	500
(e) expenses on changing pipes for one node of relocating heat exchanger [$]:	100

Notice, that changes of splitter, e.g. a new branch and a decrease of surface area of heat exchanger are not charged. It is also of importance that relocation of heat exchanger (change (d)) requires repiping (change (e)) for one or two nodes. Hence, full cost of relocation is $600 or $700.

The retrofit superstructure applied in this example is shown in Fig. 10. It contains nodes (denoted by numbers) for potential heat exchangers. Notice that a hot/cold node can be matched with every cold/hot one.

Hence, there are numerous potential heat exchangers though the number of nodes does not seem very large. The superstructure contains one split for two branches at the cold stream. It was proposed from problem analysis using a simple heuristic: split this stream which has relatively large *CP* value, particularly in comparison with streams of "opposite" kind.

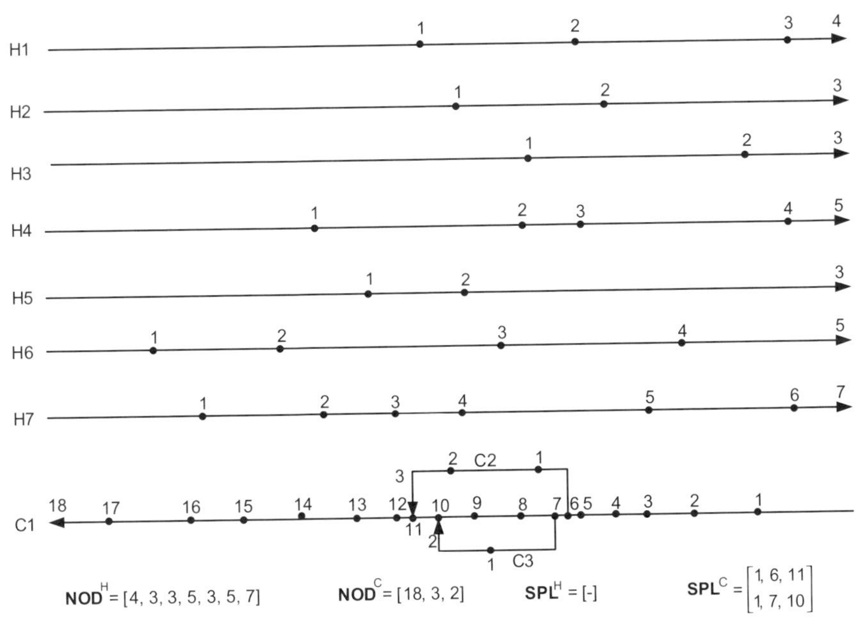

Figure 10. Graphical and matrix representation of retrofit HEN superstructure in example 1.

The problem was solved with additional constraint — we assumed that the total number of heat exchangers should not be higher than that in existing HEN. This constraint is useful if one wants to eliminate purchase of new heat exchangers. The solution space for the structure optimization defined by the superstructure in Fig. 10 embeds all possible combinations of 8-element set of connections: hot node — cold node from the set of all 464 connections between 23 hot nodes and 16 cold nodes. The number of combinations is huge and amounts to 5.0×10^{16}. The decision variables in parameter optimization level are heat transfer areas and split ratios of the splitter at stream C1. Heat loads of coolers and heaters are dependent variables calculated as the differences between enthalpy change of stream at given and calculated outlet temperatures.

Some important parameters and features of optimization at both levels are listed below.

Structure optimization level

Initial population was generated randomly from existing network. The population consisted of 40 members and we used 300 generations. The probabilities for genetic operators applied were:

1. hot node relocation	0.12
2. cold node relocation	0.12
3. clone of **SM**	0.04
4. multiple-crossover	0.4
5. exchange of hot node	0.16
6. exchange of cold node	0.16

Parameter optimization level

At second level the population contained 20 members and was processed within 100 generations. Initial population was generated randomly keeping the given ranges for parameters. Intermediate population consist of the same number of members as parent population (parameter $u\_mod = 1$). Two operators were sufficient — simple crossover with probability 0.8 and uniform mutation with probability 0.2. Computation time was of order of 11 hours at processor Intel Pentium III 1 GHz. The approach calculated the

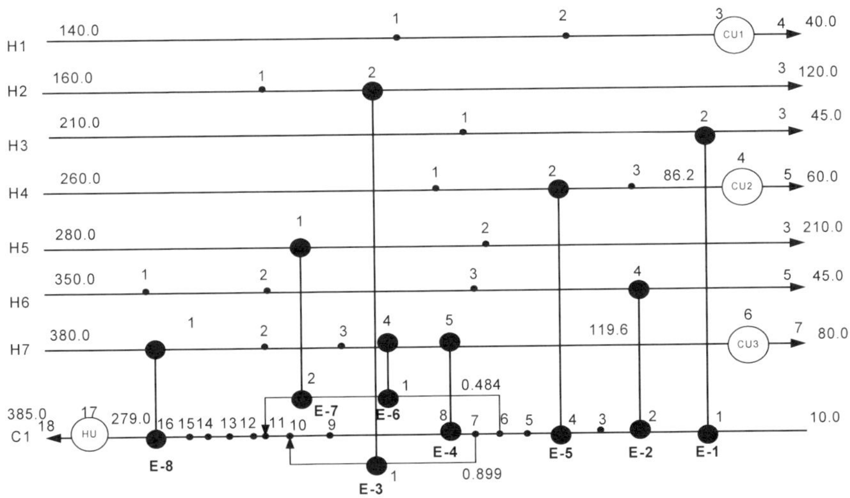

Figure 11. HEN structure for optimal solution in example 1.

Table 3. Heat exchangers areas and structural matrix for optimal solution in example 1.

Heat exchanger	SM				Area [m^2]	Added area [m^2]
1	3	2	1	1	228.4	0.0
2	6	4	1	2	604.0	9.0
3	2	2	3	1	2493.3	749.3
4	7	5	1	8	207.4	74.4
5	4	2	1	4	618.3	13.3
6	7	4	2	1	1259.4	0.0
7	5	1	2	2	1506.1	0.0
8	7	1	1	16	1569.8	477.8

network shown in Fig. 11 — flow rates in branches of splitting streams are shown as well.

The network has operation cost (goal function) of 4 382 421 \$/year. It meets the investment condition as the expenses are \$312 057. Also, the upper limit on heater load is met since this heat load is 65.9 MW. Five heat exchangers have been relocated and two splitters have been created at C1. The cost parameters of the HEN are given in Table 4.

Table 4. Cost parameters for solution in Example 1.

Cost	Value
total investment costs [$]	312 057
cost of added area in existing heat exchangers [$]	308 957
cost of relocation [$]	2 500
cost of repiping [$]	600
operation cost [$/rok]	4 382 421

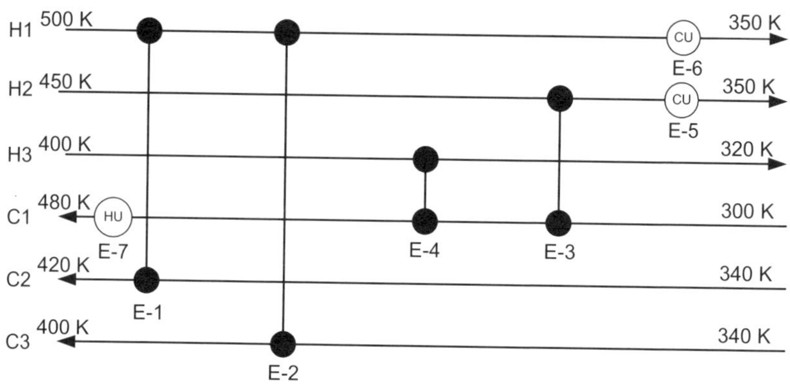

Figure 12. Structure of existing network for example 2. E-1 to E-4 — process heat exchangers, E-5, E-6 — coolers, E-7 — heater.

Example 2

This case study was addressed in Yee and Grossmann (1987), Ciric and Floudas (1989, 1990). The existing network with 4 process heat exchangers, 2 coolers and 1 heater is shown in Fig. 12. Parameters of streams, together with cost of utilities are presented in Table 5. Finally, Table 6 involves parameters of heat exchangers. The network presented in this example consumes 360 kW of heating utility and 800 kW of cooling utility. It gives operation cost of 44 800 $/year.

For design we assumed $\Delta T_{\min}$ of 9 deg. The energy recovery targets are: $Q^{\mathrm{HU}}_{\mathrm{min,glob}} = 0\,\mathrm{kW}$ and $Q^{\mathrm{CU}}_{\mathrm{min,glob}} = 442.3\,\mathrm{kW}$; it means that this is the so called threshold problem (requiring only one kind of utility). Investment cost minimization is the objective under the condition that operation cost

Table 5. Streams data and cost parameters for the Example 2.

Stream	T_{in} [K]	T_{out} [K]	CP [kW/K]	h [kW/m^{2}K]	Cost [$/kW year]
H1	500	350	10.01	1.6	
H2	450	350	12.00	1.6	
H3	400	320	8.01	1.6	
HU	540	540	—	1.6	80
C1	300	480	9.00	1.6	
C2	340	420	10.00	1.6	
C3	340	400	8.00	1.6	
CU	300	320	—	1.6	20

Table 6. Heat exchangers areas and structural matrix of existing network in example 2.

Heat exchanger	SM				Area [m^2]
1	1	1	2	3	12.50
2	1	2	3	2	23.50
6	2	2	1	2	45.06
4	3	2	1	1	33.09
5	2	4	5	1	11.49
6	1	3	4	1	5.39
7	4	1	1	3	5.75

should not be higher than 8 890 $/year. The expenses on investment are calculated using existing cost as the basis.

In addition to the possible modification changes we add here the change of utility heat exchanger (heater or cooler) into process heat exchanger. The cost of such change is equivalent to relocating of heat exchanger. The formulas for calculating cost of retrofit changes taken from Ciric and Floudas (1990) are as follows:

(a) cost of area added to the existing heat exchanger [$]: $300\,\Delta A$
(b) area dependent cost of a new heat exchanger [$]: $1200\,A^{0.6}$
(c) fixed cost of a new heat exchanger [$]: 4 000

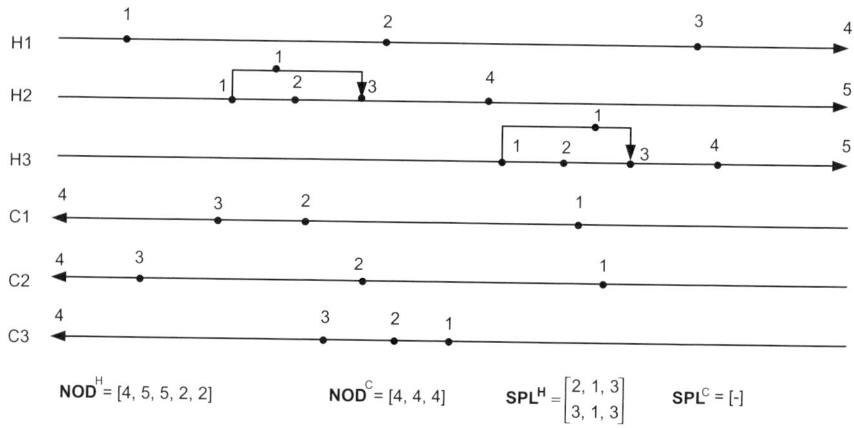

Figure 13. Graphical and matrix representation of retrofit HEN superstructure in example 2.

(d) expenses on relocating heat exchanger [$]: 300
(e) expenses on changing pipes for one node of relocating heat exchanger [$]: 100

Similarly to example 1 total cost of relocating includes also repiping for one or both nodes, and hence equals \$400 or \$500. The retrofit superstructure with two new splitters is shown in Fig. 13. The existing network has 4 process heat exchangers. Assuming that two of three utility heat exchangers are allowed to change to process heat exchangers — total number of process heat exchangers is 6. The solution space for structure optimization contains all possible combinations of 6 elements set of connections: hot node — cold node, from 81 elements set of all connections for 9 hot and 9 cold nodes. Hence, the number of all combinations is $3.0 \cdot 10^9$. Similarly to example 1 the loads of potential heaters and coolers at the end of streams are treated as dependent variables and calculated on the basis of temperature differences: required outlet temperature and calculated outlet temperature. One heater or cooler comes from the existing network. Decision variables are heat transfer surface areas of 6 process heat exchangers and flow rate in branches of the splitter at C1. To optimize topology at the upper level we employed the population of 40 members. Number of generations was set at 300.

For parameter optimization the population of 20 members and 80 generations were used. CPU time was of order 10 hours at processor Intel

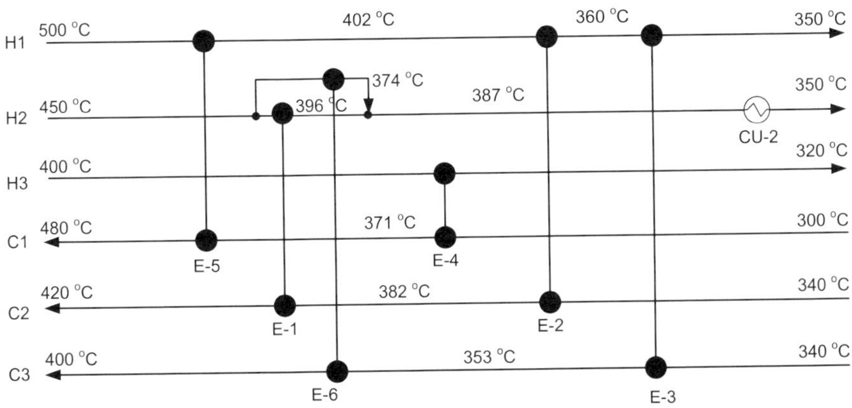

Figure 14. Final solution — network structure for example 2.

Table 7. Optimal values of heat exchanger areas and split ratio for example 2.

Apparatus	Area [m^2]
E-1	22.66
E-2	25.68
E-3	14.74
E-4	33.04
E-5	48.90
E-6	14.21
CU-2	9.53
Split	0.41

Pentium III 1 GHz. The genetic operators and their probabilities were identical to those in example 1. The best generated network has investment cost \$8 308 and is shown in Fig. 14. Table 7 gives parameters of heat exchangers. The network has one splitter at H2 with split ratio equals 0.41. The cost parameters of the solution are shown in Table 8. It is important to note that the hot utilities were eliminated. The comparison with the best literature solution calculated in Ciric and Floudas (1990) by the deterministic optimization method (General Benders Decomposition) is also shown in Table 8. Notice, that our network is better in terms of the goal function and, also, features smaller operation cost.

Table 8. Results for Example 2.

	Our solution	GBD*
total investment cost, [$]	8308	10450
cost of added area, [$]	6108	8250
relocation cost, [$]	1500	1500
repiping cost, [$]	700	700
area increase of existing heat exchanger, [m^2]	18.83	24.31
number of heat exchangers with increased area	4	5
number of new branches	1	0
annual operation cost, [$/year]	8766	8986

* denotes solution by Ciric and Floudas (1990) calculated by GBD algorithm

This case study was also solved for standard heat exchangers. The allowable standard heat transfer areas were generated randomly from given ranges. Due to space limitation we will give a brief description of major points. The solution differs from that for non-standard apparatus in regards to both topology and parameters. The goal function was worse but still better than in Ciric and Floudas (1990) for non-standard heat exchangers. At structure optimization level we applied population of 40 members, while for parameter optimization 20 members were sufficient. Similar proportions was used for number of generations: 300 generations for structural optimization and 60 generations for parameter optimization level. It is of importance that CPU time for standard apparatus case was smaller than for non-standard one and amounts to about 8 hours.

This illustrates our claim that GA is an efficient tool for highly combinatorial problems. It also shows that designing HENs with standard heat exchangers should be only done with GA or similar meta-heuristic techniques.

6. Summary

The approach for retrofitting HEN has been developed. It is based on two-level optimization where at first level structures are generated and at the second one the parameters of heat exchangers and splitters are optimized. GA are applied at both levels.

The novel concept of retrofit superstructure has been developed. This concept relies primarily on stream superstructure and, then, on match superstructure. The generation of the superstructure is easy and makes use of problem understanding of the user. The representation for the superstructure and the structures generated from it is compact and allows easy coding and efficient computation. The algorithm for generating structures ensures that they all are structurally feasible and that there exists unique relation: structure — superstructure. The original GA method was applied to solve the NLP model for parameter optimization of HEN with fixed structure. It uses two populations: parent and intermediate ones for creating a children population. Special scheme of solution of the NLP model has been developed to deal efficiently with nonlinear constraints of superstructure optimization model. The solution procedure allows solving directly equality constraints as linear equations or sets of simultaneous linear equations. Due to this scheme the use of penalty terms for most of equalities was omitted.

It is of significance that the optimization model at the second level can be of MINLP type if standard heat exchangers are to be considered. The tests proved that such case is easier to handle by GA than a HEN consisting of non-standard apparatus. It is the opposite effect than with the use of deterministic optimization approach. The disadvantage of GA-based retrofit is relatively long computation time. This causes its application to large scale problems difficult since calculation of "good" network could require too large time.

The investigations are planned to enhance computations at the second level by some mechanisms such as a solution of properly linearized model with a deterministic optimization method.

Problems and Exercises

Exercise 1

Consider the simplest heat recovery problem for two streams:

— hot stream H should be cooled down from T_1^{H} to T_2^{H}
— cold stream C should be heated up from T_1^{C} to T_2^{C}

Table 7. Stream data for Exercise 1.

Stream	T_1 [°C]	T_2 [°C]	CP [kW/K]	h [kW/m^{2}K]	Cost [$/kW year]
H	250	35	20	0.8	
C	40	200	30	0.6	
HU	300	299	—	1	100
CU	15	20	—	0.6	30

Enthalpy change can be determined from:

$$\Delta I = CP \cdot (T_2 - T_1)$$

(a) What is the maximum heat load $Q_{\max}$ that can be exchanged (recovered from hot stream H)?
What is minimum load on heating utility (Q^{HU}) and cooling utility (Q^{HU})? Notice that $Q_{\max} Q^{\mathrm{HU}}$, Q^{CU} depend on minimum temperature approach ($\Delta T_{\min}$) defined as the minimum permissible difference between temperature of hot and cold streams.
Calculate $Q_{\max}$, Q^{HU} , Q^{CU} for $\Delta T_{\min}$ varying from 0.1 to 40. Draw the plots: $Q_{\max}$, Q^{HU}, Q^{CU} vs. $\Delta T_{\min}$.
In order to check your results notice that for $\Delta T_{\min} = 10°\mathrm{C}$ the results are:

$$Q_{\max} = 4000\,\mathrm{kW}; \quad Q^{\mathrm{HU}} = 800\,\mathrm{kW}; \quad Q^{\mathrm{CU}} = 300\,\mathrm{kW};$$

(b) Could you apply such small $\Delta T_{\min}$ as 0.1 deg in practice? To answer let think if $\Delta T_{\min}$ influence other costs of heat exchange? To check this calculate heat transfer area for some $\Delta T_{\min}$ values from equation:

$$A = \frac{Q}{U \cdot \Delta T_{\mathrm{m}}},$$

where ΔT_{m} is logarithmic mean temperature difference calculated from:

$$\Delta T_m = \frac{(T^{\mathrm{H}}_{\mathrm{in}} - T^{\mathrm{C}}_{\mathrm{out}}) - (T^{\mathrm{H}}_{\mathrm{out}} - T^{\mathrm{C}}_{\mathrm{in}})}{\ln \frac{T^{\mathrm{H}}_{\mathrm{in}} - T^{\mathrm{C}}_{\mathrm{out}}}{T^{\mathrm{H}}_{\mathrm{out}} - T^{\mathrm{C}}_{\mathrm{in}}}}.$$

Assume that U can be determined from equation:

$$\frac{1}{U} = \frac{1}{h^{\mathrm{H}}} + \frac{1}{h^{\mathrm{C}}}.$$

Notice that area A influences fixed cost according to:

$$INV = a + b \cdot A^{c},$$

where $a = 8600$; $b = 700$; $c = 0.83$.

Draw the plot A versus $\Delta T_{\min}$.

(c) Make the plot of total annual cost (TAC) versus $\Delta T_{\min}$.

$$TAC = Cost^{HU} Q^{HU} + Cost^{CU} Q^{CU} + \frac{(1 + ROR)^{PL}}{PL} \cdot INV$$

where: rate of return — ROR $= 10\%$; plant life — PL $= 6$ years; $Cost^{\mathrm{HU}}$ and $Cost^{\mathrm{CU}}$ (unit prices of utilities) get from Table 7. What is your optimum $\Delta T_{\min}$?

Hint: practitioners apply $\Delta T_{\min}$ of order 11°C (~20°F).

Remark: Such the computation for multiple H, C are called HEN targeting. You are asked to refer to textbook by Smith (2005) for explanations.

Exercise 2

Heat exchanger network (HEN) design problem is usually formulated as follows:

> Given n^{H} hot streams and n^{C} cold streams with data such as in the Table 8 of exercise 1 (i.e. inlet and required outlet temperatures and heat capacity flow-rate of each stream). Design HEN that minimizes total annual cost.

To solve it the designer should choose matches (streams to be matched in heat exchangers) and loads on matches. Also structure of streams flow has to be calculated, i.e. location of splitters and split ratios.

The problem is complex and requires specialized software to achieve good solution.

Without such aids the designer has to use heuristics and/or simplifications. The crucial decision is what streams should be matched and in what sequence. This could be found by solving the optimal assignment problem. Appropriate heuristics can be employed, too.

Table 8. Stream data for Exercise 2.

Stream	T_1 [°C]	T_2 [°C]	CP [kW/K]
H1	250	35	25
H2	200	90	15
H3	150	35	20
C	40	250	50
HU	300	299	—
CU	15	20	—

Consider the case of several hot streams and one cold stream. Here we limit to three hot streams and one cold stream of parameters from the Table 8. Let assume that the goal is the minimum heat of utilities that have to be applied to reach outlet temperatures. Furthermore, assume serial structure of matches, i.e. the arrangement without splitters. Finally, assume that a pair of streams can be matched only one time.

Such assumptions make the use of the optimal assignment problem formulation possible.

Question 1: Formulate optimal assignment problem for n^H hot streams and n^C cold streams.

The next question is whether there exist a heuristic on matching streams? To help in answering this question we will assume that heat load on match is determined with the rule: "exchange as much heat as thermodynamically possible". This means that we should reach minimum achievable temperature approach (minimum temperature approach) in each match. For this case assume minimum temperature approach $\Delta T_{\min}$ equal to 11°C.

Apply to the data the following matching rules:

(a) hottest hot stream with hottest cold stream,
(b) hottest hot stream with coldest cold stream,

and calculate the network, i.e. loads of utilities to reach the final temperatures.

Notice that rule (a) or (b) allows to calculate 3 matches from 9 possible combinations — see number of possible matches in examples 1 and 2 in Sec. 5 of the chapter.

What rule has produced the best results? Could you explain why? Is this rule general enough to be applied for maximizing heat recovery in designing HENs?
Help: rule a) gives $Q^{\mathrm{HU}} = 3714.5\,\mathrm{kW}$; $Q^{\mathrm{CU}} = 2539.5\,\mathrm{kW}$;

Symbols*

A	heat transfer area of heat exchanger
Cp	stream heat capacity
CP	stream heat capacity flow-rate, $CP = GCp$
ΔA	area increase
ΔT_m	mean logarithmic temperature difference
DVV	decision variables vector in parameter optimization
FP	fitness function
G	stream flow-rate
h	heat transfer coefficient
NA	total number of heat exchangers in HEN
NB	number of all branches in hot/cold streams
NGEN	total number of generations
NOD	vector of nodes
NOPER	number of genetic operators
NPOP	size of parent and offspring population
NPOPgen	size of sub-population in optimization by GA
NS	number of side branches in hot/cold streams
p_gen	probabilities of using genetic operator
$p0$	selection probabilities from population POP0
$p1$	selection probabilities from population POP1
Q	heat load
SM	structural matrix
SPL	split matrix
U	overall heat transfer coefficient

Superscripts

H	hot process stream
C	cold process stream
HU	hot utility (stream)
CU	cold utility (stream)

Subscripts

in	inlet
out	outlet
*	Symbols used for description of GA algorithm are explained in point 4.

Abbreviations

EA/GA	evolutionary/genetic algorithms
HEN	heat exchanger network
MINLP	mixed-integer nonlinear programming
NLP	nonlinear programming

Acknowledgment

The authors would like to give sincere thanks to Prof. A. Kraslawski (Lappeenranta University of Technology, Finland) for his advices and comments on presentation of the material for the final revision of the chapter.

References

Ahmad, S., Poley, G.T. (1990). Debottlenecking of heat exchanger networks. *Heat Recovery Systems and CHP*, **10**, pp. 369–385.

Androulakis, I.P. and Venkatasubramanian, V. (1991). A genetic algorithmic framework for process design and optimization. *Computers and Chemical Engineering*, **4**, pp. 217–228.

Asante, N.D.K. and Zhu, X.X. (1996). A new method for automated retrofit of heat exchanger networks, *Conference, CHISA'96*, H7.1.

Asante, N.D.K. and Zhu, X.X. (1997). An Automated and interactive approach for heat exchanger network retrofit, *Chemical Engineering Research and Design*(*Transactions of the Institution of Chemical Engineers*), **75**, pp. 349–360.

Babu, B.V. and Munawar, S.A. (2007). Differential evolution strategies for optimal design of shell-and-tube heat exchangers. *Chemical Engineering Science*, **62**, pp. 3720–3739.

Barbaro, A.B. and Bagajewicz, M.J. (2005). New rigorous one-step MILP formulation for heat exchanger network synthesis. *Computers and Chemical Engineering*, **29**, pp. 1945–1976.

Bergamini, M.L., Scenna, N.J. and Aguirre, P.A. (2007). Global optimal structures of heat exchanger networks by piecewise relaxation. *Industrial and Engineering Chemistry Research*, **46**, pp. 1752–1763.

Biegler, L.T., Grossmann, I.E. and Westerberg, A.W. (1997). *Systematic Methods of Chemical Process Design*, Prentice Hall PTR, New Jersey.

Björk, K.-M. and Westerlund, T. (2002). Global optimization of heat exchanger network synthesis problems with and without the isothermal mixing assumption. *Computers and Chemical Engineering*, **26**, pp. 1581–1593.

Bochenek, R. and Jeżowski, J. (1999). Adaptive random search approach for retrofitting flexible heat exchanger networks. *Hungarian Journal of Industrial Chemistry*, **27**, pp. 89–97.

Bochenek, R. and Jeżowski, J. (2008). Optimization with evolutionary algorithms — calculation procedure and applications. *Chemical and Process Engineering*, **29**, pp. 179–189.

Bochenek, R., Jeżowski, J. and Bartman, S. (2007). Optimization of heat exchanger network with fixed topology by genetic algorithms. *Chemical Engineering Transactions*, **12**, pp. 195–200.

Bochenek, R., Jeżowski, J. and Jeżowska, A. (2005). Genetic algorithms optimization and its use in chemical engineering. *Studia Universitatis Babes-Bolyai, Seria Chemia 2*, Anul L, pp. 73–83.

Bochenek, R., Jeżowski, J.M. and Jeżowska, A. (2001). Multi-product batch plant optimization using genetic algorithms. *28th International Conference of Slovak Society of Chemical Engineering*, Tatranske Matliare.

Briones, V. and Kokossis, A.C. (1999). Hypertargets: a Conceptual Programming approach for the optimization of industrial heat exchanger networks — II. Retrofit design. *Chemical Engineering Science*, **54**, pp. 541–561.

Chaudhuri, P.D. and Diwekar, U.M. (1997). An automated approach for the optimal design of heat exchangers. *Industrial and Engineering Chemistry Research*, **36**, pp. 3685–3693.

Ciric, A.R. and Floudas, C.A. (1989). A retrofit approach for heat exchanger networks. *Computers and Chemical Engineering*, **13**(6), pp. 703–715.

Ciric, A.R. and Floudas, C.A. (1990). A mixed integer nonlinear programming model for retrofitting heat-exchanger networks. *Industrial and Engineering Chemistry Research*, **29**, pp. 239–251.

Dolan, W.B., Cummings, P.T. and Le Van, M.D. (1989). Process optimization via simulated annealing: application to network design. *AIChE Journal*, **35**, pp. 725–736.

Dolan, W.B., Cummings, P.T. and Le Van, M.D. (1990). Algorithmic efficiency of simulated annealing for heat exchanger network design. *Computers and Chemical Engineering*, **14**, pp. 1039–1050.

Floudas, C.A. (1995). *Nonlinear and Mixed-Integer Optimization. Fundamentals and Applications*, Oxford University Press, New York.

Frausto-Hernadez, S., Rico-Ramirez, V., Jimenez-Gutierrez, A. and Hernandez-Castro, S. (2003). MINLP synthesis of heat exchanger networks considering pressure drops effects. *Computers and Chemical Engineering*, **27**, pp. 1143–1152.

Furman, K.C. and Sahinidis, N.V. (2002). A critical revue and annotated bibliography for heat exchanger network synthesis in the 20th century. *Industrial and Engineering Chemistry Research*, **41**, pp. 2335–2370.

Gundersen, T. (1989). Retrofit process design research and applications of systematic methods. *Conference: Foundations of Computer-Aided Process Design*, Snowmass Village, Colorado, USA.

Jeżowski, J., Bochenek, R. and Poplewski, G. (2007). On application of stochastic optimization techniques to designing heat exchanger and water networks. *Chemical Engineering and Processing: Process Intensification*, **46**, pp. 1160–1174.

Kotjabasakis, E. and Linnhoff, B. (1986). Sensitivity tables for the design of flexible processes (1) — How much contingency in heat exchanger networks is cost-effective? *Chemical Engineering Research and Design*, **64**, pp. 197–211.

Lewin, D.R. (1998). A generalized method for HEN synthesis using stochastic optimization — II. The synthesis of cost-optimal network. *Computers and Chemical Engineering*, **22**, pp. 1387–1405.

Lewin, D.R., Wang, H. and Shalev, O. (1998). A generalized method for HEN synthesis using stochastic optimization — I. General framework and MER optimal synthesis. *Computers and Chemical Engineering*, **22**, pp. 1503–1515.

Lin, B. and Miller, D.C. (2004). Solving heat exchanger network synthesis problems with Tabu Search. *Computers and Chemical Engineering*, **28**, pp. 1451–1464.

Ma, X., Yao, P., Luo, X. and Roetzel, W. (2008). Synthesis of multi-stream heat exchanger network for multi-period operation with genetic/simulated annealing algorithms. *Applied Thermal Engineering*, **28**, pp. 809–823.

Michalewicz, Z. and Fogel, D.B. (2002). *How to Solve It: Modern Heuristics*, Springer-Verlag, Berlin.

Nielsen, J.S., Hansen, M.W. and Joergensen, S.B. (1996). Heat exchanger network modeling framework for optimal design and retrofitting. *Computers and Chemical Engineering*, **20**, pp. S249–S254.

Őzcelik, Y. (2007). Exergetic optimization of shell and tube heat exchangers using a genetic based algorithm. *Applied Thermal Engineering*, **27**, pp. 1849–1856.

Pettersson, F. (2005). Synthesis of large-scale heat exchanger network synthesis problems using a sequential match reduction approach. *Computers and Chemical Engineering*, **29**, pp. 993–1007.

Pettersson, F. and Söderman, J. (2007). Design of robust heat recovery systems in paper machines. *Chemical Engineering and Processing: Process Intensification*, **46**, pp. 910–917.

Ponce-Ortega, J.M., Serna-Gonzales, M. and Jimenez-Gutierrez, A. (2007). Heat exchanger network synthesis including detailed heat exchanger design using genetic algorithms. *Industrial and Engineering Chemistry Research*, **46**, pp. 8767–8780.

Ravagnani, M.A.S.S. and Caballero, J.A. (2007). Optimal heat exchanger network synthesis with the detailed heat transfer equipment design. *Computers and Chemical Engineering*, **3**, pp. 1432–1448.

Ravagnani, M.A.S.S., Silva, A.P., Arroyo, A.A. and Constantino, A.A. (2005). Heat exchanger network synthesis and optimization using genetic algorithms. *Applied Thermal Engineering*, **25**, pp. 1003–1017.

Seider, W.D., Seader, J.D. and Lewin, D.R. (2004). *Product and Process Design Principles: Synthesis, Analysis, and Evaluation,* 2nd Edition, John Wiley and Sons Inc., New York.

Selabs, R., Kizilkan, O. and Reppich, M. (2006). A new design approach for shell-and-tube heat exchangers using genetic algorithms from economic point of view. *Chemical Engineering and Processing*, **45**, pp. 268–275.

Shenoy, U.V. (1995). *Heat Exchanger Network Synthesis*, Gulf Publishing Co., Houston.

Smith, R. (2005). *Chemical Process Design and Integration*, John Wiley and Sons, Chichester, UK.

Sorsak, A. and Kravanja, Z. (2002). Simultaneous MINLP synthesis of heat exchanger networks comprising different exchanger types. *Computers and Chemical Engineering*, **26**, pp. 599–615.

Sorsak, A. and Kravanja, Z. (2004). MINLP retrofit of heat exchanger networks comprising different exchanger types. *Computers and Chemical Engineering*, **28**, pp. 235–251.

Tayal, M.C. and Fu, Y. (1999). Optimal design of heat exchangers: A genetic algorithm framework. *Industrial and Engineering Chemistry Research*, **38**, pp. 456–467.

Xie, G.N., Sunden, B. and Wang, Q.W. (2008). Optimization of compact heat exchangers by a genetic algorithm. *Applied Thermal Engineering*, **28**, pp. 895–906.

Yee, T.F. and Grossmann, I.E. (1987). Optimization model for structural modifications in the retrofit of heat exchanger networks. *Conference: Foundations of Computer-Aided Process Operations*, Park City, Utah, pp. 653–662.

Yee, T.F. and Grossmann, I.E. (1991). A screening and optimization approach for the retrofit of heat-exchanger networks. *Industrial and Engineering Chemistry Research*, **30**, pp. 146–162.

Yoon, S.G., Lee, J. and Park, S. (2007). Heat integration analysis for an industrial ethylbenzene plant using pinch analysis. *Applied Thermal Engineering*, **27**, pp. 886–893.

Chapter 17

ANT COLONY OPTIMIZATION FOR CLASSIFICATION AND FEATURE SELECTION

V. K. Jayaraman

Evolutionary Computing Group
Center for Development of Advance Computing
Pune University Campus, Pune-411008, India
jayaramanv@cdac.in

P. S. Shelokar, P. Shingade, B. D. Kulkarni[†],
B. Damale and A. Anekar

Chemical Engineering & Process Development Division
National Chemical Laboratory, Pune-411008, India
[†]*bd.kulkarni@ncl.res.in*

1. Introduction

Classification tasks form an important class of problems in process engineering and a number of methods such as decision trees (Chang, Lin and Chang, 2002), statistical (Hsiung and Himmelblau, 1996), neural networks (Venkatasubramanian, Vaidyanathan and Yamamoto, 1990) and rule based (van Kampen *et al.*, 1997; Özyurt *et al.*, 1998) techniques have been applied to tackle these types of problems.

Recently Dorigo and coworkers (Dorigo, Di Caro and Gambardella, 1999) have developed a heuristic methodology known as *Ant Colony Optimization* (ACO) for solving combinatorial optimization problems.

ACO is a cooperative search technique that mimics the foraging behavior of real life ant colonies. Ant algorithms are inspired by the techniques employed by real life ants to optimize the shortest route from their nest to the food source and vice-versa. It is well known (Beckers, Deneubourg and Goss, 1992) that ants have the capability to deposit pheromone and use their pheromone trails as an indirect medium for communication of information among them. When an isolated ant comes across some food source it deposits pheromone on that location. Other randomly moving ants in the neighborhood can detect and get attracted to this pheromone rich trail thereby simultaneously enhancing the trail by depositing their own pheromone. More and more ants follow the pheromone rich trail and the probability of the trail being followed by other ants is further enhanced by increased trail deposition. It is this auto catalytic process that helps the ants to establish the shortest route. ACO has been applied to solve the process optimization problems such as flowshop scheduling (Jayaraman *et al.*, 2000), dynamic control of bioreactors (Jayaraman *et al.*, 2001) and multiple objective refinery design (Summanwar *et al.*, 2002).

The present chapter describes the application of ant colony optimization (ACO) as a machine learning technique, viz., *AntzClass* (Shelokar, Jayaraman and Kulkarni, 2004; Parpinelli, Lopes and Freitas, 2002) for discovery of classification rules in process data. We further employ another ACO based algorithm, viz., *AntzFeat* (Patil *et al.*, 2008; Gupta, Jayaraman and Kulkarni, 2006) for selecting most informative input features in process data and hybridize it along with *AntzClass* for simultaneous classification and feature selection. In *AntzClass* algorithm, software ants generate rules by using heuristic information and pheromone communication. At each iteration, the rules constructed by different ants are evaluated and the rule with highest prediction quality is denoted as a discovered rule, which represents information extracted from the environment (database). Instances correctly covered by the discovered rule are removed from the training set, and a new iteration is started. This process is repeated for as many iterations as are necessary to find rules covering almost all cases in the training set. At this point the algorithm has a collection of discovered rules, which can be utilized to identify class labels of new examples or can become integral part of an expert system.

The rules discovered by *AntzClass* are generally of the following form:

$$\text{IF} \;<\text{antecedent}>\; \text{THEN} \;<\text{consequent}>\;.$$

The antecedent (conditions) part of rule consists of terms combined by using logical operator as: term_1 AND term_2 AND.... A term is nothing but a combination of descriptors, mathematical operators and domain values of the descriptors (e.g. outlook = sunny). The consequent part of rule is a class label. The rule will assign this class label to those cases, which satisfy antecedent part of the rule. A comparative evaluation of the performance of *AntzClass* with tree based J4.8 and PART algorithm (available in WEKA, *http://cs.waikato.ac.nz/ml/weka/*, an open source machine learning software) has also been provided. The comparison is based on the predictive accuracy and the simplicity of discovered rules for well-known datasets in UCI Machine Learning Repository (*http://archive.ics.uci.edu/ml/datasets.html*). Further, *AntzClass* is applied for multiple fault detection in Tennessee Eastman process, a benchmark case study in process engineering.

In many classification problems, like cancer identification (employing gene expression profiles) and quantitative structure activity relationship (QSAR) etc. selection of relevant input features becomes very important. Advances in microarray technology, have enabled us to simultaneously measure the expression levels of thousands of genes in a tissue sample using a single experiment. It is well known that these gene expression patterns have been found to be of immense use in cancer classification (Golub *et al.*, 1999; Nutt *et al.*, 2003; Alizadeh *et al.*, 2000; West *et al.*, 2001). Microarray data have been successfully used to classify tissue samples. These classification techniques can be multi-class (into various sub-types of cancer) classification (Peng *et al.*, 2003; Paul and Iba, 2005; Chai and Domeniconi, 2004). In these problems, the number of expression profiles of genes (features) is quite large as compared to the number of tissue samples. This makes it difficult to devise a prediction model with high accuracy. Only a subset of the expression profiles may be informative and the rest may be redundant. The presence of many redundant attributes may act as a noise during model building process and may considerably reduce the classification performance. Due to this fact, classification methods are usually applied in conjunction with feature selection methods (Weston *et al.*, 2001; Zhang *et al.*, 2006; Paul and Iba, 2005; Kumar *et al.*, 2005). In

QSAR problems a model is built by correlating a set of physicochemical, molecular or electronic descriptors to the activity of a particular compound. We are particularly interested in solving binary classification problems in which, given a set of experimentally determined descriptors, one needs to identify whether the compound is active or inactive against a particular target/disease. In all these problems we can calculate thousands of descriptors and only a subset of features will be informative.

Feature selection can be categorized into Filter, Wrapper, and Embedded approaches (Huiqing, Jinyan and Limsoon, 2002). Filter methods generally use statistical correlations to evaluate the relevance of each feature (or feature subset), and are independent of the classification algorithm. Various filter approaches like SNR, t-statistic and other metrics have been applied for the purpose of gene selection in microarray data (Huiqing, Jinyan and Limsoon, 2002). Wrapper methods employ a classification algorithm to select a feature subset with highest classification accuracy. Embedded methods perform feature subset selection during the training of the classifier. Since the wrapper and embedded approaches involve the use of a classification algorithm, they are computationally more intensive but perform better than the filter methods.

Selection of a small subset of informative features from the pool of features is a combinatorial optimization problem. For example, the problem of selecting a subset of $R = 10$ features from $F = 100$ features has $(F!/(R!*(F-R)!) \sim= 17\text{E} + 12)$ possible solutions. An exhaustive search technique is computationally intractable and as a result various search techniques have been used in the literature, for traversing the search space to obtain near optimal solution. These techniques can be grouped into heuristic and randomized searches (Saeys, Inza and Larrañaga, 2007). Heuristic search methods like forward selection and backward elimination (Pudil *et al.*, 1994) intelligently search through the search space for the optimal solution without having to evaluate all the subsets, and select a reasonably high quality subset. Randomized searches probabilistically traverse the search space looking for optimal solutions. Various wrapper methods utilize these search techniques (Blanco, *et al.*, 2004; Jirapech-Umpai and Aitken, 2005; Li *et al.*, 2001). In this chapter, a new wrapper method is proposed for selection of most informative features that can support a classifier to produce maximum classification in cancer prediction study (Fig. 1). For the purpose of selecting informative features, a novel ACO based algorithm

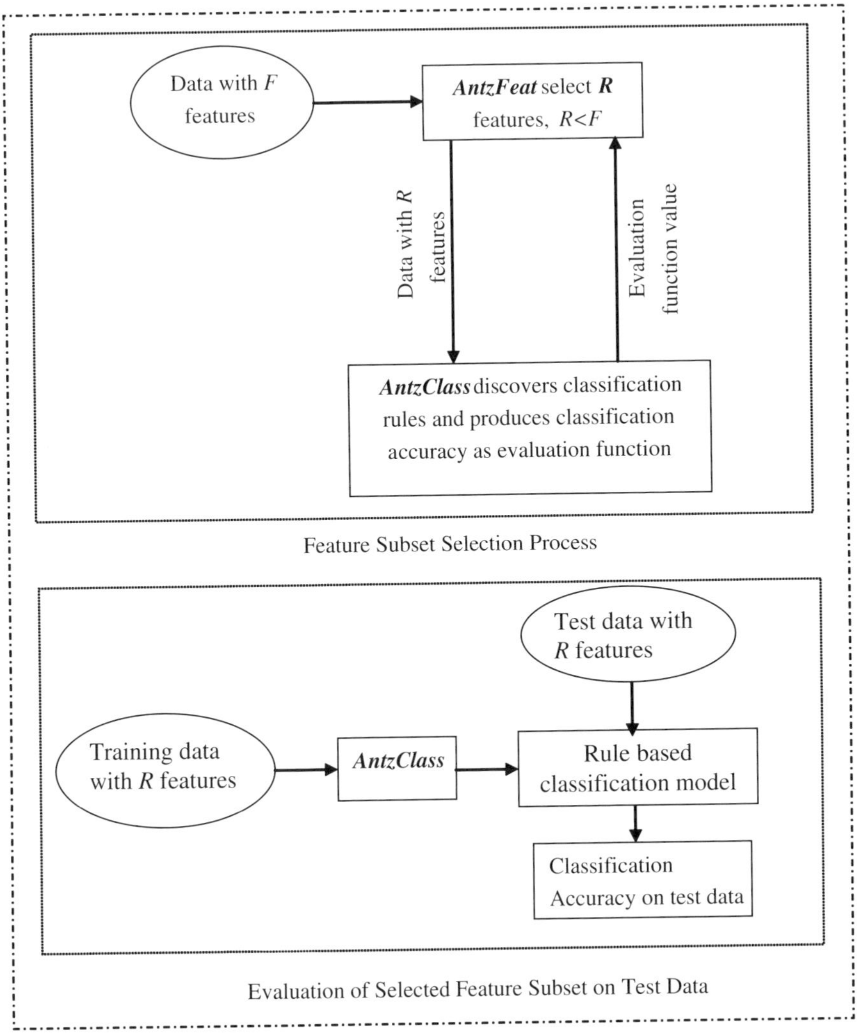

Figure 1. Wrapper method for feature selection using *AntzFeat* and *AntzClass* algorithms.

viz. *AntzFeat* (Patil *et al.*, 2008; Gupta, Jayaraman and Kulkarni, 2006) is utilized which employs software ants to generate a subset of features. Each ant incrementally generates a subset of features of known size by utilizing information about individual features in the form of pheromone trail and the ability of individual feature to correctly predict class label. The subsets generated by the ants in *AntzFeat* algorithm are subsequently employed as input to *AntzClass* algorithm. *AntzClass* then employs the methodology described in the earlier section to develop classification rules with supplied subsets of features. With the optimally evolved set of rules the ratio of number of correctly predicted cases to the total number of training cases is computed for each of the input feature subsets. These values are then returned as fitness values to *AntzFeat* module. Subsequently, *AntzFeat* algorithm employing the pheromone mediated search produces new subsets of input features, which are again passed to *AntzClass* algorithm. The cycle is repeated for a few generations until the best subset of input features providing the maximum classification accuracy is obtained. The proposed wrapper approach provides a large reduction in the number of informative features, compared to similar analyzes by generating biologically relevant minimal subsets of genes. Application of this wrapper method is illustrated with two data sets, viz., colon data and QSAR (quantitative structure activity) data.

In this chapter we first introduce an ant algorithm for discovery of classification rules and feature selection, which were developed and used, in our group. Further, application of these methods is illustrated using some benchmark datasets and results are discussed. Finally some potential advantages and drawbacks of these methodologies are briefly described in the concluding remarks.

2. Ant Algorithm for Discovery of Classification Rules (*AntzClass*)

The present form of ant colony optimization as a classification algorithm, viz. *AntzClass* (Shelokar, Jayaraman and Kulkarni, 2004; Parpinelli, Lopes and Freitas, 2002) handles only categorical (discrete) attributes. Thus, real (continuous) attributes need to be discretized as a data-preprocessing step. The simplest discretization technique is binning which, divides an attribute range into equal length intervals. More rigorous supervised and unsupervised discretization techniques have also been employed (Dougherty, Kohavi and Sahami, 1995). Binning is an example of unsupervised technique which does not take into account class labels whereas the supervised methods, such as, entropy-based methods discretize the feature space based on class label information. Let us illustrate the procedure with a simple example of a dataset (Table 1) consisting of two attributes and fourteen examples/cases representing two classes. For attribute X_1, using three equal interval partitioning we get three domains $D_{11}(\leq 76)$, $D_{12}(> 76$

Table 1. Example of discretization of continuous feature space.

		Data set (14 cases; 2 attributes)													
		1	2	3	4	5	6	7	8	9	10	11	12	13	14
Continuous data	X_1	90	85	88	75	73	70	69	77	74	80	80	77	86	76
	X_2	90	95	83	101	85	75	70	100	75	85	75	95	80	85
Discretized data	X_1	D_{13}	D_{13}	D_{13}	D_{11}	D_{11}	D_{11}	D_{11}	D_{12}	D_{11}	D_{12}	D_{12}	D_{12}	D_{13}	D_{11}
	X_2	D_{22}	D_{23}	D_{22}	D_{23}	D_{22}	D_{21}	D_{21}	D_{23}	D_{21}	D_{22}	D_{21}	D_{23}	D_{21}	D_{22}
Class		2	2	1	1	1	2	1	2	1	1	1	1	1	2

and ≤ 83) and $D_{13}(> 83)$. Attribute X_2 can similarly be discretized into appropriate domains $D_{21}(\leq 80)$, $D_{22}(> 80$ and $\leq 90)$ and $D_{23}(> 90)$. The present form of the *AntzClass* handles only categorical attributes. Each of these domains will henceforth be denoted as a term. More specifically the term S_{ij} refers to the jth domain value of ith attribute. To facilitate the implementation of the algorithm to solve the process engineering problems having continuous attributes, we represented domains of each attribute by numbers $1, 2, \ldots$ etc. The real values of each attribute will be represented by any of its domain value. With this background information, we now outline the ant algorithm for discovery of classification rules in the following four sub-sections namely: generation of classification rules, probability calculation, heuristic function, and rule pruning procedure.

3. Generation of Classification Rules

To construct a rule each ant starts with an empty rule. A term S_{ij} to be added in the current partial rule depends on (1) a heuristic function evaluated on the basis of information content of candidate terms and (2) the amount of pheromone trail associated with each term. For example, dataset in Table 1 has 14 cases defined by two attributes. Let us first, for example, impose the condition that the antecedent part of current partial rule should cover at least some cases, say, 3 in the dataset (Higher value of this algorithmic parameter may result into generation of few rules having those specific terms with high quality. While low value of parameter may result into generation of many rules having redundant terms). Suppose ant has selected first term as D_{11} and second term as D_{21}. From Table 1, we can say that term D_{11} covers 6 cases and term D_{21} covers 5 cases. While terms D_{11} and D_{21} jointly cover at least 3 cases satisfying the required condition. Thus, the antecedent part of this rule consists of terms D_{11} and D_{21}. Considering both terms the antecedent part now covers three cases (6,7 and 9) out of which cases 7 and 9 representing class $c_1$while case 6 belongs to class c_2. In the covered cases, the number of instances of class c_1 is in majority. Thus, the consequent part for this rule is assigned the class label c_1. Finally this rule generated by the ant is given as: IF $X_1 = D_{11}$ AND $X_2 = D_{21}$ THEN class c_1. The ant terminates the process of rule construction if it has already visited all the attributes. All the ants follow the same procedure to build the rules. The predictive power of a given rule can be expressed by the following equation (Lopes, Coutinho and Lima, 1998):

$$Q = \left(\frac{TP}{TP + FN}\right)\left(\frac{TN}{FP + TN}\right). \tag{1}$$

As different terms are added employing stochastic probabilities and the nature of heuristic functions (like the one considered here) may not take the interdependency of the attributes, the predictive power of a rule at this stage may not be optimal. Therefore pruning of the rule can help to increase the predictive power and simplicity of the rule. The rule pruning procedure is briefly explained in a separate section. The rule having the best predictive power among the constructed rules by all ants is considered as the rule discovered by the system during current iteration. This rule is stored in a set of discovered rules while the other rules are discarded. Further, all the cases from the training set correctly covered by this discovered rule are removed and only the remaining uncovered cases are considered for further rule discovery in the next iteration. The software ants are deputed to discover more rules as per the above-mentioned procedure. The rule discovery is continued until almost all the cases in the training set are covered. Thus at the end of the process, we have several discovered rules at hand. These rules are stored in the order of their discovery. A default rule is added at the end of the rule list. It has an empty antecedent with consequent as a class that represents majority of left out cases in the training set (cases not covered by any of the discovered rules).

This discovered rule set is then used on test cases that were not a part of training examples. The rules are tried on test cases in the order of discovery and the first rule having the terms in the antecedent part matching with that of the test case is employed to assign the class label to the test case. The system will apply default rule on the test case if any of the rules from the ordered rule list is not able to classify the test case.

4. Probability Calculation

Initially all the terms S_{ij} are assigned the same pheromone trail value which is usually inversely proportional to the total number of terms. As mentioned earlier the probability with which a term S_{ij} can be selected is given by the following normalized equation:

$$\Pr_{ij} = \frac{\tau_{ij}(t) \cdot \eta_{ij}}{\sum_k^F \sum_l^{d_k} \tau_{kl}(t) \cdot \eta_{kl}}, \quad i, k \in I. \tag{2}$$

The heuristic function value η_{ij} is the measure of quality of the term S_{ij}. For higher values of τ_{ij} and η_{ij}, term S_{ij} will have higher probability of being selected. The problem dependent heuristic function used in this study is given in the heuristic information section. The pheromone trail τ_{ij} is an indirect form of communication between ants. After every iteration after completion of discovery of rules by all ants the best rule is selected as the discovered rule at that iteration. At the end if each iteration the pheromone values of all terms are updated. The amount of pheromone deposited by a given ant depends on the quality of rule discovered and is given by the following expression:

$$\tau_{ij}(t+1) = \begin{cases} (1-\rho)\lfloor \tau_{ij}(t) + Q \rfloor; & (i,j) \in \text{ to the rule terms} \\ (1-\rho)\tau_{ij}(t); & (i,j) \notin \text{ to the rule terms} \end{cases}. \tag{3}$$

At any given iteration, different ants construct rules with differing qualities and depending upon the quality of terms selected by different ants, differing amounts of pheromones will be deposited at different terms. As a consequence the selection of terms would be greatly influenced by the quality. The extent of importance of term S_{ij} representing the domain j of an attribute i is thus collectively decided by the qualities of rules discovered by different ants which employ that term in the rule. The indirect communication capabilities of different ants thus influence the learned desirability of choosing a given term.

The procedure described so far is concerning development of antecedent part of a rule. The consequent part of rule can be obtained by calculating the predictive power of rule for each class label using Eq. (1). The class label that maximizes the quality of a given rule is assigned as the consequent part of that rule.

5. Heuristic Information

The heuristic function η_{ij} considered in this study is given by Cover and Thomas (1991). It provides a measure of the amount of information (entropy) associated with term S_{ij}. A higher value of η_{ij} makes the term S_{ij} more relevant for classification. For each term S_{ij} of the form $X_i = D_{ij}$, where X_i is the ith attribute and D_{ij} is a value of jth domain of attribute i, its information content is given by the following equation.

$$\text{info } W_{ij} = -\sum_{i=1}^{n}((P(c|X_i = D_{ij})) \cdot (\log_2 P(c|X_i = D_{ij}))), \quad (4)$$

where, info W_{ij} is a measure of the quality of term S_{ij} with respect to its ability to improve the predictive accuracy of rule. A higher value of entropy of a given term represents a more uniform distribution of cases among classes and thus reducing the predictive power and quality of the term. Thus, terms having smaller info W_{ij} values will have higher probability of being selected (Eq. (2)). The entropy of terms obtained by Eq. (4) can be further normalized to fall in the range $0 \leq \text{info } W_{ij} \leq \log_2(N)$. The equation used to normalize info W_{ij} is given as follows:

$$\eta_{ij} = \frac{\log_2(C) - \text{info } W_{ij}}{\sum_i^F \sum_j^{d_i} \log_2(C) - \text{info } W_{ij}} \quad (5)$$

Absence of the term S_{ij} in the training set indicates that the partition W_{ij} is empty and the term S_{ij} has zero predictive power (info $W_{ij} = \log_2(F)$ and $\eta_{ij} = 0$). Also, if the number of cases covered by the term S_{ij} in partition W_{ij} corresponds to the same class, then Eq. (5) assigns maximum predictive power to term S_{ij} (info $W_{ij} = 0$).

6. Rule Pruning Procedure

In the well known that in decision tree C4.5 algorithm, pruning of constructed tree is often carried out to avoid data over-fitting, increase predictive power of discovered rules, improve simplicity of rules etc. The rule pruning procedure adopted is similar to one proposed by Quinlan (1993) but with different criterion to measure the quality of rule.

Once an agent develops a rule, it is immediately sent to the rule pruning procedure. The pruning method is iterative and begins by considering the full rule. It provisionally takes off one term at a time from the rule, which results in a new rule. The quality of the resultant rule is calculated by using Eq. (1). The pruning of rule may assign a different class to the pruned rule. Any term whose removal enhances the predictive power of the rule is readily removed from the rule. This completes an iteration of the rule pruning. The procedure is terminated when one of the following conditions hold true (i) single term remained in the antecedent part of rule, or (ii) further removal of any term from the rule will not improve the quality of rule.

The steps involved in the algorithm are depicted in the flowchart shown in Fig. 2.

7. Ant Algorithm for Feature Selection (*AntzFeat*)

Let $F = \{X_1, X_2, X_3, \ldots, X_N\}$ be the complete set of F features, and $R \in F$ is a subset of R features, *where* $R < F$. In our implementation of ACO for feature selection, viz., *AntzFeat*, (Patil *et al.*, 2008; Gupta, Jayaraman and Kulkarni, 2006) the path followed by every ant can be seen as a subset of R features, where every feature represents a discrete state in the path. Thus, features subset selected by ant k can be written as R_k. At any time, every state (i.e. feature X_i) in the path has a pheromone value associated with it, which is represented by $\tau(X_i)$. Every feature also has a heuristic function value $\eta(X_i)$ associated with it, which is a measure of quality of an individual feature. Initially the pheromone level of all the features is set as $\tau_0 = 1/F$. The heuristic function is a prior knowledge of the relative importance of individual feature, which is calculated using information gain implemented in Weka (Witten and Frank, 2005). Information Gain

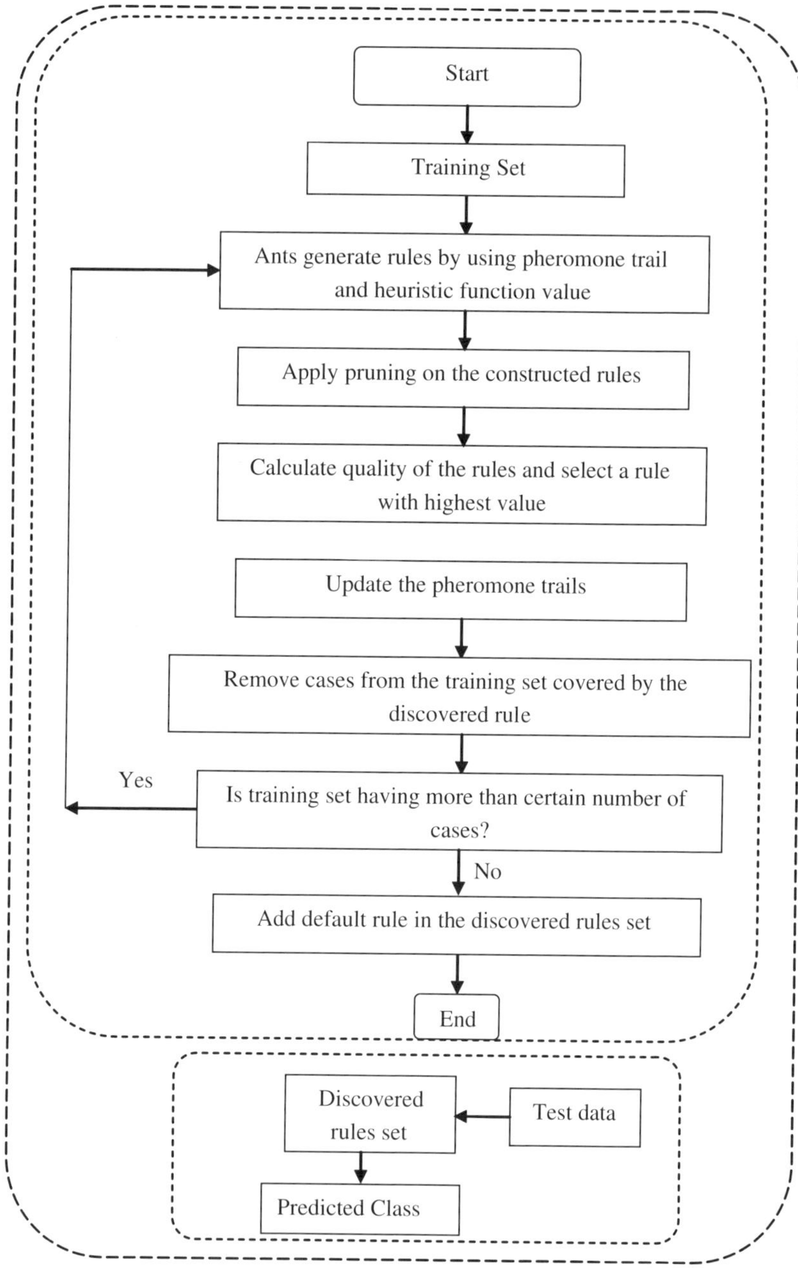

Figure 2. Flow chart of the steps involved in *AntzClass* system.

evaluates the effectiveness of a feature X_i in classifying by measuring the reduction in entropy as:

$$
\begin{aligned}
H(C) &= -\sum_{c\in C} p(c)\,\log_2 p(c) \\
H(C|X_i) &= \sum_{X_j\in D_{ij}} p(D_{ij}) \sum_{c\in C} p(c|D_{ij})\,\log_2 p(c|D_{ij})
\end{aligned}
. \quad (6)
$$

The importance of a feature is the increase in information about the class (or reduction in entropy of the class) given by, $I(C, X_i) = H(C) - H(C|X_i)$. Obviously, a feature with a higher value of information gain will be more desirable, and will have a higher heuristic function value. The features were ranked according to the information gain values, and the rankings were used as the heuristic function value after scaling from 0 to 1. Incorporating the heuristic function allows us to bias the algorithm so as to selectively favor a feature, thereby improving its chance to get selected in the best feature subset.

Ant k uses the following state transition rules to select a specified number of features for its solution R_k as:

$$
X_i = \begin{cases} \max\{\tau(X_i)\eta(X_i)^\beta\} & \text{if } (q < q_0) \quad \text{Exploitation} \\ \dfrac{\tau(X_i)\eta(X_i)^\beta}{\sum_{X_i\in I_k} \tau(X_i)\eta(X_i)^\beta} & \text{otherwise} \quad \text{Biased Exploration} \end{cases}, \quad (7)
$$

where I_k represents the set of features which are not the part of partially generated solution p_k, β is a parameter that represents the relative importance of the heuristic function over pheromone value. A value of $\beta = 1$ implies that the pheromone value and the heuristic function are given equal importance. A suitable value of β should be selected accordingly. In the probabilistic selection process, q_0 parameter allows us to choose a term based on absolute values of pheromones, instead of cumulative probability based term selection. This is a tunable parameter and we need to carry out simulations and we found the optimal values of q_0 is in the range [0.65, 0.75], q is a uniform random number in the range [0, 1]. The heuristic function value obtained by filter ranking procedure carries prior ranking information about the features while pheromone concenrations represents to the knowledge gained by the ants through indirect communication by the traversing the search space over time. The product of these two, in a way can be said to represent the total available dynamic knowledge about the features. At every step, the ant chooses between the two modes of feature selection, namely, exploitation and biased exploration mode, whose relative importance is given by the parameter q_0. Through the mode of exploitation the ants select the features having maximum values of the product of the pheromone and heuristic function, thereby fully exploiting the available knowledge. In biased exploration mode, the selection process is probabilistic and higher the pheromone level and heuristic value of a given feature implies higher chance of the feature to get selected. This mode enables ants to explore newer paths as well. A high value of q_0 will ensure the selection of known "good" features, while keeping the exploration ability of the ants less. A low value of q_0 will force the ants to keep exploring newer paths. The subset of features p_k generated by ant k needs to be evaluated to gauge its fitness. Sending the subset of features as input to the rule ant classifier *AntzClass* and estimating the classification accuracy can readily facilitate this. All the ants generate their solutions (different subsets of features) and corresponding fitness values are calculated using the above-mentioned process. After iteration all the ants update pheromone trails as:

$$\tau_i(t+1) = (1-\alpha) \times \tau_i(t) + \alpha \times \sum_k \Delta\tau_i^k, \tag{8}$$

where α is an evaporation rate in the range [0,1]. $\Delta\tau_i^k$ is the pheromone value given as:

$$\Delta\tau_i^k = \begin{cases} Q_k & \text{if } X_i \in p_k \\ 0 & \text{else} \end{cases}, \tag{9}$$

where Q_k is the classification accuracy of the *AntzClass* with the input as subset of features selected by the best ant in *AntzFeat*. Thus at every iteration of the *AntzFeat* the global best solution p_{best} (best subset of input features) stored corresponding to maximum fitness value and *AntzClass* algorithm is run with this best subset to discover classification rules set on the training data. Thus at every iteration of *AntzFeat* we need to run *AntzClass* to evaluate the fitness. After running the algorithms for a fixed number of iterations the software ants produce the best subset of features having the best information content. The rules developed by the *AntzClass* with this subset will thus have the maximum classification accuracy. It should be noted that, *AntzClass* and *AntzFeat* are two different methods and do not share pheromone trails information or heuristic information and any algorithm parameters. The application of ant algorithm for feature selection is illustrated by solving two case studies.

8. Results and Discussion

8.1. Performance study of AntzClass

The performance of *AntzClass* was tested on eight well-known publicly available datasets shown in Table 2 using the customary 10-fold cross-validation approach. Illustration of rule generation for classification of 3 Iris flower species (setosa, virginica and versicolor) based on 4 attributes (sepal length, petal length, petal width and sepal width) is given in Fig. 3. The results are reported in Table 3. The CPU time required by *AntzClass* algorithm for rule mining on datasets (shown in Table 2) is given in Table 4. Illustration of rule generation for one another data set, viz., wine data set taken from UCI machine learning repository (*www.ics.uci.edu/~mlearn/MLRepository.html*) is shown in Fig. 4.

The performance of *AntzClass* was also compared with PART and J4.8 algorithms (Witten and Frank, 2005). These algorithms have produced nearly 86% average classification accuracy over all the datasets. *AntzClass*

Table 2. Illustrative data sets for the study.

Dataset	Instances	Attributes	Attribute type: Real	Attribute type: Categorical	Class
Iris	150	4	4	0	3
Wine	178	13	13	0	3
Thyroid Gland	215	5	5	0	3
Pima Indian	768	8	8	0	2
Wisconsin Diagnosis Breast Cancer (WDBC)	569	30	30	0	2
Wisconsin Prognosis Breast Cancer (WPBC)	198	32	32	0	2
Wisconsin Breast Cancer (WBC)	699	9	9	9	2

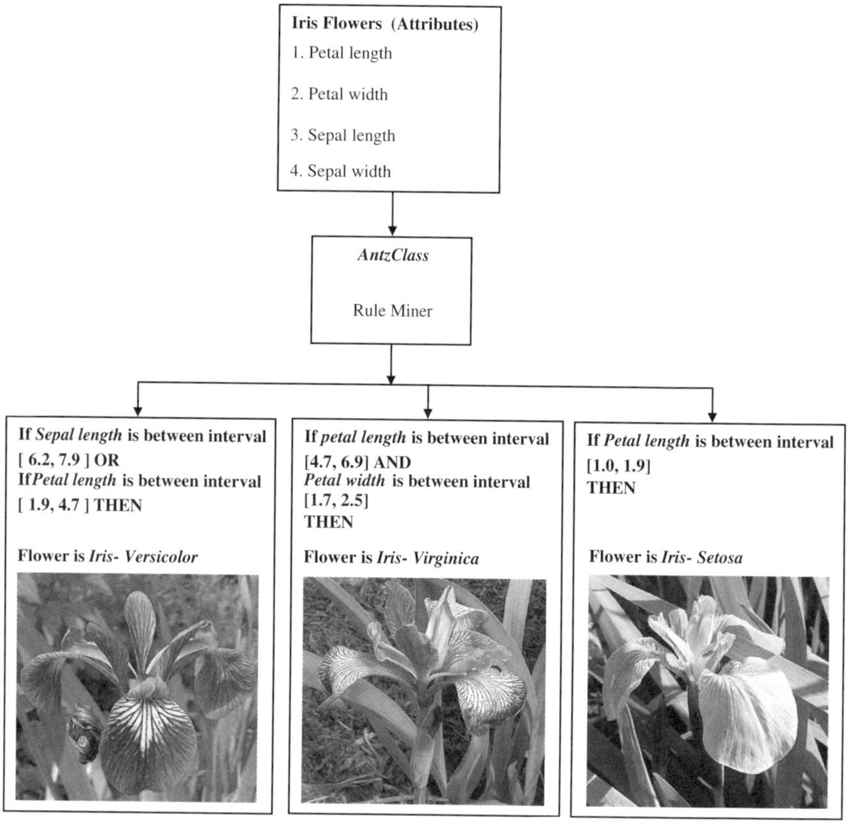

Figure 3. Illustration of rules generation using Iris data.

and PART have generated at an average around 3 rules with 2 terms over eight datasets. While J4.8 has generated average nearly 5 rules with 3 terms over five datasets. The results show that performance of *AntzClass*, is comparable to PART and J4.8.

During this study several experiments were performed with different set of parameter values and the results are reported using the following optimal set of parameter values: number of ants = 50, maximum number of iterations = 500, evaporation rate = 0.01, minimum cases to be covered by a rule = 8–10; maximum cases remained uncovered = 5.

We now illustrate the application of *AntzClass* for Tennessee Eastman Process (TEP), a benchmark problem in process engineering.

Table 3. Performance comparison study of *AntzClass* algorithm.

	% Accuracy			Avg. no. of rules			Avg. no. of terms		
Data	AntzClass	Part	J48	AntzClass	Part	J48	AntzClass	Part	J48
Iris	93.33	92.66	92.99	3.7	3.8	4.3	1.62	1.23	2.42
Wine	91.41	93.46	94.08	6.4	4.4	5	2.25	1.32	2.84
Thyroid	95.41	94.46	93.12	6	4	8.3	1.6	2.1	4.62
Pima	75.01	74.6	74.06	7.7	7.8	18.8	2.38	2.03	5.57
Hepatitis	64.45	67.5	63.75	0.4	1.0	0.8	1.8	3.9	1.4
WBDC	93.8	93.5	93.2	1	0.7	1.3	1.9	1.3	2.4
WBPC	75.76	73.2	73.71	1.1	.5	1.6	2.2	0.7	3
WBC	95.24	95.46	96.05	1.2	1.1	1.1	3	2.1	2
Overall	85.55	86.01	85.12	3.44	2.91	5.15	2.1	1.84	3.03

Table 4. CPU time for discovery of classification rules by *AntzClass*.

Data	Processing time (sec)
Iris	0.0
Wine	3.0
Thyroid	0.0
Pima	1.0
Hepatitis	1.0
WBDC	6.4
WBPC	4.8
WBC	1.0
Overall	2.15

Table 5. Performance comparison of *AntzClass* on TEP process data.

Classifier	% Misclassification on test data
AntzClass	19.4
J4.8	18.1
PART	17.7

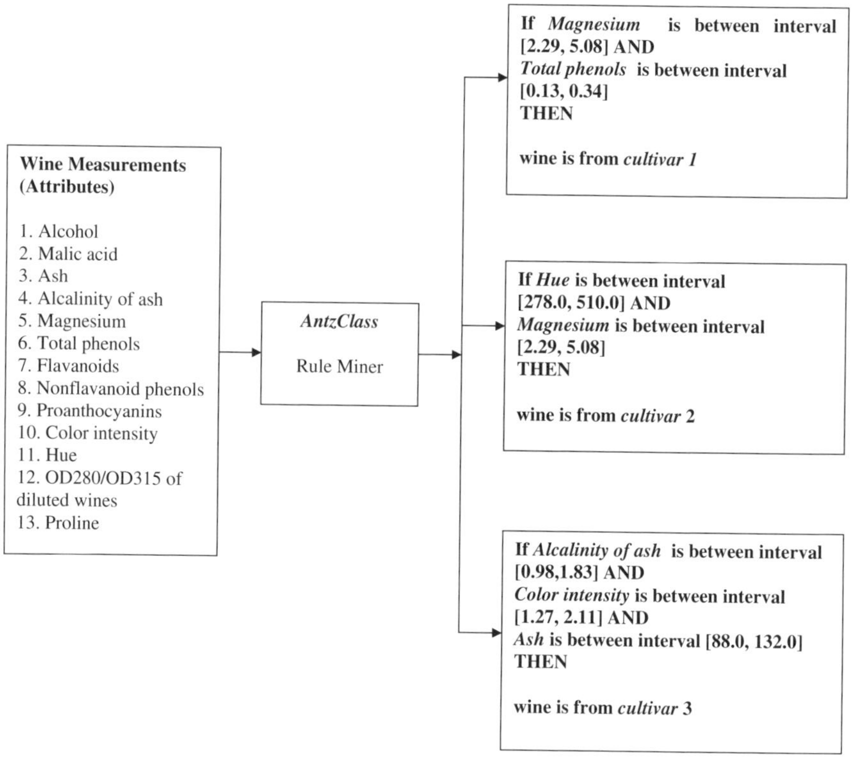

Figure 4. Illustration of rules generation using wine data.

8.2. Tennessee eastman process (TEP)

In this study, we employed plant-wide control structure (Downs and Vogel, 1993; Lyman and Georgakis, 1995) to simulate the closed loop process to generate the data for different faults. TEP simulator can generate 21 types of different faults entering in the process due to various disturbances. Once the fault enters the process, it affects almost all state variables in the process. For multiple fault identification, two faults viz. step change in the reactor cooling water inlet temperature and random variation in the reactor cooling water inlet temperature were considered. Third fault was due to random variation in D feed temperature (Chiang *et al.*, 2004). It was found that there is considerable overlap amongst the reactor faults. After introduction of third fault, the degree of overlap increased further which in a way made the dataset a challenging task for an algorithm to learn. For each fault, we generated 480 observation as training data and 960 as testing data. Thus, for three faults, training data and testing data contain 1440 and 2880 observations respectively. Each observation contains 52 process variables. However, only variable 9 (reactor temperature) and variable 51 (reactor cooling water valve position) are important in terms of distinguishing the three classes (Chiang *et al.*, 2004). *AntzClass* has produced an accuracy of 80.6% on the test set. We applied PART and J4.8 on the test data, which produced accuracy of 82.3% and 81.9% respectively. The classification rate produced by *AntzClass* is somewhat inferior to that of PART and J4.8. However, *AntzClass* generated simpler rules (9 rules with 11 terms, shown in Fig. 5) as compared to the rules obtained by PART (12 rules with 26 terms) and J4.8 (26 rules with 51 terms).

8.3. Performance study of wrapper approach (AntzFeat-AntzClass)

In this study, we applied both *AntzClass* and *AntzFeat* synergistically to simultaneously classify and select an optimal subset of features. *AntzClass* discovers rules using the training data while *AntzFeat* iteratively obtains the

Rule1:	IF 43.527 < A2 <= 49.229 AND 120.35 < A1 <= 120.43 THEN class 1
Rule2:	IF 43.527 < A2 <= 49.229 AND 120.43 < A1 <= 120.74 THEN class 3
Rule3:	IF 39.809 < A2 <= 43.527 AND 120.43 < A1 <= 120.74 THEN class 3
Rule4:	IF 120.43 < A1 <= 120.74 THEN class 3
Rule5:	IF 43.527 < A2 <= 49.229 THEN class 3
Rule6:	IF 120.11 < A1 <= 120.35 THEN class 3
Rule7:	IF 33.663 < A2 <= 39.809 THEN class 3
Rule8:	IF 120.35 < A1 <= 120.43 THEN class 2
Rule9:	Default class 1

class 1	step change in the reactor cooling water inlet temperature
class 2	step change random variation in the reactor cooling water inlet temperature
class 3	random variation in D feed temperature
A1	reactor temperature
A2	reactor cooling water valve position

Figure 5. Rules generated by *AntzClass* using Tennessee Eastman Process data.

best subset of features. The combined algorithm was tested on the Colon cancer data and a benchmark QSAR data. Parameters of *Antzclass* were applied as: number of ants = 50; maximum number of iterations = 100, evaporation rate = 0.01, minimum cases to be covered by a rule = 8; maximum cases remained uncovered = 5. While parameters of *Antzfeat* were set as: number of ants = 20; maximum number of iterations = 50, evaporation rate = 0.01, number of features to be selected = 100.

8.4. Colon data

Gene expression data were obtained on Affymetrix human 6000 arrays. This data contains expression level of 2000 genes in 22 normal and 40 tumor colon tissues (Alon *et al.*, 1999). The genes were chosen to give the highest minimal intensity across all samples. The data was pre-processed by transforming the raw intensities to base 10 logarithmic values and standardizing each sample to zero mean and variance one. We omitted 5 samples (N34, N36, T30, T33, T36), which were contaminated as suggested by Li *et al.* (2001). We have applied the distribution of training and test set given by Li *et al.* (2001) where first 40 samples were used for training and remaining 17 samples for testing. We applied *AntzFeat* algorithm to select 100 most informative genes. *AntzClass* algorithm employed those 100 genes and correctly classified 13 samples out of 17 test samples with misclassification of 4 samples. GA/kNN method produced misclassification of 1 sample out of 17 test samples. Shena *et al.* (2007) applied a wrapper method with particle swarm optimization (PSO) for feature selection and SVM for classification. They employed 50 samples for training and 12 samples as test set. PSO selected 4 genes that produced 91.67% classification accuracy using SVM.

8.5. QSAR data

This QSAR data set (Hattotuwagama *et al.*, 2005) was obtained from the CoEPrA website (Comparative Evaluation of Predictive Algorithms, http://www.coepra.org). In this study, QSAR and 3D QSAR experiments were carried out to quantify the binding affinities of genes in H2, a multigene cluster of mouse MHC, containing three major gene classes. The first class of these genes is located in the H2-L, H2-D, H2-K, Qa, and H2-T18 regions, the second in the H2-I region, and the third in the H2-S region. From the 152 peptides taken for this study, those with a pIC50exp between 5.010 and 7.748 were labeled as inactive, and those with a higher binding affinity, pIC50exp between 7.793 and 8.403 were labeled as active. Thus, the dataset consisted of 77 peptides in the negative class (inactive) and 75 in the positive class (active). For subset with 100 features, *AntzClass* algorithm discovered classification rules on training data. The discovered rules when applied on test data correctly classified 65 samples out of 76 test samples. Oloff and Muegge (2007) reported similar results with their kScore algorithm for this data set available at CoEPrA website.

Overall, this is one of the first applications in which a hybrid classification cum feature selection is carried out by the same metaheuristic paradigm, viz., Ant Colony Optimization. A small simulation study of this wrapper approach on two datasets (selecting 100 features out of the attributes ranging from 2000–5867) has shown promising results. An exhaustive study on such tasks using metaheuristics could be interesting to explore their viability in different machine learning applications.

9. Conclusions

Ant colony optimization, a novel heuristic paradigm based on the foraging behavior of real life ants was employed to develop two data mining algorithms, viz., *AntzClass* and *AntzFeat*. Given a set of instances, *AntzClass* produces an optimal set of rules to classify the instances into different groups. *AntzFeat* algorithm extracts a subset of most informative features to enable accurate classification. In fact *AntzFeat* synergistically combines with *AntzClass* for simultaneous classification and feature extraction. The performance of these methods was tested on well-known benchmark case studies. The results indicate that *AntzClass* algorithm is quite competitive to state-of-the-art PART and J4.8 algorithms and can be a useful tool in machine learning systems for knowledge discovery in database. Although preliminary results do not show much improved performance to its counterparts in terms of classification accuracy, *AntzClass* has certain merits, for example, it applies evolutionary search as against greedy search and it can be easily parallelized. Considering the merits of both metaheuristics and induction algorithms, a promising approach could be the use of *AntzClass* to discover small disjunct rules while J4.8 produces relatively large disjunct rules. The novel hybrid approach (*AntzFeat* for feature selection and *AntzClass* for classification rule discovery) based on ant colony optimization also compares very well with the existing methods.

Acknowledgment

Author PSS acknowledges a partial financial support received under Fast Track Scheme for Young Scientist from the DST, New Delhi, India.

Nomenclature

Arabic symbols

F	Number of attributes, Eq. (2).
d_k	Number of domains for attribute k, Eq. (2).
X_i	ith attribute value in domain D_{ij}.
D_{ij}	jth domain value of attribute X_i, $j = 1, 2, \ldots, d_i$.
I	Domain of attributes not yet been used, Eq. (2).
C	Number of classes, Eq. (4).
S	Total number of terms $= \sum_k^n d_k$
W_{ij}	Partition where attribute $X_i = D_{ij}$, Eq. (4).
$\|W_{ij}\|$	Total number of cases in partition W_{ij} where attribute $X_i = D_{ij}$, Eq. (4).
c	Class attribute, Eq. (4).
$P(c\|X_i = D_{ij})$	Empirical probability of observing class w conditional on having observed $X_i = D_{ij}$, Eq. (4).
FN	Number of cases is not covered by the rule but that have same class to the rule consequent.
FP	Number of cases covered by the rule but that have class different from the rule consequent.
TN	Number of cases is not covered by the rule and that have class different from the rule consequent.
TP	Number of cases covered by the rule and that have same class to the rule consequent.
ACO	Ant Colony Optimization
GA	Genetic Algorithm

PSO	Particle Swarm Optimization
AntzClass	Ant Algorithm for Classification
AntzFeat	Ant Algorithm for Feature Selection
SVM	Support Vector Machine
kNN	*k*-Nearest Neighbors

Greek symbols

$\tau_{ij}(t)$	Pheromone at iteration t in the position (i, j), Eq. (2).
ρ	Pheromone evaporation rate in the range [0,1], Eq. (3).
η_{ij}	Value of heuristic function for term S_{ij}, Eq. (5).
α	Weighting parameter in Eq. (6)
β	Weighting parameter in Eq. (6)

References

Alizadeh, A.A., Eisen, M.B. *et al.* (2000). Distinct types of diffuse large B-cell lymphoma identified by gene expression profiling. *Nature*, **403**, pp. 503–511.

Alon, U., Barkai N., Notterman, D.A., Gish, K., Ybarra, S., Mack, D. and Levine, A.J. (1999). Broad patterns of gene expression revealed by clustering analysis of tumor and normal colon tissues probed by oligonucleotide arrays. *Proc Natl Acad Sci USA*. **96**, pp. 6745–6750.

Beckers, R., Deneubourg, J.L. and Goss, S. (1992). Trails and U-turns in the selection of the shortest path by the ant *Lasius niger*. *Journal of Theoretical Biology*, **159**, pp. 397–415.

Blanco, R., Larranaga, P., Inza, I. and Sierra, B. (2004). Gene selection for cancer classification using wrapper approaches. *Int. J. Pattern Recognit. Artif. Intell.*, **18**, pp. 1373–1390.

Chai, H. and Domeniconi, C. (2004). An evaluation of gene selection methods for multi-class microarray data classification. *Proceedings of the 2nd European Workshop on Data Mining and Text Mining for Bioinformatics*, pp. 7–14, Pisa, Italy.

Chang, S.-Y., Lin, C.-R. and Chang, C.-T. (2002). A fuzzy diagnosis approach using dynamic fault trees. *Chemical Engineering Science*, **57**(15), pp. 2971–2985.

Chiang, L.H., Kotanchek, M.E. and Kordon, A.K. (2004). Fault diagnosis based on Fisher discriminant analysis and support vector machines. *Computers and Chemical Engineering*, **28**, pp. 1389–1401.

Cover, T.M. and Thomas, J.A. (1991). *Elements of Information Theory*. John Wiley, New York.

Dorigo, M., Di Caro, G. and Gambardella, L.M. (1999). Ant algorithms for discrete optimization. *Artificial Life*, **5**, pp. 137–172.

Dougherty, J., Kohavi, R. and Sahami, M. (1995). Supervised and unsupervised discretization of continuous features. In Prieditis, A. and Russell, S., *Proceedings of the 12th international conference Machine Learning*. San Francisco, CA, Morgan Kaufmann.

Downs, J.J. and Vogel, E.F. (1993). A plant-wide industrial-process control problem. *Computers and Chemical Engineering*, **17**, pp. 245–255.

Furey, T.S., Cristianini, N., Duffy, N., Bednarski, D.W., Schummer, M. and Haussler, D. (2000). Support vector machine classification and validation of cancer tissue samples using microarray expression data. *Bioinformatics*, **16**, p. 10.

Golub, T.R., Slonim, D.K., Tamayo, P., Gaasenbeek, C.H.M., Mesirov, J.P., Coller, H., Loh, M.L., Downing, J.R., Caligiuri, M.A., Bloomfield, C.D. and Lander, E.S. (1999). Molecular classification of cancer: Class discovery and class prediction by gene expression monitoring. *Science*, **286**, pp. 531–537.

Gupta, A., Jayaraman, V.K. and Kulkarni, B.D. (2006). Feature Selection for Cancer Classification using Ant Colony Optimization and Support Vector Machines. In: Sanghamitra Bandyopadhyay, Ujjwal Maulik and Jason T L Wang (ed.) *Analysis Of Biological Data: A Soft Computing Approach, World. Scientific, Singapore*, **3**, pp. 259–280.

Hattotuwagama, C.K., Doytchinova, I.A. and Flower, D.R. (2005). In silico prediction of peptide binding affinity to class I mouse major histocompatibility complexes: A comparative molecular similarity index analysis (CoMSIA) study. *J. Chem. Inf. Model.*, **45**, pp. 1415–1423.

Hsiung, J.T. and Himmelblau, D.M. (1996). Detection of leaks in a liquid-liquid heat exchanger using passive acoustic noise. *Computers and Chemical Engineering*, **20**(9), pp. 1101–1111.

Huiqing, L., Jinyan, L. and Limsoon, W. (2002). A comparative study on feature selection and classification methods using gene expression profiles and proteomic patterns. *Genome Informatics*, **13**, pp. 51–60.

Jayaraman, V.K., Kulkarni, B.D., Karale, S. and Shelokar, P. (2000). Ant colony framework for optimal design and scheduling of batch plants. *Computers and Chemical Engineering*, **24**(8), pp. 1901–1912.

Jayaraman, V.K., Kulkarni, B.D., Gupta, K., Rajesh, J. and Kusumaker, H.S. (2001). Dynamic optimization of fed-batch bioreactors using the ant algorithm. *Biotechnology Progress*, **17**, pp. 81–88.

Jirapech-Umpai, T. and Aitken, S. (2005). Feature selection and classification for microarray data analysis: Evolutionary methods for identifying predictive genes. *BMC Bioinformatics*, **6**, p. 148.

van Kampen A.H.C., Ramadan, Z., Mulholland, M., Hibbert, D.B. and Buydens, L.M.C. (1997). Learning classification rules from an ion chromatography database using a genetic based classifier system. *Analytica Chimica Acta*, **344**, pp. 1–15.

Kulkarni, A., Jayaraman, V.K. and Kulkarni, B.D. (2005). Knowledge incorporated support vector machines to detect faults in Tennessee Eastman Process. *Computers and Chemical Engineering*, **29**, pp. 2128–2133.

Kumar, R., Jayaraman, V.K. and Kulkarni, B.D. (2005). An SVM classifier incorporating simultaneous noise reduction and feature selection: Illustrative case examples, *Pattern Recognition*, **38**(1), pp. 41–49.

Li, L., Weinber, C.R., Darden, T.A. and Pederse, L.G. (2001). Gene selection for sample classification based on gene expression data: Study of sensitivity to choice of parameters of the GA/KNN method. *Bioinformatics*, **17**(12), pp. 1131–1142.

Lopes, H.S., Coutinho, M.S. and Lima, W.C. (1998). An evolutionary approach to simulate cognitive feedback learning in medical domain. *Genetic algorithms and Fuzzy Logic Systems: Soft Computing Perspectives* (pp. 193–207) World Scientific.

Lyman, P.R. and Georgakis, C. (1995). Plant-wide control of the Tennessee Eastman problem. *Computers and Chemical Engineering*, **19**, pp. 321–331.

Nutt, C.L., Mani, D.R., Betensky, R.A., Tamayo, P., Cairncross, J.G. *et al.* (2003). Gene expression-based classification of malignant gliomas correlates better with survival than histological classification. *Cancer Res*, **63**, pp. 1602–1607.

Oloff, S. and Muegge, I. (2007). kScore: A novel machine learning approach that is not dependent on the data structure of the training set. *Journal of Computer-Aided Molecular Design*, **21**(1–3), pp. 87–95.

Özyurt, B., Sunol, A.K., Çamrudan, M.C., Mogili, P. and Hall, L.O. (1998). Chemical plant fault diagnosis through a hybrid symbolic-connectionist machine learning approach. *Computers and Chemical Engineering*, **22**(1–2), pp. 299–321.

Parpinelli, R.S., Lopes, H.S. and Freitas, A.A. (2002). An ant colony algorithm for classification rule discovery. Abbas, H., Sarker, R. and Newton, C., *Data mining: A Heuristic Approach* (pp. 191–208). London, UK, Idea group publishing.

Patil, D., Raj, R., Shingade, P., Jayaraman, V.K. and Kulkarni, B.D. (2008). Feature selection and classification employing hybrid ACO-Random forest methodology. *Combinatorial Chemistry and High Throughput Screening* (*accepted*).

Paul, T.K. and Iba, H. (2005). Gene selection for classification of cancers using probabilistic model building genetic algorithm. *Biosystems*, **82**(3), pp. 208–225.

Peng, S., Xu, Q., Ling, X.B., Peng, X., Du, W. and Chen, L. (2003). Molecular classification of cancer types from microarray data using the combination of genetic algorithms and support vector machines. *FEBS Letters*, **555**(2), pp. 358–362.

Pudil, P., Ferri, F.J., Novovicova, J. and Kittler, J. (1994). Floating search methods for feature selection with nonmonotonic criterion functions. *Proc. Int. Conf. Pattern Recognition*, vol. **1**, pp. 279–283.

Quinlan, J.R. (1993). *C4.5: Programs for machine learning.* San Francisco, CA, Morgan Kaufmann.

Saeys, Y., Inza, I. and Larrañaga, P. (2007). A review of feature selection techniques in bioinformatics. *Bioinformatics*, **23**(19), pp. 2507–2517.

Shelokar, P.S., Jayaraman, V.K. and Kulkarni, B.D. (2004). An ant colony classifier system: Application to some process engineering problems. *Computers and Chemical Engineering*, **28**, pp. 1577–1584.

Shena, Q., Shia, W.-M., Konga, W. and Yea, B.-X. (2007). A combination of modified particle swarm optimization algorithm and support vector machine for gene selection and tumor classification. *Talanta*, **71**, pp. 1679–1683.

Summanwar, V.S., Shelokar, P.S., Jayaraman, V.K. and Kulkarni, B.D. (2002). Ant colony framework for process optimization: Unconstrained and constrained problems with single and multiple objectives. Luus, R., *Recent Developments in Optimization and Optimal Control in Chemical Engineering* (pp. 67–87). Trivandrum, India: Research Signpost.

Venkatasubramanian, V., Vaidyanathan, R. and Yamamoto, Y. (1990). Process fault detection and diagnosis using neural networks-I. *Steady-state processes. Computers and Chemical Engineering*, **14**(7), pp. 699–712.

West, M., Blanchette, C., Dressman, H., Huang, E., Ishida, S., Spang, R., Zuzan, H., Olson, J.A. Jr., Marks, J.R. and Nevins, J.R. (2001). Predicting the clinical status of human breast cancer by using gene expression profiles. *Proc. Natl. Acad. Sci.*, **98**(20), pp. 11462–11467.

Weston, J., Mukherjee, S., Chapelle, O., Pontil, M., Poggio, T. and Vapnik, V. (2001). Feature selection for SVMs. *Advances in Neural Information Processing Systems*, **13**, pp. 668–674.

Witten, I.H. and Frank, E. (2005). *Data Mining: Practical Machine Learning Tools and Techniques*, Morgan Kaufmann.

Zhang, X., Lu, X., Shi, Q., Xu, X.Q., Leung, H.C., Harris, L.N., Iglehart, J.D., Miron, A., Liu, J.S. and Wong, W.H. (2006). Recursive SVM feature selection and sample classification for mass-spectrometry and microarray data. *BMC Bioinformatics*, **7**, p. 197

Exercises

(1) UCI Machine Learning Repository currently maintains more than 100 datasets under classification task as a service to the machine learning community. The available datasets are from different fields of science and engineering. Students can download some of the datasets, such

as, Dermatology, Parkinsons from the link: *www.ics.uci.edu/~mlearn/MLRepository.html*

Task 1. Find a set of parameter values of *AntzClass* algorithm that can produce best classification performance on a given dataset.

Task 2. Vary the parameter 'minimum number of cases' and observe its effect in terms of classification accuracy, number of rules generation and number of terms selected.

Task 3. Validate the performance of discovered rules on test data. Make test data by arbitrarily selecting 1/3rd of a dataset and remaining 2/3rd as training data.

Perform all the above three tasks on both the datasets.

(2) Apply *AntzFeat* algorithm to select 5, 10 and 15 features that produce best classification accuracy on Dermatology and Parkinsons datasets.

Acknowledgement

Author VKJ acknowledges financial support received under grant no: SR/S3/CE/050/2009 from the DST, New Delhi, India. Author PSS acknowledges a partial financial support received under Fast Track Scheme for Young Scientist from the DST, New Delhi, India.

Chapter 18

CONSTRAINT PROGRAMMING AND GENETIC ALGORITHM

Prakash R. Kotecha, Mani Bhushan and Ravindra D. Gudi*

Department of Chemical Engineering
Indian Institute of Technology Bombay
Powai, Mumbai-400 076, India
**ravigudi@iitb.ac.in*

1. Introduction

Genetic Algorithms (GA) have found wide acceptance in Chemical Engineering (Yee *et al.*, 2003; Tarafder *et al.*, 2005, 2007; Agrawal *et al.*, 2007) because of their simplicity, and ability to determine optimal or near optimal solutions for both single and multi-objective, non linear optimization problems. In this chapter, we present another AI based technique, Constraint Programming (CP), which has been finding increasing use in Chemical Engineering. Recent applications of CP in Chemical Engineering involving combinatorial optimization include the design of sensor networks based on fault diagnosis perspective (Kotecha *et al.*, 2007), design of robust and reliable sensor network (Kotecha *et al.*, 2008a), design of precise and reliable sensor networks (Kotecha *et al.*, 2008b), short term batch scheduling of chemical plants (Maravelias and Grossmann, 2004; Roe *et al.*, 2005), scheduling of job shop plants (Jain and Grossmann, 2001), and multistage scheduling problems (Harjunkoski and Grossmann, 2002). In this chapter, we compare CP and GA techniques for solving some of these combinatorial optimization problems. Both these techniques share some common features despite having completely different working principles. In the next section, we give a brief background of CP along with its working principle and demonstrate its working on a simple single objective problem. We present some of its unique features along with discussing its similarities and differences with GA followed by a brief description of its extension to multi-modal and multi-objective optimization (m3o) problems. We subsequently present two case studies (a job shop scheduling problem and a sensor network design problem) widely discussed in the Chemical Engineering literature and demonstrate the performance of both GA and CP on each of these problems. We finally conclude the chapter by summarizing the results and present possible future extensions.

2. Constraint Programming

Constraint Programming is an intelligent, non gradient, tree based, implicit enumerative search technique. It has been developed by the Artificial Intelligence and Computer Science Community for solving Constraint Satisfaction Problems (CSPs) arising in these fields. It has found applications in diverse areas (Van Hentenryck, 2002) such as computer graphics, software engineering, databases, hybrid systems, finance, circuit design, etc. CP has been found to be particularly successful in areas involving combinatorial optimization such as planning, resource allocation and scheduling. A brief description of CP is presented next.

2.1. Working principle

CP is a domain reduction technique relying primarily on constraint propagation for reducing the domain of the variables. CP solves an optimization problem by transforming it to a CSP. A CSP is a feasibility problem that is similar to an optimization problem except for the fact that it does not have an objective function. Hence, any solution satisfying the set of constraints is an acceptable solution as there is no notion of optimality. Consider a single objective optimization problem as shown in formulation SO.

$$\textbf{SO} \quad \begin{array}{ll} \textit{Optimize} & f(\mathbf{x}) \\ \text{s.t.} & G(\mathbf{x}) \end{array}, \tag{1}$$

where $f(\mathbf{x})$ is a multivariable single objective function that needs to be optimized subject to the set of constraints represented by $G(\mathbf{x})$. The set of constraints represented by $G(\mathbf{x})$ can directly include, without any transformation, constraints involving the non-equality operator ($\neq$), implication constraints or disjunctive constraints. The optimization problem in SO can be converted to a CSP (or feasibility problem) as shown in SF

$$\mathbf{SF} \qquad \begin{array}{ll} \textit{Solve} & G(\mathbf{x}) \\ & F = f(\mathbf{x}) \end{array}. \tag{2}$$

The problem statement in SF is a feasibility problem and as mentioned earlier any set of decision variables which satisfies the set of constraints represented by $G(\mathbf{x})$ is a solution. The feasibility problem in SF ignores the objective function and instead adds a new variable (F) evaluating the value of the objective function of the original problem SO. The following constraint is also added to SF after determining every feasible solution to ensure that the value of the objective function progressively improves throughout the exploration of the search space.

$$F^{k+1} \lhd F^{k}, \tag{3}$$

where F^k, F^{k+1} indicates the value of the objective function of the kth and $(k+1)$th feasible solution and $\lhd$ is the dominant operator (Deb, 2001). To keep the formulation generic, the dominant operator has been used to avoid the distinction between a maximization problem and a minimization problem. For maximization problems, $\lhd(\unlhd)$ will represent the $>$ $(\geq)$ operator and for a minimization problem, it will represent the $<$ $(\leq)$ operator. The initial value of the variable (F), F^0, is set to $-\infty(\infty)$ for a maximization (minimization) problem. In this chapter, we will be using the term "suboptimal" solution to denote solutions that are feasible but do not correspond to the global optima. Figure 1 shows the working principle of CP to solve a single objective optimization problem and additional details can be obtained from literature (Lustig and Puget, 2001; Grossmann and Biegler, 2004; Roe *et al.*, 2005).

The first step of the CP algorithm is to propagate the set of constraints to eliminate infeasible values from the domain thus reducing the potential search space. However, this step eliminates only those values in a domain

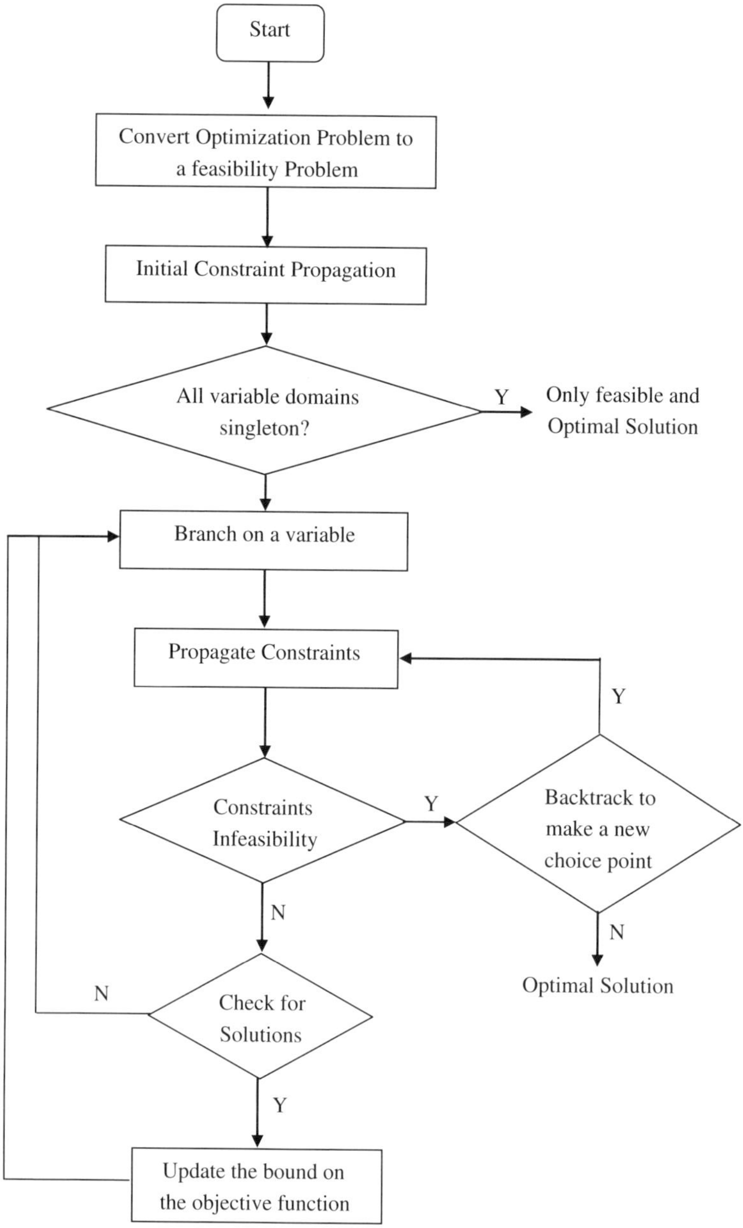

Figure 1. CP algorithm for solving an optimization problem.

that can be identified to be infeasible without making any choice on other decision variables (i.e. fixing any other variable arbitrarily to a particular value from its domain). If this step leads to a singleton set for all the decision variables, then the problem has only one feasible solution which then corresponds to the optimal solution. However, if the reduced domains of all variables are not singleton, a choice point is made, i.e. a decision variable is selected and arbitrarily assigned a value from its (latest updated) domain. The constraints are once again propagated to possibly reduce the domain of other variables. If necessary, additional choice points are made on the other variables during the search till either the domain of any one of the decision variable becomes null or the domain of all variables become singleton. It should be noted that constraint propagation is performed after every choice point. If the domain of any decision variable becomes null, the choice point is said to have led to a failure and backtracking is performed to revise the latest choice point. This procedure of choosing choice point and propagating constraints is continued till the whole search space is explored. It should be noted that on realizing a failure, the latest choice point corresponding to this failure cannot be removed from the domain of the variable. This is because this failure may have been reached for a particular set of choice points. However a choice point can be removed from the domain of a variable if all subsequent choice points on all the other variables lead to a failure. At any point, a feasible solution is said to be obtained if the domain of all variables are singleton. Once a feasible solution is obtained, the value of F is updated and the procedure is continued till the complete search space is explored. However, if every choice point leads to a failure, the problem is an infeasible problem. In literature there are intelligent ways of selecting a choice point leading to fewer failures (Jaffar and Maher, 1994; Grossmann and Biegler, 2004; Roe *et al.*, 2005; Rossi *et al.*, 2006). We next discuss a simple example to illustrate the working principle of CP.

2.2. Example

We now demonstrate the working of CP on a small example to illustrate its superior modeling power compared to traditional optimization techniques and guarantee of global optimality. Consider the following two variable constrained single objective Integer Programming (IP) optimization problem (SOE).

$$\max x_1 - x_2 \tag{4}$$

$$\text{s.t.} \quad x_1 \neq x_2; \tag{5}$$

$$x_1 < 2 \Rightarrow x_2 > 4; \tag{6}$$

$$x_1 + x_2 < 5; \tag{7}$$

$$x_1 = \{0, 1, 2, 3, 4, 5\}; \quad x_2 = \{0, 1, 2, 3, 4, 5\}. \tag{8}$$

It can be seen that the constraints involve non-equality, implication and strict inequalities. The use of CP enables the handling of the problem in SOE without any modification of the constraints to the form $G(\mathbf{x}) \leq 0$. As mentioned earlier, the first step is to convert the optimization problem to a feasibility problem as shown in SFE (refer to Fig. 2). From Fig. 2, it can be seen that the initial constraint propagation reduces the domains of variables x_1 and x_2. The value 5 is removed from the domain of x_1 and x_2 as it would violate constraint (7). The values {0, 1} are removed from the domain of x_1 since they would lead to violation of constraints (6) and (7). Thus, the smallest feasible value of x_1 is 2 and hence any value of x_2 greater than 3 would violate constraint (7). It can be seen that the initial constraint propagation removes a large number (but not all) of infeasible values. Figure 2 shows the remaining steps in CP to converge to the globally optimal solution of 4. It can be seen from step 5 that the choice point of $x_1 = 2$ and $x_2 = 1$ leads to a failure. However, the value of 1 is not removed from the domain of x_2 because other combinations of x_1 with $x_2 = 1$ may not violate any constraint. However, at the end of step 5, the value of 2 is removed from the domain of x_1 because all combinations of x_2 with $x_1 = 2$ have been explored and found to lead to failures. Thus, we see that CP can intelligently search the whole space without requiring/determining any gradient information or enumerating the whole search space. The case studies presented later in this chapter will reinforce the power of constraint propagation even for complex problems.

2.3. Features of CP

In this section, we list some of the advantages and disadvantages of CP along with a brief mention of other CP features (Kotecha *et al.*, 2007;

Example:

SOE

$$\max \quad x_1 - x_2$$

$$s.t \quad x_1 \neq x_2; \quad x_1 < 2 \Rightarrow x_2 > 4; \quad x_1 + x_2 < 5; \quad x_1 = \{0,1,2,3,4,5\}; \quad x_2 = \{0,1,2,3,4,5\}$$

SFE

$$Solve \quad x_1 \neq x_2; \quad x_1 < 2 \Rightarrow x_2 > 4; \quad x_1 + x_2 < 5; \quad x_1 = \{0,1,2,3,4,5\}; \quad x_2 = \{0,1,2,3,4,5\}$$

$$F = x_1 - x_2;$$

	Operation	Domain		Description	Best Sol
		x_1	x_2		
Step 1	Initial Domains	{0,1,2,3,4,5}	{0,1,2,3,4,5}	Initialization	$F^0 = -\infty$
Step 2	Initial Constraint Propagation	{2,3,4}	{0,1,2}	Select a choice point as the domains of the variables are not singleton	$F^0 = -\infty$
Step 3	Choice Point $x_1 = 2$; Propagating Constraints	2	{0,1}	Select a choice point for x_2 as its domain is not singleton	$F^0 = -\infty$
Step 4	Choice Point $x_2 = 0$	2	0	Feasible Solution; Update the bound on the objective function; Select another choice point as the whole search space has not been explored	$F^1 = 2$
Step 5	Choice Point $x_2 = 1$	2	1	Failure; Backtrack to select a new choice point	$F^1 = 2$
Step 6	Constraint Propagation	{3,4}	{0,1}	Select a choice point as the domains of the variables are not singleton	$F^1 = 2$
Step 7	Choice Point $x_1 = 3$	3	{0,1}	Select a choice point for x_2 as its domain is not singleton	$F^1 = 2$
Step 8	Constraint Propagation	3	0	Feasible Solution; Update the bound on the objective function; Select another choice point as the whole search space has not been explored	$F^2 = 3$
Step 9	Choice Point $x_2 = 1$	4	0	Feasible solution; As the whole space has been explored, this is the optimal solution	$F^3 = 4$

Figure 2. Demonstration of CP on a two variable optimization problem.

Kotecha *et al.*, 2008a). Some of the important benefits of CP over traditional mathematical programming techniques are listed under the following categories:

(i) *Modeling*: The expressive modeling ability of CP can be used to develop compact models with minimal efforts. For example, traditional mathematical programming techniques require the constraints of an optimization problem to be of the form $G(\mathbf{x}) \leq 0$. However, CP permits a richer description of the problem and does not impose any such restriction. As mentioned earlier, it can directly include, without any transformation, constraints involving the non-equality operator ($\neq$), implication constraints or disjunctive constraints. Also, there is no restriction on the variables to be non negative which is in contrast to some existing techniques such as the popular simplex method that uses additional artificial variables to incorporate this non negativity constraint. Additionally, CP modeling environments have problem specific constructs which result in compact models thereby enabling efficient constraint propagation leading to significant reduction in computational efforts.

(ii) *Optimality*: Unlike mathematical programming techniques, CP does not rely on integer relaxations or gradients but instead uses constraint propagation to reduce the domain of the variables. Thus, it avoids convergence to suboptimal (or local) solutions irrespective of the nature (convex/non-convex) of the problem. Hence, CP can guarantee globally optimal solutions for single objective problems and globally optimal pareto points for multi-objective optimization problem irrespective of the combinatorial and convex nature of the problem.

(iii) *Realizations*: The existence of multiple solutions (realizations) is a typical characteristic of both single and multi-objective combinatorial problems. The importance of determining such realizations has been discussed in literature (Tarafder *et al.*, 2007). Such realizations can be determined using CP without repeatedly re-solving the optimization problems.

(iv) *Next best solutions*: The search strategy of CP can be modified so as to determine K-best feasible solutions (K being an user defined parameter) of a single objective optimization problem instead of determining the globally optimal solution. Depending on the requirement of the designer, the set of distinct or non-distinct K-best feasible solutions can be determined.

(v) *Soft-constraints/soft-objectives*: In certain cases, the enumerative nature of CP can be appropriately used to solve problems involving either soft constraints or soft objectives (Kotecha *et al.*, 2008a).

Inspite of the above advantages, as with any other technique, CP also suffers from certain drawbacks as mentioned below

(i) Though CP enables richer modeling compared to the traditional gradient based mathematical techniques, it nevertheless requires an explicit optimization formulation for the propagation of constraints. This is in contrast to GA which has the capability of handling constructive constraints.
(ii) Most traditional optimization techniques use some mathematical property such as convexity or bounds obtained by appropriate relaxations to confirm local/global optimality of the obtained solution. However, since CP is purely a search based technique, it may require considerable amount of computational resource even after the determination of an optimal solution to confirm this optimality.

Comparative works: In the literature, CP and Integer Programming (IP) have been compared for a wide range of problems like the modified generalized assignment problem (Darby-Dowman *et al.*, 1997), the template design problem (Proll and Smith, 1998), the progressive party problem (Smith *et al.*, 1996), the change problem (Heipcke, 1999), the short term batch scheduling problem (Maravelias and Grossmann, 2004), the job shop problem (Jain and Grossmann, 2001) and the sensor network design problems (Kotecha *et al.*, 2008c). However, to the best of our knowledge, there have been no comparisons of CP with GA. As will be explained later, CP and GA share some interesting common features that help in addressing complexity issues in optimization problems. However, they differ in their search mechanisms and it would be interesting to analyze their relative performances on some benchmark problems. In this chapter, we make such a comparison of GA with CP for two representative problems reported in the literature and list the advantages and drawbacks of each technique.

Hybrid CP based methods: Recent developments in the field of combinatorial optimization include the development of hybrid methods and integration schemes (Hooker *et al.*, 1999; Van Hentenryck, 2002) which use the complementary properties of the traditional IP optimization techniques and CP to efficiently handle combinatorial problems. Most of these schemes have been developed for scheduling problems (Jain and Grossmann, 2001; Harjunkoski and Grossmann, 2002; Maravelias and Grossmann, 2004; Roe *et al.*, 2005) wherein the original problem is decomposed into a MILP Master problem and a CP subproblem leading to a large reduction in computational efforts.

CP tools: Some of the available CP solvers associated with different optimization software are ILOG Solver (ILOG, Last Accessed: July 2008), Kalis (DashOptimization, Last Accessed: July 2008), CHIP (Dincbas *et al.*, 1988), ECLiPSe (Wallace *et al.*, 1997). CP has been quite popular for solving scheduling problems and hence a large number of scheduling specific developments have been incorporated in various CP tools. The ILOG Scheduler is one such example that has been specially built for efficient modeling and solving of scheduling problems. It provides a large number of constructs that can be used to easily model scheduling problems.

Till recently, CP was primarily used for determining either a feasible solution or determining one single optimal solution for single objective optimization problems. However, recent works (Kotecha *et al.*, 2007, 2008a) have used CP to determine globally optimal pareto solutions and efficiently solve m3o problems to global optimality. To the best of our knowledge, CP is the only technique that can provide guarantee on the global optimality of the pareto front even for non linear Integer Programming problems. In the following sections, we briefly describe some of the CP based techniques that have been proposed in literature for the determination of multiple optimal solutions and the determination of pareto optimal fronts along with their realizations. A detailed discussion and specific observations on each of these techniques and their variants can be obtained from literature (Kotecha *et al.*, 2008d).

2.4. Determination of multiple optimal solutions

A two step CP based strategy has been used in literature (Kotecha *et al.*, 2007) to determine the multiple optimal solutions of a single objective optimization problem.

The two steps of this strategy are (i) the determination of a single optimal solution, and (ii) the determination of all feasible solutions of an appropriate feasibility problem. The first step is as explained in the Sec. 2.1. The second step involves the transformation of the optimization problem into a CSP whose solution will correspond to the optimal solution of the optimization problem. Consider the optimization problem in (1). Let the optimal solution of this problem be $f^{optimal}$. The second step involves modifying the optimization problem in (1) to

$$\begin{aligned} \textit{Solve} \quad & G(\mathbf{x}) \\ & f(\mathbf{x}) = f^{optimal} \end{aligned} \quad . \tag{9}$$

It is easy to observe that any feasible solution to (9) will also be an optimal solution to (1) due to the constraint $f(\mathbf{x}) = f^{optimal}$ and all the feasible solutions of (9) will correspond to the total multiple optimal solutions of (1). The individual steps in the above strategy are summarized in Fig. 3.

This strategy has the following advantages over the traditional techniques for determining multiple optimal solutions

(a) it does not require re-solving the optimization problem for determining every optimal solution.
(b) it does not require *a priori* specification of the total number of multiple optimal solutions.
(c) it does not require the construction and addition of any cuts.
(d) it does not alter the nature (linear to non-linear) or increase the dimensionality of the problem even in the presence of integer variables.

Though this strategy involves solving the problem twice, the computational effort required in the second step will be comparatively less due to the presence of the constraint $f(\mathbf{x}) = f^{optimal}$. As will be demonstrated in the case study, this constraint is used during constraint propagation and can considerably reduce the search space. The above strategy can be easily modified to determine only a pre-defined number of optimal solutions.

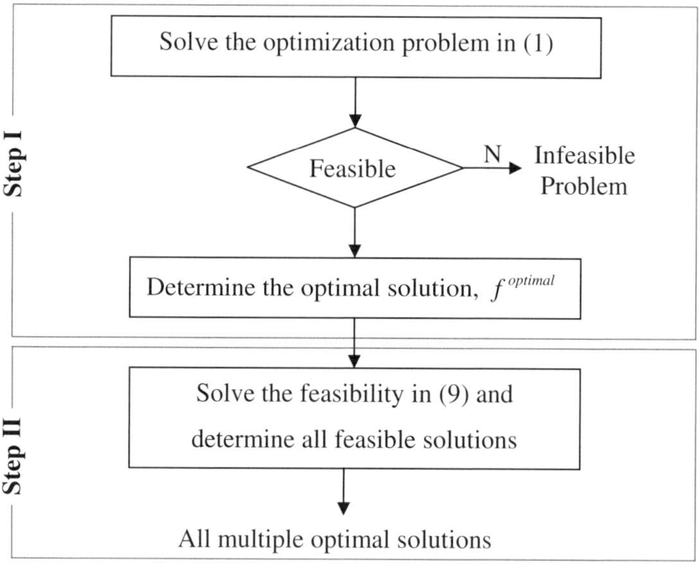

Figure 3. Two step strategy for determining multiple optimal solutions.

2.5. Determination of pareto fronts along with multiple realizations

CP has been successfully used to determine globally optimal pareto fronts along with their realizations. All the techniques proposed in literature for the determination of pareto fronts using CP harness the ability of CP to efficiently solve feasibility problems. The initial work proposed (Kotecha *et al.*, 2007) used the ability of CP to determine all feasible solutions without the addition of integer cuts. The strategy involved the determination of all feasible solutions followed by sorting these feasible solutions to identify the non dominated solutions. This strategy was demonstrated on the benchmark Tennessee Eastman (TE) problem for the design of sensor networks for efficient fault diagnosis. However, this technique requires the determination of all feasible solutions and can be computationally intensive. Subsequently, a two-step strategy for the efficient determination of pareto optimal solutions has been proposed (Kotecha *et al.*, 2008d). Similar to the previous approaches, this strategy also solves a feasibility problem but in addition employs integer cuts to ensure that inferior points are not explored. These integer cuts are designed based on the principle of pareto optimality and are added after the determination of each feasible solution ensuring that future solutions are non dominated. As this strategy does not require the determination of all feasible solutions, it can lead to potential computational benefits. This technique has been demonstrated on the benchmark TE problem for the design of sensor networks for efficient fault diagnosis. The two step strategy has also been used for the design of robust and reliable sensor networks (Kotecha *et al.*, 2008a) and for the design of reliable and precise sensor networks (Kotecha *et al.*, 2008b). We now present a brief description of this two step strategy and detailed information can be obtained from the cited literature (Kotecha *et al.*, 2008d).

Two step strategy: MO and MF are the corresponding multi-objective counterparts of SO and SF. The problem statement represented in MO is a multi-objective optimization problem with m multivariable objectives.

$$\textbf{MO} \quad \begin{array}{ll} \textit{Optimize} & f_j(\mathbf{x}); \quad j = 1, 2, \ldots, m \\ s.t & G(\mathbf{x}) \end{array}. \tag{10}$$

MF represents the feasibility problem of MO with each of the objectives evaluated. $F$is a $m$dimensional vector whose jth element corresponds to the value of the jth objective function.

$$\textbf{MF} \quad \begin{array}{ll} \textit{Solve} & G(\mathbf{x}) \\ & F_j = f_j(\mathbf{x}); \quad j = 1, 2, \ldots, m \end{array}. \tag{11}$$

This strategy consists of two distinct steps: (i) the determination of pareto optimal points, and (ii) the determination of realizations (pareto optimal solutions). The first step involves the conversion of the optimization problem to a feasibility problem (MO to MF) and determining a single feasible solution. Further exploration of the search space is performed with the additional criterion that every subsequent solution should be better than the current solution in at least one of the m objectives. Mathematically, this criterion can be represented by the following Type I cut

$$\textbf{Type I} \qquad \bigvee_{j=1,2,\ldots,m} (F_j \lhd F_j^t), \tag{12}$$

where F_j^t represents the jth objective function corresponding to the tth feasible solution, $\vee$ represents the commonly used OR operation in Boolean logic and $\lhd$ indicates the dominant operator. The successive addition of

Type I cut is continued till either the problem becomes infeasible or the search space is completely explored. The set of pareto points can then be determined by a simple post-optimal sorting of these Type I cuts. The second step of this strategy involves the construction of a single Type II cut corresponding to each Type I cut as

$$\textbf{Type II} \qquad \left\{ \bigwedge_{j=1,2,\ldots,m} (F_j = F_j^i) \right\} \vee \left\{ \bigvee_{j=1,2,\ldots,m} (F_j \lhd F_j^i) \right\}. \quad (13)$$

The T Type II cuts (one for each Type I cut) are subsequently added to the feasibility problem and all solutions of this modified feasibility problems are determined. This solution set directly corresponds (does not require any sorting) to all the pareto optimal points along with all possible realizations for each of the pareto point. The two step strategy to determine globally optimal pareto fronts/pareto solutions is summarized in Fig. 4.

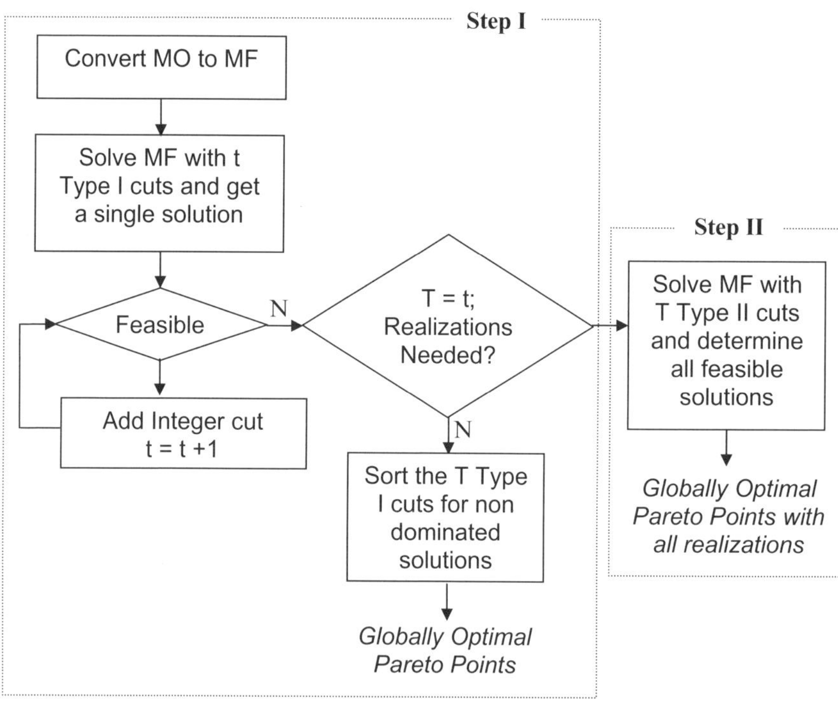

Figure 4. Two step strategy to determine globally optimal pareto solutions.

2.6. Comparison of CP and GA

In this section, we compare and contrast both the CP and GA techniques which have originated from different backgrounds. CP was developed by the Computer Science & Artificial Intelligence community whereas the stochastic GA technique finds its roots in the theory of evolution. From an application point of view, GA has found relatively more applications in Chemical Engineering than CP whereas CP has found considerable applications in scheduling problems.

Similarities between CP and GA: The following are some of the similarities between GA and CP:

(i) Both these techniques fall under the paradigm of AI. These techniques are independent of the geometry of the problem and do not make use of any gradient information.
(ii) Both these techniques can be used for solving both single objective optimization and multi-objective optimization to determine the pareto-optimal solutions (along with realizations) in a single run.
(iii) Both these techniques offer high modeling flexibility and do not restrict the designer to specify constraints in the form of $G(x) \leq 0$. Neither do these techniques require the decision variables to be non negative.
(iv) Both these techniques can determine multiple realizations and K-best solutions without the use of any integer cuts and without re-solving the problem.
(v) Both these techniques can prove to be computationally intensive as these do not have a bound on the objective function due to the absence of relaxations. This is evident from the fact that these techniques very often continue the exploration of the search space to prove the optimality (or satisfy the convergence criteria) despite attaining the optimality.

Differences between CP and GA: CP is a deterministic technique in nature and hence its results are easily reproducible whereas GA is a stochastic technique and the results are reproducible only if the same set of random numbers is generated. Other prominent differences between CP and GA can be categorized into three subdivisions and are summarized in Table 1. We would like to emphasize that the distinctions reported in this table are between the most commonly used implementations of CP and GA.

Table 1. Differences between CP and GA.

Sl. no.	Parameter	Constraint programming	Genetic algorithm
I. Modeling			
1	Handling of constraints	CP predominantly relies on the constraints to reduce the domain of the variables and determine the optimal solutions	GA predominantly relies on the fitness function and ignores the constraints during its principal operations such as crossover and mutation thereby possibly generating large number of infeasible solutions
2	Search space	The search space is predominantly decided by the set of constraints and is intelligently explored. As the constraints are used for domain reduction, defining a wider domain for variables domain is acceptable as redundant values can get eliminated	The search space is defined by the domain of variables with no role for constraints. Hence the domain of variables has to be chosen appropriately failing which many infeasible solutions can get generated (These infeasible solutions have an inferior fitness value and may get eliminated at a later stage). In recent implementations of GA, this drawback has been overcome by using the repair operator (Michalewicz, 1994; Duda, 2005)
3	Replicates	Works with a single solution and no replicates (solutions identical both in the objective function and decision variables) are ever created/re-visited	Works with a population of solutions and replicates may get created
4	Understanding, ease of implementation	Well understood technique but constraint propagation techniques need to be effectively implemented	Though some attempts have been made, GAs are not yet mathematically well understood. Nevertheless, operations like crossover and mutation are relatively simple to implement

(*Continued*)

Table 1. (*Continued*)

Sl. no.	Parameter	Constraint programming	Genetic algorithm
5	Modeling capabilities	Superior modeling power than the traditional optimization techniques but nevertheless requires an explicit optimization framework	GA can handle constructive constraints and does not require the constraints to be written in terms of explicit equations thereby making it suitable for a wide range of optimization problems
II. Computational Performance			
1	Progression of objective function	The objective function progressively improves throughout the search	Though NSGA-II is designed to progressively improve, it can nevertheless lose globally optimal pareto points due to the distance crowding operator (in cases where the size of population is less than the number of solutions in the best front obtained from the pooling of parents and offspring)
2	Parameter sensitivity	Since the search space of CP is primarily decided by constraints, a change in the parameters involved in the constraints can vary the search space of the problem. This can lead to variation in time required by CP to attain optimality. Hence, CP may be viewed as a technique susceptible to parametric changes	The search space of GA is determined by the domain of the decision variables as specified by the user. This domain may not change due to small variations in the parameters. Hence, the time required by GA to explore a given fraction of search space (as specified by choice of number of generations and population size) may be less sensitive to parametric changes
3	Number of runs required	Single run enables the determination of globally optimal solutions (both for multiple and single objective optimization problems)	Multiple runs are required (in terms of different set of random numbers) to verify/improve optimality

(*Continued*)

Table 1. (*Continued*)

Sl. no.	Parameter	Constraint programming	Genetic algorithm
III. Optimality			
1	User defined parameters	No user defined parameters (the choice points can be arbitrarily selected with no effect on optimality but may affect the time required for optimality)	A considerable number of user defined parameters like size of population, number of generations, mutation probability, crossover probability need to be carefully selected for reaching the optimal solution.
2	Global optimality	Guaranteed optimality for linear and non linear IP problems, irrespective of convexity	No guarantee on the optimality for even well structured problems
3	Realizations	All realizations in single and multi objective optimization can be guaranteed to be obtained	No guarantee of determining all realizations in either single or multi objective optimization. The maximum number of realizations that can be obtained is restricted by the population size
4	Stopping criterion	Well defined stopping criterion i.e. exploration of the complete search space	Arbitrary stopping criteria (such as fixed number of generations or fixed amount of time)

3. Case Studies

In this section, we evaluate both CP and GA approaches using two case studies taken from the Process Systems Engineering literature. The first case study is a single objective optimization problem involving scheduling of tasks in a job shop while the second case involves the design of sensor networks with multiple objectives. Both the techniques are implemented on a 3.20 GHz Pentium IV CPU with 1 GB RAM and using Windows XP Operating System. The CP solver used is ILOG SOLVER 6.1 and ILOG SCHEDULER 6.1 with ILOG OPL Studio (Van Hentenryck, 2002; ILOG, 2003) as the modeling language. The GA is implemented using an in house developed elitist code on MATLAB 7.5. Unless otherwise mentioned, the following GA parameters have been used: The crossover probability is set to 0.9. The mutation probability is set to $1/n_d$, where n_d is the actual number of decision variables. The population size is fixed at $20n_d$. The random numbers are generated using the *rand* function in MATLAB with the seed value set to zero.

3.1. Single objective optimization: Job shop scheduling

In this section, we will be using a job shop scheduling (Jain and Grossmann, 2001) combinatorial optimization problem for comparing CP and GA regarding their capability to (i) efficiently model a problem, (ii) determine the globally optimal solution, and (ii) determine all the multiple optimal solutions.

3.1.1. Problem definition

This problem involves the processing of a given set of orders (I) on a set of dissimilar parallel machines (M). For each order, a release date and due date is specified. The processing of an order i on any machine m can start only after its release date and has to be completed on or before its due date. Each pair of order and machine has a different cost (c_{im}) and processing time (p_{im}) associated with them. It should be noted that more than one order can be processed on a machine. The two types of decision variables in this problem are (i) assignment of an order to a machine, and (ii) the start time of each order on the assigned machine subject to the following three constraints: (i) the processing of an order cannot start before its release date, (ii) the processing of the order should end before the due date, and (iii) if more than one order is processed on a single machine, then the time of processing of any two orders should not overlap i.e. a second order cannot start before the completion of the first order. The exact number of variables and constraints are dependent on the number of orders and machines in a given problem.

3.1.2. CP formulation

This problem has been addressed in literature (Jain and Grossmann, 2001) and four different explicit optimization formulations, namely (i) MILP formulation, (ii) CP formulation, (iii) combined MILP-CP formulation and, (iv) MILP/CP hybrid model have been proposed. The discrete time CP scheduling model presented here was proposed in literature (Jain and Grossmann, 2001) and is based on the ILOG's OPL modeling language which has a wide range of constructs for scheduling problems. These constructs make the modeling task easier leading to compact models and also have efficient constraint propagation schemes to reduce the computational effort. ILOG OPL considers the set of orders to be processed as activities that have to be completed and the set of machines available to process these orders as resources. Each activity is associated with a start time, processing time, and end time. The CP model is as given in F1:

F1

$$\min \sum_{i \in I} C_{iz_i}, \tag{14}$$

$$\text{s.t} \quad i.start \geq r_i \quad \forall i \in I, \tag{15}$$

$$i.start \leq d_i - p_{z_i}, \tag{16}$$

$$i.duration = p_{z_i} \quad \forall i \in I, \tag{17}$$

$$i\, requires T \quad \forall i \in I, \tag{18}$$

$$activityHasSelected\,\mathrm{Re}\,source(i, T, t_m) \Leftrightarrow z_i = m \; \forall i \in I, m \in M, \tag{19}$$

$$z_i \in M \quad \forall i \in I, \tag{20}$$

$$i.start \in \Im \quad i \in I, \tag{21}$$

$$i.duration \in \Im \quad i \in I. \tag{22}$$

The machine selected to process order i is represented by the variable subscript z_i. Equation (14) is the objective function to determine the cost of a feasible schedule. The start time of the order is given by "*i.start*" and constraint (15) ensures that the start time of any order is not earlier than its release time. Constraint (16) ensures that the order is completed before its corresponding due date. The term p_{z_i} is machine dependent and corresponds to the processing time of order i on the selected machine m. The appropriate value of p_{z_i} is determined using constraint (17). ILOG OPL provides the definition of a special type of resource known as "*unary resource*".

The uniqueness of this type of resource is that they can be used by only one activity at any given point of time. This is identical to the problem at hand and hence the machines are defined as "*unary resources*" represented by T. Constraint (18) enforces that any order i requires a machine/resource from the set T. In addition to the assignment of an order to a specific machine, this constraint also makes sure that the orders on the same machine are sequenced appropriately. The construct "*activityHas-SelectedResource*( )" in constraint (19) is a boolean construct used to assign the machine processing order i to the variable z_i. The variable z_i is assigned the value of the machine m if the machine t_m processes the order i. Constraints (20)–(22) define the domains of the variables z_i, the start time and the durations of the orders respectively. Additional details on various constructs available in ILOG OPL Studio can be obtained from the cited literature. We now discuss some issues related with the GA implementation before presenting the results.

3.1.3. GA implementation

In this chapter, we have made use of an elitist GA. The set of decision variables and constraints is as specified in the earlier section on problem definition.

Decision variables: The two types of decision variables are (i) the machine processing order $i\,(x_{im})$, and (ii) the start time of the order $i\,(ts_i)$. As mentioned earlier, specifying appropriate domains in GA is very critical to its performance. The range of x_{im} is the integer values between 1 and the total number of machines m whereas ts_i can take any value between the release and the due date of order i.

Constraints: In addition to the constraints stated earlier, we have also included the following constraint as it was observed to significantly help GA achieve better solutions

$$\sum_{i\in I} p_{im}x_{im} \leq \max_i\{d_i\} - \min_i\{r_i\} \quad \forall m \in M. \tag{23}$$

This constraint ensures that the total processing time for all the orders assigned to a machine is less than the difference of the latest due date and the earliest release date of orders on that machine. This constraint is used

(Jain and Grossmann, 2001) to tighten the LP relaxation of the MILP model. To aid in the determination of appropriate penalty for constraint violation, all the three types of constraints are normalized by transforming them to the form (Deb, 2001)

$$g(x) \geq 0. \tag{24}$$

Then the penalty associated with a constraint c, w_c is calculated as follows

$$w_c = \begin{cases} |g(x)|, & \textit{if } g(x) < 0 \\ 0, & \text{otherwise} \end{cases}. \tag{25}$$

The above equation ensures that the penalty is zero for all feasible solutions giving them better fitness value and higher rate of survival.

Fitness function: The fitness function is used while deciding the candidate solutions for the next generation. For a feasible solution, the fitness function is given as

$$f_{feasible} = \sum_{i \in I} c_{im} x_{im}, \tag{26}$$

where c_{im} is the cost associated with processing order i on machine m. However, an infeasible solution has to be penalized and thus the fitness function is given as,

$$f_{infeasible} = \sum_{i \in I} c_{im} x_{im} + \sum_{c \in C} w_c + \sum_{i \in I} \max_m \{C_{im}\}. \tag{27}$$

The first term on the RHS of (27) corresponds to the cost of processing all the orders on the selected machines. The second term corresponds to the penalty associated with each of the constraint while the last term is an upper bound on the cost. The term $\max_m\{C_{im}\}$ denotes the cost of processing an order i on a machine with the maximum cost. The third term is necessary to ensure that an infeasible solution does not have a better fitness function value than a feasible solution.

GA parameters: The number of generations in GA is restricted to 500 with the rest of parameters as mentioned earlier. We now present the comparative results for the two techniques.

3.1.4. Data

Jain and Grossmann (Jain and Grossmann, 2001) considered 5 different problems of varying sizes (as determined by the number of orders and machines). For each problem size, two sets of data were considered to demonstrate that the difficulty of solving a scheduling problem can vary significantly with data. The processing times in the first data set are longer and have fewer feasible schedules and the total cost of processing all the tasks is higher. All the required parameters for these 10 instances (5 problems with 2 data sets each) can be obtained from the online supplement available with (Jain and Grossmann, 2001). However, we will restrict ourselves to only 6 instances by solving problems of three different sizes. In particular, we have used three problems labeled as Problem 1, Problem 4 and Problem 5 (Jain and Grossmann, 2001). To maintain consistency, we will also refer to these problems as Problem 1, Problem 4 and Problem 5. The dimensionality of these three problems is given in Table 2.

All the required parameters for these 6 instances are given in the Appendix I. In the rest of the chapter, we have used labels to refer to each of the six instances. For example, the label P1S1 indicates the first data set of Problem 1 while P4S2 indicates the second data set of Problem 4.

3.1.5. Results

Table 3 shows the comparison of the problem sizes for the six instances between CP and GA. It can be seen that as the problem size increases from P1 to P5, the number of variables and constraints increases significantly in the CP formulation as compared to that in GA. This shows the superior modeling power of GA which does not require an explicit optimization formulation as opposed to CP. However, it should be noted that a compact

Table 2. Details of orders and machines for the three problems.

Problem	Number of orders	Number of machines
Problem 1	3	2
Problem 4	15	5
Problem 5	20	5

Table 3. Problem statistics for six different instances.

Problem label	Problem Statistics: Number of orders (n)	Problem Statistics: Number of machines (m)	Problem[$]	Constraint Programming*: Number of variables	Constraint Programming*: Number of constraints	Genetic Algorithm (Number of generations = 500): Number of variables ($n_d = 2n$)	Number of constraints[†] ($n + 2m$)	Population size ($20n_d$)	Size of search space ($m^n \prod_{i \in I} (d_i - r_i)$)
P1S1	3	2	Problem 1 Set 1	15	21	6	7	120	19040
P1S2	3	2	Problem 1 Set 2	15	21	6	7	120	19040
P4S1	15	5	Problem 4 Set 1	120	105	30	25	600	$4.675 * 10^{31}$
P4S2	15	5	Problem 4 Set 2	120	105	30	25	600	$4.675 * 10^{31}$
P5S1	20	5	Problem 5 Set 1	160	140	40	30	800	$3.843 * 10^{42}$
P5S2	20	5	Problem 5 Set 2	160	140	40	30	800	$3.843 * 10^{42}$

[$] As classified in (Jain and Grossmann, 2001).
[*] As indicated by ILOG Scheduler 6.1 and ILOG Solver 6.1.
[†] Bound constraints on the individual variable are not counted. Also, the constraint that the start time of any order cannot be less than the release date is not included because this constraint is always satisfied as the start time is always selected between the release date and the due date.

model representation need not always translate to a reduced computational time. The interested reader can also refer to literature (Jain and Grossmann, 2001) wherein the problem sizes for a MILP formulation are given. The size of the MILP increases even more drastically than CP. As is obvious, there is no difference in the problem size or search space between the two instances of a particular problem. From Table 3, it can also be seen that whereas the search space for GA increases exponentially from Problem 1 to Problem 5, the population size has not been increased proportionately. Hence the ratio of number of solutions evaluated (calculated as number of generations × size of population) to the size of search space in GA decreases from P1S1 to P5S2.

Table 4 shows the computational performance of both CP and GA for determining the globally optimal solution along with the number of multiple optimal solutions. For Problem 1, it can be seen that both GA and CP are able to determine the globally optimal solutions in a small amount of time. For P1S1, both GA and CP are able to determine all the three multiple optimal solutions. However, GA is able to determine only two of the three globally optimal solutions for P1S2.

For P4S1, CP is able to determine the globally optimal solution of 115 in less than 111 seconds whereas GA is not able to determine the globally optimal solution even after the completion of 500 generations (102.5 seconds). The best solution reported by GA is 123. Similar results are also observed for problem P4S2. However, the time taken for CP to determine the globally optimal solution is significantly less for this instance of the data set (P4S2). This shows the susceptibility of CP to the change in the data of a problem. It can also be seen that these two instances of Problem 4 are characterized by a large number of realizations and CP is able to determine them in a reasonable amount of time. Due to a large number of realizations, we have restricted ourselves to determining only 10,000 globally optimal solutions. However for P4S1, GA is not able to determine more than one solution (best) and for P4S2 it is able to determine only 22 solutions with an objective of 109 in 500 generations. As mentioned earlier during CP discussion, the second step in CP (i.e. the solution of the CSP while determining multiple optimal solutions) is solved efficiently as 10,000 solutions are discovered in a reasonable time. This case study also shows the possibility of GA not being able to determine all realizations with a fixed population size.

Table 4. Computational performance for the six different instances.

Problem label	Constraint Programming[a]				Genetic Algorithm[b]				
	Time	Objective	Time for realizations[#]	No. of realizations	Time for optimal solution	Objective	Time for best sol	No. of realizations (multiple best sols)	Time for 500 generations
P1S1	~0	26	~0	3	0.4	26	NA	3	12.76
P1S2	~0	18	~0	3	0.62	18	NA	2	13.62
P4S1	110.52	115	10.06	10,000[m]	NA	123[s]	1.22	1	102.50
P4S2	3.8	102	15.32	10,000[m]	NA	109[s]	64.25	22	80.14
P5S1	103.47	169[s]	104.64	100[m]	NA	378[i]	53.99	4	138.45
P5S2	535.5	140	99.67	100[m]	NA	157[s]	1.51	8	118.60

[a] indicates the time required on ILOG CP Solver 6.1 and Scheduler 6.1.
[b] indicates the time required by a elitist GA on MATLAB 7.5.
[#] indicates the time taken by CP to determine the multiple optimal solutions (only Step II and does not include the time for determining the optimality in Step I)
[m] indicates the time reported to determine the specified number of solutions.
[s] Suboptimal Solution.
[i] Infeasible Solution; NA indicates that GA has not determined the globally optimum solution and hence the notion of time for globally optimum solution does not exist.

For example, if we assume that GA is able to determine the globally optimal solution, it still cannot determine all the 10,000 realizations because the population size is restricted to 600. A higher population size increases the possibility of determining many more realizations but increases the computational burden.

For P5S1, the optimal solution has been reported (Jain and Grossmann, 2001) as 158. This solution was determined by using a hybrid approach as a standalone CP was not able to determine the optimal solution in 19 hrs (using ILOG OPL Studio 2.1, ILOG Scheduler 4.4 and ILOG Solver 4.4 single processor versions on a dual processor SUN Ultra 60 workstation, refer (Jain and Grossmann, 2001)). Hence, we decided not to determine the globally optimal solution. As can be seen from Table 4, a solution of 169 is obtained in less than 104 seconds. While executing CP, it was found that for further improvement to 168, CP required 1449.11 seconds. In contrast to CP, GA is not able to determine even a single feasible solution in 500 generations. The best objective reached is 378 and this corresponds to an infeasible solution because the maximum possible cost for a feasible solution is $209\left(=\sum_{i \in I} \max_m \{C_{im}\}\right)$. To determine the effect of the number of generations, GA was executed for 10,000 generations. However, no improvement in the objective was observed indicating that even a single feasible solution has not been determined in 10,000 generations. For P5S2, CP determines the global optimal solution of 140 in 535.5 seconds whereas GA converges to a suboptimal solution of 157 in around 1.51 seconds. However, CP was seen to achieve the same objective function value of 157 in 1.72 seconds. As in Problem 4, this problem also has a large number of realizations and due to the complexity of the problem, we have reported the time taken to determine only 100 realizations. For P5S2, GA is able to determine only 8 solutions with a suboptimal cost value whereas CP is able to determine 100 solutions which are globally optimal.

The reasons for the inability of GA to determine the globally optimal solution in four of these six instances can possibly be that the search space is very large for these four instances and GA explores only a small portion of this search space. Moreover, the set of constraints is not used in exploring the search space. In contrast, CP explores the complete search space intelligently with the help of constraints and is able to determine the globally optimal solutions along with their realizations. Thus, we see that despite the larger problem sizes due to explicit modeling, CP is able to efficiently determine the globally optimal solutions compared to GA.

We next discuss the performance of GA for the above six instances of the problem. In particular, we show the improvement in the objective function value and the number of multiple solutions that are determined in each generation. We also show the number of feasible solutions in each generation along with the time required for each generation.

Figure 5 shows the improvement in the objective function for GA as the evolution progresses for all these three problems and their instances. Due to the elitist nature, the objective function is bound to improve (decrease in this case) monotonically. From Fig. 5, it can be seen that GA converges for all the six problems in less than 200 generations.

For P1S1, GA determines the globally optimal solution (26) in the first generation whereas for P1S2 it determines the globally optimal solution (18) in the 9th generation. For P4S1, GA converges to a suboptimal solution of

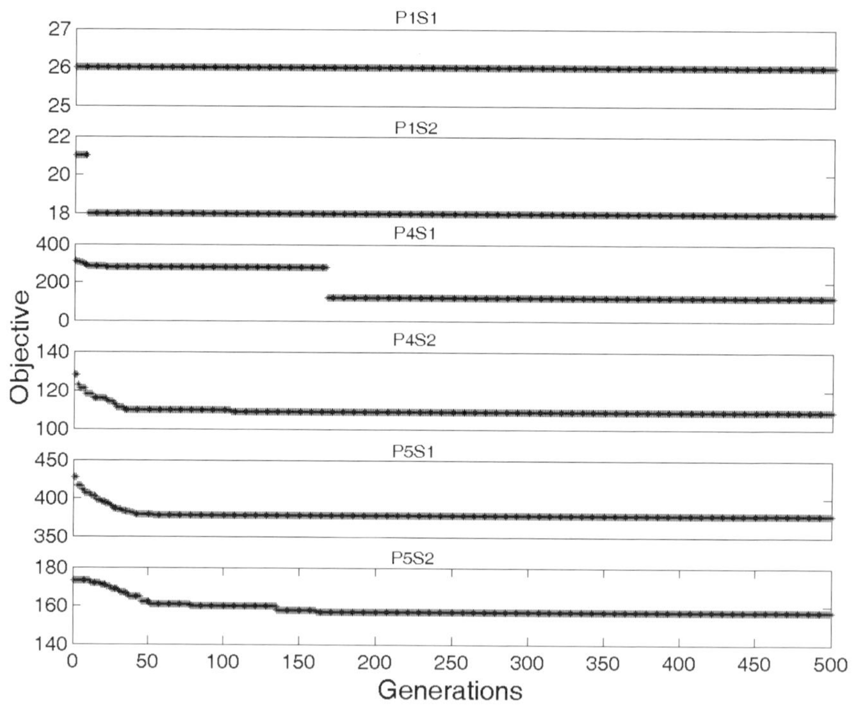

Figure 5. Improvement in the objective function with generations in GA.

123 in the 168th generation while for P4S2, GA converges to a suboptimal solution of 109 in the 105th generation. For P5S1, GA converges to a solution with an objective of 378 which is an infeasible solution as discussed earlier. This can also be seen from Fig. 6 which shows the total number of feasible solutions in a generation. For P5S2, GA converges to a suboptimal solution of 157 in the 162nd generation.

Figure 6 shows the number of feasible solution in each generation. It is obvious that the number of feasible solutions cannot be greater than the population size in a particular generation. The number of feasible solutions should increase monotonically with every generation because of the elitist nature. It is to be noted that due to the stochastic nature of GA, the number of feasible solutions need not remain constant even after the determination of the optimal solution because an infeasible (or inferior) solution from the set of population can be replaced by a feasible (better) solution.

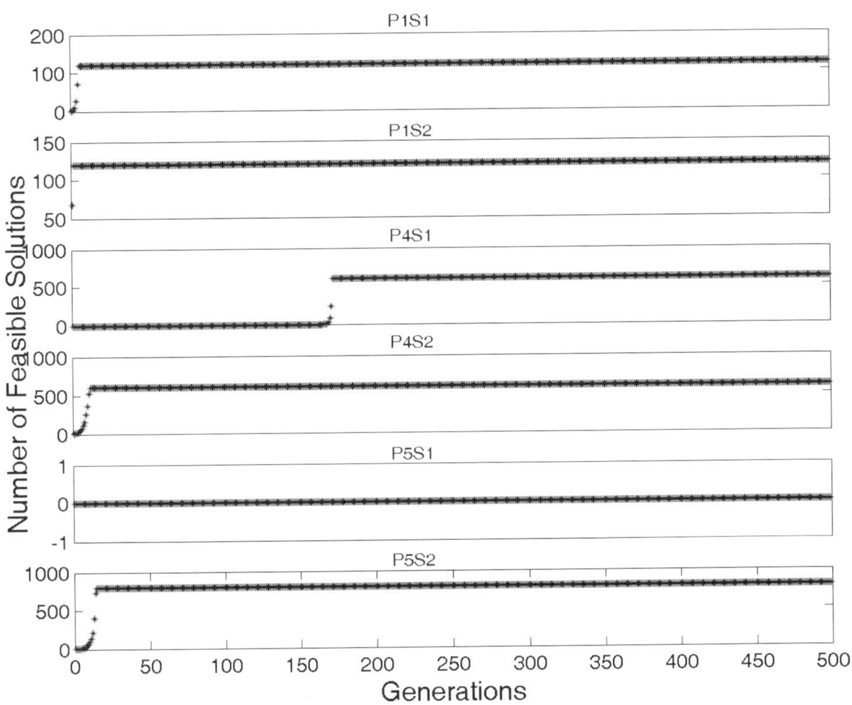

Figure 6. Number of feasible solutions in each generation in GA.

For Problem 1, the population size is 120 and an equal number of feasible solutions are obtained in 6th and 2nd generations for P1S1 and P1S2 respectively. It takes 6 generations for P4S1 to determine feasible solution equaling the size of population (600) whereas 12 generations are required for P4S2. For P5S1, as expected the number of feasible solutions is always zero whereas it takes 15 generations to determine 800 feasible solutions for P5S2. However, many of these solutions may be replicates of one another.

Figure 7 shows the number of current best solutions in each generation. The number of realizations (of the current best solution) need not increase monotonically because there can be an improvement in the objective function which may cause a decrease in the number of realizations. However, the number of multiple optimal solutions can only monotonically increase after the globally optimal solution (or the best solution) is reached because of the elitist nature. In all the cases, GA appears to have converged from the

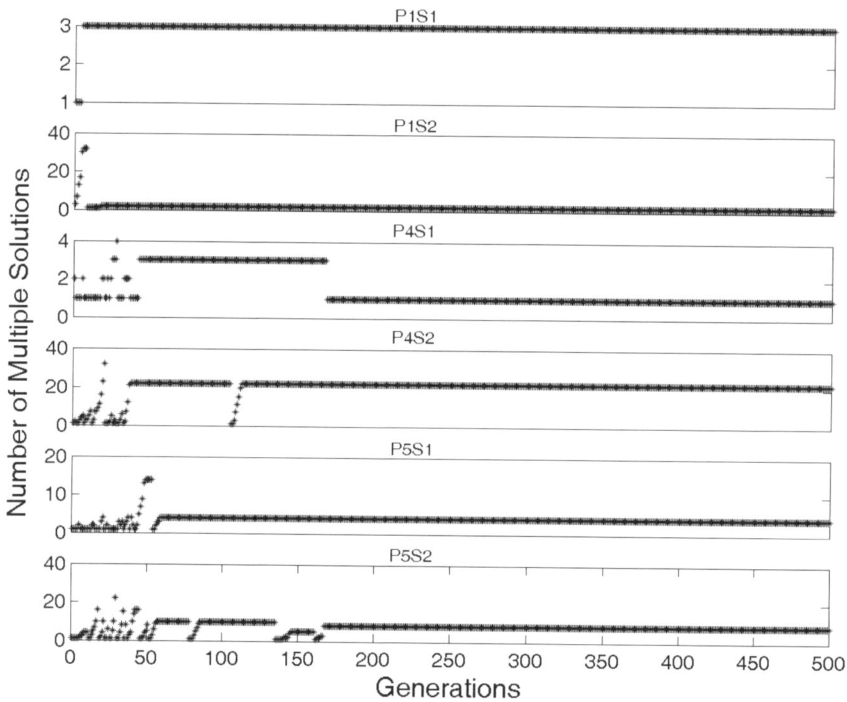

Figure 7. Number of realizations (best) in each generation in GA.

perspective of multiple best solutions. The number of generations required for convergence from the perspective of multiple optimal solutions will never be less than the number of generations for convergence from the perspective of the objective function value. This is because, GA is a stochastic search strategy and all the realizations need not be obtained in the same generation in which the best solution is determined.

All the globally optimal realizations (3) are obtained for P1S1. For P1S2, the number of realizations initially increases and suddenly decreases due to an improvement in the objective function (as can be seen in Fig. 5). However, only 2 of the 3 globally optimal realizations are determined using GA. Similarly for P4S1, the number of realizations generally increases in the beginning and then decreases due to an improvement in the objective function. Only one solution is obtained with the best objective value (123). For P4S2, a total of 22 realizations are obtained for an objective of 109. For P5S1, a total of 4 solutions are reported by GA with an objective of 378. However, all of them are infeasible (also seen in Fig. 6) because the objective value is greater than the maximum possible value of 209 for a feasible solution. For P5S2, 8 solutions are obtained with an objective value of 157 in the 168th generation.

The computational time required for each generation is shown in Fig. 8. As expected, the time for P1S1 (and P1S2) is less than P4S1 (and P4S2). This is because these problems have a larger population size (600 against 120). Similar, observations can be made for P5S1 and P5S2.

As mentioned earlier, the stochastic nature of GA makes it very susceptible to the selection of random numbers. All the six instances of the problem were implemented with 100 different sets of random number (generated by initializing the seed in the *rand* function from 1 to 100 in MATLAB). The objective function at the end of the 500 generations for every seed is shown in Fig. 9 and the number of realizations at the end of the 500 generations is shown in Fig. 10. As can be seen from Fig. 9, the objective function for P1S1 always remains at 26. Except for P1S1, the best objective function for all the other problems does not remain constant. Table 5 shows various statistics including the best possible solution for the six instances.

From Table 5 (and Fig. 9), it is interesting to note that GA is able to determine feasible schedules with an objective of 165 for P5S1. It may be recalled that GA was not able to determine even a single feasible solution

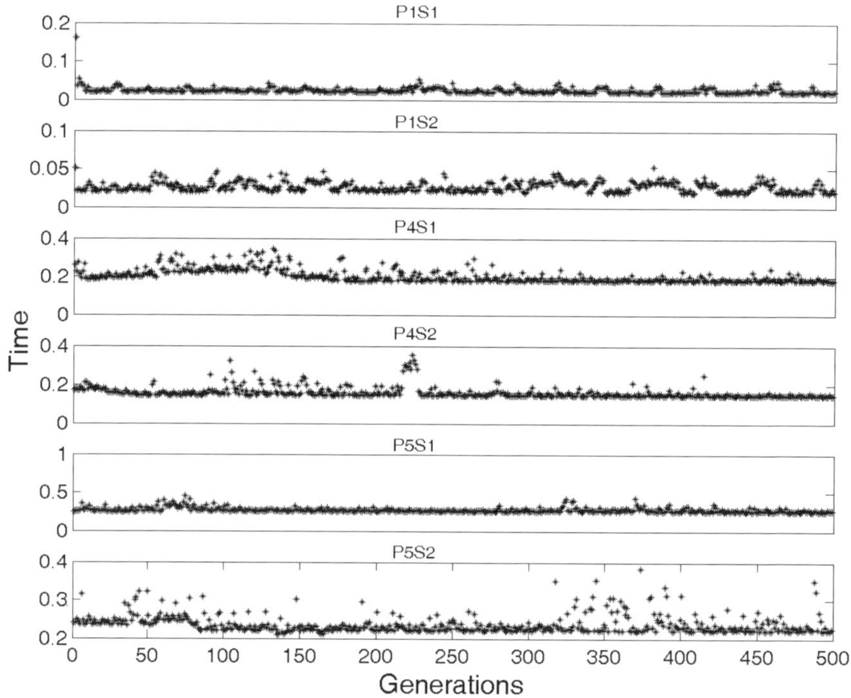

Figure 8. Time required for each generation.

when the random numbers were generated with the seed set to zero. On the other hand, for some values of seed, GA is not able to determine even a single feasible solution for Problem P4S1 (as the maximum value is 283 while the maximum possible cost for a feasible solution is 154).

3.2. Multi-objective optimization

In this section, we present a multi-objective optimization problem taken from the PSE literature and compare the performance of CP and GA on it. We only provide preliminary details about the problem and a complete description can be obtained from (Bhushan *et al.*, 2008).

3.2.1. Problem definition

This design problem involves the selection of the variables (along with the number of sensors for each of the selected variable) to be measured

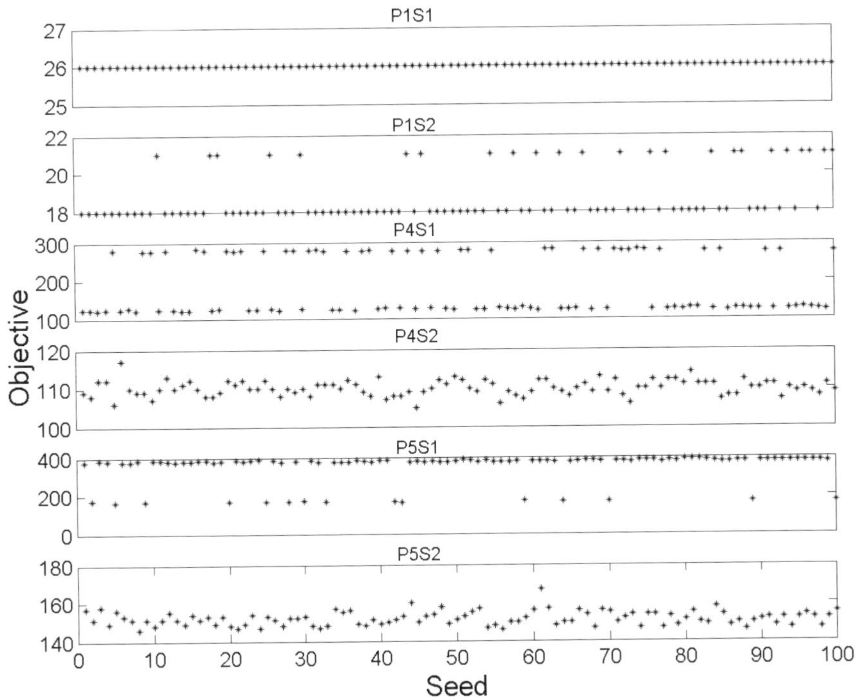

Figure 9. Objective function value for various sets of random number.

in the presence of uncertainties so as to aid in efficient fault diagnosis. These uncertainties can be in either the fault occurrence or sensor failure probabilities or in both these data. An explicit MILP formulation with four different objectives namely (i) minimization of nominal system unreliability of detection (U), (ii) maximization of robustness to the uncertainties in the probability data (ϕ), (iii) maximization of the network distribution (N), and (iv) minimization of the sensor network cost (or the maximization of the cost saving (x_s)) has been proposed in literature (Bhushan *et al.*, 2008). However, these objectives were considered in various lexicographic orderings and hence the trade-offs between them were not fully evaluated.

3.2.2. CP formulation

This problem was reformulated into an explicit CP model (Kotecha *et al.*, 2007) using the expressive power of CP leading to a significant reduction

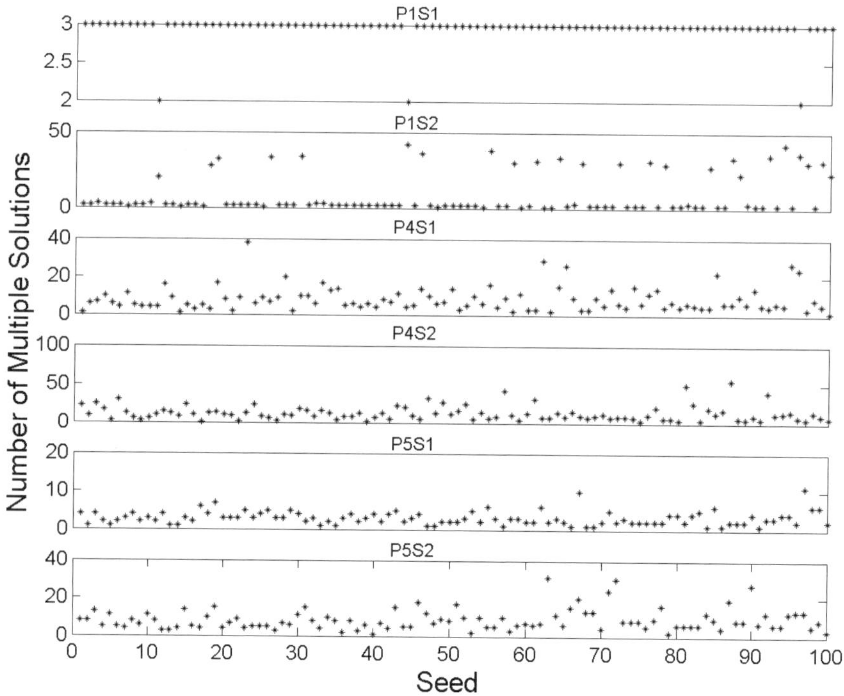

Figure 10. Number of realizations for various sets of random number.

Table 5. Statistics on the six instances with various sets of random number.

Problem tag	Mean	Standard deviation	Global optima	Minimum (Best) solution among the 100 seeds	Maximum (Worst) solution among the 100 seeds
P1S1	26	0	26	26	26
P1S2	18.72	1.2812	18	18	21
P4S1	185.67	75.7262	115	120	283
P4S2	109.96	1.9387	102	105	117
P5S1	349.26	75.4334	158	165	389
P5S2	151.63	3.5990	140	146	167

in the dimensionality of the optimization problem. In this chapter, we will be using the explicit CP based optimization formulation (reproduced in F2) used in literature (Kotecha *et al.*, 2008d). To be consistent with literature and without loss of generality, we will fix the value of one of the objectives (minimization of nominal system unreliability of detection (U)) to its optimal value and determine the tradeoffs between the other three objectives.

$$\begin{aligned} \text{Max} \quad & [\phi, N, x_s] \\ & U = U^{Optimal'} \end{aligned} \tag{28}$$

$$\text{s.t} \quad \sum_{j=1}^{n} c_j x_j + x_s = C^*, \tag{29}$$

$$U = \log_{10}(f_i) + \sum_{j=1}^{n} b_{ij} x_j \log_{10}(s_j) + \phi_i, \quad \forall i \in (I_f \cup I_s), \tag{30}$$

$$U_i = \log_{10}(f_i) + \sum_{j=1}^{n} b_{ij} x_j \log_{10}(s_j), \quad \forall i \in I \backslash (I_f \cup I_s), \tag{31}$$

$$U = \max_{\forall i \in I}(U_i), \tag{32}$$

F2

$$\phi_i^* = \phi_{fi}^*, \quad \forall i \in (I_f \backslash I_s), \tag{33}$$

$$\phi_i^* = \phi_{si}^*, \quad \forall i \in (I_s \backslash I_f), \tag{34}$$

$$\phi_i^* = \phi_{fi}^* + \phi_{si}^*, \quad \forall i \in (I_s \cap I_f), \tag{35}$$

$$\left.\begin{aligned} \phi_i < \phi_i^* \Rightarrow \tilde{\phi}_i = \phi_i \\ \phi_i \geq \phi_i^* \Rightarrow \tilde{\phi}_i = \phi^* \end{aligned}\right\}, \quad \forall i \in (I_s \cup I_f), \tag{36}$$

$$\phi = \min_{\forall i \in (I_f \cup I_s)} (\tilde{\phi}_i), \tag{37}$$

$$\phi_{si}^* = -\sum_{j \in J_s} B_{ij} (\log_{10} s_j) x_j, \quad \forall i \in I_s, \tag{38}$$

$$N = \sum_{j=1}^{n} \min(x_j, 1), \tag{39}$$

$$(U_i, U) \in \mathbb{Z}^-; \; (x_j, x_s, \phi, \phi_i^*, \phi_i, \phi_{si}^*, N) \in \mathbb{Z}^+, \tag{40}$$

where I is the set of all faults, I_f is the set of faults with uncertain occurrence probabilities, and I_s is the set of faults that affect variables which can only be measured by sensors with uncertain failure probabilities. Note that the ith fault is characterized as uncertain if (i) its fault occurrence probability is not known accurately ($i \in I_f$), (ii) failure probability of any sensor available for measuring the variables affected by that fault (the corresponding b_{ij} is 1; b_{ij} is the (i, j)th element of the cause-effect matrix B) is uncertain ($i \in I_s$), or (iii) both i and ii hold ($i \in (I_s \cap I_f)$). In Eq. (30), ϕ_i is the slack in the detection unreliability constraint for the ith uncertain fault. A nonzero ϕ_i ensures that the system detection unreliability is robust to changes in the data used to estimate the detection unreliability of the ith uncertain fault. ϕ_i^* is the corresponding value of the slack that is necessary to ensure complete robustness of the system detection unreliability to the uncertain data used to calculate the detection unreliability of fault i, and is required because the upper values of the probabilities are bounded by 1. Depending on the type of uncertainty in the faults (cases i, ii, or iii), Eqs. (33)–(35) can be used to determine the value of ϕ_i^*. The objective ϕ represents the network's ability to tolerate uncertainty in the probability data without affecting the nominal system detection unreliability and is defined as the minimum of all those ϕ_i which are below their corresponding ϕ_i^* values. Equation (36) ensures that if ϕ_i is less than ϕ_i^*, then the maximum meaningful value of the slack ($\tilde{\phi}_i$) is ϕ_i. On the other hand, if $\phi_i \geq \phi_i^*$ then the maximum meaningful value of the slack is ϕ^*. The values of ϕ^* and ϕ_{fi}^* can be determined *a priori* (Bhushan *et al.*, 2008) by the following equations

$$\phi^* = \max \left\{ \begin{array}{l} \max\limits_{i \in I_f \setminus I_s} (\phi_{fi}^*),\ \max\limits_{i \in I_s \setminus I_f} \left(-\sum\limits_{j \in J_s} B_{ij} (\log_{10} s_j) x_j^* \right), \\ \max\limits_{i \in I_s \cap I_f} \left(\phi_{fi}^* - \sum\limits_{j \in J_s} B_{ij} (\log_{10} s_j) x_j^* \right) \end{array} \right\}, \tag{41}$$

$$\phi_{fi}^* = -\log_{10} f_i, \quad \forall i \in I_f. \tag{42}$$

In the above equation, x_j^* is an upper bound on x_j and corresponds to the maximum number of sensors that can be installed on the jth variable on the basis of available cost and is given by $\lfloor C^*/c_j \rfloor$ (where $\lfloor x \rfloor$ indicates rounding off x to the nearest integer not greater than x). In F2, while ϕ denotes the robustness to uncertainty in the probability data, N is the network distribution and is a measure of robustness to modeling errors. Network distribution is defined as the total number of measured variables and is calculated as shown in Eq. (39). The variable x_s in the objective function represents the cost saving and is equal to the difference in the available and the used cost (Eq. (29)).

3.2.3. GA implementation

We have used the NSGA — II technique (Deb, 2001) to determine the pareto optimal fronts along with their realizations. The various decision variables and constraints are as discussed below

Decision Variables: From the formulation F2, it can be seen that a large number of decision variables such as $x_j, x_s, \phi, \phi_i^*, \phi_i, \phi_{si}^*, N$ are added to express the problem in an explicit optimization framework. However, the only decision variables necessary in this problem are the number of sensors that need to be deployed on each variable i.e. x_j. As mentioned earlier, all these variables are bounded between 0 and x_j^* and hence the search space for these problems is $\prod_{j=1}^{n} x_j^*$.

Constraints: The three types of constraints that define the problem are

(i) the cost constraint represented by (29),
(ii) the value of ϕ should be positive,
(iii) the set of constraints shown below (corresponding to Eqs. (31) and (32))

$$U \geq \log_{10}(f_i) + \sum_{j=1}^{n} B_{ij} x_j \log_{10}(s_j) + \phi_i, \quad \forall i \in (I_f \cup I_s). \tag{43}$$

The constraints are normalized as explained in the previous case study. Equations (30) are used to determine the values of ϕ_i whereas Eqs. (33)–(35) are used to determine ϕ_i^*; The values of ϕ_i are subsequently used to determine the meaningful slacks $\tilde{\phi}_i$ using Eq. (36) while Eq. (37) is used to evaluate the robustness (ϕ) of the system. The network distribution is calculated using Eq. (39).

GA parameters: The number of generations in GA is set to 2000 while the rest of the parameters are as specified earlier. As discussed earlier, CP is known to determine the globally optimal pareto front along with their realizations and hence the true pareto can be estimated *a priori*. Thus at any generation, the distance of the pareto front (Q) determined by GA to the true pareto front (P^*) (determined by CP) can be estimated using the generational distance (GD) (Deb, 2001) given as:

$$GD = \frac{\left(\sum_{i=1}^{|Q|} d_i^p\right)^{1/p}}{|Q|}. \tag{44}$$

For $p = 2$, the parameter d_i is the Euclidean norm (in the objective space) between the solution $i \in Q$ and the nearest member of the true pareto front

$$d_i = \min_{k=1}^{|P^*|} \sqrt{\sum_{m=1}^{M} (f_m^{(i)} - f_m^{*(k)})^2} \tag{45}$$

where $f_m^{*(k)}$ is the mth objective function value of the kth member of P^*. Ideally, due to the elitist nature of NSGA-II, the generational distance should monotonically decrease with the progress of the evolution.

3.2.4. Data

In this chapter, we will be using the Tennessee Eastman Problem for comparing the performance of CP and GA to solve multi-objective optimization problems. The TE flowsheet (shown in Appendix II) has been widely discussed in Process Systems Engineering literature as a benchmark problem. For the sake of brevity, further details of the TE problem are not presented here and the interested reader is referred to the literature (Downs and Vogel, 1993; Bhushan *et al.*, 2008). It has $50(= n_d)$ potential measurements and 33 faults (16 bidirectional and 1 unidirectional; given in Appendix II). The cost of the sensors, the sensor failure probability, and fault occurrence probability data are taken from literature (Kotecha *et al.*, 2007; Bhushan *et al.*, 2008). All these data have been reported in Appendix II. Similar to literature (Kotecha *et al.*, 2007; Bhushan *et al.*, 2008), we have considered that the sensor failure probabilities of sensors 3 and 4 and occurrence probabilities of faults 1 and 9, are not known accurately. We present the results for the case when the maximum available monetary resource (C^*) is 1,000 units (Kotecha *et al.*, 2007).

3.2.5. Results

In this section, we compare the performance of both CP and GA on various criteria for the above case study.

Model statistics: Table 6 shows the number of constraints and decision variables for both the CP and GA formulation. As expected, the GA formulation is more compact than the CP formulation and hence the number of variables and constraints are comparatively fewer.

Pareto points: It can be seen from Table 6 that CP is able to determine 17 globally optimal pareto points whereas GA is able to determine only 14 of the pareto points. However, all these 14 pareto points discovered by GA are part of the globally optimal pareto front. The details of the pareto points determined by CP and GA are given in Table 7. All these 14 globally optimal pareto points are obtained in the 489th generation. The three points that are not determined by GA are with the tag CO, CP and CQ (marked in bold).

Realizations: From Table 6, it can be seen that CP discovers all the possible 102 realizations on the global pareto front whereas GA is able to determine only 49 (at the end of 2,000 generations). The number of realizations of the three pareto points not discovered by GA is 16 and hence it can be seen that GA is able to determine 49 realizations out of 86 realizations for the remaining 14 pareto points. Details of the realizations associated with each pareto point are given in Table 7.

Table 6. Model statistics and computational performance of CP and GA.

Problem statistics								Time (secs)		
CP		GA		Pareto points		Pareto solutions			GA	
No. of vars	No. of cons	No. of vars	No. of cons	CP	GA	CP	GA	CP[a]	Time for best front[b]	Time[c]
81	37	50	14	17	14	102	49	10	3915.3	5175.6

[a] indicates the time required on ILOG CP Solver 6.1
[b] indicates the time required on MATLAB 7.5 for determining the best front i.e. time for 489 generations
[c] indicates the time required on MATLAB 7.5 for 2,000 generations

Table 7. Details of pareto points.

Constraint programming					Genetic algorithm				
	Pareto points					Pareto points			
Tag	ϕ	N	x_s	No. of realizations	Tag	ϕ	N	x_s	No. of realizations
CA	0	1	900	2	GA	0	1	900	1
CB	0	2	800	9	GB	0	2	800	5
CC	0	3	700	16	GC	0	3	700	6
CD	0	4	600	14	GD	0	4	600	6
CE	0	5	500	6	GE	0	5	500	4
CF	0	6	400	1	GF	0	6	400	1
CG	0	7	250	3	GG	0	7	250	1
CH	0	8	100	3	GH	0	8	100	1
CI	3	2	600	1	GI	3	2	600	1
CJ	3	3	500	5	GJ	3	3	500	4
CK	3	4	400	10	GK	3	4	400	6
CL	3	5	300	10	GL	3	5	300	7
CM	3	6	200	5	GM	3	6	200	5
CN	3	7	100	1	GN	3	7	100	1
CO	**6**	**2**	**200**	**1**					
CP	**6**	**3**	**100**	**5**					
CQ	**6**	**4**	**0**	**10**					

Computational time: In this work, CP and GA have been implemented on different platforms; hence exact comparison of their computation times cannot be made. However, to get an approximate idea, we have reported the time taken by CP and GA to determine their respective pareto fronts in Table 6. It can be seen that CP is able to determine the pareto front in 10 seconds whereas GA takes 3915.3 seconds to identify 14 of the 17 globally optimal pareto points. The actual time taken by GA to complete 2,000 generations is 5175.6 seconds. Hence for this case study, the computational performance of CP is superior to GA.

From Fig. 11(a), it can be seen that GA is able to find 1,000 feasible solutions in 44 generations. Also from Fig. 11(b), it can be seen that the time for each generation is high initially and decreases as the number of feasible solution increases. This behavior can be attributed to the fact that in the absence of infeasible solutions, there is no computational effort required for determining the penalties associated with the constraints.

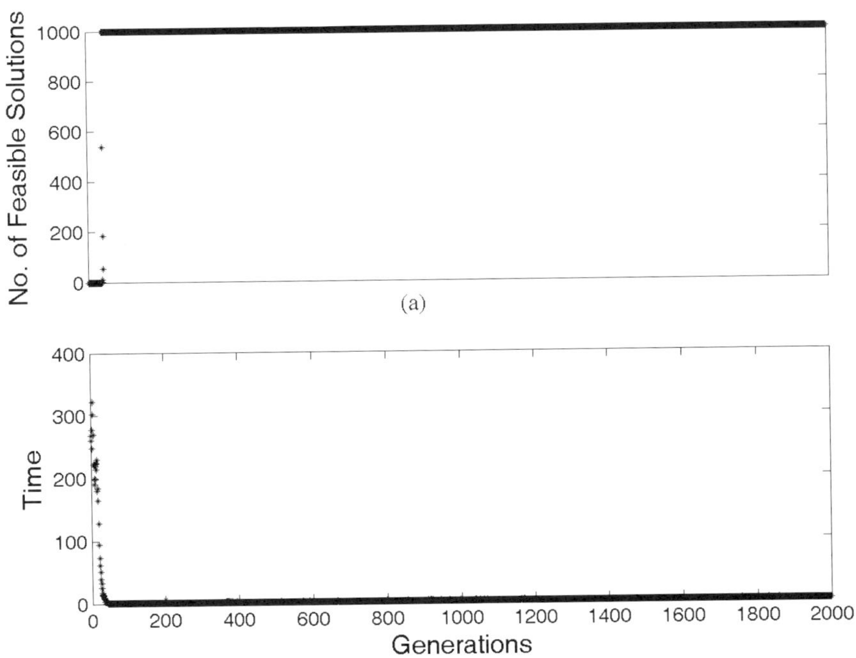

Figure 11. (a) Number of feasible solutions and (b) time taken for each generation.

It can be noted that the number of true pareto points determined by GA should always increase monotonically with the generations whereas the number of pareto points in the best front need not follow this behavior. This is due to the fact that a pareto point can be discovered (in a current generation) which could be dominating more than one another pareto point obtained till the previous generation, causing a decrease in the number of pareto points. However, a true pareto point can never be dominated and hence the number of true pareto points has to increase monotonically with the generations. Figure 12 shows the number of pareto points reported by GA and also the number of true pareto points in this set of pareto points. It can be seen that the number of true pareto points increase monotonically whereas the number of pareto points does not display such behavior. In the 482nd generation, 14 pareto points are discovered of which 10 are globally pareto optimal points. However, in the 489th generation, 14 pareto points are determined which are all globally pareto optimal.

(a)

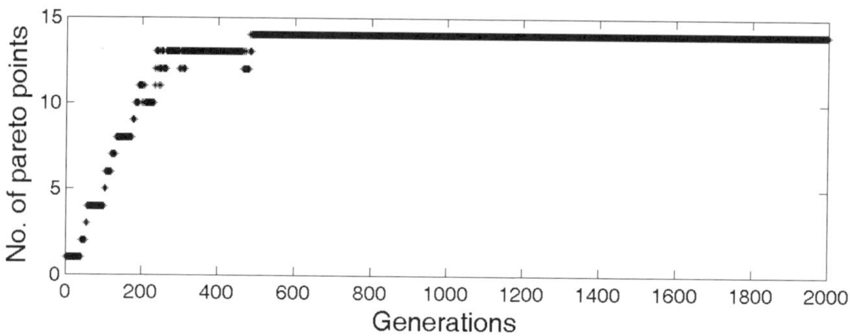

(b)

Figure 12. (a) Number of true pareto points and (b) Number of pareto points in each generation.

From Fig. 13(a), it can be seen that the generational distance from the true pareto front monotonically decreases with the generations and finally settles at zero. This should not be used to interpret that the true globally optimal pareto front is determined because the GD will be zero even if GA determines only m pareto points (all of these being present in the true pareto front) out of a total of $n(n > m)$ true pareto points. In the current case, the value of n is 17 and the value of m is 14. Also, since all these 14 points are part of the globally optimal pareto front, the GD becomes zero. This anomaly is due to the fact that the definition of GD (as in (44) and (45)) only takes into account the distance of a pareto point (determined by GA) to the nearest true pareto point. It does not take into account two other important factors namely (i) the number of true pareto points that have been determined by GA, and (ii) the number of realizations discovered by GA compared to the total number of true realizations.

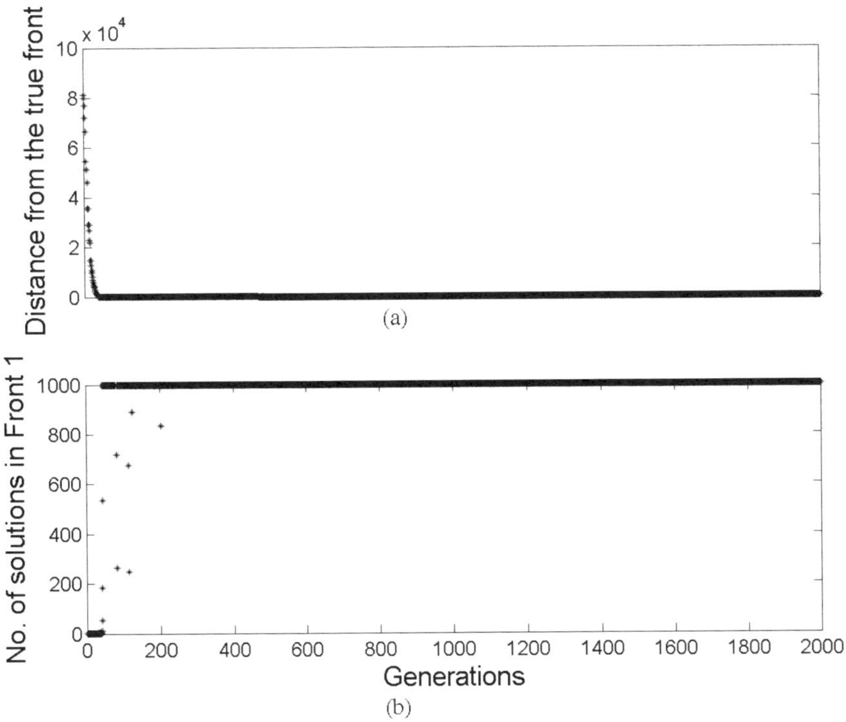

Figure 13. (a) Distance from the true and front and (b) Number of solutions in the best front in each generation.

Figure 14 shows the number of pareto solutions determined by GA and also the number of true pareto optimal solutions among those determined by GA. Ideally, the number of true pareto solutions should increase (for reasons mentioned for the true pareto points) whereas the number of pareto solutions can either increase or decrease. However, in Fig. 14(a) the number of true pareto solutions is not increasing monotonically. This can happen under the following scenario: The set of solutions passed on to a subsequent generation includes all the pareto optimal points of the combined population consisting of parent and offpsring population at the current generation. However, if the number of pareto optimal points (in the combined population) is greater than the population size, the crowding distance metric (Deb, 2001) is used to determine which of the current set of pareto optimal solutions are passed to the next generation. This crowding distance metric

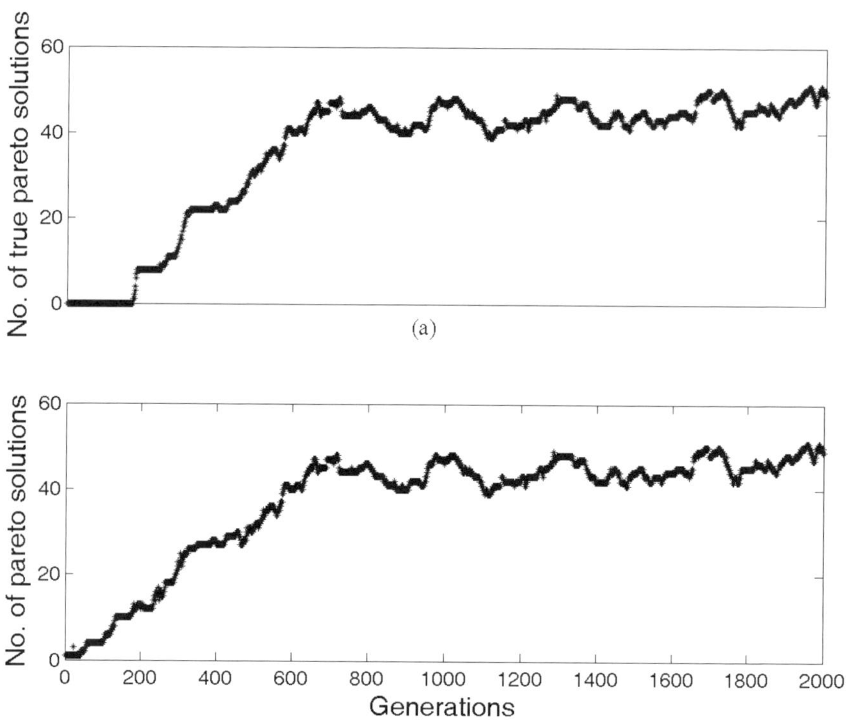

(b)

Figure 14. (a) Number of true pareto solutions and (b) Number of pareto solutions in each generation.

however need not necessarily rank a globally optimal pareto point higher than a locally pareto optimal point and hence this strategy can even eliminate a globally optimal pareto point/solution. This is the reason for the non monotonic increase of the number of true pareto optimal solutions. However, this anomaly cannot happen if the number of solutions in the best front of the current generation is less than the population size. As can be seen from Fig. 13(b), the number of solutions in the best front is equal to 1,000 (the population size). It should be noted that the loss of globally optimal solutions can happen even for the true pareto optimal points in Fig. 12(a). However, for this case study, such phenomenon does not happen for the number of pareto points possibly due to the existence of a large number of realizations.

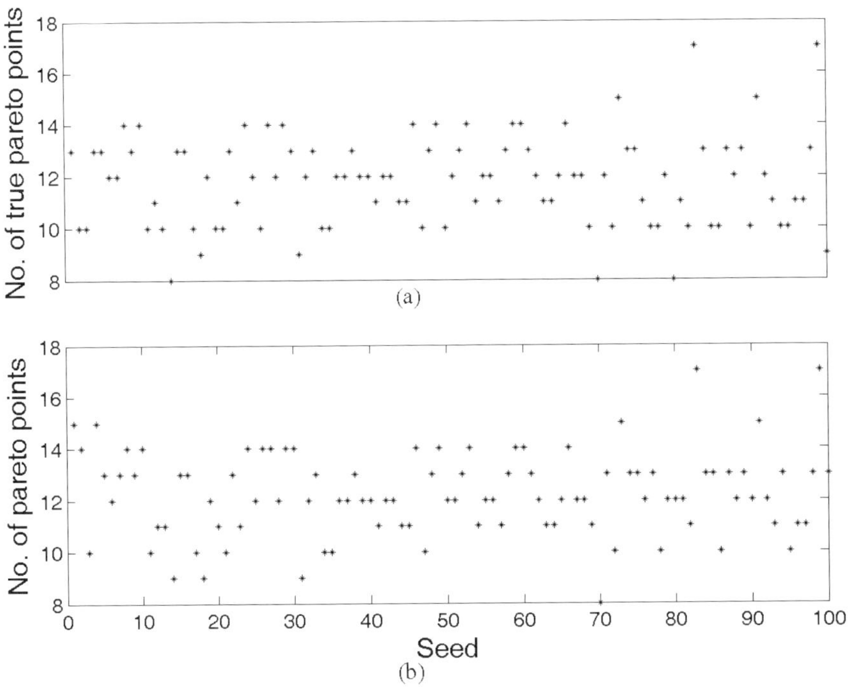

Figure 15. (a) Number of pareto points and (b) Number of true pareto points for various sets of random number.

As in the previous case study, we solved this multi-objective optimization problem with 100 different sets of random numbers (generated by initializing the seed in the *rand* function between 1 and 100 in MATLAB). The number of generations for each set of random number was set to 1,000. Figure 15 shows the number of pareto points as well as the number of true pareto points determined by GA at the end of 1,000 generations for each set of random numbers. It can be seen that GA was able to determine all the 17 globally optimal pareto points in two instances (with seed values 83 and 99). As seen in Fig. 16, these two instances are characterized by 59 and 62 pareto optimal solutions. It should be evident that these pareto optimal solutions will also be globally pareto optimal. However, there is also one instance wherein GA is able to determine only 8 pareto points (all of them globally optimal) with 13 realizations.

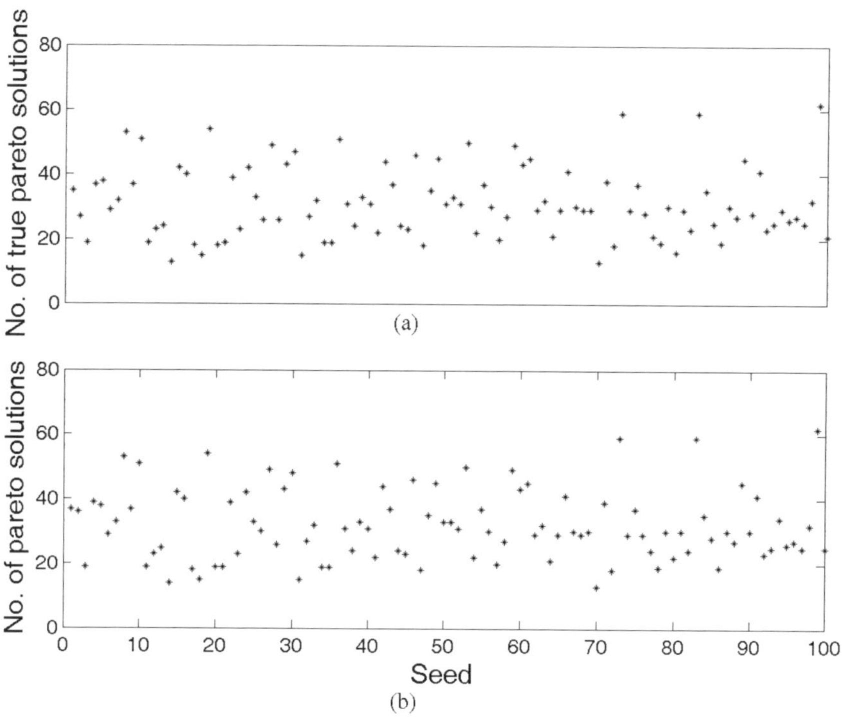

Figure 16. (a) Number of pareto solutions and (b) Number of true pareto solutions for various sets of random number.

Figure 17 shows the distance of the pareto front determined by GA to the globally optimal pareto front. It can be seen that the GD is zero in most of the instances. However, as indicated earlier, it should not be used to interpret that the complete pareto optimal front has been determined. The maximum GD is 6.6622 where GA determined 13 pareto points (with 85 as the seed). Among these 13 pareto points, only 10 are present in the globally optimal pareto front. Thus, we see that GA can possibly determine globally optimal pareto points and solutions but is dependent on the set of random numbers that are used during the various GA operations.

It can be seen from the above case study that CP is able to solve the multi-objective optimization more efficiently than GA. However, it should be mentioned that CP requires the problem to be specified in an explicit optimization framework and this requirement restricts its usage to a limited set of problems. On the other hand, GA does not require explicit optimization

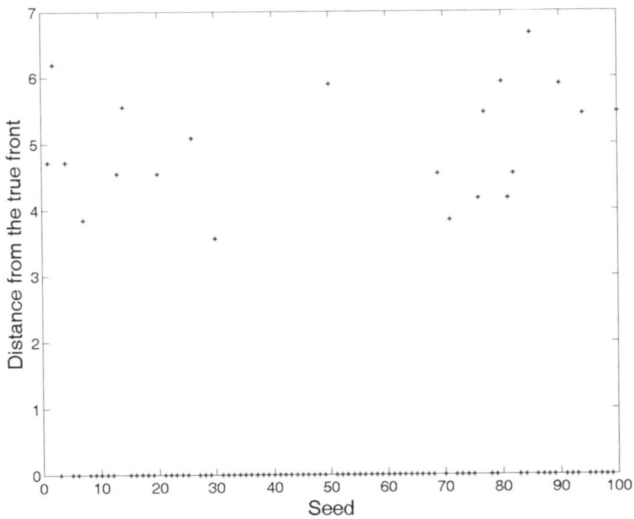

Figure 17. Distance of the pareto front from the true front for various sets of random number.

formulation and hence addresses a wider set of problems, but nevertheless does not guarantee the global optimality of the solution.

4. Conclusions

In this chapter, we have compared the performance of CP and GA in terms of their capabilities to efficiently model the problems, determine the optimal solutions and the realizations for both single objective and multi objective optimization problems. GA was able to model the problem more compactly (less number of variables) than CP as it does not require posing the problem in an explicit optimization framework. However, CP was able to efficiently determine globally optimal solutions and realizations because of its powerful constraint propagation mechanism. The strengths of GA and CP need to be used in a single framework and it may be worthwhile to explore such hybrid strategies. For example, as seen in some instances (especially P5S1), CP can require a considerable amount of time for a small improvement in the objective function. In such cases, a CP-GA based hybrid strategy can be used wherein CP is used to quickly determine some feasible solutions which can be used to populate the initial population in GA. Such a strategy can help the designer to discover better solutions in a relatively lesser time without any guarantee on the optimality of the solutions.

References

Agrawal, N., Rangaiah, G.P., Ray, A.K. and Gupta, S.K. (2007). Design stage optimization of an industrial low-density polyethylene tubular reactor for multiple objectives using NSGA-II and its jumping gene adaptations. *Chemical Engineering Science*, **62**(9), pp. 2346–2365.

Bhushan, M., Narasimhan, S. and Rengaswamy, R. (2008). Robust sensor network design for fault diagnosis. *Computers and Chemical Engineering*, **32**(4–5), pp. 1067–1084.

Darby-Dowman, K., Little, J., Mitra, G. and Zaffalon, M. (1997). Constraint logic programming and integer programming approaches and their collaboration in solving an assignment scheduling problem. *Constraints*, **1**(3), pp. 245–264.

DashOptimization. http://www.dashoptimization.com. (Last Accessed: July 2008).

Deb, K. (2001). *Multi-Objective Optimization Using Evolutionary Algorithms*. New York, NY, USA, John Wiley and Sons, Inc.

Dincbas, M., Van Hentenryck, P., Simonis, H., Aggoun, A., Graf, T. and Bertier, F. (1988). The constraint programming language CHIP. *International Conference on 5th Generation Computer Systems*. FGCS-88., Tokyo, Japan.

Downs, J.J. and Vogel, E.F. (1993). A plant-wide industrial-process control problem. *Computers and Chemical Engineering*, **17**(3), pp. 245–255.

Duda, J. (2005). Lot-Sizing in a Foundry Using Genetic Algorithm and Repair Functions. Evolutionary Computation in Combinatorial Optimization, *Lecture Notes in Computer Science*, Springer Berlin/Heidelberg, **3448/2005**, pp. 101–111.

Grossmann, I.E. and Biegler, L.T. (2004). Part II. Future perspective on optimization. *Computers and Chemical Engineering*, **28**(11), pp. 1193–1218.

Harjunkoski, I. and Grossmann, I.E. (2002). Decomposition techniques for multistage scheduling problems using mixed-integer and constraint programming methods. *Computers and Chemical Engineering*, **26**(11), pp. 1533–1552.

Heipcke, S. (1999). Comparing constraint programming and mathematical programming approaches to discrete optimisation — the change problem. *Annals of Operations Research*, **50**(6), pp. 581–595.

Hooker, J., Ottosson, G., Thorsteinsson, E.S. and Kim, H. (1999). On integrating constraint propagation and linear programming for combinatorial optimization. *Proceedings of the 16th National Conference on Artificial Intelligence* (*AAAI-99*), Orlando, Florida, USA, The AAAI Press/The MIT Press.

ILOG (2003). *ILOG OPL Studio 3.7 Language Manual*.

ILOG. www.ilog.com. Last Accessed: July 2008.

Jaffar, J. and Maher, M.J. (1994). Constraint Logic Programming: A Survey. *Journal of Logic Programming*, **19/20**, pp. 503–581.

Jain, V. and Grossmann, I.E. (2001). Algorithms for hybrid MILP/CP models for a class of optimization problems. *INFORMS Journal on Computing*, **13**(4), pp. 258–276.

Kotecha, P.R., Bhushan, M. and Gudi, R.D. (2007). Constraint programming based robust sensor network design. *Ind. Eng. Chem. Res.*, **46**(18), pp. 5985–5999.

Kotecha, P.R., Bhushan, M. and Gudi, R.D. (2008a). Design of robust, reliable sensor networks using constraint programming. *Computers and Chemical Engineering*, **32**(9), pp. 2030–2049.

Kotecha, P.R., Bhushan, M. and Gudi, R.D. (2008b). Unified approach for the design of robust, reliable and precision sensor networks. *Foundations of Computer-Aided Process Operations (FOCAPO) 2008*, Cambridge, Massachusetts, USA.

Kotecha, P.R., Bhushan, M. and Gudi, R.D. (2008c). Comparison of mathematical programming and constraint programming for the design of sensor networks. *International Conference on Emerging Technologies and Applications in Engineering, Technology and Sciences* (*ICETAETS*), Rajkot, Gujarat, India.

Kotecha, P.R., Bhushan, M. and Gudi, R.D. (2008d). Efficient optimization strategies with constraint programming. *AIChE Journal*. Submitted for publication.

Lustig, I.J. and Puget, J.F. (2001). Program does not equal program: Constraint programming and its relationship to mathematical programming. *Interfaces*, **31**(6), pp. 9–53.

Maravelias, C.T. and Grossmann, I.E. (2004). A hybrid MILP/CP decomposition approach for the continuous time scheduling of multipurpose batch plants. *Computers and Chemical Engineering*, **28**(10), pp. 1921–1949.

Michalewicz, Z. (1994). *Genetic Algorithms + Data Structures = Evolution Programs*, Berlin, Springer-Verlag.

Proll, L. and Smith, B. (1998). Integer linear programming and constraint programming approaches to a template design problem. *INFORMS Journal on Computing*, **10**, pp. 265–275.

Roe, B., Papageorgiou, L.G. and Shah, N. (2005). A hybrid MILP/CLP algorithm for multipurpose batch process scheduling. *Computers and Chemical Engineering*, **29**(6), pp. 1277–1291.

Rossi, F., Beek, P.V. and Walsh, T., (eds.) (2006). *Handbook of Constraint Programming*, Elsevier Publications.

Smith, B.M., Brailsford, S.C., Hubbard, P.M. and Williams, H.P. (1996). The progressive party problem: Integer linear programming and constraint programming compared. *Constraints*, **1**(1/2), pp. 119–138.

Tarafder, A., Rangaiah, G.P. and Ray, A.K. (2005). Multiobjective optimization of an industrial styrene monomer manufacturing process. *Chemical Engineering Science*, **60**(2), pp. 347–363.

Tarafder, A., Rangaiah, G.P. and Ray, A.K. (2007). A study of finding many desirable solutions in multiobjective optimization of chemical processes. *Computers and Chemical Engineering*, **31**(10), pp. 1257–1271.

Van Hentenryck, P. (2002). Constraint and integer programming in OPL. *INFORMS Journal on Computing*, **14**(4), pp. 345–372.

Wallace, M., Novello, S. and Schimpf, J. (1997). ECLiPSe: A platform for constraint logic programming. *ICL Systems Journal*, **12**(1), pp. 159–200.

Yee, A.K.Y., Ray, A.K. and Rangaiah, G.P. (2003). Multiobjective optimization of an industrial styrene reactor. *Computers and Chemical Engineering*, **27**(1), pp. 111–130.

Appendix I

Scheduling Data for Case Study

The following is the data used for the jobshop scheduling case study discussed in this chapter. Data for three different problems are given here labeled as P1, P4 and P5. Each of these problems has two sets of processing time making a total of 6 data instances. These are labeled as P1S1, P1S2, P4S1, P4S2, P5S1 and P5S2.

Table A.1.1. Scheduling data of problem 1 for case study.

			Cost on machine	
Order (i)	r_i	d_i	1	2
1	2	16	10	6
2	3	13	8	5
3	4	21	12	7

		Durations	
Order (i)	Machine (m)	P1S1	P1S2
1	1	10	5
	2	14	7
2	1	6	3
	2	8	4
3	1	11	5
	2	16	7

Table A.1.2. Scheduling data of problem 4 for case study.

			Cost on machine				
Order (i)	r_i	d_i	1	2	3	4	5
1	2	33	10	6	8	9	9
2	3	34	8	5	6	7	7
3	4	31	12	7	10	11	10
4	5	33	10	6	8	9	8
5	10	34	8	5	6	7	7
6	1	34	12	7	10	11	10
7	2	33	12	10	11	12	11
8	4	25	9	5	7	9	8
9	10	38	10	6	8	9	8
10	1	37	8	5	6	7	6
11	5	30	15	9	12	14	13
12	2	20	13	7	10	12	11
13	4	32	9	5	6	8	7
14	6	20	10	6	8	10	9
15	2	25	8	5	6	7	7

		Durations (p_{im})				Durations (p_{im})	
Order (i)	Machine	P4S1	P4S2	Order (i)	Machine	P4S1	P4S2
1	1	10	5	9	1	2	4
	2	14	7		2	4	6
	3	12	6		3	3	6
	4	11	5		4	2	5
	5	13	6		5	3	5
2	1	6	3	10	1	7	2
	2	8	4		2	14	5
	3	7	3		3	11	3
	4	6	3		4	8	2
	5	7	4		5	10	3
3	1	11	2	11	1	8	2
	2	16	4		2	16	3
	3	13	3		3	12	2
	4	11	2		4	10	2
	5	12	3		5	11	2
4	1	6	3	12	1	3	2
	2	12	6		2	6	6
	3	8	4		3	5	4
	4	7	3		4	4	3
	5	8	4		5	5	3

(*Continued*)

Table A.1.2. *(Continued)*

Order (i)	Machine	Durations (p_{im}) P4S1	Durations (p_{im}) P4S2	Order (i)	Machine	Durations (p_{im}) P4S1	Durations (p_{im}) P4S2
5	1	10	2	13	1	4	1
	2	16	4		2	10	3
	3	12	3		3	7	3
	4	12	2		4	5	2
	5	13	2		5	6	2
6	1	7	1	14	1	2	2
	2	12	3		2	4	5
	3	10	2		3	4	5
	4	8	2		4	3	2
	5	9	2		5	3	3
7	1	10	1	15	1	7	4
	2	13	2		2	14	7
	3	10	1		3	13	6
	4	11	1		4	10	4
	5	12	1		5	11	5
8	1	4	2				
	2	10	5				
	3	8	4				
	4	5	3				
	5	6	3				

Table A.1.3. Scheduling data of problem 5 for case study.

Order (i)	r_i	d_i	Cost on machine 1	2	3	4	5
1	2	33	10	6	8	9	9
2	3	34	8	5	6	7	7
3	4	31	12	7	10	11	10
4	5	33	10	6	8	9	8
5	10	34	8	5	6	7	7
6	1	34	12	7	10	11	10
7	2	33	12	10	11	12	11

(Continued)

Table A.1.3. (*Continued*)

Order (i)	r_i	d_i	Cost on machine 1	2	3	4	5
8	4	25	9	5	7	9	8
9	10	38	10	6	8	9	8
10	1	37	8	5	6	7	6
11	5	30	15	9	12	14	13
12	2	20	13	7	10	12	11
13	4	32	9	5	6	8	7
14	6	20	10	6	8	10	9
15	2	25	8	5	6	7	7
16	3	34	9	5	7	9	8
17	3	37	10	6	8	9	8
18	7	38	8	5	6	7	6
19	6	32	15	9	12	14	13
20	0	30	13	7	10	12	11

Order (i)	Machine	Durations (p_{im}) P5S1	P5S2	Order (i)	Machine	Durations (p_{im}) P5S1	P5S2
1	1	10	5	4	1	6	3
	2	14	7		2	12	6
	3	12	6		3	8	4
	4	11	5		4	7	3
	5	13	6		5	8	4
2	1	6	3	5	1	10	2
	2	8	4		2	16	4
	3	7	3		3	12	3
	4	6	3		4	12	2
	5	7	4		5	13	2
3	1	11	2	6	1	7	1
	2	16	4		2	12	3
	3	13	3		3	10	2
	4	11	2		4	8	2
	5	12	3		5	9	2
7	1	10	1	14	1	2	2
	2	13	2		2	4	5
	3	10	1		3	4	5
	4	11	1		4	3	2
	5	12	1		5	3	3

(*Continued*)

Table A.1.3. (*Continued*)

Order (i)	Machine	Durations (p_{im}) P5S1	Durations (p_{im}) P5S2	Order (i)	Machine	Durations (p_{im}) P5S1	Durations (p_{im}) P5S2
8	1	4	2	15	1	7	4
	2	10	5		2	14	7
	3	8	4		3	13	6
	4	5	3		4	10	4
	5	6	3		5	11	5
9	1	2	4	16	1	3	2
	2	4	6		2	8	4
	3	3	6		3	7	3
	4	2	5		4	5	2
	5	3	5		5	6	3
10	1	7	2	17	1	6	3
	2	14	5		2	12	6
	3	11	3		3	10	4
	4	8	2		4	7	3
	5	10	3		5	8	4
11	1	8	2	18	1	2	2
	2	16	3		2	8	4
	3	12	2		3	6	3
	4	10	2		4	13	2
	5	11	2		5	4	2
12	1	3	2	19	1	4	1
	2	6	6		2	7	3
	3	5	4		3	6	2
	4	4	3		4	5	2
	5	5	3		5	5	2
13	1	4	1	20	1	5	1
	2	10	3		2	7	2
	3	7	3		3	7	1
	4	5	2		4	6	1
	5	6	2		5	6	1

Appendix II

The Data for Case Study

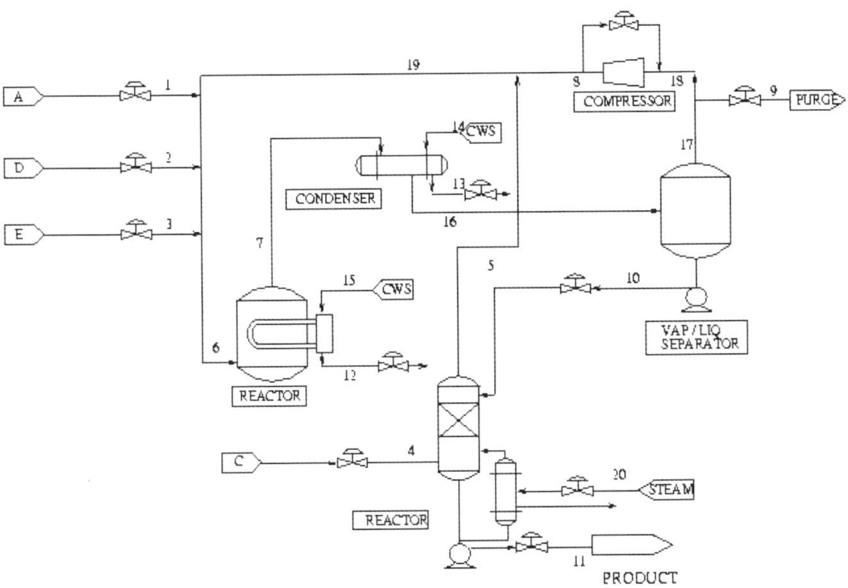

Figure A.2.1. Schematic of the Tennessee Eastman problem.

Table A.2.1. Sensor cost and failure probabilities for potential measurements.

	Var	log s_j	Cost		Var	log s_j	Cost		Var	log s_j	Cost		Var	Log s_j	Cost		Var	log s_j	Cost
1	P_r	−3	100	11	$y_{G,6}$	−3	800	21	$y_{H,7}$	−3	800	31	$x_{E,10}$	−3	700	41	$x_{H,r}$	−3	700
2	P_s	−3	100	12	$y_{H,6}$	−3	800	22	$y_{A,8}$	−3	800	32	$x_{F,10}$	−3	700	42	F_{10}	−3	200
3	P_m	−3	100	13	F_7	−3	300	23	$y_{B,8}$	−3	800	33	$x_{G,10}$	−3	700	43	F_{11}	−3	200
4	F_6	−3	300	14	$y_{A,7}$	−3	800	24	$y_{C,8}$	−3	800	34	$x_{H,10}$	−3	700	44	T_s	−2	500
5	$y_{A,6}$	−3	800	15	$y_{B,7}$	−3	800	25	$y_{D,8}$	−3	800	35	$x_{G,11}$	−3	700	45	VLr_{Xm}	−2	150
6	$y_{B,6}$	−3	800	16	$y_{C,7}$	−3	800	26	$y_{E,8}$	−3	800	36	$x_{H,11}$	−3	700	46	VLr_e	−4	100
7	$y_{C,6}$	−3	800	17	$y_{D,7}$	−3	800	27	$y_{F,8}$	−3	800	37	$x_{D,r}$	−3	700	47	VLs_{Xm}	−2	150
8	$y_{D,6}$	−3	800	18	$y_{E,7}$	−3	800	28	$y_{G,8}$	−3	800	38	$x_{E,r}$	−3	700	48	VLs_e	−4	100
9	$y_{E,6}$	−3	800	19	$y_{F,7}$	−3	800	29	$y_{H,8}$	−3	800	39	$x_{F,r}$	−3	700	49	VLp_{Xm}	−2	150
10	$y_{F,6}$	−3	800	20	$y_{G,7}$	−3	800	30	$x_{D,10}$	−3	700	40	$x_{G,r}$	−3	700	50	VLp_e	−4	100

Table A.2.2. Types of faults and their occurrence probability.

Fault No.	Description	$\log f_j$	Fault No.	Description	$\log f_j$
1,9	F_1 high, low	−2	16,25	$VL_{r_m,bias}$ high, low	−2
2,10	F_2 high, low	−2	17,26	$VL^{set}_{r_m,bias}$ high, low	−2
3,11	F_3 high, low	−2	18,27	$VL_{rVP,bias}$ high, low	−2
4,12	F_4 high, low	−2	19,28	$VL_{s_m,bias}$ high, low	−2
5,13	F_8 high, low	−2	20,29	$VL^{set}_{s_m,bias}$ high, low	−2
6,14	F_9 high, low	−2	21,30	$VL_{sVP,bias}$ high, low	−2
7,15	T_r high, low	−1	22,31	$VL_{p_m,bias}$ high, low	−2
8	C_d low	−2	23,32	$VL^{set}_{p_m,bias}$ high, low	−2
			24,33	$VL_{pVP,bias}$ high, low	−2

Chapter 19

SCHEMES AND IMPLEMENTATIONS OF PARALLEL STOCHASTIC OPTIMIZATION ALGORITHMS: APPLICATION OF TABU SEARCH TO CHEMICAL ENGINEERING PROBLEMS

B. Lin* and D. C. Miller†

Department of Chemical Engineering
Rose-Hulman Institute of Technology
5500 Wabash Avenue, Terre Haute, IN 47803, USA
**bao.lin@flsmidth.com*
†*david.miller@netl.doe.gov*

1. Introduction

Stochastic optimization algorithms (SOA) have increasingly been used to solve challenging problems in the chemical industry. The long computation times associated with sequential implementations of these algorithms often limit their applicability for solving large-scale problems. A variety of parallel strategies have been proposed to speedup SOA (Crainic and Toulouse, 1998; Cung *et al.*, 2001; Resende *et al.*, 2001); however, many of these approaches require specialized hardware and compilers. In addition, they

*Present address: FLSmidth Automation, 3225 Schoenersville Road, Bethlehem, PA18017, United States
†Present address: U.S. Department of Energy, National Energy Technology Lab (NETL), 3610 Collins Ferry Road, PO Box 880, Morgantown, WV 26507-0880, United States.

require the programmer to pay careful attention to details associated with the parallelization to synchronize communication among processes. The result has been the wide-spread notion that parallel implementations are too costly, both in terms of time and resources.

This is changing with the increasing availability of multi-core processors. These new processors put parallel computing hardware in the hands of most people with a personal computer. A multi-core processor is essentially a single chip with two or more CPU's embedded within it. Indeed, as of mid-2008, most PC's have at least 2 cores, and higher end workstations are available with 4 to 8 processor cores. In addition, networked clusters of computers can be built and maintained less expensively due to the increased availability of open source operating systems and network tools. Thus, hardware expense should no longer hinder the use of parallel SOA.

The time and effort required to develop and debug parallel code is also undergoing a revolution. Open Multi-Processing (OpenMP), which was originally developed for shared-memory multi-processor machines provides a way to parallelize sequential code through the use of simple compiler directives (Dagum and Menon, 1998). When originally introduced, it required expensive compilers and would only run on specialized hardware. Since the hardware constraint has been significantly relaxed with the widespread availability of multi-core computers, more compilers are being made OpenMP compliant, including GCC, the GNU Compiler Collection. Thus, the monetary expense of creating parallel implementations of SOA has been significantly reduced. Through the examples in this chapter, we will argue that the time requirements for developing an OpenMP-based parallel implementation are minimal compared to the increased performance achieved when running a parallel SOA.

Another, complementary approach to parallelizing sequential code involves the Message Passing Interface (MPI) standard, which is especially useful when processors are distributed among different machines. These machines could be distributed across a large geographic area and connected via the internet. More often, they are in close proximity and connected via an ultrafast, high bandwidth network. Unlike OpenMP, which is based entirely on compiler directives, MPI requires the programmer to explicitly control the flow of information between nodes, increasing the programming complexity and difficulty of debugging. The greater flexibility of MPI allows computation to potentially be more widely distributed.

Since most processors now have at least two cores, OpenMP can readily be used to speedup computations relative to the previous single core systems. To further increase parallelization, MPI can be combined with the OpenMP code to spread the computation across a number of nodes within a cluster. Each node may have multiple processors, each of which may have multiple cores. Multiprocessor nodes are generally referred to as symmetric multiprocessor (SMP) nodes. A cluster of such SMP machines is sometimes referred to as a CLUster of MultiProcessors (CLUMP) (Hsieh, 2000). Because memory is shared among all CPU's within a given node, each node is amenable to parallelization through OpenMP. Although MPI could be used to parallelize code within a node, He and Ding (2002) have noted several disadvantages. Hence, MPI is best used to facilitate parallelization across nodes.

Parallelization of SOA has been an active area of research. A wide range of implementations have been reported, including parallel GA (Gordon *et al.*, 1993; Negro *et al.*, 2004; Baños *et al.*, 2006; Tang *et al.*, 2007), parallel SA (Hiroyasu *et al.*, 2002; Debudaj-grabysz and Rabenseifner, 2005), parallel Ant Colony Optimization (PACO) (Cioni, 2005; Manfrin *et al.*, 2006), parallel Particle Swarm Optimization (PPSO) (Belal and El-Ghazawi, 2004; Koh *et al.*, 2006) and parallel tabu search (Porto and Ribeiro, 1995; Toulouse *et al.*, 1998; Jamesa *et al.*, 2008). The purpose of this chapter is to provide general approaches for parallelizing SOA on modern computational hardware. Parallel implementations of an SOA are used to illustrate these approaches and to demonstrate the performance improvements that can be achieved.

The remainder of this chapter is organized as follows. The second section describes several schemes for developing parallel SOA. Section 3 reviews sequential tabu search (TS), the SOA used to illustrate the parallelization approaches. Section 4 presents two case studies, Heat Exchanger Network (HEN) synthesis and Computer Aided Molecular Design (CAMD). Section 5 describes the implementation of parallel TS using OpenMP. Section 6 describes a hybrid MPI/OpenMP implementation for solving larger problems that require more computational resources than may be available on a single multi-core workstation.

2. Parallel Schemes of Stochastic Optimization Algorithms

A sequential SOA, such as SA, GA, or TS, generally consists of the following steps to search for an optimal solution.

```
/* A general SOA module */
1 Initialization
2 While termination criterion is not satisfied
  2.1 Generate neighbor solutions
  2.2 Evaluate neighbors
  2.3 Select the best neighbor solution
  2.4 Additional operations, i.e. elite list
      management, etc.
  2.5 Update the optimal solution
3 End while
4 Output
```

Parallelization of some of these steps can reduce the total computation time. Some of the time saved can be used to increase the number of neighbor solutions, thus increasing the likelihood of locating the true optimal solution. Toulouse *et al.* (1996) proposed three strategies to create parallel meta-heuristic procedures: *functional decomposition*, *multi-search threads* and *domain decomposition*.

- Functional decomposition parallelizes the process of finding the best neighbor, i.e. steps 2.1 through 2.3 of the SOA module. As shown in Fig. 1, each thread deals with a subset of neighbor solutions.
- Multi-search threads consist of multiple instances of the entire algorithm executing in parallel as shown in Fig. 2.
- Domain decomposition divides the search space into smaller sub-domains. A master-slave framework is often used in which the master partitions the domain and assigns the sub-domains to slave processors.

Functional decomposition follows the same search procedure as a sequential SOA and directly reduces the computation time by parallelizing computationally intensive loops. When using multi-search threads, each thread may start from a different initial solution and use different parameter settings. If information is properly shared across threads, this approach can result in improved solution quality in addition to reduced computation time.

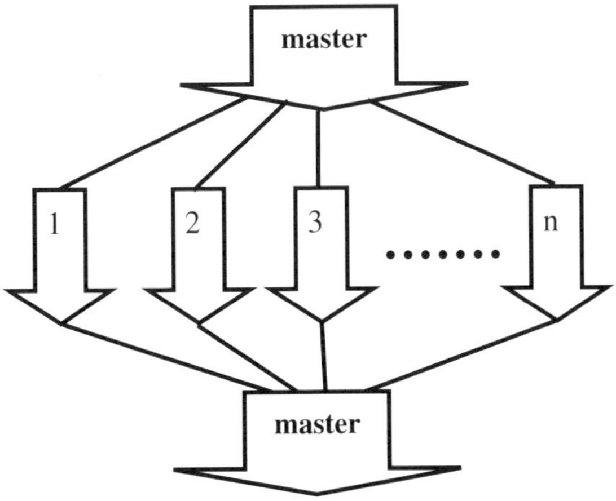

Figure 1. Functional decomposition strategy in which steps in the algorithm are assigned to processors $1, 2, \ldots, n$.

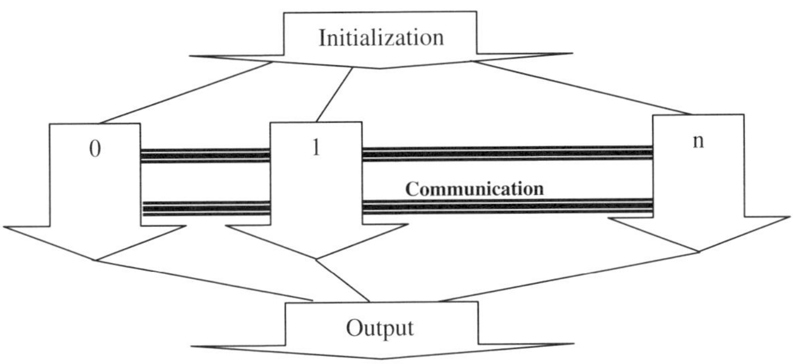

Figure 2. Multi-search threads strategy with communication in which the complete SOA is executed in parallel on processors $1, 2, \ldots, n$.

Domain decomposition tends to be problem specific and cannot be easily generalized to solve other problems. Hence, functional decomposition and multi-search threads are investigated in this study.

Parallel SOA can also be classified according to the pattern of their search (Crainic *et al.*, 1997). The two categories are defined by their cardinality: *1-control* and *p-control*. *1-control* represents the approach where one processor executes the search while assigning numerically intensive work to other processors. This is the case in functional decomposition. In *p-control* case, the search process is controlled by more than one processor. Each processor performs its own search, communicating with other processors. This is the case with multi-search threads.

Parallel implementations can be further classified based on whether they begin with a single initial solution or a different initial solution in each thread. These are referred to single-point or multi-point point strategies. In addition, each thread can either employ the same search parameters (same strategy) or different search parameters (different strategy). Combining these approaches results in four categories: single initial point, single strategy (SPSS); single initial point, different strategies (SPDS); multiple initial points with the same strategy (MPSS); and multiple initial points with different strategies (MPDS). MPSS is used in the multi-search threads implementation in this chapter.

3. Tabu Search Algorithm

TS is a meta-heuristic approach that guides a neighborhood search procedure to explore the solution space in a way that facilitates escaping from local optima (Glover and Laguna, 1997). It starts from a randomly generated initial solution. A set of neighbor solutions, $N(x)$, are constructed by modifying the current solution, x. The best one among them, x', is selected as the new starting point, and the next iteration begins. Memory, implemented with *tabu lists*, is employed to escape from locally optimal solutions and to prevent cycling. Following each iteration, the tabu lists are updated to keep track of the search process. This memory allows the algorithm to adapt to the current status of the search so as to ensure that the entire search space is adequately explored and to recognize when the search has become stuck in a local region. *Intensification* strategies are employed to search promising areas more thoroughly. At the same time, *diversification* strategies are employed to broadly search the entire feasible region, thus helping to avoid becoming stuck in local optima. Finally, *aspiration criteria* are employed to override the tabu lists in certain cases. The flow chart of the serial TS algorithm is shown in Fig. 3.

The performance of TS is highly dependent on the parameter settings. The key parameters of TS are (1) the number of neighbor solutions, (2) the number of iterations, and (3) the length of tabu lists. The reader is referred to the TS chapter in this book and Lin and Miller (2004b) for a

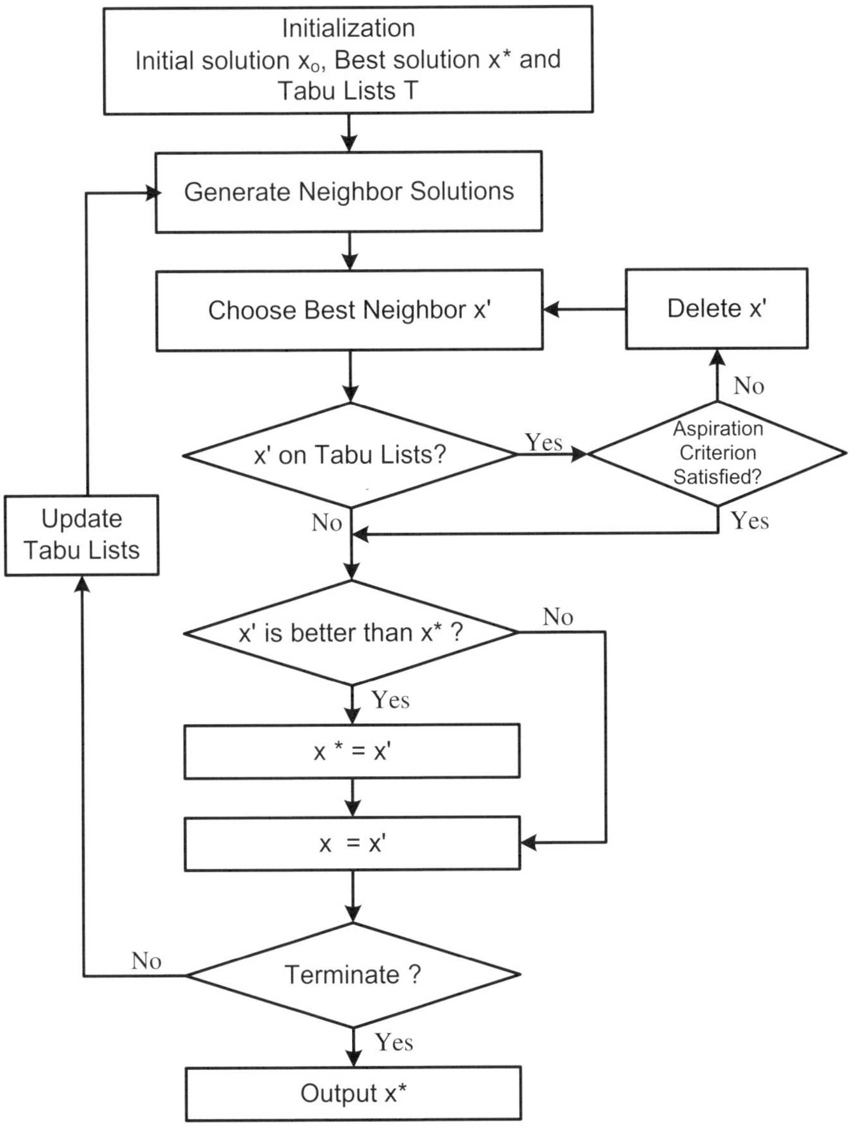

Figure 3. Flow chart of the TS algorithm.

detailed description of the algorithm and recommendations for parameter settings.

Although TS was originally proposed for combinatorial optimization problems in Operations Research, it has been applied to a wide range of problems in science and engineering, including design automation of VLSI systems (Morales *et al.*, 2000), Artificial Neural Network optimization (Sexton *et al.*, 1998), DNA sequencing (Blazewicz *et al.*, 2004) and restructuring protein structure (Paluszewski *et al.*, 2006). TS has also seen increasing application in process systems engineering as described in the TS chapter of this book. Previously reported parallel TS algorithms are predominately based on MPI (Crainic *et al.*, 1995; Porto and Ribeiro, 1995; Toulouse *et al.*, 1996; Niar and Freville, 1997; Talbi *et al.*, 1997; Porto *et al.*, 2000).

4. Test Cases

Two example problems are used to evaluate the effective speedup associated with the various parallelization schemes. They are a small HEN synthesis problem and a CAMD problem.

4.1. HEN synthesis

Heat exchanger network (HEN) synthesis problems have attracted significant research efforts because of the large savings that can be achieved by reducing energy costs. The stage-wise formulation of HEN synthesis problems results in a mixed integer nonlinear programming (MINLP) problem. This type of problem is often discussed in the literature when comparing solution techniques (Yee and Grossmann, 1990; Lewin *et al.*, 1998; Furman and Sahinidis, 2001). The objective is to minimize the annualized cost expressed as the sum of utility costs, fixed charges for each heat exchanger and an area-based cost for each heat exchanger. The area of a heat exchanger is a highly nonlinear function of the temperature difference and the heat load.

The HEN synthesis case used in this chapter is from Adjiman *et al.* (2000). It involves two hot streams, two cold streams, one hot utility and one cold utility. Stream data and parameters are given in Tables 1 and 2. There are 12 possible stream matches in the two stages. The global optimal solution consists of 6 heat exchangers with an annualized cost of $154,997.

Table 1. Stream data of the HEN synthesis case study.

Stream	T_{in} (K)	T_{out} (K)	FC_p (kW/K)
Hot1	650	370	10.0
Hot2	590	370	20.0
Cold1	410	650	15.0
Cold2	350	500	13.0
Stream	680	680	
Water	300	320	

Table 2. Parameters for the HEN synthesis case study.

$CF_{i,j}$	\$5500/yr	$CF_{i,CU}$	\$5500/yr	$CF_{j,HU}$	\$5500/yr
$C_{i,j}$	\$150/yr	$C_{i,CU}$	\$150/yr	$C_{j,HU}$	\$150/yr
$U_{i,j}$	0.5 kW/m^{2}K	$U_{i,CU}$	0.5 kW/m^{2}K	$U_{j,HU}$	0.8333 kW/m^{2}K
C_{HU}	\$80/kW.yr	C_{CU}	\$15/kW.yr	ΔT_{mapp}	10 K

The objective function is:

$$\begin{aligned}
\min_{T,Q,\Delta T,z} & \sum_{i\in HP} C_{CU} Q_{CU,i} \\
& + \sum_{j\in CP} C_{HU} Q_{HU,j} + \sum_{i\in HP}\sum_{j\in CP}\sum_{k\in SI} CF_{i,j} z_{i,j,k} \\
& + \sum_{i\in HP} CF_{i,CU} z_{CU,i} + \sum_{j\in CP} CF_{j,HU} z_{HU,j} \\
& + \sum_{i\in HP}\sum_{j\in CP}\sum_{k\in SI} \frac{C_{i,j} Q_{i,j,k}}{U_{i,j}\left(\Delta T_{i,j,k}\,\Delta T_{i,j,k+1}\frac{(\Delta T_{i,j,k}+\Delta T_{i,j,k+1})}{2}\right)^{\frac{1}{3}}} \\
& + \sum_{i\in HP} \frac{C_{i,CU} Q_{CU,i}}{U_{CU,i}\left(\Delta T_{CU,i}(T_{out,i} - T_{in,CU})\frac{(\Delta T_{CU,i}+T_{out,i}-T_{in,CU})}{2}\right)^{\frac{1}{3}}}. \\
& + \sum_{j\in CP} \frac{C_{j,HU} Q_{HU,j}}{U_{HU,j}\left(\Delta T_{HU,j}(T_{in,HU} - T_{out,j})\frac{(\Delta T_{HU,j}+T_{in,HU}-T_{out,j})}{2}\right)^{\frac{1}{3}}}
\end{aligned} \tag{1}$$

Variables of Eq. (1) are defined as follows.

Constants

SI — the set of temperature intervals or stages
HP — the set of hot process streams
CP — the set of cold process streams

Continuous Variables

$T_{i,k}$ — temperature of hot stream i at the temperature location k
$T_{j,k}$ — temperature of cold stream j at the temperature location k
$Q_{i,j,k}$ — heat changed between hot stream i and cold stream j in stage k
$Q_{CU,i}$ — heat changed between hot stream i and cold utility
$Q_{HU,j}$ — heat changed between cold stream j and hot utility
$\Delta T_{i,j,k}$ — the temperature approach for the match of hot stream i and cold stream j at the temperature location k
$\Delta T_{CU,i}$ — the temperature approach for the match of hot stream i and cold utility
$\Delta T_{HU,j}$ — the temperature approach for the match of cold stream j and hot utility

Binary Variables

$z_{i,j,k}$ — the existence of the match between hot stream i and cold stream j
$z_{CU,i}$ — the existence of the match between hot stream i and cold utility
$z_{HU,j}$ — the existence of the match between cold stream j and hot utility

4.2. CAMD

CAMD techniques have been developed to reduce the time and cost of trial-and-error experiments by identifying promising candidate molecules based on predicted properties. This example uses correlations based on topological indices to predict properties of interest. Topological indices take molecule connectivity into consideration to produce more accurate property predictions than possible with simple group contribution methods (Raman and Maranas, 1998; Siddhaye *et al.*, 2000). This example is based on the design of a molybdenum catalyst from a set of 10 basic groups shown in Table 3.

Table 3. Basic group definition.

	1	2	3	4	5	6	7	8	9	10
Group	–Mo–	$-\overset{\vert}{\text{Mo}}-$	$-\overset{\vert}{\text{Mo}}=$	S=	O=	–O–	–NH2	–NH–	–CH2–	–Cl
δ	2	3	3	1	1	2	1	2	2	1
δ^v	0.05128	0.07895	0.10811	0.66667	6	6	3	4	2	0.77778
$N_{\text{Max},i}$	2	2	2	2	2	2	2	2	2	2

Note: The value of $N_{\text{Max},i}$ denotes the maximal number of the group i allowed in a molecule. For this problem, the maximal number of groups is limited by: $\sum_{i=1} N_{\text{Max},i} = 20$.

Table 4. Zero order, first order and second order simple connectivity indices.

0th order	$^{0}\chi = \sum_{i=1} w_i \delta_i^{-\frac{1}{2}}$	$w_i = 1$ if group i exists in the molecule $= 0$ otherwise
1st order	$^{1}\chi = \sum_{(i,j);j>i} a_{ijk}(\delta_i\delta_j)^{-\frac{1}{2}}$	$a_{ijk} = 1$ if group i is bonded with j with k type of bond $= 0$ otherwise
2nd order	$^{2}\chi = \sum_{(i,j,k);k>j>i} y_{ijk}(\delta_i\delta_j\delta_k)^{-\frac{1}{2}}$	$y_{ijk} = 1$ if group i is bonded with group k through group j $= 0$ otherwise

Molecule structure is expressed with a hydrogen-suppressed graph. Topological indices are calculated with the reciprocal square root of atomic connectivity indices of basic groups as shown in Table 4 (Kier and Hall, 1976). Properties of known compounds are used to develop a correlation based on these indices.

The objective function is defined as minimizing the sum of the normalized deviation of the estimated value from the desired properties, i.e. density:

$$\sum_i \left| \frac{P_i^{\text{predict}} - P_i^{\text{target}}}{P_i^{\text{target}}} \right|, \tag{2}$$

where the correlation derived for density is:

$$\begin{aligned} \rho = {} & -55351 + 75800^0\chi - 7663^0\chi^v + 40901^1\chi + 1784^1\chi^v \\ & - 72046^2\chi - 607^2\chi^v - 24695(^0\chi)^2 + 649^0\chi^0\chi^v. \\ & - 12271(^1\chi)^2 - 65.4(^1\chi^v)^2 - 1793(^2\chi)^2 + 8.9(^2\chi^v)^2 \\ & + 72323^1\chi^2\chi \end{aligned} \tag{3}$$

The target density value is 70,050 kg/m^3.

Structural constraints are added to ensure feasibility of the obtained molecules. These include a valence balance for each basic group and a connectivity constraint to ensure a single molecule. Other constraints include bounds on the variables and property values. Up to two Mo atoms can be included in the molecule, and a maximum of 20 atoms can be a part of any molecule.

As described by Lin *et al.* (2005), neighbor generation for CAMD can exploit the structural constraints of the problem instead of relying on a purely mathematical formulation. As such, an initial molecular structure is generated by starting with a random molecular group selected from those available for the problem and adding additional groups until its connectivity constraints are met. Only atoms that meet the valence constraints are available to be added. This continues for each group that has been added until all the connectivity constraints for the molecule have been met. To generate neighbors, several operators are employed which act on this initial molecule. These operators include replacing one group with another, inserting a group between two existing groups, deleting a group that is connected to two other groups, swapping two groups, moving a group from one location to another, and a combination of the above. In addition, operators can call for rebuilding a side chain, the main chain or the complete molecule by following a procedure similar to that for generating the initial solution. The operator employed is selected at random for each neighbor generated to ensure diversity.

5. OpenMP Tabu Search

OpenMP parallelism is based on compiler directives, which are embedded in a program written with FORTRAN, C or C++. Since the directives are simply taken as comments in the case of non-OpenMP compliant compilers, directive-based parallelism has the advantage that the same code can be used on systems without parallel capabilities. Several references are available to support programming in OpenMP including the OpenMP Architecture Review Board (openmp.org) and Chandra *et al.* (2001).

The syntax for C and C++ is as follows: `#pragma omp...`, where the keyword "omp" denotes an OpenMP compiler directive. OpenMP also provides a runtime library, whose prototype is defined in the file "omp.h". The library is used to control parameters in the parallel region. In addition, shared memory systems enable the shared variables data type. Thus, data distribution among threads is easily implemented, and creating a multithreaded program is straightforward.

OpenMP follows a fork-join structure as shown in Fig. 4. When the program arrives at the beginning of a parallel region, the master thread

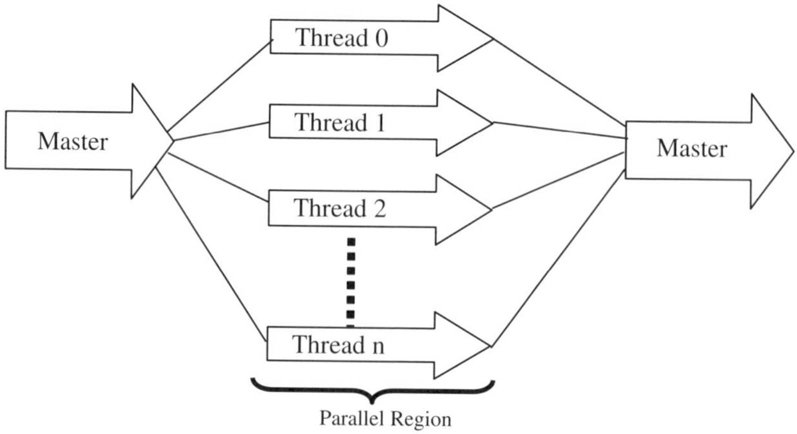

Figure 4. Fork-Join structure of OpenMP.

initializes a group of threads executing in parallel, which join at the end of the parallel region. For example, in the case of parallelizing a `for` loop, each slave thread is assigned a number of iterations to perform. An implicit barrier exists at the end of the parallel region. When a thread finishes its assignment, it waits until all other threads complete theirs. Then the barrier is released. All slave threads disappear and the master thread resumes. There can be any number of fork-join structures existing independently in one program. Such a structure can be added whenever a specific part of the program could benefit from parallelization. Thus, OpenMP provides an incremental approach for parallelization.

5.1. Functional decomposition

In functional decomposition, all the neighbors generated for an iteration are independent from one another; thus, the process of generating and evaluating neighbor solutions can be parallelized to save computation time. The loop-level parallelism of OpenMP enables the functional decomposition strategy to be readily implemented. The master process initiates a group of threads at the beginning of each iteration. Each thread deals with part of the complete set of candidate solutions. The best neighbor among all the threads is selected at the end. Communication is implemented using shared variables instead of a broadcast which would be used with MPI.

OpenMP assigns iterations to each thread in a continuous range called a *chunk*. Chunk size refers to the number of iterations each chunk contains. There are two chunking schemes within the loop level parallelization of OpenMP, *static* and *dynamic*. In the static case, the assigned iterations are determined at the beginning of a loop. In a dynamic schedule, the assignment of iterations can be changed during execution. In this case, a thread will be assigned additional iterations after finishing its initial set. For each type of schedule, only a few statements need to be added to the source code.

First, the header file "`omp.h`" needs to be included.

```
#include <omp.h>
```

Second, a `parallel for` statement is added just before the loop, defining a work-sharing construct in the source code. For the static schedule:

```
for (k = 0; k < M; k++)
  #pragma omp parallel for private(var. list)\
    shared(var. list)
  for (j = 0; j < N_Neigh; j++)
    Generate-Neighbors ();
  #pragma omp parallel for private(var. list)\
    shared(var. list)
  for (j = 0; j < N_Neigh; j++)
    Evaluate-Neighbors ();
  ⋮
```

where M is the number of iterations for each run; N_{Neigh} is the number of neighbors generated and evaluated at each iteration.

For the dynamic schedule, the statement is:

```
#pragma omp parallel for private(var. list)\
    shared(var. list) schedule(dynamic,chunksize)
```

The parameter, `chunksize`, determines the number of iterations that will be dynamically assigned to each thread. Although it is straightforward to add compiler directives, it is important to define the scope for each variable. An appropriate scope for each datum must be assigned to ensure that the loops are parallelized correctly, e.g. unintended data sharing will lead to a *race condition* and unexpected results.

5.2. Multi-search threads

In the multi-search threads strategy, the master thread initializes a group of threads after obtaining an initial solution. Since computation results demonstrate the superior performance of a cooperative search compared to an independent thread strategy (Aiex *et al.*, 1998), a cooperative pattern is implemented in the multi-search threads scheme, with communication occurring synchronously at predetermined times. At each communication interval, results from each thread are compared to determine the best solution so far. This solution is then used by all the threads to continue the search. At the end of the program, the best solution among all threads is output as the final result.

Several compiler directives are required to coordinate the communication process. The simplified implementation is as follows:

```
    Initialize();
    #pragma omp parallel private (var. list)\
         shared (var. list)
    for (k = 0; k < MaxIter; k++)
         Generate_neighbors();
         Evaluate_neighbors();
         Update_tabulist();
         if (Synchronization==1) /* Communication
              Interval Arrived*/
         #pragma omp critical
/* Mutual exclusion to guarantee only one thread
   has access to shared data*/
         if (GlobalOpt > ThreadOpt)
              Update_GlobalOpt();
         else
              Update_ThreadOpt();
         #pragma omp barrier
      /* Event synchronization, each thread waits
         for other threads to arrive */
```

```
    #pragma omp flush
    /* Make sure each thread has consistent
       view of memory*/
Output_final_solution();
```

5.3. Results

The effectiveness of the functional decomposition strategy depends on the computational sensitivity of the part of the program that is parallelized. In the case of the HEN synthesis problem, generating neighbors is more time-consuming than evaluating the objective function. Thus, if the program only parallelizes neighbor evaluation, very little speedup is seen. In fact parallelizing just neighbor generation provides nearly as much speedup as parallelizing both neighbor generation and evaluation. In general, it is recommended to parallelize both, since the same compiler directives can be used for both functions.

As described previously, functional decomposition of an OpenMP program allows the loop-level parallelization to be either static or dynamic. The HEN case study was used to compare the speedup performance of both the dynamic and static scheduling approaches. Figure 5 shows the dependence of speedup on chunk size, scheduling approach and number of threads. Smaller chunk sizes give better speedup than larger ones. In all

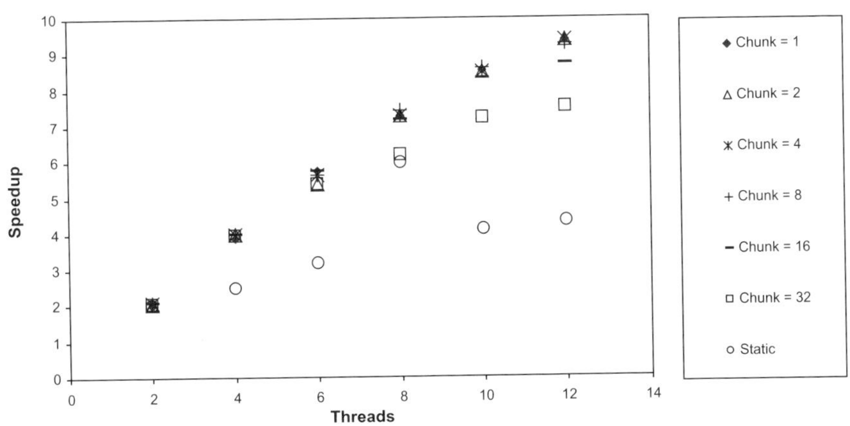

Figure 5. Speedup of HEN synthesis with functional decomposition scheme of different chunk sizes.

cases, dynamic scheduling achieves better speedup performance than the static approach.

Since the chunk size determines how many times the distribution is coordinated, a small chunk size should require additional time for coordination among threads. Thus, it might be expected that smaller chunk sizes would result in longer overall computation times; however, the benefit of the additional coordination significantly outweighs the costs of providing the coordination. Thus, a small chunk size tends to provide the maximum performance benefit. Care should be taken if choosing a chunk size of one, since it could lead to excessive dynamic scheduling in certain circumstances.

The OpenMP TS for the HEN examples was originally implemented on a Sun Enterprise 4500 workstation with 9 GB of shared memory and 12 CPUs (ultra-sparc 450 Mhz) located on 6 CPU modules, which are connected with buses (Sun Microsystems, 2001). The compiler was Guide for C/C++ (version 3.9) from Kuck and Associates (KAI). The number of neighbor solutions generated at each iteration, N_{Neigh}, was 1024 based on the guidelines given in Lin and Miller (2004b). Each set of parallel parameters were evaluated based on 50 runs, each consisting of 100 iterations. The fully documented code for this example is available on the CD accompanying this book.

6. Hybrid MPI/OpenMP Tabu Search

Several models have been developed for parallel programming within a CLUMP environment (Cappello and Richard, 1999). These include pure MPI, software supported OpenMP, and a hybrid MPI/OpenMP approach. Since MPI runs on both shared and distributed memory systems, parallel programs implemented with pure MPI can be directly implemented in CLUMP environment. As mentioned previously, additional programming overhead is inevitable. Since the system bus is shared for intra-node communication, communication must be coordinated to avoid data collisions. In addition, the overlap between intra-node and inter-node message passing degrades the performance of pure MPI implementations.

Parallel programs using pure OpenMP are able to run in a CLUMP environment with additional software support, which provides a software-based distributed shared memory system (SDSM), such as SCASH

(Ojima *et al.*, 2003). The original OpenMP code is translated during run-time; however, additional overhead comes from indirect reference. A hybrid memory model (HMM) uses OpenMP for intra-node parallelism and MPI for inter-node communications. The HMM provides a straightforward way of parallel programming and takes full advantage of both MPI and OpenMP. Figure 6 shows a hybrid parallel TS scheme that has been implemented using a HMM approach.

The framework of the hybrid parallel TS approach is based on the multi-search threads strategy, consisting of several cooperative TS processes which run on separate nodes. Communication among the nodes is achieved with MPI function calls. Within each node, the functional decomposition strategy is implemented with OpenMP compiler directives. By taking advantage of shared variables within a node, the amount of data that must be explicitly moved is significantly reduced. Thus, the hybrid model reduces the cost of data distribution, which can be expensive for pure MPI implementations.

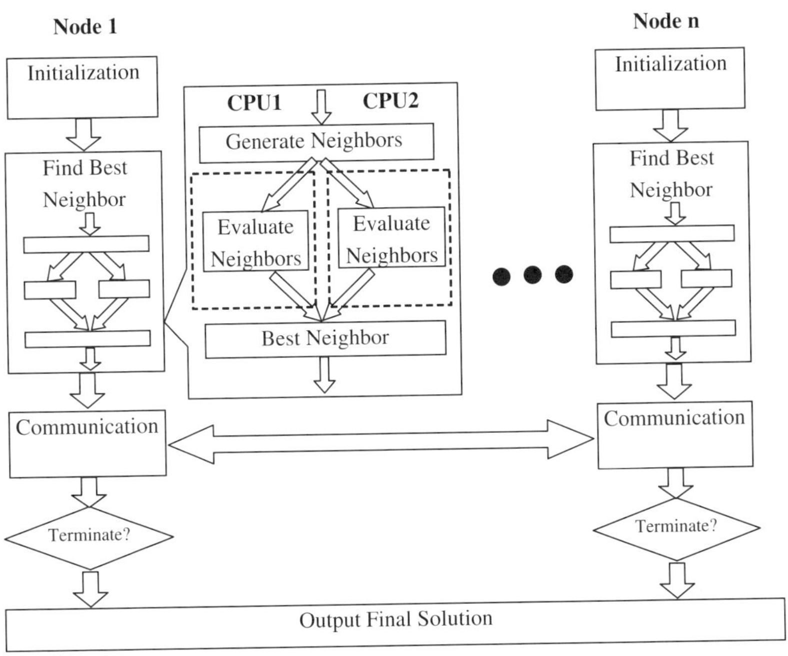

Figure 6. Flow chart of mixed MPI/OpenMP parallel TS scheme.

The implementation of the mixed parallel TS model in a CLUMP environment is as follows. First, both the header files of "`mpi.h`" and "`omp.h`" should be included in order to establish the connection to the libraries. Then, message passing procedures should be initialized on each node. Details of MPI-specific syntax and commands are provided by Snir *et al.* (1996). Within a node, OpenMP directives can then be used to parallelize certain computationally intensive loops. MPI controls the communication among the nodes. The pseudo code is as follows:

```
/* hybrid OpenMP/MPI model */
#include "mpi.h"
#include "omp.h"

MPI_Init(&argc, &argv);
MPI_Comm_rank(MPI_COMM_WORLD, &rank);
MPI_Comm_size(MPI_COMM_WORLD, &size);

omp_set_num_threads(n);
#pragma omp parallel
   /* intra-node parallel work here */

#pragma omp master
   /* inter-node communications */
   MPI_comm(Iter, Rank, var1, var2, ...);
MPI_Finalize();
```

6.1. Results

The hybrid parallel TS implementation has been run with one to five nodes, each consisting of two CPU's to solve the CAMD example. The total number of functional evaluations remains fixed. As a result, the total number of iterations is split evenly among the nodes. For example, if the total number of iterations is 1,000 with 1 node, then 200 iterations will be performed on each node when 5 are used. Figure 7 shows the average speedup based on 100 test runs. The hybrid model parallelizes the procedure to find the best neighbor at each iteration, which is the most time consuming part of the CAMD problem. Significant speedup is obtained, despite the need for

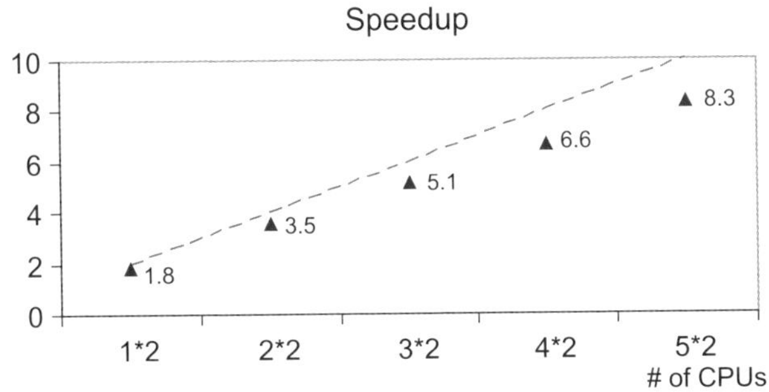

Figure 7. Speedup versus number of CPUs of hybrid parallel TS model. The dotted line indicates the hypothetical linear speedup.

communication among the nodes. The maximum speedup observed is 8.3 with 10 CPUs.

Increasing the number of nodes has the potential to increase the diversity of candidate solutions in two ways. First, at the beginning of a run, each node will generate its own initial solution. The search within that node will continue to be influenced by that initial solution until the nodes communicate and synchronize their best solution. Depending on the length of the communication interval, each node will increasingly generate neighbor solutions from different starting points, until they are again synchronized. Thus, more nodes have the potential to better explore the search space. However, since the total number of function evaluations remains the same, when there are more nodes, any given node performs fewer iterations. Within this case study, no clear relationship between solution quality and number of nodes has been observed, but empirical evidence suggests that additional nodes reduce the likelihood of becoming stuck in local optima.

The effect of the communication interval on the hybrid model has also been investigated. When the interval is set to 1, all threads update the best neighbor at each iteration, eliminating the additional diversity expected from the additional nodes. This is because all the threads will generate neighbors from the same solution after the initial iteration. Thus, a communication interval of 1 is equivalent to simply increasing the number of neighbors generated at each iteration and distributing their evaluation among all the threads. A larger communication interval enables each TS thread to potentially generate a more diverse set of neighbors from different current solutions, increasing the diversification of the search as described above; however, if the communication interval is too large, the overall speedup is somewhat reduced due to some nodes being idle while waiting for others to finish. Thus, a tradeoff exists between speedup and communication interval.

The hybrid model parallel TS scheme was implemented on an Apex® cluster with 18 nodes, each consisting of 2 Pentium-III 1GHz CPUs and 1G shared memory. All nodes are connected by a HP switch. MPICH is employed for the message passing, and the OpenMP compiler is PGCC from Portland group®. A switch is required for compiling the hybrid code:

```
pgcc -mp -o a.out a.c -lmpich
```

where `-mp` denotes the support of multiple processors on each node. Its output file then runs the same as normal MPI programs.

```
mpirun -np 3 a.out
```

6.2. Recommendations for hybrid parallel TS model

First, all MPI calls should be placed within thread sequential regions to ensure that the code is portable. When MPI calls occur within an OpenMP parallel region, the calls should be placed inside a `CRITICAL`, `MASTER` or `SINGLE` region, depending on the nature of the code. Since different threads may execute the code on successive passes, placing an MPI statement inside a `SINGLE` region is not a good practice. Second, the number of threads should be set within each MPI process using `omp_set_num_threads(n)`, because this is more portable than the environment variable of `OMP_NUM_THREADS`. Third, although the two models are mixed together, experience suggests that it is more effective to carry out the debugging and performance optimization of MPI and OpenMP separately.

7. Summary

With the increasing availability of multi-core processors and networks of workstations, a growing number of users will find parallel computing a cost effective approach to solving larger and more complex problems. As interest in parallel implementations of SOA grows, this chapter serves as a starting point for not only understanding fundamental approaches to parallelization, but also for parallelizing existing code. This chapter demonstrates two complementary approaches to parallelizing code: OpenMP and MPI. Multi-core processors provide a convenient platform for limited parallelization using OpenMP on desktop machines. OpenMP tends to be much simpler to implement, because it is based on compiler directives, while MPI requires the programmer to explicitly control the flow of information between nodes. Since most processors now have at least two cores, OpenMP can readily be used to speedup computations by approximately 70% per core. To further increase parallelization, MPI can be combined with the OpenMP code to spread the computation across a number of nodes within a cluster of workstations.

While the focus of this chapter has been on speedup, it is also important to consider how parallelization can potentially increase diversification of the search process. Under a multi-search threads approach, additional threads can decrease the likelihood of being stuck in a local optima so long as the communication interval between threads is sufficiently large to allow each thread to generate new solutions from different starting points. Thus, parallelization of SOA has the potential to improve the quality and reliability of solutions in addition to reducing computation time.

Acknowledgments

This work was partially supported by NSF grants MRI-9871133 and CTS-0224887.

References

Adjiman, C.S., Androulakis, I.P. and Floudas, C.A. (2000). Global optimization of mixed-integer nonlinear problems. *AIChE Journal*, **46**(9), pp. 1769–1798.

Aiex, R.M., Martins, S.L., Ribeiro, C.C. and Rodriguez, N.D.L.R. (1998). Cooperative multi-thread parallel tabu search with an application to circuit partitioning. *Lecture Notes in Computer Science*, **1457**, pp. 310–331.

Baños, R., Gil, C., Paechter, B. and Ortega, J. (2006). Parallelization of population-based multi-objective metaheuristics: An empirical study. *Applied Mathematical Modelling*, **30**(7), pp. 578–592.

Belal, M. and El-Ghazawi, T. (2004). Parallel models for particle swarm optimizers. *International Journal on Intelligent Cooperative Information Systems*, **4**, pp. 100–111.

Blazewicz, J., Glover, F. and Kasprzak, M. (2004). DNA sequencing — Tabu and scatter search combined. *INFORMS Journal on Computing*, **16**(3), pp. 232–240.

Cappello, F. and Richard, O. (1999). *Intra-Node Parallelization of MPI Programs with Openmp.*

Chandra, R., Dagum, L., Kohr, D., Maydan, D., McDonald, J. and Menon, R. (2001). *Parallel Programming in OpenMP*, San Diego, CA, Academic Press.

Chavali, S., Lin, B., Miller, D.C. and Camarda, K.V. (2004). Environmentally-benign transition metal catalyst design using optimization techniques. *Computers and Chemical Engineering*, **28**(6), pp. 605–611.

Cioni, L. (2005). *Some Strategies for Parallelizing Ant Systems.*

Crainic, T. and Toulouse, M. (1998). *Parallel Metaheuristics. Fleet Management and Logistics.* Crainic T.G. and Laporte, G., Norwell MA, Kluwer Academic, pp. 205–251.

Crainic, T.G., Toulouse, M. and Gendreau, M. (1995). Synchronous Tabu search parallelization strategies for multicommodity location-allocation with balancing requirements. *OR Spektrum*, **17**, pp. 113–123.

Crainic, T.G., Toulouse, M. and Gendreau, M. (1997). Toward a taxonomy of parallel tabu search heuristics. *INFORMS Journal on Computing*, **9**(1), pp. 61–72.

Cung, V.-D., Martins, S.L., Ribeiro, C.C. and Roucairol, C. (2001). Strategies for the Parallel Implementation of Metaheuristics. *Essays and Surveys in Metaheuristics.* Ribeiro, C.C. Dordrecht, Kluwer, pp. 263–308.

Dagum, L. and Menon, R. (1998). Openmp: An industry-standard Api for shared-memory programming. *IEEE Computational Science and Engineering*, **5**(1), pp. 46–55.

Debudaj-grabysz, A. and Rabenseifner, R. (2005). Nesting Openmp in Mpi to implement a hybrid communication method of parallel simulated annealing on a cluster of Smp nodes. *Recent Advances in Parallel Virtual Machine and Message Passing Interface*, Heidelberg, Springer-Verlag, Berlin.

Furman, K.C. and Sahinidis, N.V. (2001). Computational complexity of heat exchanger network synthesis. *Computers and Chemical Engineering*, **25**(9–10), pp. 1371–1390.

Glover, F. and Laguna, M. (1997). *Tabu Search.* Boston, Kluwer Academic Publishers.

Gordon, V.S. and Whitley, D. (1993). Serial and parallel genetic algorithms as function optimizers. *Proceedings of the 5th International Conference on Genetic Algorithms*, Morgan Kaufmann.

He, Y. and Ding, C.H.Q. (2002). Conference on high performance networking and computing. *Proceedings of the 2002 ACM/IEEE Conference on Supercomputing, Baltimore, Maryland*, IEEE Computer Society Press.

Hiroyasu, T., Miki, M., Ogura, S., Aoi, K., Yoshida, T., Okamoto, Y. and Dongarra, J. (2002). Energy minimization of protein tertiary structure by parallel simulated annealing using genetic crossover. *Proceedings of Genetic and Evolutionary Computation Conference* (*GECCO*).

Hsieh, J. (2000). Design choices for a cost-effective, high-performance beowulf cluster. *HP Power Solutions*.

Jamesa, T., Regob, C. and Glover, F. (2008). A Cooperative parallel Tabu search algorithm for the quadratic assignment problem. *European Journal of Operational Research*, in press.

Kier, L.B. and Hall, L.H. (1976). *Molecular Connectivity in Chemistry and Drug Research*. New York, Academic Press.

Koh, B.-I., George, A.D., Haftka, R.T. and Fregly, B.J. (2006). Parallel asynchronous particle swarm optimization. *International Journal for Numerical Methods in Engineering*, **67**, pp. 578–595.

Lewin, R., Wang, H. and Shalev, O. (1998). A generalized method for hen synthesis using stochastic optimization: (I) General framework and mer optimal synthesis. *Computers and Chemical Engineering*, **22**(10), pp. 1503–1513.

Lin, B., Chavali, S., Camarda, K. and Miller, D.C. (2005). Computer-aided molecular design using Tabu search. *Computers and Chemical Engineering*, **29**, pp. 337–347.

Lin, B., Leibovici, C.F. and Jorgensen, S.B. (2008). Optimal component lumping: Problem formulation and solution techniques. *Computers and Chemical Engineering*, **32**, pp. 1167–1172.

Lin, B. and Miller, D.C. (2004a). Solving heat exchanger network synthesis problems with Tabu search. *Computers and Chemical Engineering*, **28**(8), pp. 1451–1464.

Lin, B. and Miller, D.C. (2004b). Tabu search algorithm for chemical process optimization. *Computers and Chemical Engineering*, **28**(11), pp. 2287–2306.

Linke, P. and Kokossis, A. (2003). On the robust application of stochastic optimization technology for the synthesis of reaction/separation systems. *Computers and Chemical Engineering*, **27**, pp. 733–758.

Manfrin, M., Birattari, M., Stutzle, T. and Dorigo, M. (2006). Parallel ant colony optimization for the traveling salesman problem. *Proc. of the 5th Int. Workshop on Ant Colony Optimization and Swarm Intelligence*, Université Libre de Bruxelles.

Morales, L.B., Garduño-Juárez, R., Aguilar-Alvarado, J.M. and Riveros-Castro, F.J. (2000). A parallel Tabu search for conformational energy optimization of oligopeptides. *Journal of Computational Chemistry*, **21**, pp. 147–156.

Negro, F.D.T., Ortega, J., Ros, E., Mota, S., Paechter, B. and Martín, J.M. (2004). Psfga: parallel processing and evolutionary computation for multiobjective optimisation. *Parallel Computing*, **30**, pp. 721–739.

Niar, S. and Freville, A. (1997). A parallel Tabu search algorithm for the 0–1 multidimensional knapsack problem. *International Parallel Processing Symposium*, **97**, Geneva.

Ojima, Y., Sato, M., Harada, H. and Ishikawa, Y. (2003). Performance of cluster-enabled OpenMP for the SCASH software distributed shared memory system. *3rd IEEE/ACM International Symposium on Cluster Computing and the Grid.*

OpenMP Architechture Review Board (2008) website at openmp.org.

Paluszewski, M., Hamelryck, T. and Winter, P. (2006). Reconstructing protein structure from solvent exposure using Tabu search. *Algorithms for Molecular Biology*, **1**, pp. 1–20.

Porto, S.C.S., Kitajima, J.P. and Ribeiro, C.C. (2000). Performance evaluation of a parallel Tabu search task scheduling algorithm. *Parallel Computing*, **26**(1), pp. 73–90.

Porto, S.C.S. and Ribeiro, C.C. (1995). Parallel Tabu search message-passing synchronous strategies for task scheduling under precedence constraints. *Journal of Heuristics*, pp. 207–223.

Raman, V.S. and Maranas, C.D. (1998). Optimization in product design with properties correlated with topological indices. *Comp. Chem. Eng.*, **22**(6), pp. 747–763.

Resende, M.G.C., Pardalos, P.M. and Eksioglu, S.D. (2001). Parallel metaheuristics for combinatorial optimization. *Advanced Algorithmic Techniques of Parallel Computation with Applications*, Kluwer Academic Publishers.

Sexton, R.S., Alidaee, B., Dorsey, R.E. and Johnson, J.D. (1998). Global optimization for artificial neural networks: A Tabu search application. *European Journal of Operational Research*, **106**, pp. 570–584.

Siddhaye, S.S., Camarda, K.V., Topp, E. and Southard, M.Z. (2000). Design of novel pharmaceutical products via combinatorial optimization. *Computers and Chemical Engineering*, **24**, pp. 701–704.

Snir, M., Otto, S.S., Huss-Lederman, S., Walker, D. and Dongarra, J. (1996). *MPI: The Complete Reference*, Cambridge, MA, MIT Press. Also available electronically at http://www.netlib.org/utk/papers/mpi-book/mpi-book.html.

Srinivas, M. and Rangaiah, G.P. (2007). A study of differential evolution and Tabu search for benchmark, phase equilibrium and phase stability problems. *Computers and Chemical Engineering*, **31**, pp. 760–772.

Sun Microsystems, I. (2001). *Sun Enterprise 6500/5500/4500 Systems Reference Manual.*

Talbi, E.-G., Hafidi, Z. and Geib, J.-M. (1997). Parallel Adaptive Tabu search for large optimization problems. *MIC'97–2nd Metaheuristics International Conference*, Sophia Antipolis, France.

Tang, Y., Reed, P.M. and Kollat, J.B. (2007). Parallelization strategies for rapid and robust evolutionary multiobjective optimization in water resources applications. *Advances in Water Resources*, **30**, pp. 335–353.

Teh, Y.S. and Rangaiah, G.P. (2003). Tabu search for global optimization of continuous functions with application to phase equilibrium calculations. *Computers and Chemical Engineering*, **27**, pp. 1665–1679.

Toulouse, M., Crainic, T., Sans'o, B. and Thulasiraman, K. (1998). Self-organization in cooperative Tabu search algorithms. *IEEE International Conference on Systems, Man, and Cybernetics*, San Diego, USA.

Toulouse, M., Crainic, T.G. and Gendreau, M. (1996). Communication issues in designing cooperating multi-thread parallel searches. *Meta-Heuristics: Theory and Applications*, Norwell, MA, Kluwer Academic.

Wang, C., Quan, H. and Xu, X. (1999). Optimal design of multiproduct batch chemical process using Tabu search. *Computers and Chemical Engineering*, **23**, pp. 427–437.

Yee, T.F. and Grossmann, I.E. (1990). Simultaneous optimization models for heat integration — II. Heat exchanger network synthesis." *Computers and Chemical Engineering*, **14**(10), pp. 1165–1184.

INDEX